AF530702

MICROPALEONTOLOGY AND ITS APPLICATIONS

The Editors

Dr. P. K. Kathal, Professor, Department of Applied Geology, Dr. Harisingh Gour Vishwavidyalaya, Sagar, M.P., an alumnous of the same University with vast academic and administrative experiences as the former Head, Department of Applied Geology, Director, Centre of Advance Study in Geology, Dean, Engineering Faculty, Dean Research & Development, an awardee of Japan Society for Promotion of Science (JSPS) Fellowship (1999 and 2003) and DFG Fellowship (under DAAD Programme, Germany) in 2007 to carry out advanced research in Micropaleontology at Akita University, Japan; Kumamoto University, Japan and Tubingen University, Tubingen, Germany, respectively; Member of International UNESCO Programme on the Ediacaran (Vendian) Puzzle from 2011 to 2016; and responsible for developing a field of research on comparison of Recent Foraminifera from the Indian and Japanese coasts. He has chaired sessions in international research events held at Canada (2000), Austria (2002) and India (2013, 2014 and 2015). Having obtained M. Tech. (Applied Geology), and Ph. D. (Recent Foraminifera, Micropaleontology) he is in the field of teaching and research in Micropaleontology for over 35 years with 65 research contributions in research journals of national and international repute along with the credit of bringing out the first two text books in Micropaleontology from India.

He is Associate Editor of the Journal of Foraminiferal Research, USA.

Dr. Rajiv Nigam, Retd. Deputy Director and Head of Geological Oceanography and Marine Archaeology, National Institute of Oceanography (NIO), Goa is currently working there as a Consultant. An alumnous of Lucknow University, Lucknow he obtained his M. Phil, Ph. D. and D. Sc. degrees from Aligarh Muslim University, Aligarh and NORAD Diploma in Marine Micropaleontology from Oslo University, Norway. He joined the NIO as Scientist in 1978 and superannuated from there in 2016. He not only made significant contributions in the field of Marine Geology (Micropaleontology) but also established the first 'Laboratory for Culture Studies on Foraminifera' in India at NIO. He guided 12 Ph. D. theses and published over 135 papers in research journals of National and International repute. Recipient of the 'CSIR-Young Scientist Award', 'Krishnan Gold Medal', 'National Mineral Award', 'M. R. Sahni Lecture Award' and 'Prof. S. N. Bhalla Gold Medal' of the Palaeontological Society of India and 'H. N. Siddiquie Lecture Award' of the Indian Geophysical Union. He has authored a book on Indian Monsoon, which received a cash prize award of Rs 50000/- by the Ministry of Earth Sciences, Government of India.

Dr. Abu Talib, Professor and Head, Department of Geology, Aligarh Muslim University (AMU), Aligarh obtained his M. Sc. and Ph. D. degrees from AMU and joined the Department of Geology, AMU as Lecturer in 1986. His teaching and research interests include Petroleum Geology, Palaeontology and Micropalaeontology with 30 years of research and teaching experience. He has supervised 5 Ph. D. theses and several M. Phil theses and contributed over 35 papers in national and international research journals and has completed three research projects sponsored by UGC, DST and a petroleum company. He is a Life Member, Palaeontological Society of India, Lucknow; Indian Society for Earth, Planetary and Environmental Sciences, New Delhi; and remained an elected Member, Executive Council, Palaeontological Society of India, Lucknow in 2014-15.

MICROPALEONTOLOGY AND ITS APPLICATIONS

A Festschrift for Septuagenarian Scholar Professor Satyendra Nath Bhalla

Editors

Professor Pradeep K. Kathal
Department of Applied Geology
Dr. Harisingh Gour Vishwavidyalaya
(A Central University)
Sagar, 470 003 M.P., India

Dr. Rajiv Nigam
Retd. Deputy Director,
Geological Oceanography
National Institute of Oceanography
Dona Paula, Goa-403004, India

Professor Abu Talib
Head, Department of Geology
Aligarh Muslim University
Aligarh 202 001, U.P., India

Published by
SCIENTIFIC PUBLISHERS (INDIA)

Jodhpur –

5 A, New Pali Road
P.O. Box 91
Jodhpur - 342 001 INDIA

Delhi –

• 4806/24, Ansari Road
Daryaganj
New Delhi - 110 002 INDIA

ISBN: 978-93-86652-09-6

Visit the Scientific Publishers (India) website at
http://www.scientificpub.com

Printed in India

वाडिया हिमालय भूविज्ञान संस्थान
(भारत सरकार के विज्ञान एवं प्रौद्योगिकी विभाग का एक स्वायत्तशासी संस्थान)
३३, जनरल महादेव सिंह मार्ग, देहरादून - २४८००१ (उत्तराखण्ड)
WADIA INSTITUTE OF HIMALAYAN GEOLOGY
(An Autonomous Institution of Dept of Science & Technology, Govt of India)
33, General Mahadeo Singh Road, Dehra Dun - 248001 (Uttarakhand)

प्रो. अनिल कुमार गुप्ता
निदेशक
एफ एन ए एस सी, एफ ए एस सी, एफ एन ए, एफ टी डब्लू ए एस
जे.सी. बोस नेशनल फैलो

Prof. Anil K. Gupta
Director
FNASc, FASc, FNA, FTWAS
J.C. Bose National Fellow

Foreword

Micropaleontology has played an important role in understanding paleoceanographic and paleoclimatic evolution of the Ocean basins. In the late sixties and early seventies, importance of microfossils was realized in reconstructing deep-sea water masses. During the 1960's and 1970's, a new era of Micropaleontology developed into applied discipline, i.e. applied micropaleontology. Since then this area of research has led to major information explosion! The most text-books suffered from either faster obsolescence or considerable revision. Many new research journals exclusively devoted to applied micropaleontology or to even the ecological parameters of microorganisms as the living representatives of the fossilized microorganisms, were launched. Accent on publication of the thematic mostly multi-authored volumes appeared. The present edited volume is also one of the results and upshot studies in the field of Mircopaleontology and has been planned as a festschrift for a septuagenarian scholar and my revered teacher Professor S.N. Bhalla, Former Chairman, Department of Geology, Aligarh Muslim University, Aligarh and a legendary figure in the field of Micropaleontology.

The volume is covering results of the 'cutting edge research' dealing with Foraminifera, Ostracoda, Radiolaria, Diatoms, Micro-vertebrates, Micro-algae etc. by well known academicians and researchers from across the world.

The Editors of the volume, Professor Pradeep K. Kathal, Dr. Rajiv Nigam and Professor Abu Talib are acclaimed researchers in Micropaleontology. Prof. Kathal has already contributed two text books on the applications of microfossils, which have been received well by the academic fraternity.

All the papers in the volume were reviewed by the reviewers and editors and invariably scrutinized and approved before being finally accepted for publication, which I am sure, will motivate other workers towards the fascinating world of microfossils.

(Professor Anil K. Gupta)

Dehra Dun, 15th March 2017

Tel (Off) : +91 135 2525101, FAX : +91 135 2625212, 2525200, E-mail : director@wihg.res.in

Contents

Preface

This volume, titled **"Micropaleontology and its Applications"** contains invited papers from eminent scholars of academic and research institutions from across the world on the diverse groups of microfossils, like - Foraminifera, Ostracoda, Radiolaria, Pollen and Spores, Algae/Coralline Algae, Diatoms, Microbialites and Microstructures of Dinosaur Eggs.

Ever since Micropaleontology developed into a new era of applied discipline in the 1960'-70's this area of research has generated exponential explosion of literature! The prevailing text-books therefore required extensive revision to reveal the current state of research on applied micropaleontology. New research journals have appeared covering applied micropaleontology or to even the ecological parameters of microorganisms as the living representatives of the fossilized microorganisms thereafter. This has lead to an accent on publication on the thematic multi-authored volumes. The present volume, also an outcome of the 'cutting edge research' in the field of Mircopaleontology, is being dedicated to our esteem teacher Professor Satyendra Nath Bhalla, former Chairman, Department of Geology, Aligarh Muslim University, Aligarh and a legendary figure in the field of Micropaleontology.

We are thankful to Prof. A. K. Gupta, Director, Wadia Institute of Himalayan Geology (WIHG), Dehra Dun for scripting the 'Foreword' of the book and the referees of the research papers who painstakingly reviewed the papers promptly. We also acknowledge indebtedness to Professor M. B. Hart, Earth and Environmental Sciences, Plymouth University, Plymouth, United Kingdom; Professor Pamela Hallock, College of Marine Science, University of South Florida, USA; Professor Jacob Jhon, Department of Environmental Biology, Curtin University of Technology, Perth, Australia; and Prof. Michal Kucera, Bremen University, Bremen, Germany for their timely and learned suggestions in improving the contents. Authors thanks Professor K.L. Srivastava, Jai Narain Vyas University, Jodhpur for support.

We thank all the contributors to this book for their time and efforts in preparing their papers, We are indebted to Head, Department of Applied Geology, for providing infrastructure facilities, Professor J. K. Mishra Department of Library and Information Science, and Dr. Jemini Khateek, Woman Scientist (UGC), Mr. Navneet Ranjan and Miss Paishnee Patel Research Scholars, Department of Applied Geology, Dr. Harisingh Gour Vishwavidyalaya, Sagar for their help in different ways.

Last but not the least, we thank Shri Tanay Sharma, Managing Director, Scientific Publishers (India), Jodhpur and New Delhi for taking personal interest in publishing this volume with the sense of meeting the dead-line.

Professor P. K. Kathal
Dr. Rajiv Nigam
Professor Abu Talib

15th March 2017

Professor Satyendra Nath Bhalla – a legendary micropaleontologist

Professor Satyendra Nath Bhalla was born on 1st July, 1940 at Basti, Uttar Pradesh, India. His mother, Mrs. Vaishno Devi, was a home maker and father, Shri B. N. Bhalla was Superintendent of Police. His father was a very religious person and known as '*Maharaj Jee*' to his disciples who still gathered on *makar sankranti* day every year at his Samadhi at Kanpur. Prof. Bhalla inherited disciplined and poised life-style of his father.

He had his early education at Calvin Taluqedars College in Lucknow and obtained his B.Sc. and M.Sc. (Geology) degrees from Aligarh Muslim University, Aligarh. He joined the Department of Geology of Punjab University, Chandigarh, to pursue research in Micropaleontology, under the supervision of the famous Indian micropaleontologist Professor S. B. Bhatia. He often mentioned that the other person who had inspired him was Prof. M. R. Sahni.

Prof. Bhalla was awarded his Ph. D. in 1964 and he joined, as Lecturer in the Department of Geology, Aligarh Muslim University, Aligarh. He established the Micropaleontology Lab in the Department and started pioneering research on Jurassic foraminifera of Kutch and within a few years the Micropaleontology Lab was elevated as a leading research centre on Recent Foraminifera from the Indian Coasts as well as Jurassic foraminiferal research in India. During his academic career as a Professor (1987 to 2002), the Department of Geology got fresh impetus in academic and research endeavors under his Chairmanship from 1990-94.

Prof. Bhalla has a distinguished research career of more than half a century! His main research focused on the taxonomy and distribution of Recent and Jurassic foraminifera of India but he also applied microfossils for dating, palaeoenvironmental analysis, palaeogeographic interpretation, pollution studies, and petroleum exploration. During his research career, he worked on other groups of microfossils, *viz.*, shark teeth and ostracods as well as stratigraphic studies. He was the first to carry out pioneering work on the east coast of Gondwana and to predict the possibility of finding oil and gas in on-shore as well as off-shore areas of Andhra Pradesh, eastern India, as far back as 1970, which was confirmed by the Oil and Natural Gas Commission, Government of India, during their drilling operations in this region in 1982. He was invited by the Elsevier Science Publishers, Holland, in 1983 to contribute Chapter 10 (India) for the book, "Phanerozoic Geology of the World: II. The Mesozoic". Prof. Bhalla published 105 research papers and supervised six Ph. D. and several M. Phil. theses.

He was invited by several institutions of different countries, including the British Council, London on Visitorship in 1971 and 1975 for doing advanced research on foraminifera and marine environment; Royal Society, London, to carry out higher research and discuss problems of mutual research interest with colleagues in the U. K. in 1994; German Academic Exchange Service, Bonn, in 1972, 1975, 1995 and 1997 for advance research on marine microfossils in Germany; the Museum d'Historie Naturelle, Geneva, Switzerland, in 1985 to do advanced research on Jurassic foraminifera; and the Japan Society for Promotion of Science in 1987 to deliver lectures at different Universities and Institutes in Japan and to do higher research on Recent benthic foraminifera of the Pacific Ocean. In addition, he visited and carried out research in the U.K., Germany, France, Holland, Switzerland, U.S.A., Canada, Japan, Al-Ain (UAE), Kuwait, Muscat, Nepal.

Prof. Bhalla has a long association with micropaleontology movement in India, *i.e.* Indian Colloquium on Micropaleontology and Stratigraphy (ICMS). In 1976, he successfully convened the 'V-ICMS' at the Aligarh Muslim University, Aligarh. He was the President of the 'XVII-ICMS' at Ujjain. He was felicitated at the 'XVI-ICMS' at Goa, 1998 and in 2005 at Lucknow for significant contributions in the field of Micropaleontology and Stratigraphy and received the 'Lifetime Achievement Award' at the 'XXII-ICMS' held at Tiruchirapalli in 2009. The Government of India appointed him as Chairman of the Research Advisory Council, Birbal Sahni Institute of Palaeobotany, Lucknow in 2011.

In recognition of his contribution to the field of micropalaeontology, the Palaeontological Society of India, Lucknow constituted a National Award, the 'Prof. S. N. Bhalla Gold Medal in Micropalaeontology' in 2003. The award, in the form of a citation and a medal, is given once every two years for outstanding contributions to Micropalaeontology.

He retired from the Aligarh Muslim University in 2002 from the service but kept his interest alive in the subject and to date he is actively participating in conferences, writing research papers and counseling and advising research scholars on various problems of micropalaeontology.

His passion, devotion and suggestions in the field of Micropaleontology in general, and taxonomy in particular, are unparalleled! He is a meticulous researcher and excellent teacher besides, being an innovative thinker with modesty and concern for others. He is a soft spoken person with high moral and social values and a valued and trusted friend, philosopher and guide to his associates.

By joining hands with all the contributors and reviewers to this book, we wish him all the best for the future. It is indeed a matter of pride for all his research students to be associated with Prof. Bhalla not only in the field of Micropaleontology but also in the various pursuits of life. We feel honoured to dedicate this book- 'Micropaleontology and its Aapplication' to the subject discipline builder Prof. S. N. Bhalla.

Pradeep K. Kathal
Rajiv Nigam
Abu Talib

Contributors

Abbas, S. M.
Senior Geologists,
KDMIP, Oil & Natural Gas Corporation,
Dehradun, India

Anjum, Shabana Irshad
Palaeontology Lab, K.D.M. Institute of Petroleum Exploration,
Oil & Natural Gas Corporation,
Dehradun, India

Ansari, Ainul H.
Department of Geology,
Aligarh Muslim University,
Aligarh, India

Bharti, S. R.
Centre of Advanced Study in Geology,
Banaras Hindu University,
Varanasi, India

Campagna, S.
Department of Environmental Biology,
Curtin University of Technology,
Perth, W. Australia

Chakraborty, Arindam
Birbal Sahni Institute of Palaeosciences,
Lucknow, India

Chhabra, N. L.
University of Lucknow,
Lucknow, India

Deepti
Department of Geology,
Kumaun University, Nainital, India

Dhiman, Harsha
Department of Geology,
Centre for Advanced Studies in Geology,
University of Delhi, Delhi, India

Fauth, John E.
Department of Biology,
University of Central Florida,
Orlando, USA

Gandhi, M. Suresh
Department of Geology,
University of Madras,
Chennai, India

Gaur, K. N.
Former Head,
Department of Geology,
D. S. College, Aligarh, India

Ghosh, Amit
Birbal Sahni Institute of Palaeosciences,
Lucknow, India

Govindan, Abiraman
Asian Biostratigraphy Services,
Chennai, India

Hallock, Pamela
College of Marine Science,
University of South Florida,
St. Petersburg,
FL, USA

Hart, M. B.
Earth & Environmental Science,
School of Geography,
Plymouth University,
Plymouth, UK

Hussain, S. M.
Department of Geology,
University of Madras,
Chennai, India

Irshad, Roohi
Department of Geology,
Aligarh Muslim University,
Aligarh, India

Iwao, Kenji
Akajima Marine Science Laboratory,
Zamami-son,
Okinawa, Japan

Jain, Sreepat
Department of Applied Geology,
Adama Science & Technology,
University,
Adama, Ethiopia

Jauhri, A. K. (Deceased)
Department of Geology,
University of Lucknow,
Lucknow, India

Jeshma, P.
Department of Geology,
University of Madras,
Chennai, India

Jha, Neerja
Birbal Sahni Institute of Palaeosciences,
Lucknow, India

John, Jacob
Department of Environmentology
Curtin University of Technology
Perth, W. Australia

Kaithwar, Amrata
Micropaleontology Lab,
CSIR-National Institute of
Oceanography,
Goa, India

Kathal, P. K.
Department of Applied Geology,
Dr. Harisingh Gour Vishwavidyalaya,
Sagar, India

Kishore, S.
Department of Botany,
University of Lucknow,
Lucknow, India

Krassilov, Valentin A.
Paleontological Institute,
Profsoyuznaya, Moscow, Russia and
The Institute of Evolution,
University of Haifa,
Israel (Deceased)

Kurtarkar, Sujata R.
Micropaleontology Lab,
CSIR-National Institute of
Oceanography, Goa, India

Linshy, V. N.
408, Technocrats Apartments,
Gurgaon, India

Manjima, M.
Department of Geology,
University of Madras,
Chennai, India

Mehrotra, Naresh Chandra
Birbal Sahni Institute of Palaeosciences,
Lucknow, India

Moses, Elizabeth Fisher
Science, Engineering and Technology
Division, Howard Community College,
Columbia, USA and College of Marine
Science, University of South Florida,
Petersburg, USA

Nigam, Rajiv
Micropaleontology Lab,
CSIR-National Institute of
Oceanography, Goa, India

Pande, Kanchan
Department of Earth Sciences,
Indian Institute of Technology,
Mumbai, India

Prakasan, Megha
Department of Geology,
Aligarh Muslim University,
Aligarh, India

Prasad, G. V. R.
Department of Geology,
Centre for Advanced Studies,
University of Delhi,
Delhi, India

Rao, N. Rajeshwara
Department of Applied Geology,
University of Madras,
Chennai, India

Rogers, Karyne
National Isotope Centre,
GNS Science,
Lower Hutt, New Zealand

Sanyal, Prasanta
Department of Earth Sciences,
Indian Institute of Science Education and Research (IISER),
Kolkata, India

Saraswat, R.
Micropaleontology Lab,
CSIR-National Institute of Oceanography,
Goa, India

Saraswati, Pratul Kumar
Department of Earth Sciences,
Indian Institute of Technology, Bombay,
Mumbai, India

Sharma, G. K.
Department of Geology,
Kumaun University,
Nainital, India

Singh, A. D.
Centre of Advanced Study in Geology,
Banaras Hindu University,
Varanasi, India

Singh, A. P.
Department of Geology,
Mizoram University,
Aizwal, India

Singh, Asmita
Department of Earth Sciences,
Indian Institute of Technology,
Mumbai, India

Singh, Gaurav Kumar
Department of Applied Geology,
Dr. Harisingh Gour Vishwavidyalaya,
Sagar, India

Singh, S. K.
Department of Botany,
Charak Institute of Education,
Lucknow, India

Singh, V. K.
Department of Geology,
National Institute of Technology (NIT),
Raipur, India

Singh, Y. P.
Department of Geology,
D. S. College,
Aligarh, India

Sujata, R.
Micropaleontology Lab,
CSIR-National Institute of Oceanography,
Goa, India

Sukhrie, Thejasino
Micropaleontology Lab,
CSIR-National Institute of Oceanography,
Goa, India

Talib, Abu
Department of Geology,
Aligarh Muslim University,
Aligarh, India

Taukulis, Fiona E.
Department of Environmental Biology,
Curtin University of Technology,
Perth, W. Australia

Tharun, R.
Department of Geology,
University of Madras,
Chennai, India

Tiwari, R. P.
Department of Geology,
Mizoram University,
Aizwal, India
(Present position: Vice Chancellor,
Dr. Harisingh Gour Vishwavidyalaya,
Sagar, India)

Verma, K.
Paleontology Division I CHQ,
Geological Survey of India,
Kolkata India

Verma, Vishal
Bakaner,
Dhar, M. P., India

Veronica, S. Campagna
Department of Environmental Biology,
Curtin University of Technology,
Perth, W. Australia

Wasim, S. M.
Department of Geology,
D. S. College, Aligarh, India

Watkinson, M. P.
Earth & Environmental Science,
School of Geography,
Plymouth University,
Plymouth, UK

Micropaleontology and its Applications
Pradeep K. Kathal, Rajiv Nigam & Abu Talib (Editors)
Scientific Publishers (India), 2018, 1-13 pp.

Micropalaeontology and Palaeoecology of the Kallankuruchchi Limestone Formation, Cauvery Basin, S.E. India

M.B. Hart* and M. P. Watkinson

School of Geography, Earth & Environmental Sciences, Plymouth University,
Drake Circus, Plym PL4 8AA, United Kingdom
***Email:** mhart@plymouth.ac.uk

Abstract

The Kallankuruchchi Limestone Formation of the Cauvery Basin in S. E. India is exposed in a series of quarries east of Ariyalur (Tamil Nadu). The limestone, of latest Campanian to Maastrichtian age, represents a carbonate ramp that is richly fossiliferous. The fauna includes large bivalves, brachiopods and echinoderms, together with abundant foraminifera – especially *Siderolites* and *Lepidorbitoides*. The succession appears to represent a single transgressive/ regressive cycle with minor facies oscillations.

Keywords: Cauvery Basin, Kallankuruchchi Limestone Formation, micropalaeontology, palaeoecology, thin-section analysis.

INTRODUCTION

The Cretaceous succession of the Cauvery basin has been the subject of continuing research since the early studies of Blanford (1862) and the palaeontological investigations of Stioliczka (1870–3) and Kossmat (1897). The Cauvery Basin is located within the Ariyalur-Pondicherry Depression (Fig. 1) and its geological history is key to understanding the tectonic development of the eastern continental margin of India (Watkinson et al., 2007). The break-up of

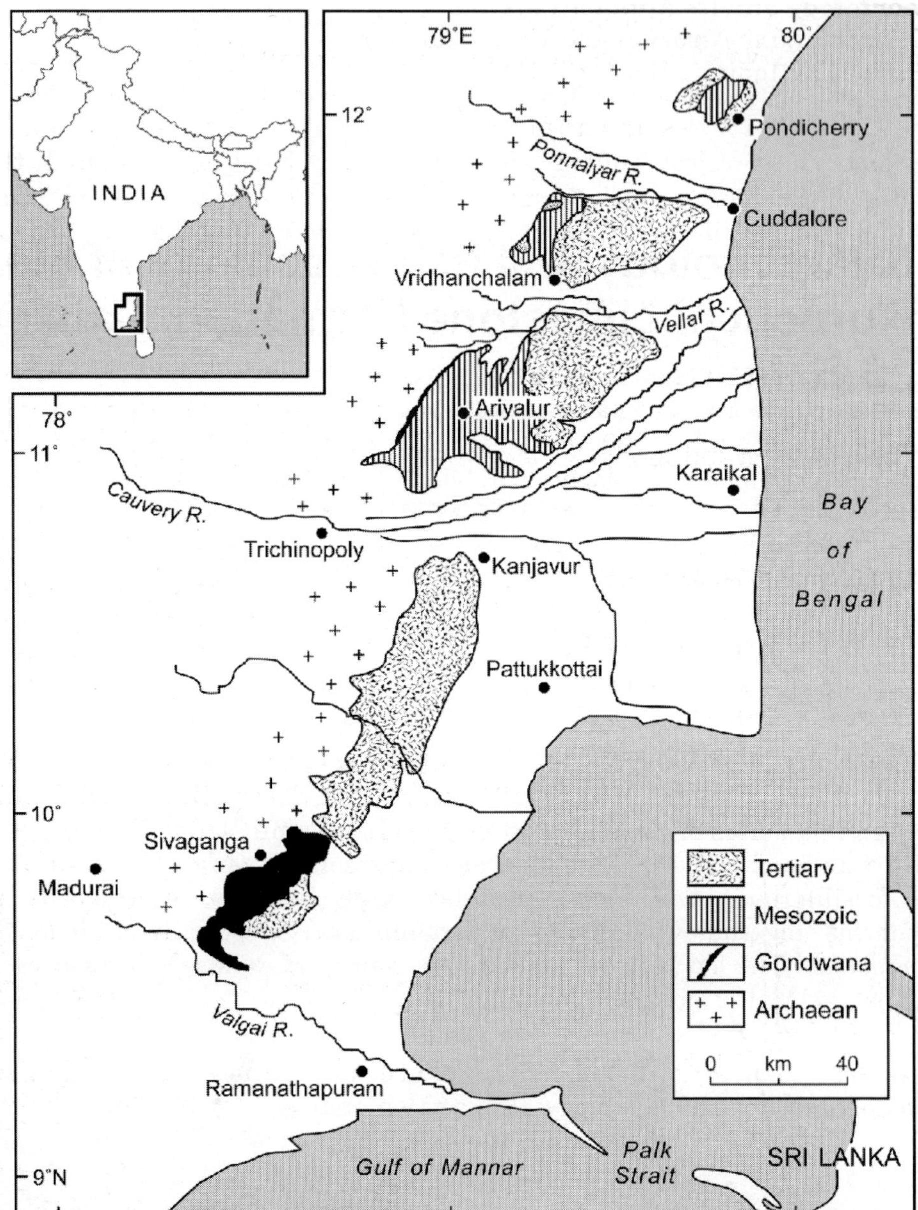

Fig.1. Geological map of the Cauvery Basin, showing the location of Ariyalur.

East Gondwanaland – which includes India, Australia and Antarctica – led to the initial opening of the Indian Ocean (Powell et al., 1988; Tarling, 1988). Major changes in plate configuration influenced the Cretaceous palaeogeography of the Indian sub-continent (Acharyya and Lahiri, 1991). In particular, the latest Jurassic – earliest Cretaceous rifting of India and Antarctica led to the formation of a number of major NE–SW trending basins in the Indian Precambrian crystalline basement (Powell et al., 1988), including the Cauvery Basin. The latter is the southernmost basin on the eastern margin of India. It follows the regional NE–SW trend, and is block-faulted with horst/graben architecture. The most northerly depression, within the basin, is the Ariyalur-Pondicherry area and this includes the well-known outcrops of the Ariyalur area. From the mid-Cretaceous, however, sedimentation in the basin appears to follow more global patterns of sea level change (Hancock, 1990; Jacquin and de Graciansky, 1998a, b; Jacquin et al., 1998).

LITHOSTRATIGRAPHY

Following the early work of Blanford (1862) there have been a number of papers dealing with the lithostratigraphy and biostratigraphy of the Cauvery Basin. This work has suffered from a plethora of local facies names, many of which have not been formally defined or used in a consistent manner. Changes in the spelling of many of the place names in the literature have also created their own problems. A comprehensive lithostratigraphy for the basin was proposed by Sundaram and Rao in 1979, but revised by the same authors a few years later (Sundaram and Rao, 1986; Sundaram et al., 2001). It was, however, Ramasamy and Banerji (1991) who attempted the formal definition of stratigraphical units and their lithological characteristics (with reference to type sections). Their work was, unfortunately, limited to the lower parts of the succession and they did not consider the Ariyalur Group. Subsequently Tewari et al. (1996) devised a more complete lithostratigraphy for the basin and this was extended into a sequence stratigraphical framework by Watkinson et al. (2007, and references cited therein).

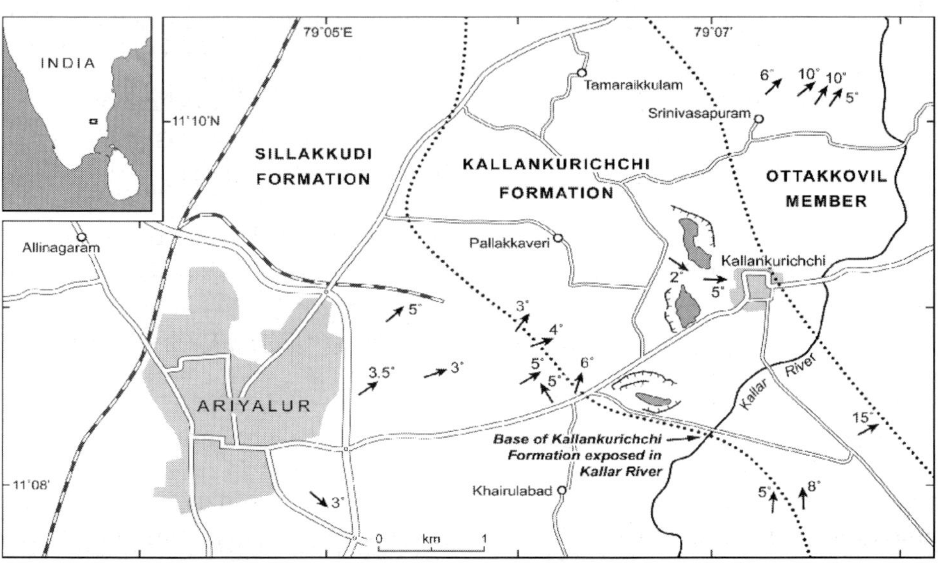

Fig.2. Field map of the outcrop of the Kallanku- ruchchi Limestone Formation east of Ariyalur. The succession shown in Figure 3 is derived from the three quarries located either side of the main east-west road. The Cretaceous strata dip very gently (3°–10°) to the NE, with the base of the formation exposed in the Kallar River.

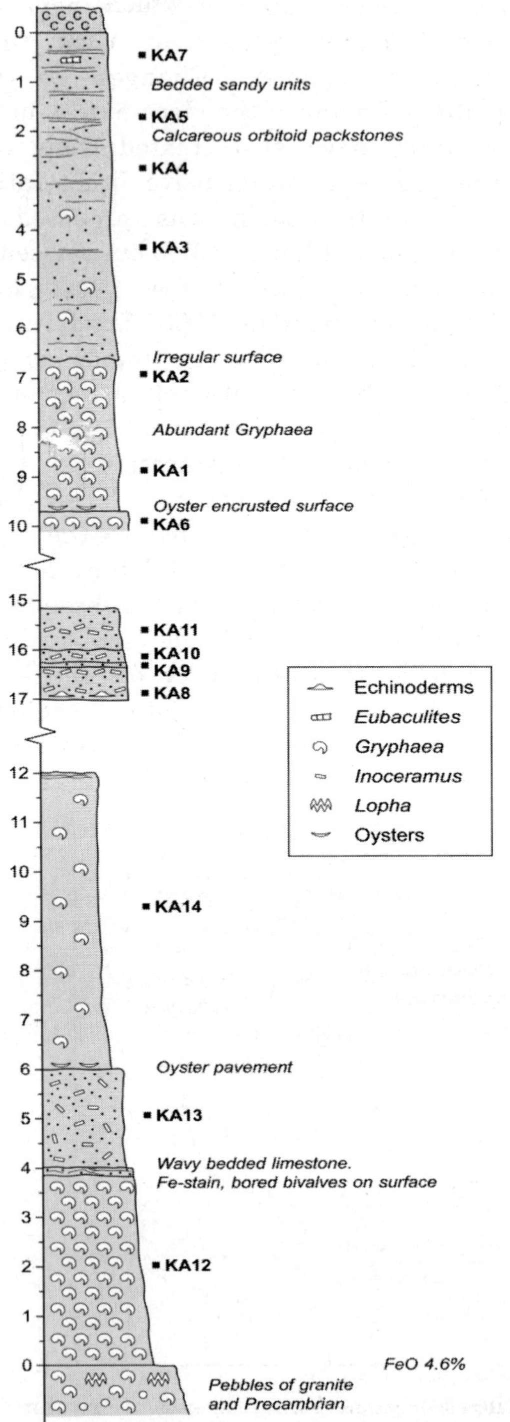

Fig.3. The succession of the Kallankuruchchi Limestone Formation, distribution of key fossils and the location of samples.

The Ariyalur Group is customarily divided into three formations: the Sillakkudi Sandstone Formation (which is rather poorly exposed in the area), the Kallankuruchchi Limestone Formation (which is well-exposed to the east of Ariyalur) and the Kallamedu Sandstone Formation (Fig. 2 and 3). Tewari et al. (1996) divided the Kallankuruchchi Limestone Formation into three members, although the relationship between the uppermost (Dherani Sandstone Member) and the overlying Ottakkovil Sandstone Member remains unclear. In the Kallar River the base of the Kallankuruchchi Limestone Formation is exposed and consists of a massive conglomerate with large cobbles/pebbles of Precambrian rocks, other exotic lithologies and large, fresh, pink feldspars. This major erosional surface, and overlying conglomerate, marks a distinctive change in the depositional record of the area, with a switch to a shallow-water, carbonate-dominated succession with a distinctive palaeontology (both macro- and micro-fossil).

The Kallankuruchchi Limestone Formation has been described by Tewari et al. (1996) as a retrograding (transgressive) carbonate ramp. The three members identified by Tewari et al. (*op. cit.*) include the lower, Tancem Limestone Member, which is a hard, yellow-coloured, coarse-grained sandy bioclastic grainstone/ packstone. The main components are larger foraminifera, bryozoans fragments and bivalve fragments – some of which may be rudist in origin.

The overlying Dalmia Biostromal Member comprises oramge-yellow, sandy, carbonate-rich mudstones that contain abundant, very large, *Gryphaea*, terebratulids (see Radulović and Ramamoorthy, 1992), *Lopha* (or *Alectronia*) and other large shell fragments. Large foraminifera, such as *Siderolites* and *Lepidorbitoides* are less common, or absent.

The uppermost Dherani Limesone Member is a yellow-coloured, coarse-grained, lepidorbitoid-rich grainstone or – in places – a lepidorbitoid rich sand. The assemblage is domnated by *Lepidorbitoides* sp. cf. *L. socialis* (Leymerie, 1851), which is associated with *Siderolites calcitrapoides* Lamarck *in* Faujas de Saint Fond, 1799. The separation of *S. calcitrapoides* from the often associated *S. laevigatus* (d'Orbigny, 1826) is almost impossible, especially in thin-section. Even in images produced by scanning electron microscopy (see Hart et al., 2000, pl. 2, figs 8, 9) show that the spaces between the characteristic elongate 'arms', that are typical of *S. calcitrapoides*, become infilled progressively until one develops the *S. laevigatus* morphotype (Renema and Hart, 2012). The macrofauna is reduced, with fragmentary *Gryphaea*, *Lopha*, *Inoceramus* and rare terebratulids. One specimen of *Eubaculites* was recorded but too fragmentary for a full identification. The overlying Ottakkovil Sandstone Member, while superficially similar, does not contain the assemblage of larger foraminifera.

KALLANKURUCHCHI LIMESTONE FORMATION

In the area east of Ariyalur, around Pallakkaveri and Kallankuruchchi, the outcrop of the limestone can be followed in a number of large quarries (Fig. 2). During the time we were undertaking fieldwork in the area, active quarry operations were constantly changing access to the succession, and this report must inevitably be viewed as a 'snapshot' of the situation at the time of sample collection and field descriptions. The succession is characterized by the presence of abundant, massive *Gryphaea*, which are associated with other bivalves (*Inoceramus*, *Cremnoceramus*, *Lopha,* etc.), echinoderms (*Hemineustes*, *Stygmatopygus*), nautiloids and other fragmentary macro-fossils. Only one, broken, example of *Eubaculites* was found adjacent to the uppermost sample KA7. In the thin-sections of the Kallankuruchchi Limestone Formation planktic foraminifera are either extremely rare or totally absent. The presence of larger foraminifera, that may well have hosted algal symbionts, is indicative of a shallow water environment in which planktic foraminifera would not be expected, especially the deeper water morphotypes. Small, juvenile planktic foraminifera are also difficult to identify as they are often missing the diagnostic characteristics seen in mature forms. *Gansserina gansseri* (Bolli, 1951) is recorded from the Kallankuruchchi Limestone Formation and is also recorded from coeval strata in deeper parts of the basin (Raju et al., 1993; Ravindran et al., 1995). This species was noted as characteristic of the 'Kallanhuruchchi Cycle' by Raju et al. (1993). Many of the breaks in the succession are bored, while others preserve distinctive oyster pavements. It is clear that many of the breaks in the succession display incipient hardgrounds, indicating distinct breaks in deposition. Samples (KA1 – KA14) were collected from discrete horizons, with a vertical stratigraphical thickness of only <5 cm. Sample KA16 was collected from the outcrop in the Kallar River, at the base of the Kallankuruchchi Limestone Formation. Thin-sections of all the samples were cut to a 30 µm standard thickness and imaged using a Nikon Eclipse LV100POL microscope with an attached Nikon DS-Fi1 digital camera using a NIS-Elements BR operating program.

INTERPRETATION

The presence of *G. gansseri* is significant as this species is a marker for the uppermost Campanian and Lower Maastrichtian (Robaszynski et al., 1984; Robaszynski and Caron, 1995). The co-occurrence of this species with *S. calcitrapoides* and *Lepidorbitoides* sp. cf. *Lepidorbitoides socialis* also indicates an early Maastrichtian age for the limestones. So-called 'primitive' *L. socialis* are recorded by Özcan and Özcan-Altiner (1999, fig. 7) in the middle of the *G. gansseri* Zone, although in the randomly oriented thin-sections used in this investigation we have not seen the indicative embryonic apparatus (Van Gorsel, 1975; Özcan and Özcan-Altiner, 1999; Dieni, 2010; Robles-Salcedo, 2013). In one thin-section (Fig. 5E) the embryonic apparatus is almost visible, and confirms that these individuals are not *Orbitoides*.

The study of thin-sections is an important, but often neglected, part of micropalaeontological investigations. In a recent review, Hart (2016) has shown some of the benefits, but also the limitations, of thin-section analysis. There is no currently accepted methodology for the 'counting' of taxa present in thin-sections in a way that is both time-effective and capable of providing suitable palaeoecological information. Point-counting is often suggested as a way forward but this tends to identify the larger taxa (more 'hits'), but these are quite obvious even in a simple, rapid, inspection. The main problem with thin-sections is that even if a large number, made from one sample, are inspected one may still fail to see a critical view of, for example, the diagnostic embryonic apparatus of a large foraminiferid. In this analysis of the Kallankuruchchi Limestone Formation six thin-sections of each sample were inspected and the biotic and abiotic components recorded as a simple rare/scarce/common/abundant check list (Table 1).

Following Jacquin and de Graciansky (1998a, b) and Jacquin et al. (1998), the erosion surface at the base of the Kallankuruchchi Limestone Formation could, therefore, correspond to the major event at the base of their sequence Cam9. This sequence has a maximum flooding surface in the early Maastrichtian, followed by a fairly dramatic fall in global sea level: this is quite comparable to the succession recorded within the limestone succession and the overlying Kallamedu Sandstone Formation. Watkinson *et al.* (2007) placed Kallankuruchchi Limestone Formation within the lower part of their 'Post-rift sequence 3'–part of their Turonian–Maastrichtian 'Postrift Megasequence'.

The community represented by the assemblage of *Gryphaea*, echinoderms and oyster-encrusted surfaces appears to be relatively shallow water in aspect. This suggestion is supported by the presence of the larger foraminifera as their modern equivalents contain algal symbionts (Renema, 2006, 2008; Renema and Hart, 2012) and occupy environments within the photic zone. In the Maastricht Chalk of Limburg (Netherlands), which is mid-Late Maastrichtian in age, equivalent genera of larger foraminifera are associated with sea grass fossils which, by analogy to modern sea grass meadows, probably indicate a water depth of <20 m at most (Hart et al., 2016).

The similarity with the Maastricht Chalk is reduced in one respect: the distinctive taxon *Omphalocyclus macro-*

pora (Lamarck, 1816) is missing from the Ariyalur succession. This species is known from the Middle East (e.g., Oman) and from North India (Vrendenburg, 1908; Hart et al., 2000), but was clearly absent from the proto-East Indian Ocean between India and Australia/Antarctica. This is probably a genuine palaeoenvironmental absence as the 'lepidorbitoid sand' represented by the Dherani Limestone Member is a facies in which this taxon would be expected.

In some of the thin-sections (e.g., Figs 5F, 6A, 6C, 6F) we have noted the presence of encrusting red algae (as distinct from free-living rhodoliths; see Hart et al., 2015). In the absence of internal layering and reproductive cells it is almost impossible to identify the genus.

Table 1. This table provides a summary of the thin-section analysis of the Kallankuruchchi Limestone Formation which are arranged in stratigraphical order. Relative abundance was recorded as: RA – rare; SC – scarce; CO – common; and AB – abundant. Relative abundance was selected over a quantitative approach, such as point-counting for the reasons outlined in Hart (2016).

Rock Sample	Red algae	*Siderolites calcitrapoides*	*Lepidorbitoides socialis*	Planispiral foraminifera	Conical foraminifera	Low trochospiral foraminifera	Trochospiral foraminifera	Bryozoa	Echinoid spines	Echinoid fragments	Bivalve fragments	Micrite	Calcite crystals	Quartz grains	Limonite	Fieldspar grains	Rock fragments
KA7	SC	SC	–	CO	CO	CO	SC	CO	SC	–	–	AB	CO	CO	CO	–	–
KA5	CO	CO	AB	SC	AB	–	SC	AB	–	–	–	AB	CO	SC	CO	–	–
KA4	AB	CO	AB	SC	CO	SC	CO	AB	CO	–	–	AB	AB	AB	CO	–	–
KA3	CO	CO	AB	CO	CO	CO	CO	CO	CO	–	–	AB	AB	CO	AB	–	–
KA2	SC	CO	CO	CO	CO	SC	SC	CO	SC	–	–	AB	AB	AB	AB	–	–
KA1	RA	AB	AB	SC	CO	RA	CO	CO	CO	–	SC	AB	AB	CO	AB	–	RA
KA6	–	–	SC	CO	CO	SC	CO	CO	SC	–	–	SC	AB	SC	CO	–	–
KA11	–	–	–	CO	SC	SC	AB	SC	CO	SC	–	AB	CO	AB	AB	SC	–
KA10	–	–	–	SC	SC	–	SC	SC	CO	–	–	SC	AB	AB	AB	SC	–
KA9	–	–	–	–	SC	–	SC	RA	CO	–	–	SC	AB	AB	AB	SC	–
KA8	SC	–	–	CO	RA	–	SC	CO	CO	–	–	AB	CO	AB	AB	SC	–
KA14	RA	–	–	SC	CO	SC	RA	SC	CO	CO	RA	AB	AB	SC	AB	–	–
KA13	–	–	–	RA	SC	CO	RA	RA	RA	RA	–	SC	SC	AB	–	–	–
KA12	SC	CO	AB	SC	SC	–	SC	SC	RA	SC	–	AB	CO	SC	AB	–	–
KA16	–	–	–	–	–	–	–	–	–	SC	–	–	AB	RA	–	–	–

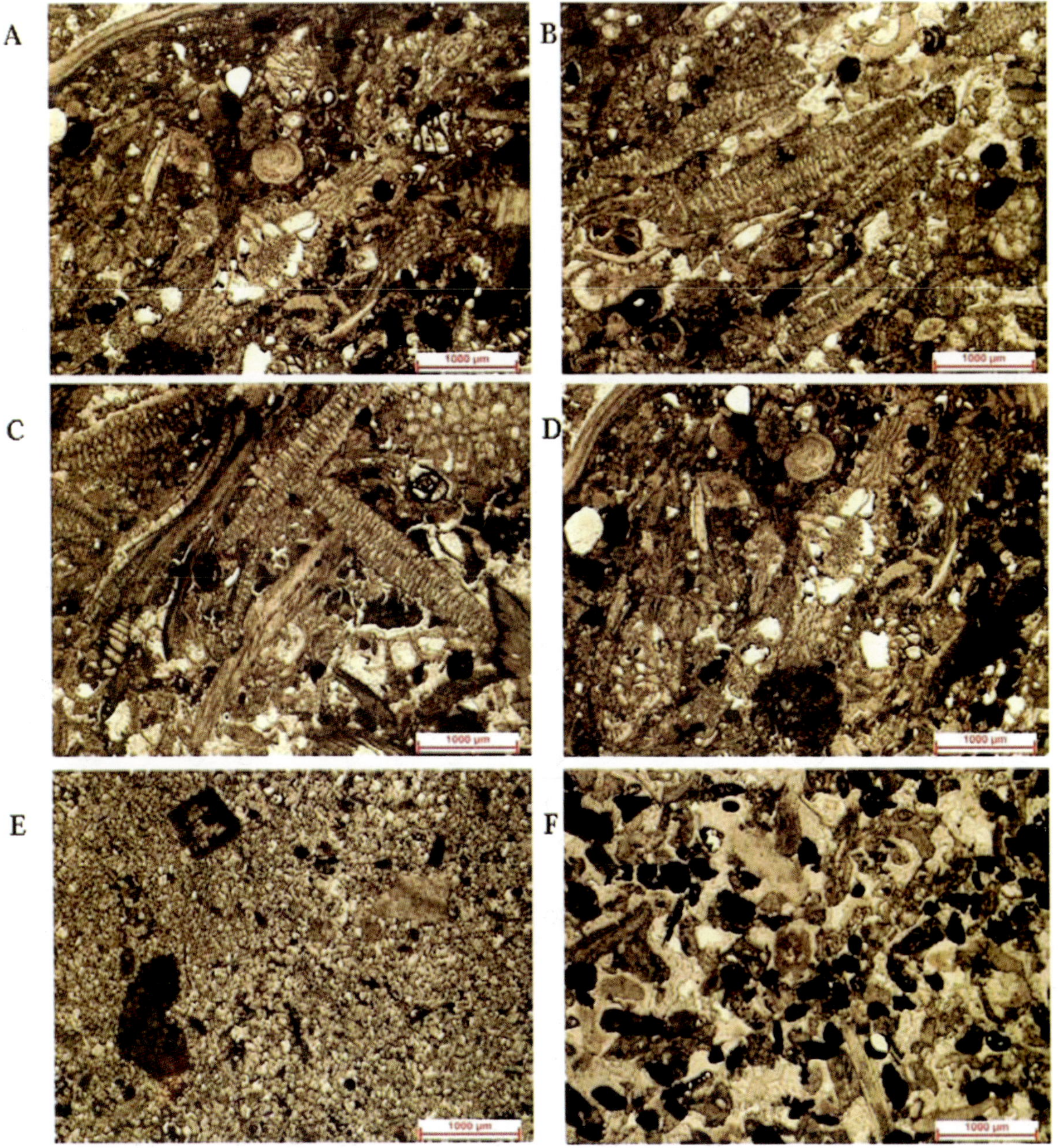

Fig.4. Thin section photomicrographs of the Kallankuruchchi Limestone Formation. **(A)**, KA12.1, *Siderolites calcitrapoides*, *Marssonella trochus* (d'Orbigny) in an highly fossiliferous bioclastic limestone; **(B)**, KA12.2, *Siderolites calcitrapoides*, *Lepidorbitoides* sp. cf. *L. socialis* (Leymerie) and *Lenticulina* sp.; **(C)**, KA12.3, *Lepidorbitoides* sp. cf. *L. socialis*, *Marssonella trochus* and *Quinqueloculina* sp.; **(D)**, KA12.4, *Siderolites calcitrapoides* and macrofaunal debris, including echinoid spines; **(E)**, KA16.2, sandstone with rare feldspar fragments and some macrofaunal debris; **(F)**, KA13.1, *Marssonella trochus* and macrofaunal debris, including echinoid spines.

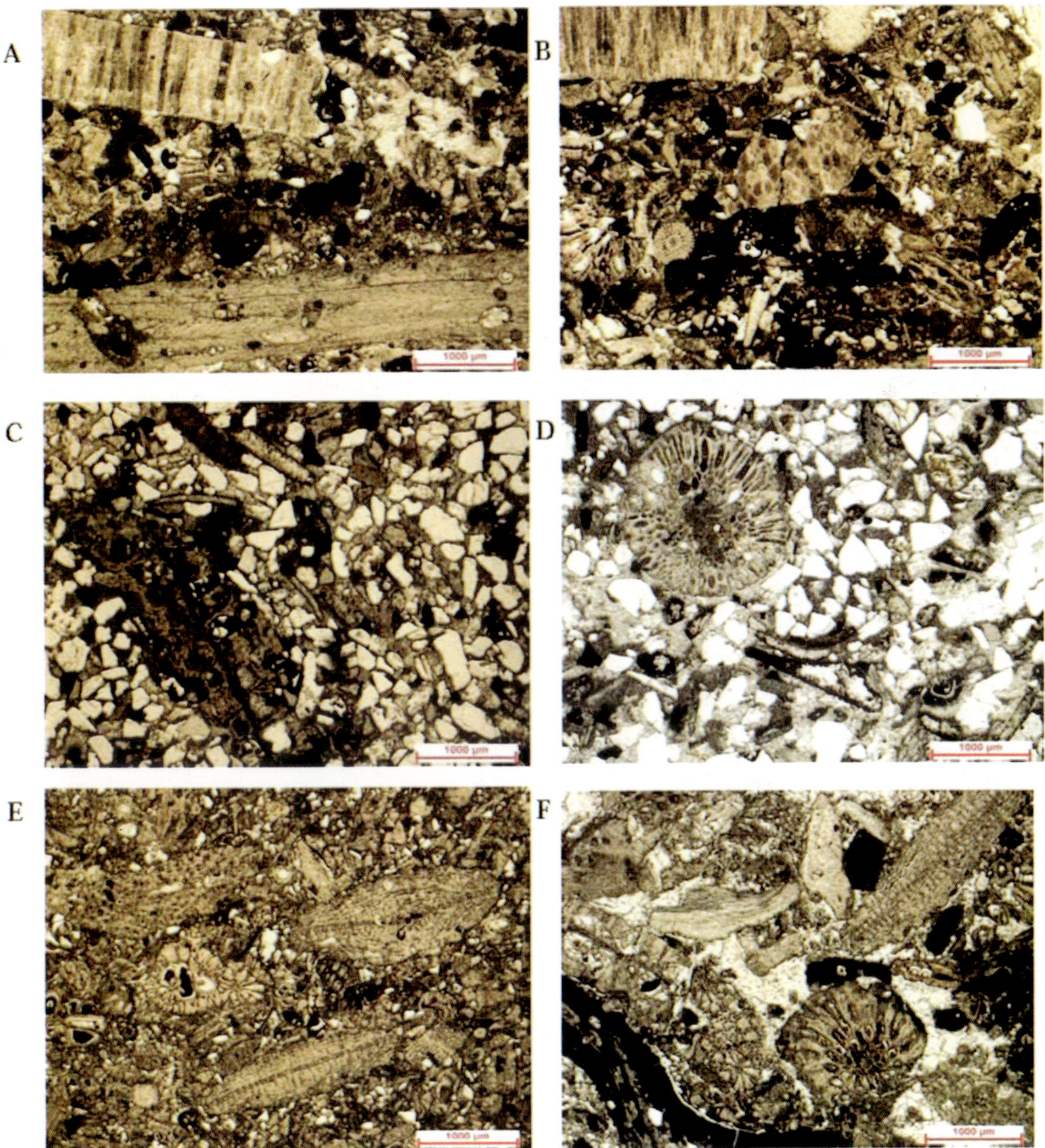

Fig.5. Thin section photomicrographs of the Kallankuruchchi Limestone Formation. Note that in all the images the 1000 µm scale bar is located in the bottom right hand corner of the image. **(A)**, KA14.1, *Inoceramus* and brachiopod shell fragments; **(B)**, KA14.2, *Inoceramus* fragments, bryozoa and echinoid spines; **(C)**, KA8.1, feldspathic sandstone with bryozoan fragments; **(D)**, KA11.1, feldspathic sandstone with bryozoan fragments and an indeterminate rotaliid; **(E)**, KA1.1a, *Siderolites calcitrapoides* and *Lepidorbitoides* sp. cf. *L. socialis*; **(F)**, KA1.2, *Siderolites calcitrapoides* and *Lepidorbitoides* sp. cf. *L. socialis* with bryozoan fragments and red algae.

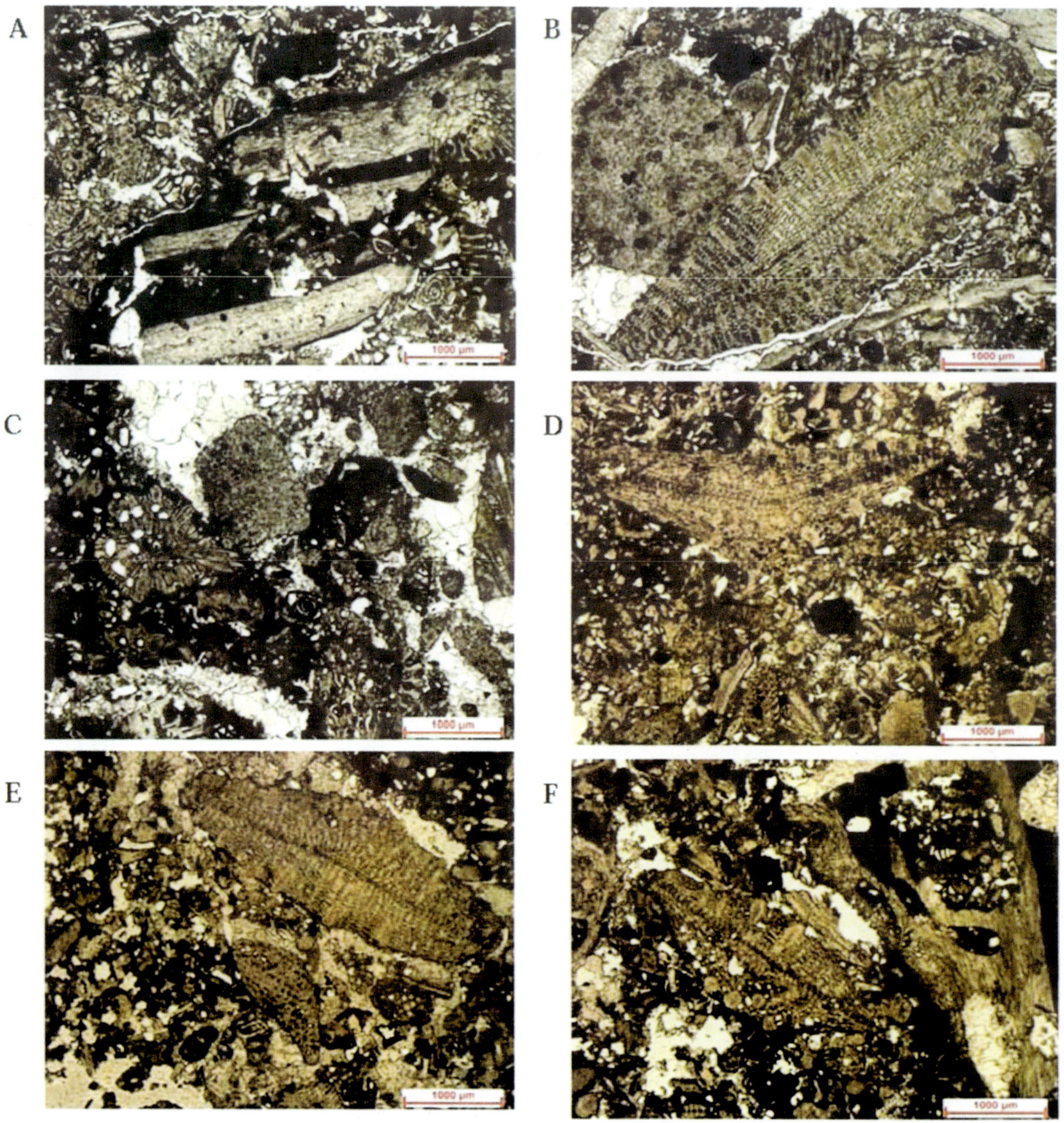

Fig.6. Thin section photomicrographs of the Kallankuruchchi Limestone Formation. Note that in all the images the 1000 µm scale bar is located in the bottom right hand corner of the image. **(A)**, KA4.1, *Marssonella trochus*, echinoid spines, an indeterminate rotaliid foraminifera and encrusting red algae; **(B)**, KA4.3, *Lepidorbitoides* sp. cf. *L. socialis* with bryozoan fragments; **(C)**, KA4.4, *Siderolites calcitrapoides*, *Marssonella trochus*, *Quinqueloculina* sp., bryozoan fragments and red algae; **(D)**, KA5.1, *Siderolites calcitrapoides* and *Lepidorbitoides* sp. cf. *L. socialis*; **(E)**, KA5.3, *Lepidorbitoides* sp. cf. *L. socialis* with bryozoan fragments; **(F)**, KA5.5, *Lepidorbitoides* sp. cf. *L. socialis* with red algae and a large bivalve shell (rudist?).

SUMMARY / CONCLUSION

The Kallankurichchi Limestone Formation has been confirmed as a latest Campanian to early Maastrichtian transgressive sequence (cycle). The distinctive assemblage of larger foraminifera, bryozoans, and large, distinctive, bivalves makes this an interesting part of the Cretaceous succession in the Cauvery Basin. Further work needs to be undertaken on the relationship with the underlying Sillakkudi Sandstone Formation and, especially, the overlying Ottakkovil Sandstone Member.

Acknowledgements: The authors acknowledge financial assistance from the British Council, the Royal Society (London), the Indian National Science Academy and Plymouth University. The accommodation provided by Dalmia Cement was much appreciated during our field visits. A large number of individuals have assisted us on visits to India, but special thanks go to the late R. Banerji, A. Govindan, C. Ravindran, S. Ramasamy, K. Ramamoorthy and R. Venkatachalapathy. James Quinn (Plymouth University) is thanked for providing the maps and diagrams used in this paper.

REFERENCES

ACHARYYA, S.K. and LAHIRI, T.C. (1991) Cretaceous palaeogeography of the Indian subcontinent; a review. Cretaceous Res., v.12, pp.3-26.

BLANFORD, H.F. (1862) On the Cretaceous and other rocks of South Arcot and Trichinopoly Districts. Mem. Geol. Surv. India, v.4, pp. 1-217 [published in 1865].

DIENI, I. (2010) Maastrichtian and Selandian decapods crustaceans from Sardinia. Boll. della Soc. Paleon. Italiana, v.49, pp.135-144.

HANCOCK, J.M. (1990) Sea level changes in the British region during the late Cretaceous. Proc. Geologists' Assoc., London, v.100, pp.565-594 [for 1989].

HART, M.B. (2016) Problem solving with microfossils: a brief review of the role of thin-section studies in micropaleontology. Gulf Coast Assoc. Geol. Societies, J. v.5, pp.111-129.

HART, M.B., BHASKAR, A. and WATKINSON, M.P. (2000) Larger foraminifera from the Upper Cretaceous of the Cauvery Basin, S.E. India. Mem. Geol. Soc. India, v.46, pp.159-171.

HART, M.B., CARTWRIGHT, K., FISK, B., SMART, C.W., CONSOLARO, C. and HALL-SPENCER, J.M. (2015) Foraminifera of the Fal Estuary (Cornwall), including taxa associated with maerl beds. Geoscience South-West England, v.13, pp.483-490.

HART, M.B., FITZPATRICK, M.E.J. and SMART, C.W. (2016) The Cretaceous/Paleogene boundary: Foraminifera, sea grasses, sea level change and sequence stratigraphy. Palaeogeogr. Palaeoclimat. Palaeoecol., v.441, pp.420-429.

JACQUIN, T. and DE GRACIANSKY, P.C. (1998a) Major transgressive/regressive cycles: the stratigraphic signature of European Basin development. *In*: DE GRACIANSKY, P.C., HARDENBOL, J., JACQUIN, T. and VAIL, P.R. (1998), Mesozoic and Cenozoic Sequence Stratigraphy of European Basins, SEPM, Spec. Pub., v.60, pp. 15-29.

JACQUIN, T. and DE GRACIANSKY, P.C. (1998b) Transgressive/regressive (second order) facies cycles: the effects of tectono-eustasy. *In*: DE GRACIANSKY, P.C., HARDENBOL, J., JACQUIN, T. and VAIL, P. R. (1998), Mesozoic and Cenozoic Sequence Stratigraphy of European Basins, SEPM, Spec. Pub., v. 60, pp. 31-42.

JACQUIN, T., DARDEAU, G., DURLET, C., GACIANSKY, P.C. and HANTZPERGUE,

P. (1998) The North Sea Cycle: an overview of 2nd-order transgressive/regressive facies in Western Europe. *In*: DE GRACIANSKY, P.C., HARDENBOL, J., JACQUIN, T. and VAIL, P.R. (1998), Mesozoic and Cenozoic Sequence Stratigraphy of European Basins, SEPM, Spec. Pub., v.60, pp. 445-466.

KOSSMAT, F. (1897) The Cretaceous deposits of Pondicherry. Rec. Geol. Surv. India, v.30, pp.51-110.

ÖZCAN, E. and ÖZCAN-ALTINER, S. (1999) The genus *Lepidorbitoides*: evolution and stratigraphic significance in some Anatolian Basins (Turkey). Rev. Micropaléont., v.42, pp.111-131.

POWELL, C.McA., ROOTS, S.R. and VEEVERS, J.J. (1988) Pre-breakup of continental extension in East Gondwanaland and the Early Cretaceous opening of the eastern Indian Ocean. Tectonophy, v.155, pp.261-283.

RADULOVIĆ, V. and RAMAMOORTHY, K. (1992) Late Cretaceous (Early Maastrichtian) brachiopods from South India. Senckenbergiana Lethaea, v. 72, pp. 77-89.

RAJU, D.S.N., RAVINDRAN, C.N. and KALYANSUNDAR, R. (1993) Cretaceous cycles of sea-level changes in the Cauvery Basin, India – a first revision. Bull. Oil Nat. Gas Comm, v.30, pp.101-113.

RAMASAMY, S. and BANERJI, R.K. (1991) Geology, petrology and systematic stratigraphy of pre-Ariyalur sequence in Tiruchirapalli District, Tamil Nadu, India. J. Geol. Soc. India, v.37, pp.577-594.

RAVINDRAN, C.N., KALYANASUNDAR, R. and RAMESH, P. (1995) Status of surface and subsurface stratigraphy of Cretaceous sequence in the Trichinopoly type area, Cauvery Basin, India. Proc. Petrotech-95, New Delhi, Tech. Trends in Petrol. Industry, v.2, pp.53-64.

RENEMA, W. (2006) Habitat variable determining the occurrence of large benthic foraminifera in the erau area (East Kalimantan, Indonesia. Coral Reefs, v. 25, pp. 351-359.

RENEMA, W. (2008) Habitat selective factors influencing the distribution of large benthic foraminifera assemblages over the Kepualauan Seribu. Marine Micropaleont., v.68, pp. 286-298.

RENEMA, W. and HART, M.B. (2012) Larger benthic foraminifera of the type Maastrichtian. In: JAGT, J.W.M., DONOVAN, S.K. and JAGT-YAZYKOVA, E.A. (eds), Fossils of the type Maastrichtian (Part 1), Scripta Geol. Spec. Issue, v.8, pp.33-43.

ROBASZYNSKI, F. and CARON, M. (1995) Foraminifères planctoniques du Crétacé: commentaire de la zonation Europe-Méditerrané. Bull. Soc. Géol. France, v.166, pp. 681-692.

ROBASZYNSKI, F., CARON, M. , GONZALEZ DONOSO, J.M., WONDERS, A.A.H. and the European Working Group on Planktonic Foraminifera (1984) Atlas of Late Cretaceous Globotruncanids. Rev. Micropaléont., v.26, pp.145-305.

ROBLES-SALCEDO, R., RIVAS, G., VICEDO, V. and CAUS, E. (2013) Paleoenvironmental distribution of larger foraminifera in Upper Cretaceous siliciclastic-carbonate deposits (Arén Sandstone Formation, South Pyrenees, Northeastern Spain). Palaios, v.28, doi : 10.2110/palo.2012.p.12-125r.

STOLICZKA, F. (1870–3) Cretaceous fauna of Southern India. The Pelecypoda with a review of all known genera of this class, fossil and recent. Palaeontologica Indica, Mem. Geol. Surv. India, ser. 1 & 2, v.1.

SUNDARAM, R. and RAO, P.S. (1979) Lithostratigraphy classification of Uttattur and Trichinopoly Groups of Upper Cretaceous rocks of Tiruchirapalli District, Tamil Nadu. Geol. Surv. India, Miscell. Pub., v.45, pp.111-119.

SUNDARAM, R. and RAO, P.S. (1986) Lithostratigraphy of Cretaceous and

Palaeocene rocks of Tiruchirapalli District, Tamil Nadu, South India. Rec. Geol. Surv. India, v.115, pp.9-23.

SUNDARAM, R., HENDERSON, R.A., AYYASAMI, K. and STILWELL, J.D. (2001) A lithostratigraphic revision of palaeoenvironmental assessment of the Cretaceous System exposed in the onshore Cauvery Basin, southern India. Cretaceous Res., v.22, pp.743-762.

TARLING, D.H. (1988) Gondwanaland and the evolution of the Indian Ocean. *In*: AUDLEY-CHARLES, M. G. and HALLAM, A. (eds), Gondwana and Tethys, Geol. Soc., London, Spec. Pub., v.37, pp.61-77.

TEWARI, A., HART, M.B. and WATKINSON, M.P. (1996) A revised lithostratigraphical classification of the Cretaceous rochs of Trichinopoly District, Cauvery Basin, Southeast India. *In*: PANDEY, J., AZMI, R.J., BHANDARI, A. and DAVE, A. (eds), Contri. XV Indian Colloq. Micropalaeont. Stratigr., pp.798-800.

VAN GORSEL, J.T. (1975) Evolutionary trends and stratigraphic significance of the Late Cretaceous *Helicorbitoides-Lepidorbitoides* lineage. Utrecht Micropalaeont. Bull. v.12, pp.1-99.

VRENDENBURG, W. (1908) The Cretaceous *Orbitoides* of India. Rec. Geol. Surv. India, v.36, pp.172-213.

WATKINSON, M.P., HART, M.B. and TEWARI, A. (2007) Cretaceous tectono-stratigraphy and the development of the Cauvery Basin, southeast India. Petrol. Geosci., v.13, pp.181-191

Micropaleontology and its Applications
Pradeep K. Kathal, Rajiv Nigam & Abu Talib (Editors)
Scientific Publishers (India), 2018, 15-39 pp.

Dynamics of Larger Benthic Foraminiferal Populations in the Upper Florida Keys Reef Tract, USA: Comparisons with other Coral Reef Bioindicators

Elizabeth Fisher Moses[1,2], Pamela Hallock*[2,3] and John E. Fauth[3]

[1]Science, Engineering, Technology Division, Howard Community College, Columbia, MD, USA
[2]College of Marine Science, University of South Florida, St. Petersburg, FL, USA
[3]Department of Biology, University of Central Florida, Orlando, FL 32816, USA
[*]**Email**: pmuller@usf.edu

Abstract

Larger benthic Foraminifera (LBF), which host algal endosymbionts, are useful bioindicators for coral reefs worldwide; their responses can indicate changes in water quality, as well as presence and intensity of photo-oxidative stress. This study examined LBF populations quarterly for two years at sites where selected parameters for scleractinian corals also were monitored. Population densities of 11 common LBF species were examined at four 6 m deep patch reefs within Biscayne National Park (BNP) and the upper Florida Keys National Marine Sanctuary (FKNMS), and at 3, 6, 9 and 18 m depths in FKNMS. Populations of *Amphistegina gibbosa*, the dominant species, were also assessed for size distribution and prevalence of bleaching. At the 6 m sites, densities of LBF ranged from 10^3–10^5 m^{-2} and were typically highest at Algae Reef (AR), which was near an intact mangrove shoreline. The 6 m site in BNP had the lowest densities of LBF overall. At the 6 m sites, incidences of partial bleaching in adult *A. gibbosa*, typically <40%, indicated chronic photo-oxidative stress. Along the depth transect, LBF densities ranged from 5×10^3–5×10^5 m^{-2}, with higher densities at the deeper sites. Overall, the bank-margin reefs appeared to provide better habitat for these foraminifers than inshore patch reefs, although bleaching prevalence was higher offshore during summer months. This apparent paradox demonstrates the importance of high water quality to these foraminifers, even though clear water can result in chronic photo-oxidative stress. Data from the LBF populations were compared with previously published lesion-recovery data for coral colonies of *Orbicella annularis* species complex, revealing the same ranking for 6 m sites as LBF abundances: highest at AR and lowest at BNP. In contrast, along the depth transect, lesion recovery was highest at the 3 m patch-reef site and lowest at the 18 m site, consistent with long-term trends of major loss of coral cover from offshore reefs. Algae Reef represented the "sweet spot" among the reefs examined, including highest coral cover and lesion recovery, as well as LBF abundances comparable to the offshore sites. Results indicate that tannins, originating from mangrove shorelines, can play an important photo-protective role in reef waters.

Keywords: *Amphistegina*, bleaching, coral lesions, marine protected area, *Orbicella*.

INTRODUCTION

Quantifying the relative contributions of stressors acting at multiple spatial and temporal scales is a major challenge in reef conservation and management (e.g., Sammarco et al., 2007; Dustan et al., 2008). Responses of larger benthic foraminifers (LBF) can indicate changes in water quality (e.g., Cockey et al., 1996; Hallock et al., 2003; Uthicke and Nobes, 2008), as well as the presence and intensity of photo-oxidative stress, on time scales of weeks to months (Hallock et al., 2006 a,b). Reduced water quality, especially associated with increased nutrient flux, causes a shift from reef sediments dominated by symbiont-bearing, calcifying organisms (e.g., shells of LBF and coral fragments) to those dominated by calcareous algae, bioerosional products, and the calcareous remains of herbivorous and detrivorous taxa (e.g., Lidz and Hallock, 2000; Uthicke and Nobles, 2008; Oliver et al., 2014). Thus, LBF and other constituents of reef sediments reflect the makeup of the benthic community and its responses to local environmental conditions.

Benthic foraminifers that host algal endosymbionts are useful bioindicators for coral-reef studies because they 1) have physiological analogies and require similar water quality as zooxanthellate corals, 2) are abundant in healthy reef ecosystems and can be collected with minimal effort or effect on reef resources, and 3) have relatively short life spans and more rapid responses to environmental stressors than do coral populations (Hallock et al., 2003; Uthicke et al., 2010; Oliver et al., 2014). This project evaluated LBF assemblages ubiquitous in western Atlantic–Caribbean reef environments (Hallock, 1988; Baker et al., 2009), focusing on detailed population assessment of *Amphistegina gibbosa* d'Orbigny, at the same time coral assemblages and lesion recovery in individual coral colonies were studied (Downs et al., 2005, 2006; Fisher et al., 2007). Simultaneous assessment of LBFs and corals allowed us to compare responses of several indicators of reef health.

An important consideration in designing studies utilizing LBF is that this group is diverse, both with respect to the foraminiferal taxa and the symbionts they host (e.g., Hallock, 1988; Lee, 2006). While *Amphistegina* (subsequently abbreviated as "*Am.*") and a few generally less-abundant genera (e.g., *Heterostegina, Peneroplis,* and *Sorites*) occur on reefs nearly worldwide, regional differences in the composition of LBF assemblages must be considered. Nevertheless, because of their algal symbioses, LBF are typically most abundant and diverse in low-nutrient environments and therefore can be used as bioindicators on reefs worldwide, for example, in Florida (USA), Indonesia, the Great Barrier Reef of Australia, Majuro, and Puerto Rico (e.g., Cockey et al., 1996, Renema and Troelstra, 2001; Uthicke and Nobes, 2008; Osawa et al. 2010; Oliver et al., 2014; respectively).

Rationale for Assessing Populations of Amphistegina

Amphistegina is the most common, algal symbiont-bearing, foraminiferal genus on reefs worldwide (e.g., Hallock, 1988; Langer and Hottinger, 1997; Weinmann et al., 2013), typically found living on coralline and filamentous algae on reef substrata, as well as on macrophytes and less commonly in sediments. The two most similar species (Caribbean: *Am. gibbosa*; Indo-Pacific: *Am. lessonii*) occur from the shallowest subtidal zones to as deep as 100 m, and tend to be most abundant between 10 m and 40 m (Hallock, 1999; Hohenegger, 2004). Field studies and laboratory experiments demonstrated that these two species are biologically and ecologically very similar (e.g., Hallock et al., 1986; Hallock, 2000a).

Distributions of *Amphistegina* spp. are controlled regionally by temperature (~12–33°C) and locally by hydrodynamics, water quality, light and substrata (e.g., Renema and Troelstra, 2001; Hohenegger, 2004; Uthicke and Altenrath, 2010; Weinmann et al., 2013). Populations of *Amphistegina* exhibit stress symptoms similar to those reported for zooxanthellate corals, including bleaching, predation, and microbial infestations (Hallock et al., 2006 a, b, and references therein). *Amphistegina* responds to environmental stress within days to weeks, thereby providing a low-cost method to quickly distinguish between the relative contributions of water quality, photo-oxidative stress, and other factors. These characteristics make *Amphistegina* spp., particularly the very abundant siblings, *Am. gibbosa* and *Am. lessonii,* powerful indicators for coral-reef conservation and management worldwide, with uses ranging from determining the suitability of reef areas for coral-restoration efforts to forecasting the likelihood of coral bleaching (Hallock et al., 2006a; Hallock, 2012).

Amphistegina spp. host diatom endosymbionts, so light is necessary for growth and calcification. However, high intensities of solar radiation, particularly shorter wavelengths (blue–ultraviolet: 290-490 nm; $\geq$ 0.10 W m^{-2}), can induce photoinhibition (Lee et al., 1980; Williams and Hallock, 2004) and loss of symbionts (Hallock et al., 1986, 1995; Talge and Hallock, 2003), while suppressing growth rates (Williams and Hallock, 2004). Bleaching was first documented in *Am. gibbosa* and *Am. lessonii* in laboratory experiments (Hallock et al., 1986), subsequently in field populations of *Am. gibbosa* on the Florida reef tract beginning in summer 1991 (see Hallock et al., 2006 a,b, and references therein), and worldwide in *Amphistegina* spp. in the 1990s (Hallock, 2000a). Loss of symbiont color is caused by degradation and digestion of the diatom endosymbionts; partial bleaching induced by photoinhibition in laboratory experiments is cytologically identical to that in field-collected specimens (Talge and Hallock, 1995, 2003).

Unlike corals, bleaching in *Am. gibbosa* is not induced by temperatures currently experienced along the Florida reef tract; bleaching prevalence typically peaks in early summer, well before maximum seasonal water temperatures (Hallock et al., 1995, 2006 a,b). Bleaching in *Am. lessonii* and other species, which was observed in January 1996 on the west Australian shelf, occurred in temperatures <25°C (Hallock, 2000a). Instead, bleaching in *Am. gibbosa* and *Am. lessonii* has been linked to solar radiation based on laboratory experiments (as cited above); timing of onset; seasonal and latitudinal trends in bleaching in field populations (Hallock et al., 1995; Williams et al., 1997); and significantly higher bleaching prevalence at sites with significantly higher water transparency (Williams, 2002; Ayoub et al., 2009).

Although mass bleaching in corals is strongly correlated with elevated temperature (Glynn, 1996; Brown, 1997; Eakin et al., 2010; many others), local and regional variability of coral bleaching often has been explained by differences in solar radiation, particularly shorter, higher-energy wavelengths that induce photo-oxidative stress (e.g., Lesser et al., 1990; Gleason and Wellington, 1993; Lesser and Farrell, 2004). Because *Amphistegina* typically bleach in response to excess radiant energy in spring and early summer (March-July), monitoring their bleaching can allow reef managers to forecast increased susceptibility of corals to photo-oxidative stress and mass bleaching if sea surface temperatures approach bleaching thresholds later in the summer (Hallock et al., 2006 a, b).

Bleaching also affects reproductive success and recruitment of *Am. gibbosa* in ways that can provide considerable information about environmental stressors (Hallock, 2000b; Fisher, 2007). Medium-to-high population densities indicate environments suitable for growth and reproduction over recent weeks to months. High population densities with high proportions of juveniles in summer months indicate favorable environmental conditions, while very low population densities dominated by juveniles indicate environmental conditions temporarily suited for survival of propagules (Alve and Goldstein, 2003, 2010) or juveniles carried into the environment by currents, but too variable for growth to adult sizes. High population densities and high incidences of extensive bleaching in adults (>50% of the population partly bleached or bleached) indicate acute photo-oxidative stress that occurred after reproduction peaked, while high population densities and lower incidences of bleaching (<50% partly bleached) indicates chronic photo-oxidative stress. Chronic stress from bleaching also is associated with increased shell breakage (Toler and Hallock, 1998), susceptibility to predation (Hallock and Talge, 1994), and endolithic infestation (Hallock, 2000a). Therefore, damage assessment can indicate how chronic stress affects the population.

Objectives of the Study

Coral reefs along the Florida Keys Reef Tract have declined for several decades (e.g., Dustan and Halas, 1987; Somerfield et al., 2008; Ruzika et al., 2013) due to both chronic and acute stressors. We used a multi-proxy approach to assess stress responses and environmental conditions at two suites of coral reefs. Our study provided a unique opportunity to examine LBF populations at sites where morphological and physiological conditions of scleractinian corals and other reef organisms concurrently were assessed (Downs et al., 2005, 2006; Fisher et al., 2007; Fisher, 2007; Moses and Hallock, 2015). The basic questions we address in this paper are a) do population densities of common LBF differ among patch reefs at similar depths along the upper Florida Keys Reef Tract; b) do morphological characteristics of populations of *Am. gibbosa* differ among these reefs; and c) do LBF population densities or d) morphological characteristics of *Am. gibbosa* differ along an onshore–offshore depth transect? We then compare LBF responses with previous studies of LBF population dynamics and with previously published characteristics of the reef sites, including differences in solar radiation, physiological responses of selected coral colonies of *Orbicella annularis* (Ellis and Solander) species complex, and the lesion regeneration rates after biopsy samples were taken from the coral colonies.

METHODS

Sampling and Assessment of LBF

We collected foraminifers at seven reef sites in the Upper Florida Keys Reef Tract: one patch reef in Biscayne National Park (BNP), as well as four patch reefs and two deeper fore-reef sites in the upper Florida Keys National Marine Sanctuary (FKNMS). The four patch-reef sites at 6 m depth represented differences in proximity to urbanized coastal development. The depth transect, off Key Largo (KL sites) also included four sites, one each at 3, 6, 9 and 18 m depth, with the KL 6m site common to both transects (Fig. 1). All sites were suitable for LBFs with respect to temperature, salinity and hydrodynamics.

Between August 2001 and February 2003, reef rubble samples were collected quarterly at the 6 m sites; quarterly sampling along the depth gradient began in October 2001. Methods of sampling LBF and assessing *Amphistegina* populations are detailed elsewhere (Hallock et al., 1995, Baker et al., 2005). Reef rubble, although not the only habitat available to LBF, is easily compared among study sites. Suitable rubble for LBFs was readily available at all sites except KL 3 m, where rubble was both hard to find and often partially buried in fine sediments.

Rubble pieces were placed in plastic bags, which were brought to the surface and placed in a shaded bucket for transport to shore. Onshore, rubble was scrubbed using a small brush to detach associated sediment, algae and meiofauna, including Foraminifera. The approximate area of bottom covered by each rubble piece was determined using image-analysis software. Residue from each scrubbed rubble sample was distributed into 150 mm diameter petri dishes, which were then placed in a culture chamber maintained between 23 and 28°C (depending on the time of year) on a 12-hour light/dark schedule at ~10 µmol photon m^{-2} s^{-1} PAR. For each sample, we removed all LBF whose behavior or color indicated a high probability of being alive (Hallock et al., 1986) and placed them into a 100 mm diameter Petri dish of seawater.

We identified to species and counted each specimen verified to be alive based on reticulopodial activity. Maximum diameter of each *Am. gibbosa* specimen was measured to the nearest 0.05 mm, and each specimen was characterized according to symbiont color and presence of damage, either from breakage, predation, endolithic infestation or deformation [see illustrations in Hallock et al. (1995) and Toler and Hallock (1998)]. Individual *Amphistegina* were visually characterized as "unbleached" (uniform brown color), "partly bleached" (specimens possessing white patches with >50% retaining symbiont color), or "bleached" (specimens with <50% symbiont color remaining) (see Hallock et al., 1995). Bleaching in *Am. gibbosa* is typically progressive; severity of bleaching increases with increasing size of the individual (Hallock et al., 1995; Williams et al., 1997). Prior to the onset of stress in 1991, all adult individuals observed were normally colored. Subsequently, in affected populations, individuals <0.5 mm diameter seldom exhibited symbiont loss whereas specimens >1.0 mm seldom were normally colored. Therefore, we only evaluated bleaching in adult *Am. gibbosa* (> 0.6 mm) as recommended by Williams (2002) and Hallock et al. (2006 a, b). If a sample contained larger numbers of individuals (>150), all were counted, but a subsample of 150–200 individuals was measured and characterized.

Ten other species of LBF encountered live in each sample also were identified to species and counted, including *Archaias angulatus* (Fitchel and Moll), *Asterigerina carinata* d'Orbigny, *Borelis pulchra* (d'Orbigny), *Broeckina orbitolitoides* (Hofker), *Cyclorbiculina compressa* (d'Orbigny), *Heterostegina depressa* d'Orbigny, *Laevipeneroplis bradyi* (Cushman), *L. proteus* (d'Orbigny), *Peneroplis pertusus* (Forskål), and *Sorites marginalis* (Lamarck). See Baker et al. (2009) for more details on identification of these taxa.

Data Analysis

Data were analyzed in two groups: (1) by sites of 6 m depth along the northeast–southwest traverse (BNP, AR, WB and KL 6 m), and (2) by sites along the depth gradient: KL 3, 6, 9 and 18 m. As noted above, the 6 m site was common to both groups (Fig. 1).

We used two-way crossed analysis of similarities (ANOSIM) to determine if densities of LBF assemblages differed significantly among sites (averaged over the entire study period) or across time (averaged across all sites). To do this, we first calculated Bray-Curtis similarity matrices for all square-root transformed LBF densities. Analyses based on square-root transformed densities were better at distinguishing among sites than log-transformed densities, which downplay the importance of dominant species (Clarke and Warwick, 2001). We used multi-dimensional scaling (MDS) to determine how sites grouped based on densities of all LBF, followed by SIMPER (similarity percentages) analyses to determine which species were primarily responsible for site groupings (Clarke and Warwick, 2001). We performed ANOSIM, MDS and SIMPER using PRIMER v.6 (Plymouth Routines in Multivariate Ecological Research PRIMER-E Ltd., Plymouth). All other statistical analyses used JMP v.3.2. (SAS Institute Inc., Cary, NC, USA), with $\alpha = 0.05$ for all hypothesis tests.

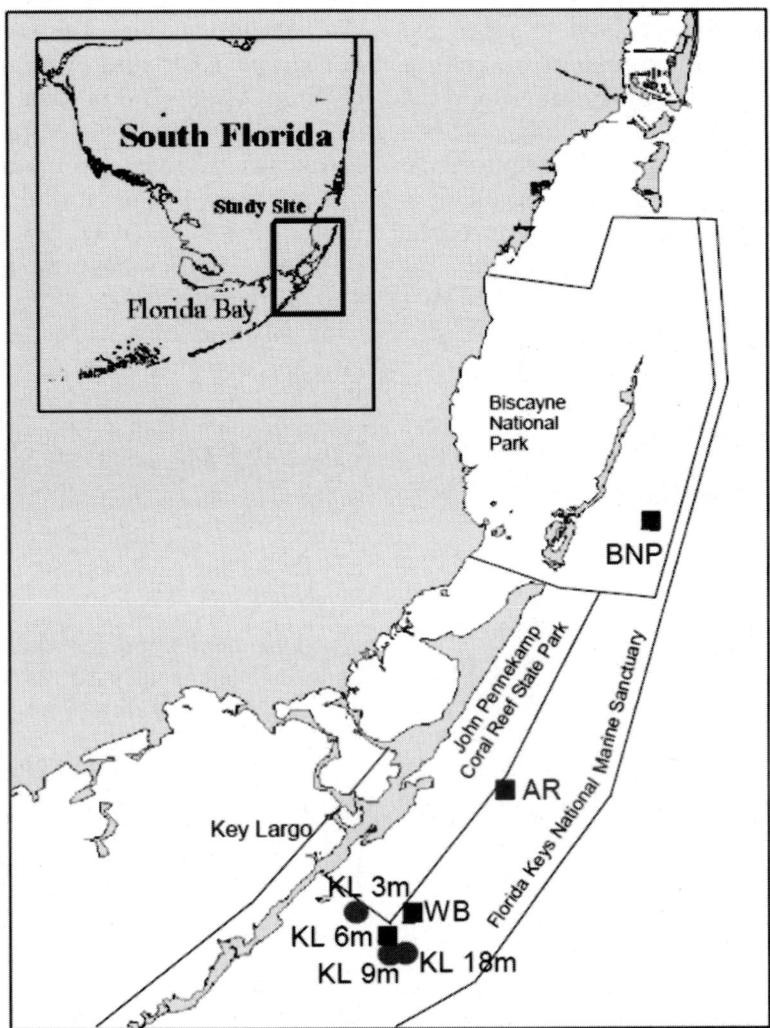

Fig.1. Map showing the seven sampling sites in the Florida Keys National Marine Sanctuary and Biscayne National Park. ■ squares designate the 6 m sites, including the four patch reefs: Key Largo (KL) 6 m (25° 0192' N, 80° 23.844' W), White Banks (WB) (25° 02.232' N, 80° 22.496' W), Algae Reef (AR) (25° 08.799' N, 80° 17.579' W), and Alina's Reef (BNP) (25° 23.185' N, 80°09.775' W). ● circles designate sites along the depth gradient, including two patch reefs, KL 3 m (25° 02.447' N, 80° 25.442' W) and KL 6 m, and two depths on one forereef, KL 9 m (25° 00.146' N, 80° 23.626' W) and KL 18 m (25°00.206' N, 80° 23.023' W).

For *Am. gibbosa* in each sample, besides mean densities we also recorded the percentage (%) of adults that exhibited bleaching (e.g., were partly bleached or bleached; Williams, 2002), % of specimens that exhibited shell damage, % juveniles (individuals <0.6 mm in diameter), and mean diameter. We used repeated-measures MANOVA to determine if significant differences existed in % juveniles and % shell damage among sites, checking model assumptions (e.g., sphericity, homogeneity of variances, normality and independence) using residual plots. Density and diameter data were log_{10} transformed to meet these assumptions. In cases where the sphericity assumption was not met, we applied a univariate (unadjusted epsilon) approach. To interpret differences detected by MANOVA, we used one-way ANOVA followed by the Tukey-Kramer Honestly Significant Difference (HSD) method.

RESULT

The 6 m Patch-Reef Sites

Combined densities of LBF from 124 samples collected at the 6 m sites ranged from approximately 10^3–10^5 individuals m^{-2} with a mean of $2.2 \times 10^4 \pm 1.6 \times 10^3$ m^{-2} (± SE hereinafter; n = 124; Fig. 2A). Densities were highest at AR and lowest at BNP. Geographic location was the primary factor affecting LBF densities (averaged across all time periods; ANOSIM2: Global R = 0.34, $p \leq 0.001$), with significant differences among all sites (Table 1). The greatest dissimilarity was between WB and BNP and the lowest between WB and KL 6 m. Densities also differed significantly with time (averaged across all 6 m sites; ANOSIM2: Global R = 0.31, $p \leq 0.001$), with the greatest dissimilarity between August 2001 and June 2002 (Table 1). SIMPER analysis based on LBF revealed AR and BNP to be most dissimilar, while WB represented a midpoint between AR and KL 6 m. Alina's Reef (BNP) was more dissimilar to all FKNMS sites than they were to each other. Those trends are illustrated in the MDS plot (Fig. 2B), with the BNP data mostly plotting to the left and the data from the three FKNMS sites plotting to the right, especially the AR data points.

Amphistegina gibbosa was the dominant LBF and primary species that distinguished all 6 m sites (Table 2), based on similarity (SIMPER) analyses. Densities of live specimens ranged from 4.8×10^2–4.4×10^4 individuals m^{-2} with a mean of 1.2×10^4 ($\pm 8.5 \times 10^2$) m^{-2}, and differed significantly among the 6 m sites and sampling dates (Fig. 3A); no site × time interactions were detected. The AR site had significantly higher densities than KL 6 m and BNP, while WB had significantly higher densities than BNP. Population densities were significantly higher in June and August 2002 than in August 2001 and February 2003. Those trends again are evident in the MDS plot (Fig. 2C).

Two other species also were identified by SIMPER analyses as contributing to dissimilarity among 6 m sites (Table 2). *Archaias angulatus* was second in abundance and the second most diagnostic species (Fig. 4A). Densities averaged $4.0 \times 10^3 \pm 5.1 \times 10^2$ individuals m^{-2}. Densities at BNP were significantly lower than at the other three sites. *Laevipeneroplis proteus* was third in importance. Densities averaged $2.5 \times 10^3 \pm 2.6 \times 10^2$ individuals m^{-2}, differing significantly among sites, seasons and their interactions (Fig. 2D). Notably, densities of *L. proteus* were significantly lower at BNP than the other sites in August and October 2001. Two other species identified as diagnostic were *H. depressa* and *B. orbitolitoides*. Densities of *H. depressa* averaged 645 ± 88 individuals m^{-2}, with significant differences among sites and with site × time interactions (Fig. 2C), generally following the pattern of highest densities at AR and lowest at BNP. Densities of *B. orbitolitoides* varied significantly among sites, seasons and their interactions, with mean densities of 452 ± 86 individuals m^{-2} (n = 124) (Fig. 2B).

Table 1. Results of two-way crossed ANOSIM comparing LBF assemblages from the 6 m sites across sampling times (Aug 2001–Feb 2003); $*p<0.05$, $**p<0.01$, $***p=0.001$.

Global Tests

	R	*p*
Site	0.34	0.001
Sampling Period	0.31	0.001

Pair wise Tests comparing differences among 6 m sites across all sampling periods

	WB	AR	BNP
KL 6 m	0.23**	0.34***	0.43***
WB		0.24**	0.47***
AR			0.46***

Pair wise Tests comparing differences among sampling periods across 6 m sites

	Oct 2001	Mar 2002	Jun 2002	Aug 2002	Nov 2002	Feb 2003
Aug 2001	0.25	0.19*	0.69***	0.45***	0.02	0.23*
Oct 2001		0.38***	0.66***	0.45**	0.10	0.43**
Mar 2002			0.30*	0.46***	0.28**	0.08
Jun 2002				0.36***	0.35**	0.36***
Aug 2002					0.20**	0.41***
Nov 2002						0.16**

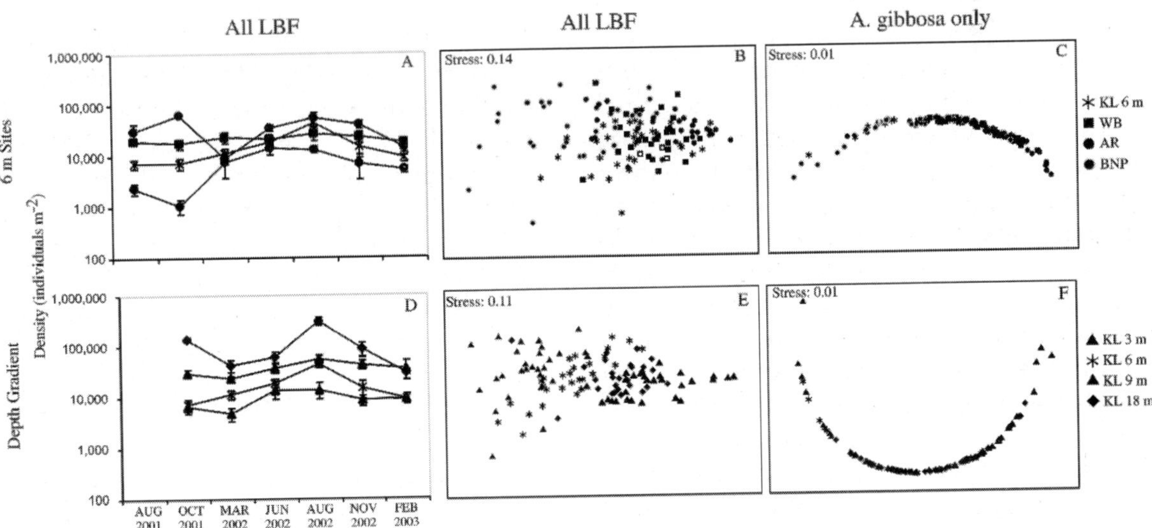

Fig.2. Comparison of sites based upon the LBF assemblages: Mean (± SE) densities of all symbiont-bearing ('larger') foraminifers along **(A)** the 6 m sites and **(D)** the depth gradient. Densities are plotted on a log scale. Multi-dimensional scaling plots illustrate the ordination of samples from the 6 m sites based on **(B)** the entire assemblage and **(C)** only *Am. gibbosa*, and from along the depth gradient based on **(E)** the entire assemblage and **(F)** only *Am. gibbosa*.

Table 2. Key LBF whose abundances differed among the 6 m sites (Aug 2001–Feb 2003), and along the depth gradient (Oct 2001–Feb 2003). Species are listed in order of importance.

Comparison	Dissimilarity	Species that together explained ≥50% dissimilarity between sites
6 m Sites		
AR–BNP	52	*Am. gibbosa, Ar. angulatus, L. proteus*
WB–BNP	49	*Am. gibbosa, Ar. angulatus, L. proteus*
KL6m–BNP	46	*Am. gibbosa, Ar. angulatus, L. proteus*
KL6m–AR	41	*Am. gibbosa, Ar. angulatus, L. proteus*
WB–KL6m	36	*Am. gibbosa, Ar. angulatus, L. proteus*
WB–AR	36	*Am. gibbosa, Ar. angulatus, L. proteus*
Depth Transect		
KL 3m vs KL 9m	65	*Ar. angulatus, C. compresssa, Am. gibbosa, L. proteus, As. carinata*
KL3m vs KL18m	55	*Am. gibbosa, L. proteus, As. carinata, C. compressa, Ar. angulatus*
KL3m–KL6m	43	*Am. gibbosa, Ar. angulatus, L. proteus*
KL6m–KL9m	52	*Ar. angulatus, C. compressa, Am. gibbosa, L. proteus, As. carinata*
KL6m–KL18m	43	*Am. gibbosa, L. proteus, As. carinata, C. compressa, Ar. angulatus*
KL9m–KL18m	37	*Ar. angulatus, Am. gibbosa, C. compressa, L. proteus*

Populations of *Am. gibbosa* at the 6 m sites were significantly dynamic with respect to the other parameters investigated (Fig. 3). Percentage of adults exhibiting any degree of bleaching was highly variable, with an overall mean of 24 ± 1%. Incidences of partial bleaching at AR were significantly higher than at KL 6 m (Fig. 3B). Juvenile *Am. gibbosa* averaged 62 ± 1% of the population and were a significantly greater percentage of the total population in June and August 2002 than in February 2003 (Fig. 3C). Similarly, mean diameters were consistently quite small, ranging from 0.40 to 0.77 mm, with an overall mean of 0.53 ± 0.01 mm. Specimens were on average smaller in summer 2002 (June and August) than other months and significantly larger in February 2003 (Fig. 3D). The percentage of the population with damaged tests averaged 17 ± 1% (Fig. 3E), differing significantly among sites and with time but not their interactions. The BNP site had significantly higher percentages of damaged *Am. gibbosa* tests than WB. Temporally, percentages of test damage significantly increased in winter months.

Depth Transect off Key Largo

Along the depth gradient, the combined average density of all LBF was $4.3 \times 10^4 \pm 6.1 \times 10^3$ individuals m^{-2} (n = 125). Depth was the primary factor affecting LBF densities (averaged across all time periods; ANOSIM2: Global R = 0.63, $p \leq 0.001$) with significant differences across depths (Table 3). The greatest dissimilarity was between KL 3 m and 9 m, with the least dissimilarity between KL 9 m and 18 m. Densities differed significantly with time (averaged across depths; ANOSIM2: Global R = 0.31, $p \leq 0.001$) with maximum differences observed between March 2002 and June 2002 (Table 4; Fig. 2D). Those trends are illustrated in the MDS plot (Fig. 2E), with the KL 3m data mostly plotting to the left and the data from the KL 9 m site plotting to the right.

Amphistegina gibbosa was again the dominant LBF (Fig. 3F). Densities of live *Am. gibbosa* ranged from 5.5×10^2–1.1×10^5 individuals m^{-2} with an overall mean of $1.6 \times 10^4 \pm 1.7 \times 10^3$ m^{-2}. Other contributing species along the depth gradient included *Ar. angulatus, C. compressa, L. proteus* and *Asterigerina carinata* (Table 3, Fig. 4E–H). The largest differences were between KL 3 m and the deeper sites: KL 9 m (90%) and KL 18 m (74%); while KL 9 m and 18 m were most similar. Those trends are illustrated in the MDS plot (Fig. 2F), with the KL 3m data mostly plotting to the left and the data from the KL 9 m site plotting to the right.

Densities of *Am. gibbosa* differed significantly among depths (Fig. 3F) but not with time or time × depth interactions. Samples from KL 9 m and 18 m had significantly higher densities than KL 3 m and 6 m. Percentages of adult *Am. gibbosa* that were partly bleached (Fig. 3G) were significantly higher at the KL 9 m and 18 m sites (typically >30%) than at the KL 3 m and 6 m (typically <30%), with summer and early autumn highs and winter lows. The percentages of juvenile *Am. gibbosa* in the populations (Fig. 3I) differed significantly, as samples from KL 3 m and 6 m had consistently higher percentages of juveniles (both 61 ± 3%) than KL 9 m and 18 m (45 ± 4% and 43 ± 2%).

Juveniles were more prevalent in summer 2002 than in February 2003. Mean diameters of *Am. gibbosa* also differed significantly with depth and time (Fig. 3H). Specimens were typically larger at KL 9 m and 18 m (0.68 ± 0.02 mm) than at KL 3 m and 6 m (0.53 ± 0.01 mm). Seasonal differences were much more prominent at the KL 9 and 18 m sites (Fig. 3I), with the largest individuals found in late winter/early spring samples (March 2002 and February 2003). The percentages of *Am. gibbosa* with damaged tests averaged 16 ± 1% (n = 100), differing significantly only with depth (Fig. 3J). The KL 3 m site had a significantly lower percentage of damaged *Am. gibbosa* tests than more hydrodynamically exposed sites (KL 3 m: 9 ± 1% vs. KL 6 m: 16 ± 2%, KL 9 m: 18 ± 2% and KL 18 m: 21± 2%; ANOVA: $F_{3,96} = 9.3$, $p < 0.001$).

DISCUSSION

The patterns of LBF densities among our study sites were consistent with previously observed depth trends (e.g., Hallock, 1988) and susceptibility to reduced water quality (e.g., Cockey et al., 1996, Hallock et al., 2003; Uthicke and Nobes, 2008). If conditions had been optimal at all sites, mean LBF densities should have exceeded 10^4 individuals/m^2, with highest densities (~10^5) in summer months and lowest (~10^4) in winter or early spring [i.e., November through March, see Hallock et al. (1986)]. Mean densities of total LBF at our study sites were mostly >10^4 individuals/m^2, indicating generally suitable conditions in the upper FKNMS. Densities at the BNP and KL 3 m sites, however, were typically <10^4 individuals/m^2.

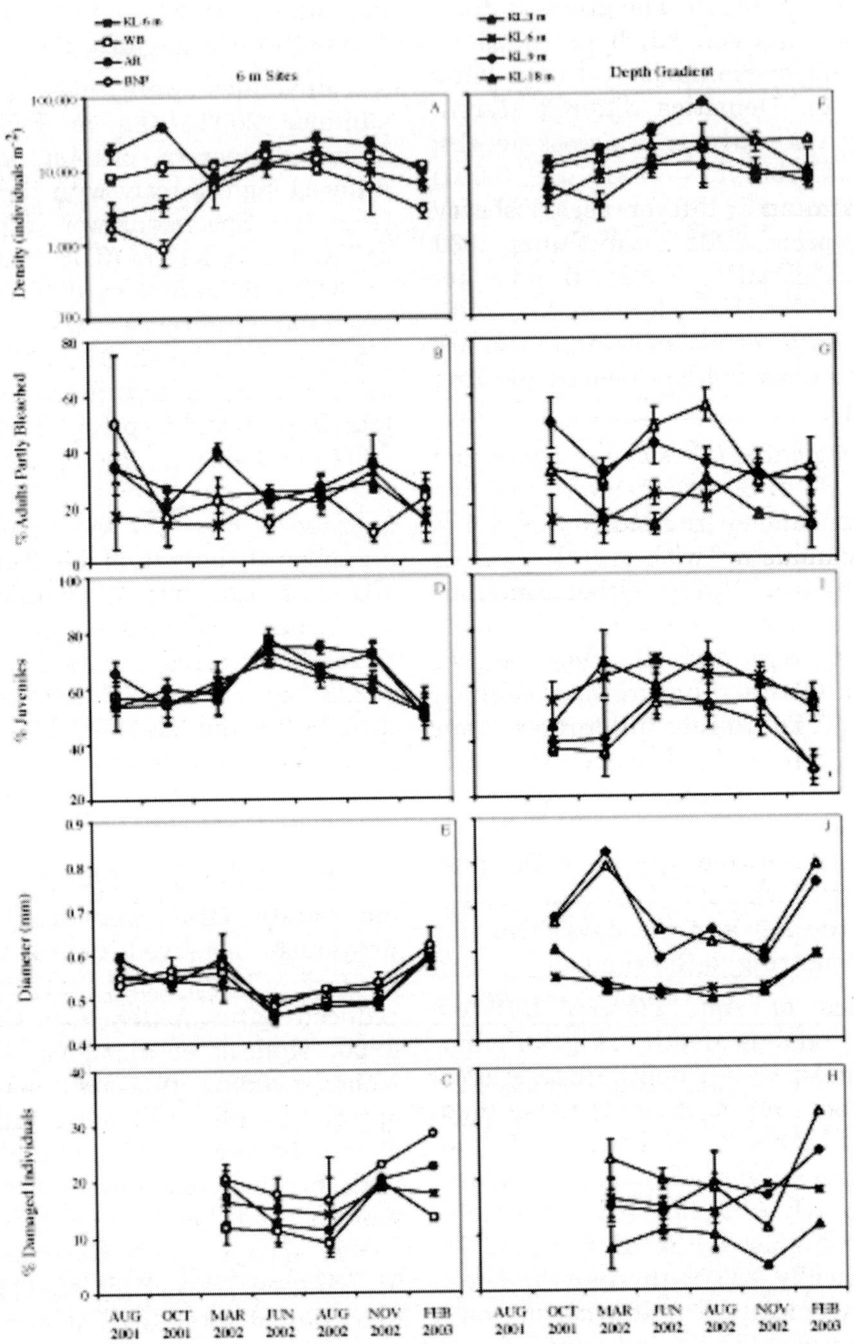

Fig.3. Population parameters (mean ± SE) of *Amphistegina gibbosa* at the 6 m sites (A–E) and the depth gradient (F–J) including **(A & F)** densities [plotted on a $\log_{10}$ scale], **(B & G)** percentages of adult *A. gibbosa* exhibiting any degree of bleaching, **(C & H)** percentage of juveniles, **(D & I)** diameter, and **(E & J)** percentages of damaged tests.

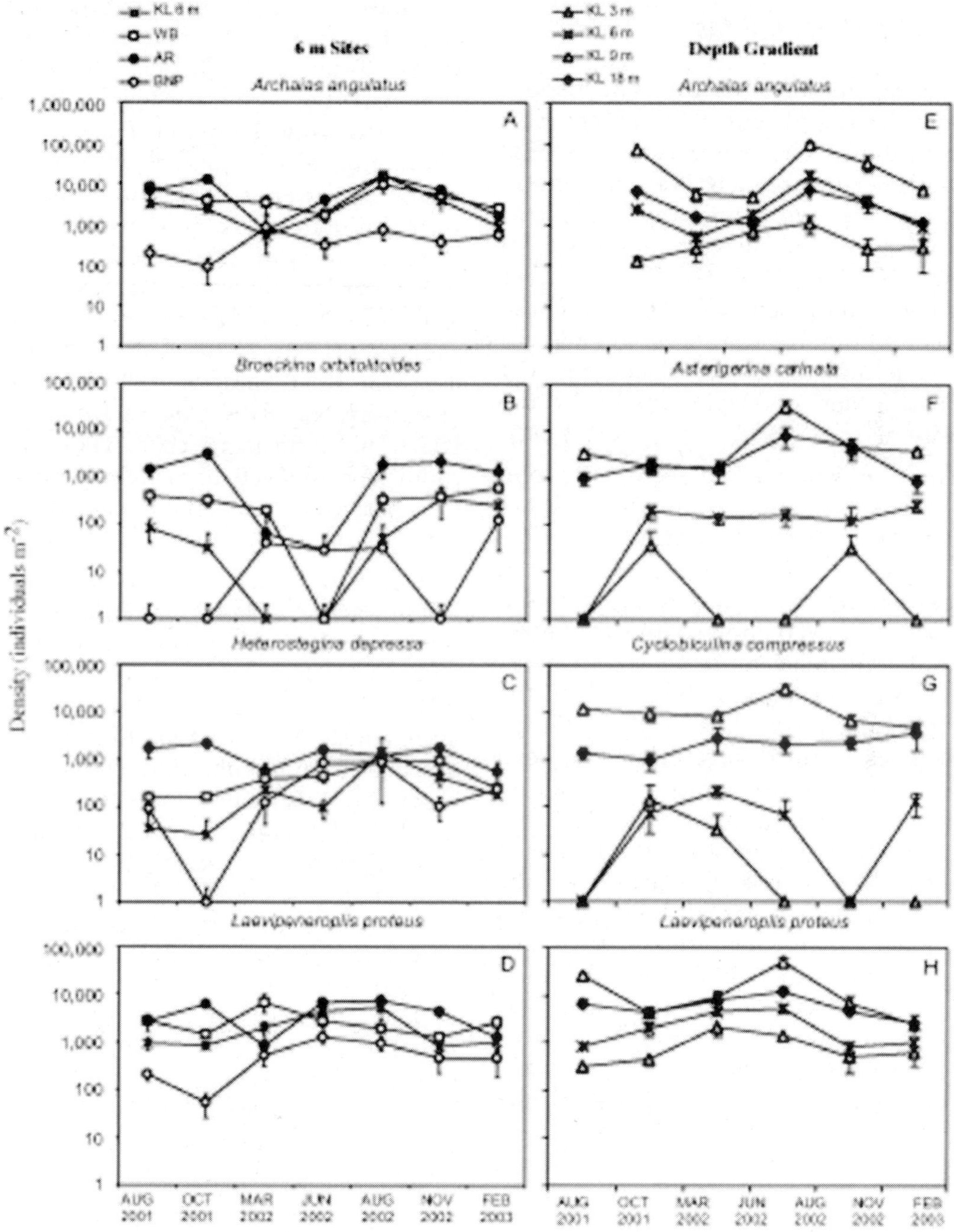

Fig.4. Mean (± SE) densities of other important species of LBF along the northeast–southwest transect **(A)** *Archaias angulatus*, **(B)** *Broeckina orbitolitoides* **(C)** *Heterostegina depressa* and **(D)** *Laevipeneroplis proteus*, and along the depth gradient **(E)** *Archaias angulatus* **(F)** *Asterigerina carinata*, **(G)** *Cyclobiculina compressus*, and **(H)** *Laevipeneroplis proteus*. Densities are plotted on a $\log_{10}$ scale. Note the difference in scale used for *Archaias angulatus* **(A & E)** relative to other LBF.

The 6 m Patch-Reef Sites

Our results for LBF and *Am. gibbosa* densities from the four 6 m patch reefs demonstrated characteristic seasonal trends, superimposed on the gradient in which densities were typically highest at Algae Reef (AR) and lowest at the site in Biscayne National Park (BNP). The other parameters assessed for *Am. gibbosa* revealed minimal differences among the sites. One unusual observation was the high percentage of shell damage at the BNP site.

The AR site is adjacent to John Pennekamp Coral Reef State Park, where most of the extensive coastline is native, intact and relatively vegetated with coastal hammock, mangroves and sea grass beds. The WB site is closer to the KL sites but still adjacent to the state park. The KL sites are located offshore from the most urbanized coastline of Key Largo, where much natural vegetation has been removed, natural topography altered to maximize waterfront properties, and the coastline armored with seawalls (Fisher et al., 2007). These sites also are along the route taken by recreational boaters and commercial dive operators to reach Molasses Reef and other heavily used outer reefs, so pollutants such as hydrocarbon combustion products may be more prevalent. In addition, larger boats regularly stir up sediments, potentially remobilizing nutrients and chemical pollutants (Kruczynski and McManus, 2002).

Williams (2002) carried out a multi-year comparison of *Am. gibbosa* populations at offshore sites at Conch and Tennessee Reefs on the Florida Reef Tract. She found differences in population responses, especially prevalence and intensity of bleaching, consistent with clearer waters at Conch Reef compared with Tennessee Reef. She further demonstrated that concentrations of colored dissolved organic matter (CDOM, commonly called tannins) were detectably higher in waters of Tennessee Reef, which were influenced by outflow from Florida Bay. Ayoub et al. (2009) compared CDOM concentrations between AR and KL 6 m, finding that significantly more ultraviolet radiation (UVR, λ320) could reach the 6 m depth at KL 6 than at AR. Moreover, the variability in UVR was ~60% greater at KL 6 m, indicating that when weather conditions allowed high light penetration, the potential for photo-oxidative stress was greater at KL 6 m.

The BNP site is closest to the urban Miami area, a large landfill, a nuclear power plant, and watershed canals that drain into Biscayne Bay. This site also is influenced by the extensive agricultural area south and west of Miami and the associated nutrients and chemicals that enter the watershed canals (e.g., Downs et al., 2005; Carnahan et al., 2008). Densities of LBF were particularly low at BNP in August and October 2001 following high rainfall associated with storms, including Hurricane Gilbert, which passed over central Florida in September 2001. Following these storms, water quality monitoring stations located at or in the vicinity of BNP showed increased sedimentation (Fisher, 2007), decreased salinity, increased turbidity, plus elevated inorganic nitrogen, organic phosphorous and chlorophyll *a* (Southeast Research Center's Water Quality Monitoring Network; http://serc.fiu.edu/wqmnetwork/) These data indicate that terrigenous stressors reached the BNP site and therefore nutrients or other land-based stressors (e.g., pesticides, PAHs) may have influenced LBF populations. Corals from the *Orbicella annularis* species complex also exhibited high stress levels in October 2001 at BNP relative to the other 6 m sites, as indicated by high levels of oxidative and xenobiotic stress that resulted in high protein turnover and denaturation (Fisher et al., 2007). Bleaching in *Am. gibbosa* also was high at BNP in August 2001.

Depth Transect off Key Largo

The depth transect off Key Largo presented much clearer differences in LBF populations among sites (Figs. 2, 3). Moreover, the trends in *Am. gibbosa* at the offshore sites (KL 9 m and 18 m) were easily interpreted based on their similarity to data sets from Conch Reef collected in the 1990s (Hallock et al., 1995; Williams et al., 1997; Williams, 2002). The two offshore sites showed typically strong seasonal variability in abundance, prevalence of bleaching, % juveniles, and mean diameters compared with those parameters at the KL 3 m and 6 m sites.

Exploring the trends in the *Am. gibbosa* populations at the KL 9 m and 18 m sites, mean diameter data tell the clearest "story", and are reinforced by data on mean densities and % juveniles. In March 2002 and February 2003, the largest mean sizes (with corresponding low % juveniles) indicate that most specimens were adults that would undergo asexual reproduction during the spring months. *Amphistegina* spp. are - *semelparous*, generally reproducing only once (Hallock, 1985). Each successful asexual reproduction by a typical adult (>1 mm diameter) can yield a few hundred offspring, with resultant drop in mean diameter and corresponding increase in percent juveniles in the population. Although alternation of asexual with sexual generations was considered typical for *Am. gibbosa* before the discovery of bleaching in 1991 (Hallock, 1985), starting in 1991, successive asexual generations were commonly observed, including production of smaller broods by smaller schizont "parents" (Harney et al., 1996). This trend can account for the continued dominance of smaller individuals through the summer and into the autumn (October 2001, June, August, and November 2002). Sexual reproduction (broadcasting of isogamous gametes) tends to occur in autumn. The sharp decline in density, especially at KL 9 m, between August and November 2002, with no corresponding drop in % juveniles and limited increase in mean diameter, reflects loss from the population of the adults that sexually reproduced. During the winter months, schizonts and agamonts (which eventually reproduce asexually) grow to adult sizes, resulting in the sharp increase in mean diameter and decrease in % juveniles in the February/March samples.

The % partly bleached adult *Am. gibbosa* were higher in summer and autumn at the KL 9 m and 18 m sites, when photo-oxidative stress was highest (Williams and Hallock, 2004; Hallock et al. 2006 a,b). Paradoxically, bleaching affected higher percentages of specimens at KL 9 and 18 m than at the shallower sites. However, Downs et al. (2002), working along the same transect, observed the same trend in bleaching and oxidative stress responses in the stony coral, *O. annularis* species complex, in the summer of 1999. Of their test colonies, none were even pale at the KL 3 m site, one of five was pale at the 6 m site, four of five were bleached at 9 m, and all were bleached at 18 m. Chlorophyll a concentrations in assessed corals dropped in May at the KL

6 m site, and most dramatically in June at the 9 m and 18 m sites, and were lowest at the 9 m site in July, recovering somewhat thereafter, while remaining low in corals from the 18 m site through September. Downs et al. (2002) reported strong positive correlations between bleaching and accumulation of oxidative-damage products, as indicated by higher chaperonin levels and protein-turnover activity. Corals that didn't bleach (i.e., at the KL 3 and 6 m sites) had the highest levels of antioxidant enzymes and small heat-shock proteins.

Two hypotheses, which are complementary, can explain the depth trends in *Am. gibbosa.* The first, which pertains equally to results for *Orbicella* at the same sites (Downs et al., 2002), involves water transparency and CDOM (i.e., tannins). Inshore waters can have as much as an order of magnitude more CDOM than the offshore sites (Ayoub et al., 2009), and increased CDOM protected both LBFs and corals from photooxidative stress. A difference in water transparency was especially evident at the KL 3 m site, which was close to a small mangrove island; corals did not bleach there and LBFs had comparatively little bleaching despite being in warm, shallow waters. The second hypothesis is based on previous observations (e.g., Hallock et al., 1995) that chronic photo-oxidative stress results in progressive bleaching with increased size. Evidence of bleaching is seldom seen in juveniles, yet, when prevalent, may affect nearly all LBF adults >1 mm diameter. Thus, because the 3 m and all 6 m sites were dominated by juveniles and mean diameters were consequently low throughout the study, *Am. gibbosa* at these sites infrequently reached size classes likely to exhibit bleaching.

Three scenarios, again based on observations in previous studies, can explain the low densities of relatively small individuals at the shallower sites. First, if recruitment on inshore patch reefs is largely from sexually-produced propagules that are transported on shelf, for example, by tidal currents (e.g., Alve and Goldstein, 2010), the initial chambers of such individuals are much smaller (Harney et al., 1996), so the individuals spend more time in the smaller size classes. At the same time, if asexually-produced individuals that recruit inshore are schizonts, they tend to reproduce at smaller sizes than agamonts (Harney et al., 1998). Finally, greater environmental variability occurs at the shallow inshore sites, including much greater differences in light intensity and spectral quality reaching the bottom (Ayoub et al., 2009, 2012). Hallock et al. (1995) found that *Am. gibbosa* specimens from Conch Reef populations under acute photo-oxidative stress tended to reproduce asexually at smaller sizes. Extensively bleached specimens that grew to larger sizes were so cytologically damaged (Talge and Hallock, 1995) that asexual reproductions were seldom successful (Hallock et al., 1995).

Bringing together all of these possibilities: even if environmental conditions temporarily kill most of the *Am. gibbosa* population at a shallow-water site, the ubiquitous availability of resistant propagules in the benthos provides recruits when environmental conditions become temporarily suitable. Moreover, the variability of the environment, especially irradiance, favors asexual reproduction as soon as the individuals become reproductively viable (~0.6–0.8

mm diameter). The result is populations made up of relatively low densities of predominantly small individuals. Considering the set of 6 m sites, the more limited variability in water transparency at AR compared to KL 6m (Ayoub et al., 2009) supports population densities at AR comparable to densities at the KL 9 and 18 m sites.

LBF as Environmental Indicators

The history of Florida's reefs over the past several decades parallels that of reefs worldwide. Over the last 50 years, water quality has declined in the Florida Keys due to changes in water-flow patterns from Florida Bay, sedimentation from boat traffic and development, and increased near-shore nutrient concentrations from local wastewaters, freshwater discharge from limestone aquifers, fertilizers, and industrial pollutants (e.g., Szmant and Forrester, 1996; Kruczynski and McManus, 2002). Changes in water quality corresponded with significant decreases in live coral cover throughout the Florida Keys (e.g., Dustan and Halas, 1987; Yentsch et al., 2002; Dustan et al., 2008) and a shift from symbiont-bearing foraminifers to heterotrophic foraminifers dominating the benthic assemblages (Cockey et al., 1996). Moreover, coral mass-bleaching events and disease outbreaks have become increasingly common along the Florida Reef Tract over those 50 years (Porter et al., 2001; Dustan et al., 2008). Similarly, bleaching, first seen on the Florida Reef Tract in summer 1991, decimated *Am. gibbosa* populations that autumn and winter (Hallock et al., 1995), and since then has been chronic to acute each summer, with related incidences of shell breakage, predation and cyanobacterial infestation (Hallock et al., 2006 a, b).

Our study was part of a larger effort to compare environmental assessments, community and physiological condition (i.e., coral regeneration rates and cellular biomarker profiles) of corals and other reef organisms at seven coral reefs (Fisher, 2007). This project provided the opportunity to determine the efficacy of using LBF abundances and characteristics of *Amphistegina* populations as bioindicators in coral reef assessment, as proposed by Hallock (2000b).

The 6 m sites were assessed using selected parameters (Fisher et al., 2007) to create a biotic health index that could be compared with the best, mean and worst regional values reported for an extensive Caribbean-wide assessment carried out by Kramer et al. (2003). The AR site and, interestingly, BNP, ranked as similar to the Caribbean means, while KL 6 m scored similar to Caribbean regional "worst" values. All study sites showed signs of stress (Fisher, 2007); low coral cover, relatively small colonies, high macro-algal abundance and low fish densities indicated suboptimal reef conditions. These and other reefs in the Upper Florida Keys Reef Tract have experienced declines in coral cover over the past four decades (e.g., Dustan and Halas, 1987; Porter et al., 2001; Dustan et al., 2008).

To assess the physiological state of the corals at the sampling sites, five *Orbicella* colonies were biopsied at each site and date that LBF samples were collected. Fisher et al. (2007) tracked the recovery of the 1–2 cm diameter lesions created when tissue samples were collected. Among the 6 m sites, the AR corals consistently had the highest lesion-regeneration rates. The lesions on biopsied colonies at AR also completely healed significantly more often than lesions at the other sites. The AR site also had the highest live coral cover, with relatively large colonies. Lesions on colonies at WB also regenerated to some extent, but many failed to heal completely, leaving the corals susceptible to fouling organisms. In contrast, KL 6 m, located offshore from the most urbanized coastline of Key Largo, had lower generation rates and overall low live coral cover. Corals from the BNP site, offshore from urban Miami, had the lowest regeneration rates. Lesions there exhibited minimal recovery with increases in lesion size often associated with seasonal increases in algae, which sometimes resulted in lesions merging.

Thus, at the 6 m sites, the overall abundances of LBF and the abundances of *Am. gibbosa*, showed the same trends as the lesion regeneration rates, which were the most direct indicator of how corals were responding to the environmental conditions that had occurred in the weeks to months prior to sampling. The biotic health index for Caribbean reefs gave a similar rank order for AR (average), WB (relatively poor) and KL 6 m (poor) sites. Despite higher coral mortality and macroalgal abundance, BNP ranked higher than WB based on larger coral colony size at BNP, which is indicative of prior reef development. Low LBF abundances, low lesion recovery rates, and high levels of cellular stress all indicated that BNP was experiencing an onset of decline, therefore these proxies appear to be better indicators of ongoing stress.

The results from the depth transect demonstrate a conundrum of Florida's reefs: inshore patch reefs have maintained higher coral cover while corals have precipitously declined on offshore reefs (e.g., Somerfield et al., 2008; Ruzika et al., 2013), while the best habitats for LBF continue to be the offshore reefs (e.g., Hallock, 1988; Baker et al., 2009). We found that the KL 3 m site had the lowest LBF abundances, the highest rates of coral lesion recovery, and by far the highest coral cover [36%, more than double the 16% at AR, the second highest of all sites sampled (Fisher et al., 2007)]. At the same time, the LBF populations at the KL 9 and 18 m sites were relatively abundant and exhibited strong seasonal trends typical of thriving populations at warm-temperate and subtropical latitudes (e.g., Williams et al., 1997; Triantaphyllou et al., 2013), while the coral lesion regeneration rates were very low at those sites, especially in 2001–02. Thus, LBF abundance trends for onshore-offshore transects are not indicating the best areas for coral growth or recovery under current environmental conditions.

Where LBF and Stony Corals Diverge

So why is KL 3 m the highest ranking site for stony corals and the worst site for LBF, while AR ranks second highest for both corals and LBF (below KL 9 m but higher than KL 18 m)? Reasons that the KL 3 m site is least favorable for LBF are likely associated with the fine sediments and generally higher turbidity (Fisher, 2007; Baker et al., 2009). Suitable rubble substratum for collecting LBF was hard to find and often covered in fine sediments. Moreover, the solar energy reaching the seafloor on a day to day basis can range 2–3 orders of magnitude, from nearly full sunlight (>1000 μmol photon m^{-2} s^{-1}) during calm, dry season conditions, to < 10 μmol photon m^{-2} s^{-1}) when winds stir the fine sediment or during the rainy season when turbid waters come from nearby land.

Most LBF are mobile to some degree, moving towards the light under low light and away from high light, but their rates of movement are limited (Zmiri et al., 1974; Sinutok et al., 2013). At least in the case of *Amphistegina* spp., optimum environmental conditions seem to be where maximum light intensities are relatively predictable on a day to day basis. Moreover, fine sediments are generally not suitable habitat; exactly why is not well understood. Are the foraminifers better adapted to more hydrodynamically active environments because their reticulopodia are better adapted to attachment when water is moving than for sorting through fine sediments for food particles? Or are other environmental parameters, such as night-time oxygen availability, less favorable for LBF in environments with limited water motion?

A similar question can be asked: Why are some coral species maintaining more stable populations at inshore patch reefs than in consistently clear offshore waters? Energetic considerations provide some hypotheses. Corals are known to produce mycosporine-like amino acids that protect them to some degree from ultraviolet radiation, especially in shallow waters (e.g., Gleason and Wellington, 1993; Ayoub et al., 2012). These sunscreens require energy to produce, but the generally high solar radiation in shallow water provides abundant energy for photosynthesis.

At the same time, inshore waters typically contain more sources of particulate organic carbon upon which corals can feed. Hughes and Grottoli (2013) showed experimentally that corals with greater potential for heterotrophy are more likely to survive and recover from bleaching (see also Levas et al., 2016). Fisher (2007) noted that the corals at KL 18 m appeared to be starved for energy; in addition to very low regeneration rates, these corals expressed very low levels of all biomarkers. Both down-regulation and up-regulation of stress proteins can indicate stress (Werner & Hinton, 1999). Mechanisms that reduce protein levels, including reduced energy (ATP) availability, can disrupt protein synthesis at multiple levels (Werner & Hinton, 1999). When a coral colony's energy reserves are depleted or its glucose supplies are reduced by loss of symbionts, the colony must utilize proteins for energy. Synthesis of stress proteins represents an additional energy burden. Our observations provide insight into why increased food supply can support higher survival rates (e.g., Hughes and Grottoli, 2013), as well as higher lesion regeneration rates (Fisher et al., 2007) for corals in near shore patch reefs.

Photo-oxidative stress causes bleaching and increased susceptibility to disease and other sources of mortality in both corals and *Amphistegina* spp. increasing maximum summer temperatures, with earlier spring warming and warmer autumn conditions, result in extended durations of warm weather and consequent high metabolic rates. This results in reduced duration of cooler waters during which corals can build their lipid reserves for summer reproduction. At the same time, depleted stratospheric ozone concentrations, though slowly recovering, are still well below pre-1978 concentrations (https://www.esrl.noaa.gov/gmd/about/ozone.html). Thus, seasonal highs during spring equinox and summer solstice have increased the potential for onset of photo-oxidative stress much earlier than the late summer temperature maxima. Partial bleaching in the *Am. gibbosa* population indicated chronic photo-oxidative stress and implicates chronic stress in the long-term decline of corals on offshore reefs along the Florida Keys Reef Tract.

Abundances of *Amphistegina* were critical to distinguishing among the 6 m sites and along the depth gradient. Incidences of bleaching and breakage in *Am. gibbosa* indicated chronic stress at all sites, with no evidence for acute photic stress during the study period. Similarly, no coral bleaching was observed during our study, but the limited lesion recovery at most sites indicated that the corals were stressed. Thus, water quality was least favorable for calcifying organisms at BNP and less than optimal at all study sites. However, the suite of stressor remains unknown. Clearly more specific diagnostic assays (e.g., cellular diagnostic biomarkers, histological studies, contaminant analysis) are needed to provide further insight into potential stressors compromising water quality at BNP and the KL 6 m site that could inform local management actions (e.g., Dustan et al., 2008; Praezeres et al., 2011, 2012). While water quality, characterized by more consistent water transparency, may have been most favorable at the two deeper sites, 9 and 18 m, partial bleaching in the *Am. gibbosa* population indicated chronic photo-oxidative stress and implicated chronic stress in the long-term decline of corals on offshore reefs along the Florida Keys Reef Tract.

What LBF Species to Assess

Our study revealed that sampling depths must be considered in using live LBF populations as indicators of reef condition. Comparisons among the 6 m sites provided very consistent results, at least when comparing LBF densities and coral lesion recovery trends. However, along the depth gradient, the patch reef at KL 3 m appeared to be in good condition based on high coral cover and lesion regeneration rates (Fisher et al., 2007), yet densities of *Am. gibbosa* were low, indicating that these foraminifers are not good indicators in very shallow, muddy, near shore environments. *Amphistegina* are most suitable for assessing reef environments at depths of ~5–10 m. *Archaias angulatus, Sorites marginalis* and *Laevipeneroplis proteus* can be abundant in low energy, near shore environments where water quality is suitable (Fujita and Hallock, 1999) and therefore may be a more useful bioindicator species there. However, low densities of all LBF (especially *Ar. angulatus*) at KL 3 m, KL 6 m, and BNP indicate that environmental conditions are limiting these populations as well.

If live LBF populations are to be used as bioindicators for coral reefs and associated coastal environments in the western Atlantic–Caribbean region, what is the most economical strategy for sampling and analyses? Is analysis of the *Am. gibbosa* population sufficient? Or is the better strategy to include *Ar. Angulatus* and *L. proteus*, which along with *Am. gibbosa* typically accounted for more than 60% of the dissimilarity among the 6 m sites we examined? Moreover, if samples are collected at depths of 20 m or more, should *Asterigerina carinata* and *Cyclorbiculina compressa* also be included? Baker et al. (2009) reported that these taxa are useful in distinguishing among the reef-margin sites. On the other hand, clearly the other LBF counted (*Borelis, Broeckina, L. bradyi, Peneroplis, Sorites*, and *Heterostegina*), were not sufficiently abundant to recommend their inclusion in future bioindicator studies from this region.

The Florida Keys Reef Tract provides a well-documented example of how dramatically coral reefs have changed over the past 50 years. Instead of extensive stands of branching corals, divers now see the occasional colony. Offshore bank reefs, which historically had as much as 50% live coral cover (e.g., Dustan and Halas, 1987; Dustan, 1999), now seldom have 10% coral cover. In contrast, many patch reefs have not exhibited the precipitous decline found on the offshore reefs, and some maintain >20% live coral cover.

CONCLUSION

1. Larger benthic foraminifers, particularly *Amphistegina gibbosa*, are sensitive indicators of both water quality and photo-oxidative stress influencing coral reefs.
2. The recommended assessment strategy is to compare sites at similar depths and especially at depths in the range of 5–10 m.
3. Densities of live LBF, and for *Am. gibbosa,* also size distributions and prevalence of partial bleaching, are key response variables for assessment.
4. Season of sample collection is also important to interpretation.
5. Tannins from intact mangrove shorelines can reduce photo-oxidative stress on nearby patch reefs.

Acknowledgements: This work would not have been possible without financial and logistical support from many individuals and agencies. Special thanks to C. Downs, W.C. Jaap, J.J. Torres, and C.M. Woodley for manuscript and field assistance. K. R. Clarke and R. Muller provided statistical assistance. E. Carnahan, R. Curry, J.C and J.F. Halas, G. Klungness, C. Moses, S. Ryan, and S. Viehman assisted in the field, and members of the Reef Indicators Lab, College of Marine Science, USF, spent countless hours of picking of foraminiferal tests. Logistical field support was provided by NOAA Florida Keys National Marine Sanctuary, NOAA National Undersea Research Center Florida Keys facility, Biscayne National Park, Everglades National Park, and Quiescence Diving Services, Inc. This work was funded, in part, by National Sea Grant, Environmental Marine Biotechnology Award No. NA86RG0052, Am.7.1 (C. M. Woodley, PI), NOAA-NURC-UNCW subcontract 2004-19B (P. Hallock Muller, PI), NOAA Hurricane Alliance Grant (T. Mason and P. Hallock Muller, PIs), and USF-College of Marine Science Sanibel-Captiva Shell Club Endowed Fellowship (to E.F. Moses). This research was conducted under permit numbers BISC-2001-SCI-0022, BISC-2002-SCI-0012 and BISC-2003-SCI-0019 from the National Park Service and FKNMS-2001-008 from the Florida Keys National Marine Sanctuary.

REFERENCES

ALVE, E. and GOLDSTEIN, S.T. (2003) Propagule transport as a key method of dispersal in benthic foraminifera. Limnol. Oceanogr., v.48, pp.2163–2170.

ALVE, E. and GOLDSTEIN, S.T. (2010) Dispersal, survival and delayed growth of benthic foraminiferal propagules, J. Sea Res. v.63, pp.36-51.

AYOUB, L., HALLOCK, P. and COBLE, P. (2008) Colored dissolved organic material increases resiliency of coral reefs by controlling exposure to UVR. Proc. 11th ICRS, Ft. Lauderdale v.1, pp.579-583.

AYOUB, L.M., HALLOCK, P., COBLE, P.G. and BELL, S.S. (2012) MAA-like absorbing substances in Florida Keys phytoplankton vary with distance from shore and CDOM: Implications for coral reefs. J. Exp. Mar. Biol. Ecol. v.420, pp.91-98.

BAKER, R.D., HALLOCK, P., MOSES, E.F., WILLIAMS, D.E. and RAMIREZ, A. (2009) Larger foraminifers of the Florida reef tract, USA: distribution patterns on reef-rubble habitats. J. Foramin. Res., v.39, pp.267–277.

BARBOSA, C.F., PRAZERES, M.F., FERREIRA, B.P. and SEOANE, J.C.S. (2009) Foraminiferal assemblage and reef check census in coral reef health monitoring of East Brazilian margin. Mar. Micropaleont., v.73, pp.62–69.

BROWN, B.E. (1997) Coral bleaching: Causes and consequences. Coral Reefs, v.16, pp.S129–S138.

CARNAHAN, E.A., HOARE, A.M., HALLOCK, P., LIDZ, B.H. and REICH, C.D. (2008) Distribution of heavy metals metals and foraminiferal assemblages in sediments of Biscayne Bay, Florida, USA. J. Coastal Res., v.24, pp.159-169.

CLARKE, K.R. and WARWICK, R.M. (2001) Change in marine communities: an approach to statistical analysis and interpretation, 2nd ed. PRIMER-E Ltd, Plymouth.

COCKEY, E., HALLOCK, P. and LIDZ, B.H. (1996) Decadal-scale changes in benthic foraminiferal assemblages off Key Largo, Florida. Coral Reefs, v.15, pp.237–248.

DE ARAUJO, H.A.B. and MACHADO, A.J. (2008) Benthic foraminifera associated with the south Bahia coral reefs, Brazil. J. Foramin. Res., v.38, pp.23–38.

DOWNS, C.A., FAUTH, J.E., HALAS, J.C., DUSTAN, P., BEMISS, J. and WOODLEY, C.M. (2002) Oxidative stress and seasonal coral bleaching. Free Rad. Biol. Med., v.33, pp.533–543.

DOWNS, C.A., FAUTH, J.E., ROBINSON, C.E., CURRY, R., LANZENDORF, B., HALAS, J.C., HALAS, J. and WOODLEY, C.M. (2005) Cellular diagnostics and coral health: declining coral health in the Florida Keys. Mar. Pollut. Bull., v.51, pp.558–569.

DOWNS, C.A., FAUTH, J.E., WETZEL, D., HALLOCK, P., HALAS, J.F., HALAS, J.C., CURRY, R. and WOODLEY, C.M. (2006) Investigating coral reef degradation at Alina's Reef in the Florida Keys: cellular physiology of white grunt (*Haemulon-plumieri*) as a biological indicator. Environ. Forensics, v.7, pp.15–32.

DUSTAN, P. (1999) Coral reefs under stress: Sources of mortality in the Florida Keys. Nat. Resour. Forum, v.23, pp.147−155.

DUSTAN, P. and HALAS, J.C.(1987) Changes in the Reef-Coral Community of Carysfort Reef, Key-Largo, Florida- 1974 to 1982. Coral Reefs, v.6, pp.91–106.

DUSTAN, P., FAUTH, J.E., PANTE, E., BANKS, K., VARGAS-ANGEL, B. and DOWNS, C.A.(2008) Using cellular diagnostics to link land-based sources of pollution with coral reef degradation in South Florida. Proc. 11th ICRS, Ft. Lauderdale, v.1, pp.502-506.

EAKIN, C.M., LOUGH, J.M. and HERON, S.F. (2009) Climate variability and change: monitoring data and evidence for increased coral bleaching stress. In: Van Oppen MH, Lough JM, eds. Coral Bleaching: Patterns, Processes, Causes and Consequences. Berlin: Springer, pp.41–67.

FISHER, E.M. (2007) Assessing the health of coral reef ecosystems in the Florida Keys at community, individual, and cellular scales. Ph.D. thesis, Univ. of South Florida, Tampa. http://scholarcommons.usf.edu/etd/705/

FISHER, E.M., FAUTH, J.E., HALLOCK, P. and WOODLEY, C.M. (2007) Lesion regeneration rates in reef-building corals (*Montastraea* spp.) as indicators of colony condition: strengths and caveats. Mar. Ecol. Progr. Ser., v.339, pp.61-71.

FUJITA, K. and HALLOCK, P. (1999) A comparison of phytal substrate preferences of *Archais angulatus* and *Sorites orbiculus* in mixed macroalgal-seagrass beds in Florida Bay. J. Foram. Res., v.29, pp.143-151.

GLEASON, D.F. and WELLINGTON, G.M. (1993) Ultraviolet-radiation and coral bleaching. Nature, v.365, pp.836–838.

GLYNN, P.W. (1996) Coral reef bleaching: Facts, hypotheses and implications. Glob. Change Biol., v.2, pp.495–509.

HALLOCK, P. (1985) Why are larger foraminifera large? Paleobiology, v.11, pp.195–208.

HALLOCK, P.(1988) Diversification in algal symbiont-bearing foraminifera: a response to oligotrophy? Rev. Paleobio., spec. v. no.2 (Benthos '86), pp.789–797.

HALLOCK, P. (1999) Symbiont-bearing foraminifera. *In*: SEN GUPTA, B.K. (Ed.), Modern Foraminifera. Kluwer Academic Publishers, Boston, pp.123–139.

HALLOCK, P. (2000a) Symbiont-bearing foraminifera: harbingers of global change. Micropal., 46(Suppl. 1), 95-104.

HALLOCK, P. (2000b) Larger foraminifera as indicators of coral-reef vitality. In: MARTIN, R.E. (Ed.), Environmental Micropaleontology. Kluwer Academic/ Plenum Publishers, New York, pp.121–150.

HALLOCK, P. and TALGE, H.K., (1994) A predatory foraminifer, *Floresina amphiphaga* n. sp., from the Florida Keys. J. Foramin. Res., v.24, pp.210–213.

HALLOCK, P., FORWARD, L.B. and HANSEN, H.J. (1986) Influence of environment on the test shape of *Amphistegina*. J. Foram. Res., v.16, pp.224–231.

HALLOCK, P., LIDZ, B.H., COCKEY-BURKHARD, E.M. and DONNELLY, K.B. (2003) Foraminifera as bioindicators in coral reef assessment and monitoring: The FORAM Index. Environ. Monit. Assess., v.81, pp.221–238.

HALLOCK, P., TALGE, H.K., COCKEY, E.M. and MULLER, R.G. (1995) A new disease in reef-dwelling foraminifera: implications for coastal sedimentation. J. Foram. Res., v.25, pp.280–286.

HALLOCK, P., WILLIAMS, D.E., TOLER, S.K., FISHER, E.M. and TALGE, H.K. (2006a) Bleaching in reef-dwelling foraminifers: a 20 year retrospective. Proc. 10th ICRS, Okinawa, pp.729–737.

HALLOCK, P., WILLIAMS, D.E., FISHER, E.M. and TOLER, S.K. (2006b) Bleaching in foraminifera with algal symbionts: implications for reef monitoring and risk assessment. Anu do Instit de Geociên., v.29, pp.108–128.

HARNEY, J.N., HALLOCK, P. and TALGE, H.K. (1998) Observations on a trimorphic life cycle in *Amphistegina gibbosa* populations from the Florida Keys. J. Foram. Res., v.28, pp.141-147.

Hohenegger, J. (2004) Depth coenoclines and environmental considerations of Western Pacific larger foraminifera. J. Foram. Res., v.34, pp.9-33.

HUGHES, A.D. and GROTTOLI, A.G. (2013) Heterotrophic compensation: A possible mechanism for resilience of coral reefs to global warming or a sign of prolonged stress? PLOS One, v.8, e81172.

KRAMER, P.A. (2003) Synthesis of coral reef health indicators for the Western Atlantic: results of the AGRRA program (1997–2000). Atoll Res. Bull., v.496, pp.1–58.

KRUCZYNSKI, W.L. and MCMANUS, F. (2002) Water quality concerns in the Florida Keys: sources, effects, and solutions. In: Porter, J.W. and Porter, K.G. (Eds.), The Everglades, Florida Bay, and Coral Reefs of the Florida Keys: An Ecosystem Sourcebook. CRC Press, Boca Raton, pp.827–881

LANGER, M.R. and HOTTINGER, L. (2000) Biogeography of selected "larger" foraminifera. Micropaleon., v.46, pp.105-126.

LEE, J. J. (2006) Algal symbiosis in larger Foraminifera. Symbiosis, v.42, pp.63–75.

LEE, J.J., MCENERY, M.E. and GARRISON, J.R. (1980) Experimental studies of larger foraminifera and their symbionts from the Gulf of Elat on the Red Sea. J. Foram. Res., v.10, pp.31–47.

LESSER, M.P. and FARRELL, J.H. (2004) Exposure to solar radiation increases damage to both host tissues and algal symbionts of corals during thermal stress. Coral Reefs, v.23, pp.367–377.

LESSER, M.P., STOCHAJ, W.R., TAPLEY, D.W. and SHICK, J.M. (1990) Bleaching in coral reef anthozoans: effects of irradiance, ultraviolet radiation, and temperature on the activities of protective enzymes against active oxygen. Coral Reefs, v.8, pp.225–232.

LEVAS, S., GROTTOLI, A. G., SCHOEPF, V., ASCHAFFENBURG, M., BAUMANN, J., BAUER, J. E. and WARNER, M. E. (2016) Can heterotrophic uptake of dissolved organic carbon and zooplankton mitigate carbon budget deficits in annually bleached corals? Coral Reefs, v.35, pp.495–506.

LIDZ, B.H. and HALLOCK, P. (2000) Sedimentary petrology of a declining reef ecosystem, Florida Reef Track (U.S.A.). J. Coast. Res., v.16, pp.675–697.

MOSES, E.F. and HALLOCK, P. (2015) Chapter 35. Coral Regeneration Assay. In Woodley, C.M. et al. (Eds.) Diseases of Coral. Wiley-Blackwell, Boston, MA, pp.473–482.

OLIVER, L.M., FISHER, W.S., DITTMAR, J., HALLOCK, P., CAMPBELL, J., QUARLES, R.L., HARRIS, P. and LOBUE, C. (2014) Contrasting responses of coral reef fauna and foraminiferal assemblages to human influence in La Parguera, Puerto Rico. Mar. Environ. Res., v.99, pp.95–105.

OSAWA, Y., FUJITA, K., UMEZAWA, Y., KAYANNE, H., IDE, Y., NAGAOKA, T., MIYAJIMA, T. and YAMANO, H. (2010) Human impacts on large benthic foraminifers near a densely populated area of Atoll, Marshall Islands. Mar. Poll. Bull., v.60, pp.1279–1287.

Porter, J.W., Dustan, P., Jaap, W.C., Patterson, K.L., Kosmynin, V., Meier, O.W., Patterson, M.E. and Parsons, M. (2001) Patterns of spread of coral disease in the Florida Keys. Hydrobiologia, v.460, pp.1–24.

PRAZERES, M.F., MARTINS, S. and BIANCHINI, A. (2011) Biomarkers response to zinc exposure in the symbiont–bearing foraminifer *Amphistegina lessonii* (Amphisteginidae, Foraminifera). J. Exp. Mar. Biol. Ecol., v.407, pp.116–121.

PRAZERES, M.F., MARTINS, S. and BIANCHINI, A. (2012) Assessment of water quality in coastal waters of Fernando de Noronha, Brazil: Biomarker analysis in *Amphistegina lessonii*. J. Foram. Res., v.42, pp.56–65.

RENEMA, W. and TROELSTRA, S.R. (2001) Larger foraminifera distribution on a mesotrophic carbonate shelf in SW Sulawesi (Indonesia). Palaegeogr. Palaeoecol. Paleocli., v.175, pp.125–146.

RUZICKA, R., COLELLA, M., PORTER, J., MORRISON, J., KIDNEY, J., BRINKHUIS, V., LUNZ, K., MACAULAY, K., BARTLETT, L. and MEYERS, M. (2013) Temporal changes in benthic assemblages on Florida Keys reefs 11 years after the 1997/1998 El Niño. Mar. Ecol. Prog. Ser., v. 489, pp.125–141.

SAMMARCO, P.W., HALLOCK, P., LANG, J.C. and LEGORE, R.S. (2007) Environmental bioindicators in coral reef ecosystems: the need to align research, monitoring, and environmental regulation. Environ. Bioindicat., v.2, pp.35–46.

SINUTOK, S., HILL, R., DOBLIN, M. and RALPH, P. J.(2013) Diurnal photosynthetic response of the motile symbiotic benthic foraminifera *Marginopora vertebralis*. Mar. Ecol. Prog. Ser., v.478, pp.127–138.

SOMERFIELD, P.J., JAAP, W.C., CLARKE, K.R., CALLAHAN, M., HACKETT, K., PORTER, J., LYBOLT, M., TSOKOS, C. and YANEV, G. (2008) Changes in coral reef communities among the Florida Keys, 1996–2003. Coral Reefs, v.27, pp.951–965.

SZMANT, A.M. and FORRESTER, A. (1996) Water column and sediment nitrogen and phosphorous distribution patterns in the Florida Keys, USA. Coral Reefs, v.15, pp.21–41.

TALGE, H.K. and HALLOCK, P. (1995) Cytological examination of symbiont loss in a benthic foraminifera, *Amphistegina gibbosa*. Mar. Micropaleontol., v.26, pp.107–113.

TALGE, H.K. and HALLOCK, P. (2003) Ultrastructural responses to bleaching in *Amphistegina gibbosa (*Foraminifera). J. Eucaryot. Microbio., v.50, pp.324–333.

TOLER, S.K. and HALLOCK, P. (1998) Shell malformation in stressed *Amphistegina* populations: relation to biomineralization and paleoenvironmental potential. Mar. Micropaleontol., v.34, pp.107–115.

TRIANTAPHYLLOU, M.V., DIMIZA, M.D., KOUKOUSIOURA, O. and HALLOCK, P. (2012) Observations on the life cycle of the symbiont-bearing foraminifer *Amphistegina lobifera* Larsen, an invasive species in coastal ecosystems of the Aegean Sea (Greece, E Mediterranean). J. Foram. Res., v.42, pp.143–150.

UTHICKE, S. and NOBES, K. (2008) Benthic foraminifera as ecological indicators for water quality on the Great Barrier Reef. Estuar. Coast Shelf Sci., v.78, pp.763–773.

UTHICKE, S. and ALTENRATH, C. (2010) Water column nutrients control growth and C:N ratios of symbiont-bearing foraminifera on the Great Barrier Reef, Australia. Limnol. Oceanogr., v.55, pp.1681–1696.

UTHICKE, S., THOMPSON, A. and SCHAFFELKE, B. (2010) Effectiveness of benthic foraminiferal and coral assemblages as water quality indicators on inshore reefs of the Great Barrier Reef, Australia. Coral Reefs, v.29, pp.209–225.

WEINMANN, A.E., ROEDDER, D., LOETTERS, S. and LANGER, M.R. (2013) Heading for new shores: projecting marine distribution ranges of selected larger Foraminifera. Plos ONE, v.8, e62182, DOI: 10.1371/journal.pone.0062182.

WERNER, I. and HINTON, D.E. (1999) Field validation of hsp70 stress proteins as biomarkers in Asian clam (*Potamocorbula amurensis*): is down regulation an indicator of stress? Biomarkers, v.4, p.473-484.

WILLIAMS, D.E. (2002) Population ecology of bleaching-stressed *Amphistegina gibbosa* in the Florida Keys (1991–1999): influence of solar radiation on reef-dwelling foraminifera. Unpubl. Ph.D. Thesis, Univ. of South Florida, Tampa.

WILLIAMS, D.E. and HALLOCK, P. (2004) Bleaching in *Amphistegina gibbosa* d'Orbigny (Class Foraminifera): observations from laboratory experiments using visible and ultraviolet light. Mar. Biol., v.145, pp.641–649.

WILLIAMS, D.E., HALLOCK, P., TALGE, H.K., HARNEY, J.N. and MCRAE, G. (1997) Responses of *Amphistegina gibbosa* populations in the Florida Keys (U.S.A.) to a multi-year stress event (1991–1996). J. Foramin. Res., v.27, pp.264–269.

YENTSCH, C.S., YENTSCH, C.M., CULLEN, J.J., LAPOINTE, B., PHINNEY, D.A. and YENTSCH, S.W. (2002) Sunlight and water transparency: cornerstones in coral research. J. Exp. Mar. Biol. Ecol., v.268, pp.171–183.

ZMIRI, A., KAHAN, D., HOCHSTEIN, S. and REISS, Z. (1974) Phototaxis and thermotaxis in some species of *Amphistegina* (Foraminifera). J. Protozool., v.21, p.133–138.

Micropaleontology and its Applications
Pradeep K. Kathal, Rajiv Nigam & Abu Talib (Editors)
Scientific Publishers (India), 2018, 41-57 pp.

Microbialites and Salt lakes of Western Australia-Salient Features, Threats and Conservation Issues

Jacob John*, Veronica, S. Campagna and Fiona E. Taukulis

Department of Environmental Biology and Agriculture
Curtin University of Technology, GPO Box U1987, Perth 6845, Western Australia
***Email**: j.john@curtin.edu.au

Abstract

The objective of this paper is to review the conservation values of the salt lakes of Western Australia, mostly based on recent unpublished data. Western Australia is noted for its extensive saline systems: both coastal and inland, the former mostly permanent and the latter temporary. Various benthic microbial communities or algal mats are the dominant primary producers in most of the coastal lakes. The Hamelin Pool in Shark Bay and Lake Clifton are two such saline coastal systems, which harbour world renowned living microbialites differing in structure and ecology. The microbialites in Lake Clifton are under severe stress due to eutrophication and drastic increase in salinity that has been apparent in recent years. Unpredictable rainfall, prolonged dry spells, a wide range of salinity and unique biota with intriguing survival strategies, are some of the salient features of the inland salt lakes of Western Australia. These palaeodrainage channels are associated with significant mineral resources that are being rapidly exploited. The biodiversity of these systems are significantly linked to a large variety of avian fauna including migratory birds. The water quality and biodiversity of the salt lakes in Western Australia have not been adequately characterised. Mining continues to have the greatest impact on the biodiversity and integrity of the inland saline systems. The use of biomonitors such as diatoms and suitable invertebrates for assessing the 'health' of the system has to be established. A case history of an inland salt lake is presented to substantiate these points.

Keywords: Microbialites, diatoms, thrombolites, stromatolites, microbial communities, inland salt lakes.

INTRODUCTION

Western Australia under tropical to temperate climatic conditions form almost 33% of the land area of Australia, and has the longest coastline of all states. Shallow embayment and hypersaline lakes are common in the coastal zone of the state. These coastal lakes were once part of the sea and got separated in relatively recent times - in the last 5000 to 10000 years. In addition to these natural wetlands, artificial lakes such as aquaculture and solar ponds in the coastal zone and deep saline mine-voids in the inland are added by excavations and mining activities. Microbial communities in the form of benthic microbial mats and the introduced anostracan *Artemia* are characteristic of coastal lakes, whereas, the endemic anostracan *Parartemia* is common in the inland salt lakes.

Inland salt lakes of Western Australia are temporary playas or saltpans shaped by their geomorphological and hydrogeological setting. Australia has the lowest percentage of rainfall as runoff, the lowest amount of runoff and the least amount of water in rivers, making it the driest inhabited continent in the world (Smith, 1998).

Spatial and temporal distribution of water is remarkably uneven in Australia. Almost 70% of the continent is arid or semi-arid where only less than 5% of the rainfall is discharged as runoff or recharge to the groundwater. Rainfall declines towards the interior arid regions where the inland salt lakes are found. The inland salt lakes are located in regions where precipitation (<200-350 mm/yr) is exceeded by evaporation (>2000 mm/yr) by several magnitudes. Yet in these extreme environments of salinity and aridity, prolific life forms are triggered by seasonal or intermittent rainfall events.

The objective of this paper is to provide a panoramic view of the unique features of Western Australia's salt lake systems based on mostly unpublished data on water quality and biota, specifically microbial communities of the coastal lakes and biota dominated by bacteria, diatoms, cyanobacteria and invertebrates of the inland salt lakes. Additionally, imminent threats to the water quality and biodiversity of these delicate ecosystems are identified. An assessment system using all physicochemical parameters as well as biomonitors to assess changes with time and any impact of human activities as part of a management package for the salt lake system is recommended.

COASTAL LAKES

Microbialites

Organosedimentary structures known as microbialites once abundant in the early stages of life on earth can be found in several coastal salt lakes of Western Australia (Fig. 1). There are two types of microbialites: stromatolites and thrombolites, the former are clearly laminated and the latter are characterised by internal clotted structures (Moore, 1987). Built predominantly by cyanobacteria and diatoms, modern analogs of microbialites are found in Hamelin Pool in Shark Bay, lake Thetis, Rottnest Islands Salt Lakes, and Lake Clifton. These modern living structures were evolved under hyposaline to hypersaline, nutrient poor environments suitable for calcium carbonate precipitation (Moore, 1987). There are only very few locations on this planet where such structures are still found. Yet the paucity of information on the processes of formation of such structures and the composition and taxonomy of associated microbial communities still persist.

Stromatolites of Hamelin Pool, Shark Bay

Located Between latitude 24°30'S and 26°45'S, 600 km north of Perth, the capital city of Western Australia, Hamelin Pool in Shark Bay is a land-locked lagoon separated from the bay by a shallow seagrass barrier bank. With an average evaporation of 2200 mm per annum and precipitation of 210 mm per annum, the salinity range of the lagoon is 55 to 70 g L^{-1}. While inactive supratidal stromatolites attract several thousands of tourists (Fig. 2), the active growing stromatolites are found in the subtidal region upto a depth of 4 m (Fig. 3). Additionally cyanobacterial mats are abundant in the entire region. It was shown that diatoms dominated the microbial communities associated with the stromatolites in species diversity compared to cyanobacteria (John 1990a, b; 1993 & 1994).

There were 79 species of diatoms in the microbial communities in Hamelin Pool. Most of the diatoms associated with the stromatolites secrete massive amount of mucilage around them and some produce profusely branched sulphated polysaccharide stalks and tubes. Prominent among them are *Brachysira aponina* and *Mastogloia cocconeiformis*. The mucilaginous tubes and stalks of these diatoms were well adapted to trapping coarse particles, thereby assisting stromatolite formation. The intertidal stromatolites are composed of particles of 125 to 250 µm in diameter, lithified by aragonite cement whereas the subtidal stromatolites are made of coarser sediment particles ranging in dimension from 500 to 5000 µm and are poorly laminated than the intertidal ones (Logan *et al.*, 1974). Winsborough and Seeler (1984) have demonstrated the role of diatoms in the formation of modern stromatolites in Mexico. While low species diversity is often associated with hypersalinity, the diatom genus *Mastogloia* was represented by 32 species in the microbial community in Hamelin Pool, forming the vast majority in the epiflora of the stromatolites (John 1990a).

Thrombolites in Lake Clifton

Lake Clifton is part of a chain of 11 salt lakes, located 120 km south of Perth, in the Yalgorup National Park, the largest conservation reserve in the Swan Coastal Plain in Western Australia (Fig. 1). Lake Clifton is well known for the largest array of thrombolites (Fig. 4) in the world and is of high conservation value. Microbialites are built by a community of microorganisms, predominantly cyanobacteria. Lake Clifton was described as one of the rare hyposaline water bodies in the world with living microbialites. The benthic microbial communities, which build these structures, are believed to have evolved in low nutrient conditions (Moore *et al.* 1983, Moore and Neil 1985, Moore 1987, Burne and Moore 1987). Additionally, there are benthic microbial mats, which are not involved in the building of microbialites. These microbial mats in the lake are also adapted to low nutrient conditions similar to those in coastal salt lakes and serve as nutrient sinks.

The salinity in Lake Clifton never exceeded that of seawater in the early 1980s, when the microbialites were described and investigated for the first time (Moore, 1987). Salinity less than that of seawater and interaction between fresh groundwater and microbial communities were considered essential for the continued growth of microbialites. Predominantly two species of hyposaline cyanobacteria, *Scytonema* sp. and *Chroococcus* sp., were identified as the cyanobacteria associated with thrombolite growth.

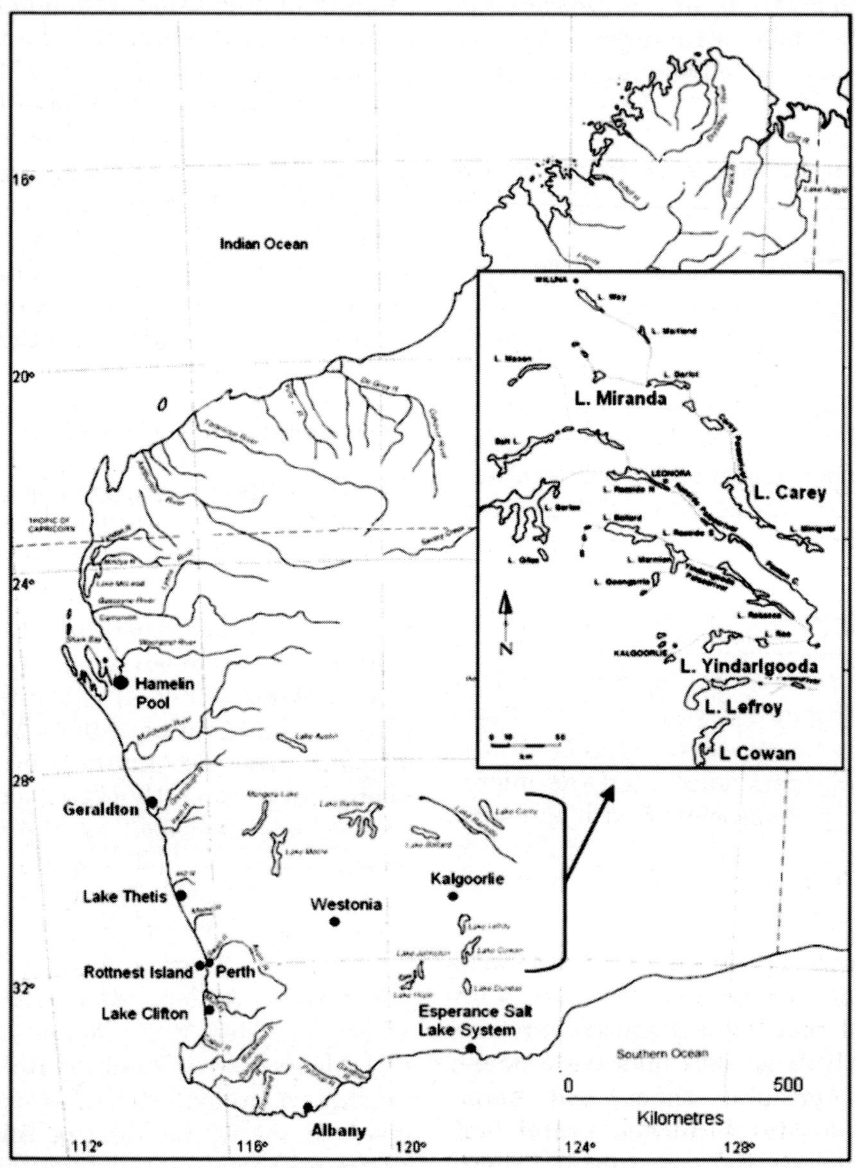

Fig.1. Map of Western Australia showing the location of coastal and inland salt lakes referred in the paper: Hamelin Pool, Lake Thetis, Rottnest Island, Lake Clifton, Esperance Salt Lakes, Lake Carey, Lake Lefroy, Lake Zot, Lake Cowan and Lake Miranda.

Fig.2. Hamelin Pool Supratidal inactive stromatolites, Shark Bay.

Fig. 3. Subtidal active stromatolites in Hamelin Pool.

Lake Clifton undergoes seasonal changes in salinity ranging from 7 to 35 g L^{-1} according to historical records (Moore *et al.,* 1983; Moore *et al.,* 1985; Moore 1987). However within the last 20 years, the salinity range has shifted to a higher level (19-80 g L^{-1}). Additionally, in 2002 and 2007 there have been massive break up of thick benthic microbial mats from the bottom of the lake to the surface, discolouring and adding nutrients to the lake. Massive fish death was also associated with the break up of these algal mats in 2007 (Fig. 5). The origin, development, composition and spatial distribution of these large benthic algal mats and their disintegration adding to the nutrient enrichment of the lake and possible impacts on the conservation of the microbialites, remain unknown.

Fig.4. Thrombolites in Lake Clifton in summer 2005.

Fig.5. Dislodged benthic algal mats and dead black bream fish in Lake Clifton in 2007.

The structure and composition of microbial communities surrounding the thrombolites have been investigated (John and Paton, 2008). The 2007 project investigated the structure of the dislodged microbial mats, exploring the origin, composition and function in relation to the ecology of the Lake Clifton. The impact of increasing salinity and eutrophication on the health and biodiversity of the microbial communities, including those associated with the thrombolites were also assessed.

The dislodged benthic microbial mats were composed of 15 species of diatoms embedded in a matrix of two species of cyanobacteria. The coccoid colonial cyanobacterium with its mucilage secretion predominantly formed the bulk of the microbial mat. A filamentous cyanobacterium, *Phormidium* sp., also formed part of the benthic mat but sparingly. Species of *Mastogloia* dominated the diatom community in the algal mat. Other diatoms found in the mat included species of *Campylodiscus, Diploneis, Navicula, Synedra, Entomoneis, Nitzschia, Amphora* and *Caloneis* many of which are known to secrete mucilage. The only remnants of animals found in the mats were ostracods, which could be identified as distinct white layers within the brownish algal mat.

As the benthic algal mat grows by addition of new layers from the top, the older layers get compressed and gradually decomposed. The mat may grow over 50 cm in thickness. The decomposition of the lower layers by bacterial activities may result in the dislodgement of the whole mats, probably induced by storm events.

The black bream death is attributed to pockets of anoxic regions due to decomposing algal mats or hypersalinity beyond the tolerance levels of this species.

The microbial communities associated with the thrombolites were dominated by a species of *Phormidium* woven into a compact mat, a coccoid bacterium and few species of diatoms such as *Entomoneis, Synedra* and *Amphora* species. The green alga *Cladophora* covered by epiphytic diatoms was also found growing profusely on the thrombolites – a clear sign of eutrophication. The salinity ranged from 29 g L^{-1} in the winter to 80 g L^{-1} in autumn. The nutrients in the standing water were at eutrophic levels.

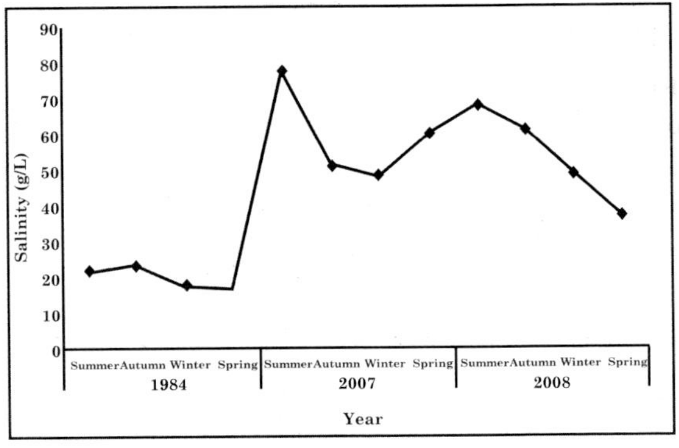

Fig.6. Salinity in Lake Clifton in 2007 and 2008 compared to 1984.

Compared with historical records of salinity and nutrients, it is clear that the lake has become eutrophic and salinity has almost doubled in the last 20 years (Figs. 6 & 7). Prolonged decline in rainfall and increased nutrient enrichment form a rapidly developing catchment area and a lack of outlet and flushing, seemed to be responsible for the degradation of the system. The cyanobacteria *Scytonema* and *Chroococcus* originally described as dominant components of the microbial communities are no more to be found. Currently there is no evidence of any growth occurring in the thrombolites. The conditions originally described as conducive to thrombolite growth no more exist today.

Climate change resulting in reduced rainfall and subsequent shortage in carbonate rich freshwater input compounded by increasing demand for fresh water due to massive development in the region are speculated causes for the doubling of the salinity of the lake. It has been also speculated that the brackish water fish black bream, introduced into the lake over 50 years ago, has altered the ecology of the lake triggering massive colonisation of *Cladophora* grazed by the fish.

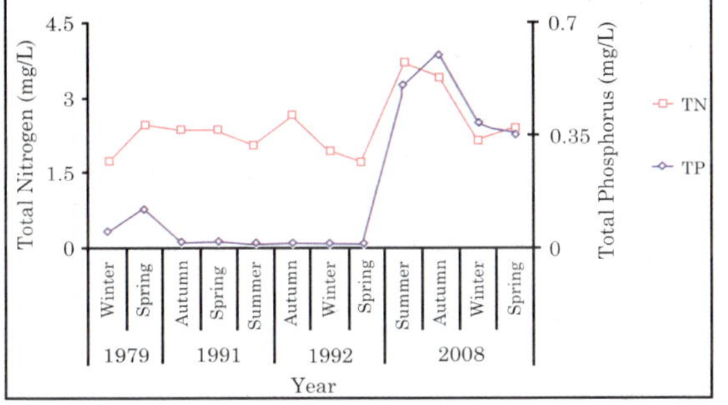

Fig.7. Total phosphorus and total nitrogen in 2008 compared to historical records (1979, 1991, 1992).

Salt Lakes of Rottnest Island

Almost 10% of Rottnest Island (area of 19 000 ha) located 18 km of the coast of Perth is occupied by salt lakes. Part of a chain of limestone islands, Rottnest was once part of the mainland of Western Australia. The islands are the result of sea level changes in the past. The salt lakes were connected to the ocean until late in the Holocene. They separated from the ocean by dune formation over 5 000 years ago. Subsequent evaporation resulted in six permanent hypersaline lakes. Microbial mats are abundant in these lakes and are the dominant primary producers. Since the arrival of Europeans, the Island has been settled first as a penal site and then as a tourist attraction. The Island is of high conservation value as it is home to unique fauna and flora. Recent developments have been impacting on the ecology of the salt lakes. Composition of the algal mats was studied by John in 1981 (unpublished data). Further studies on the algal mats in relation to water quality parameters, nutrients in particular, were conducted in 2005. The objective of the study was to compare the composition and integrity of the algal mats in four of the lakes and relate these to water quality and nutrients and *Artemia* populations.

Garden Lake had significantly lower salinity in both summer and winter (77.14 g L^{-1} in summer and 55.56 g L^{-1} in winter) compared to that of Government House Lake, Hershell Lake and Serpentine Lake (above 99 g L^{-1} in summer and above 93 g L^{-1} in winter) (Table 1).

Table 1. Salinity, Temperature and pH in lakes covered in the study.

	Salinity (g L^{-1})		Temperature (°C)		pH	
	Summer	Winter	Summer	Winter	Summer	Winter
Garden	77.14	55.56	25.5	17.74	7.8	8.21
Government	110.38	94.04	25.36	17.26	7.39	8.02
Herschell	108.62	94.06	23.78	17.84	7.81	8.20
Serpentine	99.6	93.18	23.92	17.46	7.86	8.08

The microbial mats in the salt lakes of Rottnest Island are dominated by cyanobacteria and diatoms, the former in biomass and the latter in diversity. The mats were dominated in biomass by the coccoid cyanobacterium *Aphanothece halophytica* followed by *Anacystis marina*, *Chroococcus turgidis* and *Agmenellum thermale*.

The filamentous cyanobacteria observed were *Spirulina subsalsa*, *Schizothrix calcicola* and *Microcoleus lyngbaceous*.

There were 29 species of diatoms but only species of *Navicula, Amphora, Nitzschia* and *Mastogloia* dominated the diatom assemblages. While phytoplankton are rare in these hypersaline lakes, one salt lake appropriately known as Pink Lake has abundance of *Dunaliella salina* - a green alga which appears pink due to the accumulation of ß-carotene.

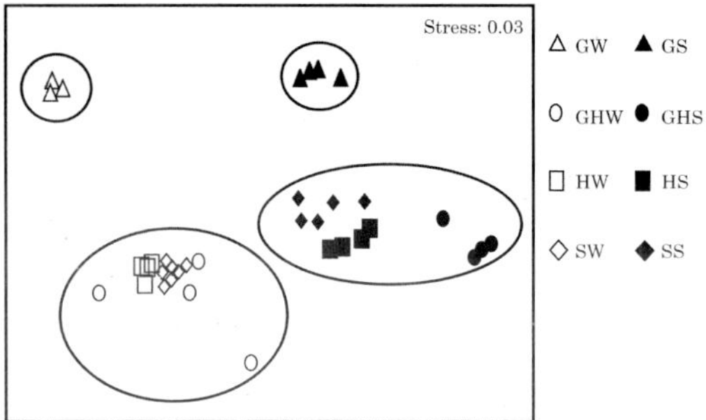

Fig.8. Biplot of water quality parameters of Rottnest Island Salt Lakes Winter (W) and Summer (S).

Nonmetric multi-dimensional scaling used to compare physicochemical data indicated that Garden Lake was separated from the other lakes in water quality parameters in both winter and summer (Fig. 8). Similarly the cyanobacterial composition of the mats was also different in Garden Lake (Fig. 9). The phosphorus and nitrogen levels in Garden Lake were much higher than those of the other lakes. *Artemia* species were much more abundant in Government House and Serpentine lakes than in Garden Lake, establishing the separation of Garden Lake in terms of nutrients and microbial communities.

The structural integrity of the algal mats in the lakes was directly related to the phosphorus levels. Garden Lake had the lowest structural integrity of algal mats and the highest phosphorus concentrations compared to other lakes. The algal mats in salt lakes originally evolved to maximise nutrient cycling by maintaining the cohesiveness of the microbial communities (Canfield and Des Marais, 1993). Low levels of phosphorus and nitrogen are a characteristic feature of lakes where microbial communities develop. A normal algal mat in a hypersaline environment forms a cohesive community to ensure efficient recycling (Canfield and Des Marais, 1993). The fact that Garden Lake had the highest phosphorus concentration and the least integrity of the microbial communities different from the other lakes indicates that eutrophication has a direct adverse impact on the benthic algal mats. Structurally the algal mats in Garden Lake had a large amount of sediment particles and were less cohesive.

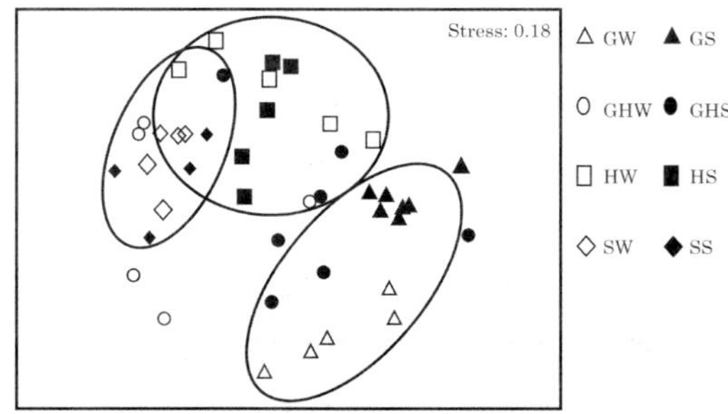

Fig.9. Biplot of composition of algal mats in Rottnest Island Salt Lakes in Winter and Summer.

The layering of the algal mats from Government House, Serpentine and Herschell lakes was also similar and far different from that of Garden Lake. The cyanobacteria and diatoms dominating the algal mats in Garden Lake were also drastically different from the other lakes. The microbial communities in Rottnest Island salt lakes reflect the water quality, nutrient status and the interactions between biota and the physicochemical environment. The benthic algal mats, which are the most predominant primary producers in the system, could be used as tools for biomonitoring the changes in the environment (John and Hay 2008). Modern stromatolites are found in Government House Lake, which appears to be healthy.

Esperance Salt Lakes

A chain of several coastal lakes exists in Esperance, a coastal town in the south of Western Australia (Fig. 1), between latitudes 33°47'S and 33°52'S and longitudes 121°47'E and 122'E. The coastal lakes are seasonally wet most of the year with a maximum depth of 2 m. The dry period may be very short. The region has a Mediterranean climate with cool dry summers and wet winters. Changes in the sea level coupled with dune formation seem to have shaped these lakes. They are of great conservation value and support 59 waterbirds including some migratory birds and two rare species. Four of the lakes are of international importance under the Ramsar Convention.

The physicochemical characteristics and algae of five of these salt lakes; Mullet Lakes, Woody Lake Reserve Marsh, Lake Wheatfield, Lake Warden and Pink Lake were investigated by Handley (2003). Out of these five only one, Pink Lake, was hypersaline with a salinity range of 87.8 to 242 g L^{-1}. All other lakes were mostly hyposaline ranging in salinity from 3 to 50 g L^{-1}. The depth and salinity of the surface water in these lakes are determined by seasons and rainfall. Pink Lake and Mullet Lake are linked with marine water, through underground and a creek, respectively.

Woody Lake Reserve Marsh was covered by the macrophyte *Lepilaena cylindrocarpa and* the charophyte *Lamprothamnium papulosum* and Lake Warden had *Lamprothamnium* meadows, with dissolved oxygen at saturation levels. There were two types of algal mats recorded in the Lakes; A pink-yellow laminated cohesive mat of 1 to 5 mm thickness in Pink Lake and a carpet-like nonlaminated mat dominated by filamentous green algae, cyanobacteria and diatoms in Woody Lake Reserve Marsh and Lake Warden. The carpet like mats emerge as the water level drops and the lakes dry up at the onset of summer. Altogether a total of 132 algal taxa were recorded from these lakes, dominated by cyanobacteria and diatoms. The diatom species in these lakes clearly show marine affinity. Pink Lake is characterised by *Dunaliella salina*, a common green alga in both coastal and inland lakes.

The threats experienced by this coastal saline system are eutrophication due to agriculture and urban developments in the surrounding land. Few streams from the catchment drain into the lakes. The Esperance salt lakes thus add to the distinctive features of the coastal lakes of Western Australia. However the inland salt lakes in the arid Western Australia while sharing certain features with the coastal lakes, maintain their unique physicochemical and biological characteristics.

INLAND SALT LAKES

Salient Features

Inland salt lakes are temporary wetlands and based upon the frequency and duration of filling events may be classified into:

1. Ephemeral: only filled after unpredictable rainfall and seldom supports macroscopic life forms for months to years.
2. Seasonal: alternatively wet and dry every year according to season, usually fills during the wet part of the

year and dries predictably, annually. Surface water persists usually long enough to complete the life stages of aquatic life.

3. Intermittent: alternatively wet and dry every year, but less frequently and regularly than seasonal wetlands; surface water persists longer than seasonal wetlands
4. Episodic: dry most of the time and very irregularly wet phases may persist for months (Boulton and Brock, 1999).

It is the variability of the water regime of temporary wetlands that acts as the major driving factor influencing the physical, chemical and biological features (Williams, 1998a,b).

The salinity in inland salt lake ranges from < 3 to 350 g L^{-1}, encompassing the whole salinity regime from fresh (< 3 gL^{-1}); Slightly saline (3-10 g L^{-1}); Saline (10-100 g L^{-1}); to highly saline (> 100 g L^{-1}). The same lake may experience these salinity ranges sequentially during a hydrocycle or at the same period at different sites of a lake. There is often great salinity variation spatially and in the duration of filling and drying cycle. Salinity may increase up to 10 times at the drying period in a hydrocycle.

Water Chemistry of the lakes is dominated by sodium and chloride ions same as in seawater and the coastal lakes. The pattern of ionic dominance is remarkably constant. Na>Mg>Ca>K for the cations and $Cl>HCO_3>SO_4$ or $Cl>SO_4>HCO_3$ for the anions; in some instances Ca may be higher than Mg (some lakes do have relatively high Ca and SO_4 (Handley 2003).

Heterogeneity

Some of the inland salt lakes are large (e.g. Lake Cowan, Lake Dundas, Lake Yindarlgooda, Lake Carey). Intersected by islands and peninsulas, combined with morphometrical differences in discharge, they are highly heterogeneous in water quality, pH, salinity, depth, temperature and sediment chemistry. The riparian zone at the margin of the lakes is dominated by various species of chenopodiaceae particularly species of *Halosarcia* (samphire), tolerant to waterlogging during flooding (Datson, 2002). Different species of *Halosarcia/Frankenia* and *Atriplex* may occupy different zones. The composition of riparian vegetation changes according to soil types and elevation from the margin. The fringing zone is followed by understory shrubs (e.g., *Acacia, Callitris*) and upper story trees (e.g. *Eucalyptus* spp.).

Vegetation often occurs in patches and the soil in between are covered by cryptogamic crust composed of the filamentous cyanobacteria *Phormidium* sp. and *Microcoleus* sp., mossplants and lichens and sometimes unicellular algae. The copious polysaccharides produced by the cyanobacteria in the crusts bind soil particles and prevent erosion.

Vast areas surrounding the inland salt lakes have been traditionally used for sheep farming and livestock. Trampling by livestock destroy the crust and facilitate erosion. Compared to the soil crusts in other parts of the world for example Florida (Hawks & Flechtner, 2002), the species diversity in the arid zone soil crust in Western Australia appears to be low in species diversity (Campagna, 2007).

Filling and Drying Periods

The hydrocycle consisting of filling and drying phases after rainfall or cyclonic events is crucial to the water quality and biodiversity of the lakes. Most of the large salt lakes do experience seasonal filling twice a year (winter and summer – characterised by ex-tropical cyclones e.g. 'Elaine' and 'Vance' in March 1999 caused 100 mm of rainfall). Prolonged dry phase followed by short wet phase has been so characteristic of some of the large inland salt lakes that there are terrestrial invertebrate fauna adapted to living in the lakebed e.g. species of wolf spider.

Salinity may be low in the early period of the filling phase. As the drying starts salinity increases and it reaches to the peak when the cycle is complete. Duration of the cycle may be up to 3 months but mostly less than two months. Depending upon the intensity or duration of the rainfall, the floodplain area may be inundated. Additionally, temporary playa lakes in the catchment of the hypersaline lakes may also be filled. These temporary water bodies are regions of high productivity and biodiversity as release of nutrients and hatching of spores and cysts are initiated. During the filling period, nutrients, biota, propagules and organic debris and particulates may be washed into the lake bed initiating biodegradation of organic materials as well as release of nutrients, triggering a chain of biological processes.

Water Quality Parameters

Most of the inland salt lakes are shallow (< 2 m) with high turbidity during filling but less during drying. Wind induced mixing is very common with virtually no stratification. Water temperature is close to the atmospheric temperature, but there is high diurnal variation. Most common range of pH is 7-9.5, but some may be acidic. Groundwater discharge along margins may be acidic, reducing the pH; mining impact may be in the form of acid drainage. Generally water is well oxygenated, (8-9 mg L^{-1}) due to wind mixing, less so during period of decomposition of large organic materials.

Total phosphorus in standing water is generally low, but relatively high at the end of drying period. However, total nitrogen is generally higher than P and higher at the end of drying period. Chlorophyll-*a* is generally low and seldom reaches eutrophic levels.

Biota: Biodiversity decreases generally as salinity increases in any inland salt lake. However the relationship between biodiversity and salinity may not be straight and direct. Decrease in biodiversity may follow drastic changes in magnitude, as salinity increases. This is illustrated by the distribution of diatoms in the inland salt lakes of Lefroy, Zot and Cowan and fresh water pools in their catchments (Fig. 10).

Generally, biota typical of low salinities have a narrow salinity tolerance. There are some biota with wide range of tolerance especially at intervals of 50 g L^{-1} to 100 g L^{-1} and 100 g L^{-1} to 200 g L^{-1}. Beyond a specific point, other factors (e.g. chance, food availability) may determine species richness other than salinity (Williams, 1998).

Halobacteria belonging to Archaebacteria (e.g. *Halobacterium*), photosynthetic and non–photosynthetic Eubacteria are present in many lakes. However, their distribution pattern and the ecological processes they are involved in remain unknown.

Common cyanobacteria found in inland salt lakes are *Schizothrix, Chroococcus, Aphanocapsa, Aphanothece, Anacystis, Synechococcus, Synechocystis, Gomphosphaeria, Oscillatoria, Spirulina, Nodularia,* and *Phormidium*. Algal mats made of *Schizothrix* are of great ecological significance in the hydric cycle. Some species of cyanobacteria have the highest salinity tolerance, up to 300 g L^{-1}.

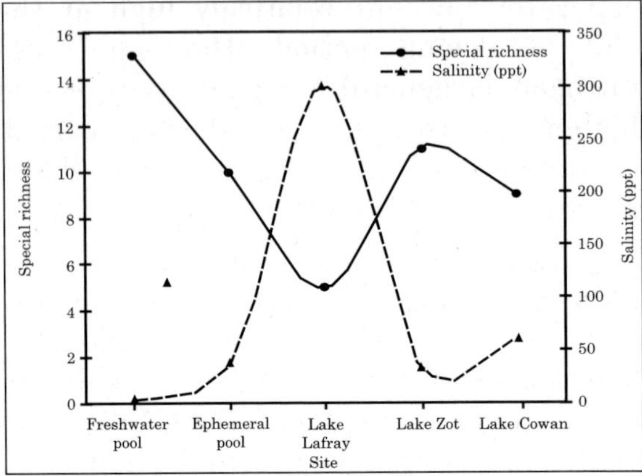

Fig.10. Distribution of diatoms in relation to salinity in three inland salt lakes and freshwater pools, in their catchments.

Green algae like *Microspora*, *Oedogonium* and *Chlorococcus* can be found up to a salinity of 10 g L^{-1}. Several strains of *Dunaliella salina* are common in these lakes. They form reddish patches in salt lakes and tolerate salinity up to 90 g L^{-1} or more. Along with cyanobacteria, some diatom species have the highest tolerance to salinity (up to 250 g L^{-1}). Beyond a salinity of 40 g L^{-1}, it is quite likely that among the algae, only certain species of diatoms and cyanobacteria survive. In shallow waters they form part of the benthic community and may get dislodged and float as pseudoplankton. True phytoplankton are rare in the temporary hypersaline lakes except species of *Dunaliella* and some nanoplanktonic flagellates.

Some species of charophytes e.g. *Lamprothamnium* spp. and *Chara* spp. can tolerate salinity up to 70 g L^{-1} and may form bulk of the macrophytes in some salt lakes. Up to 6 species of Charophytes have been collected from Kambalda region associated with the playa lakes. Two most common macrophytes are *Ruppia* and *Lepilaena* – occurring as meadows in shallow lakes forming thick beds.

A list of common invertebrates compiled from several inland salt lakes in Western Australia is presented in Table 2.

Table2. Invertebrate common representatives

Group	
Protista	Ciliates
Rotifers	*Brachionus* sp
Mollusca	*Coxiella*
Crustacea	Anostraca: e.g. *Paratemia* Copepoda: e.g. *Calamoecia* Cladocerans: e.g. *Moina*, *Daphniopsis* Ostracoda: e.g. *Autsralocypris*
Insecta	Diptera: e.g. Ephydra, *Ceratopogonidae* spp. Coleoptera: e.g. Dytiscidae, Hydrophillidae Trichoptera

Parartemia is most common in all the inland salt lakes. There are several endemic species confined to Western Australia and some not yet identified. Recently a taxonomic guide to the genus has been published (Timms, 2004). Their prolific production symbolises the food chain value of salt lakes for the endemic and migratory birds. Equally fascinating are their life-cycle involving a 'short period' of explicit life and a prolonged period of encystment- a fine example of the intricate survival strategies of saline lake biota.

A case history illustrating the impact of mining on an inland lake is presented below:

Lake Miranda – Impact of Dewatering on an Inland Lake

Lake Miranda is located about 500 km north-east of Perth in the North-eastern gold field (Fig. 1). The lake is in the arid zone of Western Australia with an annual precipitation of 250 mm/yr and evaporation of 2500 mm/yr. The lake is 200 km^2 in area and contains numerous low islands with a low relief catchment of 1500 km^2. Land use is pastoral or mining. Like other salt lakes in the area, the lake has a flat lakebed surrounded by low shrub-lands of *Halosarcia, Maireana* and *Cratystylis.*

Historically the lake was formed in early Tertiary drainage and was associated with a palaeodrainage system rich in minerals like nickel. It is a temporary lake, seasonal or intermittent and gets filled partially during storm events twice a year or once a year in summer and winter. Typically following a heavy rainfall, the lake may get partially filled up to 30 cm. As the lake dries up a thin layer of salt crust builds up on the sediment. The underground aquifer is estimated to be about 2 m below the lakebed. The lakebed consists of interbedded sand and clay overlain by dense plastic clay. These in turn are overlain by alluvial and colluvial sediments.

Mining for nickel commenced in 1999 close to the lakebed of Miranda. Dewatering of underground aquifer was used for the processing plant and dust suppression. The excess water was discharged into a suitable site.

Pre-discharge Study

As virtually no information was available on the limnology and biota of the lake, following heavy rainfall in June 2000, a study was conducted to gather basic information. Four stations were selected with a few sites within each station for sampling.

Salinity during the hydroperiod ranged from 18.7 to 25 g L^{-1} and depth of water ranged from 12.5 cm to 44 cm. The pH of the water varied from 7.14 to 9.39. The water was well oxygenated (8.2 to 9 mg L^{-1}). The relative abundance of cations was represented by the sequence of Na>Ca>Mg>K and $Cl > S0_4 > HC0_3$. Sodium formed more than 80% of the cations and Chloride formed 70% anions. Chl *a* ranged from 0.5 μgL^{-1} to 7.8 μg L^{-1}. Organic carbon, total Kjeldahl N and total P in the standing water were relatively high.

The Charophyte *Lamprothamnium* and the seagrass *Ruppia megacarpa* were the predominant macrophytes. There were mats of the filamentous green alga *Oedogonium* and a bloom of the dinoflagellate *Gymnodinium*. The planktonic diatoms *Entomoneis paludosa* and *Chaetoceros muelleri* were also part of the phytoplankton. All the algae were at reproductive stage, when sampled.

A total of eight macroinvertebrate taxa comprising of three crustaceans and five aquatic insects were collected during the sampling. Calanoid copepods attained high density at lower salinities at the floodplain sites.

Post-discharge Data

Dewatering discharge in the lake commenced in November 2000 and continued till July 2001. There was only one hydrocycle in 2001 - in January to February following a summer storm event. This was followed by a prolonged period of low rainfall. The lake dried up in October 2001, leaving a crust of salt on the floor of the lake. During the hydrocycle as the depth decreased due to evaporation, the salinity increased; a range of 1 to 33.5 cm depth was recorded. The filling phase started in January and the drying phase commenced from March.

Periphytons were collected using the substrate collector, the JJ periphytometer. Diatoms species richness declined in accordance with increase in salinity. The discharge sites continued to show decline in species diversity. There was no sampling in 2002 as the hydrocycle was very mild. However sampling resumed in 2003.

During the hydrocycle in 2003, a salinity range of 2.94 to 75.4 g L^{-1} was observed. Filamentous cyanobacteria mixed with diatoms were present in patches. The flood plain site had green filamentous algae and desmids. Some 17 species of diatoms were recorded during the period. The least species richness was found in the sites where "dewatering" was discharged.

The January and February 2001 samples were composed of rotifers, *Brachionus* sp., Chironomids and Ostracods. The community structure changed with the appearance of freshwater Copepods and Daphnids. By July, Ostracods and Chironomids were the sole members of the community. With the filling phase in April 2003, the floodplain sites had the least salinity (2.94 g L^{-1}) with an abundance of tadpoles and shield shrimps.

By April 2001, *Parartemia* was found in the lakes for the first time. Sediments samples collected from the control sites started hatching under laboratory conditions. *Parartemia* was hatched from cysts collected at two sites in April where the salinity was above 150 g L^{-1}. Further hatching of sediment samples in July and October occurred even after the lake dried up. The viability of the cysts was maintained even after depositions of the salt. In the hydrocycle in March to April 2003, *Parartemia* occurred in high abundance in sites with salinities ranging from 23.4 to 73.4 g L^{-1}.

Hydrocycle and Food Web in Lake Miranda

The food web in Lake Miranda is triggered by the filling phase resulting in the decomposition of organic materials, release of nutrients, hatching of dormant spores, cysts and germination of propagules. Flooding is crucial in inducing high diversity of biota. As the hydrocycle comes to an end with the commencement of the drying period, the biota tended to reproduce rapidly.

The surviving strategies of the biota included: 1) Short life-cycle 2) Accelerated reproductive maturity 3) Dormancy by encystment of eggs, spores or other propagates and 4) Tolerance to a wide variety of salinity.

In Lake Miranda the dewatering discharge coupled with low rainfall appears to have contributed to the accumulation of salt in the system. As salinity increased, the biodiversity tended to decrease. Phytoplankton and macrophytes characteristic of the system in the predischarge study in 2000 seem to have declined in 2001.

In spite of the hypersaline conditions and accumulation of salt crust, viable dormant forms of most benthic algae, periphyton and invertebrates, characteristic of other hypersaline systems in this area, were present in the system. The future "propagule bank" consisting of typical salt lake biota appears not to have been drastically affected by the discharge and abundance of *Parartemia* has been observed in the latest hydrocycle (2003) along with a reduced suit of the typical salt lake biota including those characteristic of floodplains of low salinity. However, effects of prolonged dewatering discharge on biodiversity is unknown.

CONCLUSION

The biodiversity of inland salt lakes are much less than that of coastal salt lakes, although both share several common species of cyanobacteria and diatoms. Handley (2003) determined that there was approximately 60% similarity between the algae of Esperance coastal lakes and Kambalda inland salt lake system which included Lake Lefroy. Different strains of *Dunaliella salina* are abundantly found in both coastal and inland salt lakes of Western Australia.

Eutrophication and increase in salinity appear to be the threats to coastal salt lakes. Doubling of salinity of Lake Clifton within the last 20 years resulting in the degradation of thrombolites remains enigmatic, while eutrophication may be explained by local land use.

Much of the inland salt lake monitoring involves comparison between 'control sites' and 'impacted' sites due to leaching of residues or dewatering discharge, e.g. Lake Yindarlgooda (Campagna 2007) and Lake Carey (John 1999). The second category of monitoring involves change over time. A pre-discharge study is followed by on-discharge and post-discharge studies.

Unpredictable rainfall patterns coupled with varying duration of the hydrocycle make comparisons between pre-discharge and post-discharge data difficult. Heavy rainfall may alleviate the impact of discharge by flushing or prolonged dry spells may intensify the impact of dewatering discharge by increased evaporation and consequent salt accumulation.

The natural and biological resources of the inland salt lakes need to be protected. It is the regular input of freshwater by rainfall events that keep the system sustainable. Any climate change resulting in rainfall decline and increase in evaporation would adversely affect the conservation of biota. Other threats to the system can be summarised as 1) Pastoral-sheep and cattle farming 2) change in local hydrology due to mining infrastructure-cause-way, roads, buildings, diversion of fresh-water flow 3) excavation and mining on the lakebed and surroundings 4) dewatering discharge on the lakebed and small playa lakes 5) Acid drainage 6) leaching of heavy metals and toxins from processing plants, settling ponds and storage facilities.

Key biota or most abundant biota such as diatoms and *Paratemia* are most suitable for assessment of health of salt lakes. As diatoms are the most dominant group of primary producers, a substrate collector such as the JJ periphytometer can be deployed at several control and impacted sites before and after the disturbance (e.g. dewatering discharge). As salinity increases with evaporation or increased discharge, sensitive species get eliminated. As diatoms occur in large numbers, they can be subjected to sophisticated statistical analyses.

The relative abundance of macroinvertebrates, specifically, juvenile to adult ratio of *Parartemia* quantified during the hydrocycle provide a powerful bioassessment tool.

Juveniles require low salinity and the cysts hatch early in the filling cycle. Toxic chemicals may have an adverse impact on the hatching of cysts. Estimation of the 'cyst bank' by sieving (wet or dry) samples of the top sediment and isolation of *Parartemia* is an effective way of determining the impact of a disturbance in a salt lake. Sediment hatching is a useful method of assessing the quality of a dry lake in terms of its potential to sustain a food chain.

Acknowledgements: We are grateful to various mining companies, consultants, and Government departments who made it possible for us to collect data from several salt lakes in Western Australia used in this paper. Special thanks to Stacey Gregory, Michelle Handley and Shane Chaplin for their contribution to the continuing study of western Australian salt lakes in Curtin University.

REFERENCES

BOULTON, A.J. and BROCK, M.A. (1999) Australian Freshwater Ecology: Processes and Management. Gleneagles Publishing, Adelaide.

BURNE, R.V. and MOORE, L.S. (1987) Microbialites: Organosedimentary Deposits of Benthic Microbial Communities. Palaios v. 2, pp. 241-254.

CAMPAGNA, V.S. (2007) Limnological Biota of Lake Yindalgoorda - An Inland Salt Lake in Western Australia. Ph. D thesis. Curtin University of Technology, Perth.

CANFIELD, D.E., MARAIS and D. J. Des (1993) Cycling of Carbon, Sulfur, Oxygen and Nutrients in a Microbial Mat. *In*: STAL, L.J. and CAUMETTE, P. (Eds.), Microbial Mats: Structure, Development And Environmental Significance. Springer-Verlag, Berlin. pp. 255-263.

CHAPLIN, S., JOHN, J., BREARLEY, D. and DUNLOP, J. (1990) A Limnological Investigation Of Lake Carey, A Large Ephemeral Lake in Western Australia. School of Environmental Biology, Curtin University of Technology, Perth.

DATSON, B. (2002) Samphires in Western Australia: a Field Guide to Chenopodiaceae Tribe Salicornieae. Department of Conservation and Land Management, Perth.

HANDLEY, M.A. (2003) The Distribution Pattern of Algal Flora in Saline Lakes in Kambalda And Esperance, Western Australia. Masters of Science Thesis. Curtin Univ., Tech., Perth.

HAWKES, C.V. and FLETCHER, V.R. (2002) Biological Soil Crust in a Xeric Florida Shrubland: Composition, Abundance and Spatial Heterogeneity of Crusts with Different Disturbance History. Microbial Ecology, v. 43, pp. 1-12.

JOHN, J. (1990a) The Diatom Flora of The Microbial Communities Associated With Stromatolites at Shark Bay, Indian Ocean, West Coast of Australia. *In*: RICARD, M., (Ed.), Ouvrage Dédie Á La Mémoire Du Professeur Henry Germain (1903-1989). Koetz Scientific, Koenigstein, pp. 97-110.

JOHN, J. (1990b) Parlibellus Panduriformis, a New Species of Diatom from Shark Bay, Western Australia. Phycologia, v. 36, pp. 38-42.

JOHN, J. (1993) Morphology and Ultrastructure of the Tubes of Mastogloia Cocconeiformis (Bacillariophyta) from Shark Bay, Western Australia. Phycologia, v. 32, pp. 88-394.

JOHN, J. (1994) Mastogloia Species Associated With Stromatolites, Shark Bay, Western Australia. Memoirs California Academy Sciences, v. 17, pp. 89-209.

JOHN, J. (1998) Diatoms: Tools for Bioassessment of River Health. Land and Water Resources Research Development Corp., Canberra.

JOHN, J. (1999) Limnology of Lake Carey with Special Reference to Primary Producers.

Salt Lake Ecology Seminar. Centre for Rehabilitation, University of Western Australia, Perth.

JOHN, J., DUNLOP, N., GOLDBERG, N., LOWE, E. and OSBORNE, J. (2000) Baseline Aquatic Biology Survey of Lake Miranda. Urs (For Sir Samuel Miles Nl), Perth.

JOHN, J. and HAY, M. (2008) Algal Mats and Artemia Populations in Hypersaline Lakes in Rottnest Island, Western Australia. *In*: Proc. V Inter. Conf. Environ. Micropal. Microbio. Meioben. University of Madras, India. pp.127-131.

JOHN, J. and PATON, J. (2008) Thrombolites of Lake Clifton, Mandurah, Western Australia. Proc. V Inter. Conf. Environ. Micropal. Microbio. Meioben. University of Madras, India, pp.131-133.

LOGAN, B.W., HOFFMAN, P. and GEBELEIN, C.D. (1974) Algal Mats, Cryptalgal Fabrics and Structures Hamelin Pool, Western Australia. Am. Assoc. Petrol. Geol., v. 22, pp. 140-194.

MOORE, L.S. (1987) Water Chemistry of the Coastal Saline Lakes of Clifton-Preston Lakeland System South-Western Australia and its Influence on Stromatolite Formation. Australian Jour. Marine Freshwater Res., v. 38, pp. 647-660.

MOORE, L.S., Knott, B. and Stanley, N. (1983) Stromatolites of Lake Clifton, Western Australia. Search, v. 14, pp. 309-314.

MOORE, L.S. and NEIL, J. (1985) Lake Clifton: a Prolific and Unusual Stromatolite Environment. Australian Microbiologist, v. 6, pp. 140.

SMITH, D.I. (1998) Water in Australia. Oxford University Press, Melbourne.

TIMMS, B.V. (2004) An Identification Guide to the Fairy Shrimps (Crustacea, Anostraca) of Australia. Cooperative Res. Centre Freshwater Ecology, Thurgoona.

TIMMS, B.V. (2005) Salt Lakes in Australia: Present Problems and Prognosis for the Future. Hydrobiologia, v. 552, pp. 1-15.

WILLIAMS, W.D. (1998a) Biodiversity in Salt Lakes. *In*: MCCOMB, A.J. and Davis, J. A. (Eds.), Wetlands For The Future. Gleneagles, Adelaide, pp. 161-166.

WILLIAMS, W.D., (1998b) Diversity and Evolution of Fauna of Dryland Wetlands. *In*: MCCOMB, A. J. and DAVIS, J. A. (Eds.), Wetlands Future. Gleneagles, Adelaide, pp. 167-172.

WINSBOROUGH, B.M. and SEELER, J.S. (1984) The Relationship of Diatom Epiflora to the Growth of Limnic Stromatolite and Microbial Mats. *In*: RICARD, M., (Ed.), Proc. 8th Inter. Diatom Symp., Koeltz Scientific, Koenigstein., pp. 395-407.

Micropaleontology and its Applications
Pradeep K. Kathal, Rajiv Nigam & Abu Talib (Editors)
Scientific Publishers (India), 2018, 59-70 pp.

$\delta^{18}O$ of Larger Benthic Foraminifera: Temperature Calibration for Shallow Marine Tropical-Subtropical Carbonates

Pratul Kumar Saraswati[1]*, Asmita Singh[1], Karyne Rogers[2], Prasanta Sanyal[3], Kanchan Pande[1] and Kenji Iwao[4]

[1]Department of Earth Sciences, IIT Bombay, Mumbai, India
[2]National Isotope Centre, GNS Science, Lower Hutt, New Zealand
[3]Department of Earth Sciences, Indian Inst. of Science Education & Research (IISER), Kolkata, India
[4]Akajima Marine Science Laboratory, Zamami-son, Okinawa, Japan
***Email:** pratul@iitb.ac.in

Abstract

In paleoclimate studies the seawater temperature is traditionally estimated on $\delta^{18}O$ of planktic and deep water benthic foraminifera. These foraminifera either are absent or do not occur commonly in shallow marine carbonates of tropical to subtropical regions. Larger benthic foraminifera (LBF) were the major contributors of such carbonates in the geologic past and continue to be a common element of the present-day carbonates in shallow sea environments. In the present study, we collected live specimens of a miliolid foraminifer (*Marginopora kudakajimaensis*) and a rotaliid foraminifer (*Amphistegina lessonii*) over different seasons for field-based calibration of $\delta^{18}O$ to estimate seawater temperatures. The $\delta^{18}O$ values were measured on whole specimens as well as on dissected chambers of the tests. The *Amphistegina* derived equation [$T\ (^{\circ}C) = 16.3 - 4.24(\delta^{18}O_f - \delta^{18}O_w)$] is indistinguishable from the mollusc derived equation by Epstein *et al.* (1953). The *Marginopora* derived equation [$T\ (^{\circ}C) = 18.8 - 4.42(\delta^{18}O_f - \delta^{18}O_w)$] is not comparable to any known foraminifer calibrations.

Key Words: Paleoclimate, Platform Carbonates, *Amphistegina*, *Marginopora*, Coral Reefs.

INTRODUCTION

During the process of calcification, the isotopes of oxygen from the ambient seawaters are incorporated in the shells of organisms through a temperature-dependent isotopic fractionation. The oxygen isotopic composition ($\delta^{18}O$ value) of biogenically precipitated calcite became a principal proxy of seawater temperatures with the introduction of the first paleotemperature equation based on mollusc shells by Epstein *et al.* (1953) and modified later by Craig (1965). The mollusc-derived equation has since been used for all types of biogenic carbonates. Later studies, however, found that organisms fractionate differently, some secrete their shells in isotopic equilibrium while others do not (see Wefer and Berger, 1991 for details). Due to microhabitat and metabolic effects (commonly termed as "vital effects") organisms produce non-equilibrium isotope fractionations and, as a result, the isotopic composition of biogenic carbonates deviate from that of inorganically precipitated calcite (Rohling and Cooke, 2003). Although metabolic processes have less effect on oxygen isotope ratios than on carbon isotopes, paleotemperature reconstructions preferably require species-specific calibrations of $\delta^{18}O$ values.

Foraminiferal calcite is the most widely used biogenic carbonate for oxygen and carbon isotope analyses in paleoclimate and paleoceanographic studies. Traditionally, most isotopic studies use planktic and deepwater benthic foraminifera for sea-surface hydrography and ocean circulation. However, shallow water, symbiont-bearing benthic foraminifera, commonly called larger benthic foraminifera (LBF), are an important constituent of benthic communities in tropical–subtropical seas. They were the major contributors of carbonate platform successions during warmer periods in the geological past (Hallock, 1985). Studies on Paleocene – Eocene carbonates, extensively developed along the margins of the Neotethys, have indicated how foraminifera played an important role in the making of carbonate platforms (Schiebner and Speijer, 2008; Khanolkar and Saraswati, 2015). The planktic foraminifera either are absent or have limited presence in platform carbonate successions and until this study, there is no $\delta^{18}O$ calibration based on LBF to estimate seawater temperatures. A preliminary study based on a smaller data set by Saraswati *et al.* (2006a) proposed a quadratic equation but it lacked essential corrections. In the present study, we take representative species from both rotaliid and miliolid groups of LBF to develop a field-based calibration of $\delta^{18}O$ values and compared them with some of the other calibrations derived from planktic and small benthic foraminifera.

MATERIALS AND METHODS

The samples for the present study were collected from Akajima, Japan (Fig. 1). The island is part of the Ryukyu Islands in southwestern Japan. The warm water Kuroshio Current flowing on the western side of the Ryukyus promotes coral reefs around many islands. Live specimens of two species, *Marginopora kudakajimaensis* and *Amphistegina lessonii* (Fig. 2), were picked at water depths of ~2m on the reef flat in different seasons. The sea-grass attached specimens and symbiont-imparted colour of the tests helped in picking live specimens. Air-dried samples were treated with a 10% solution of hydrogen peroxide and the specimens were cleaned with ethanol prior to isotope analysis.

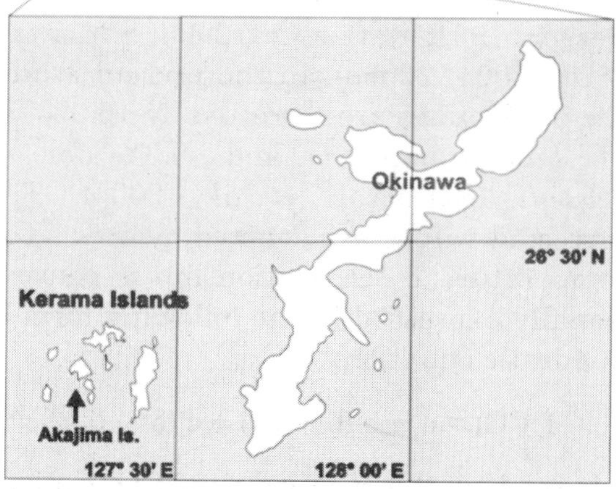

Fig.1. Location map of the sampling area.

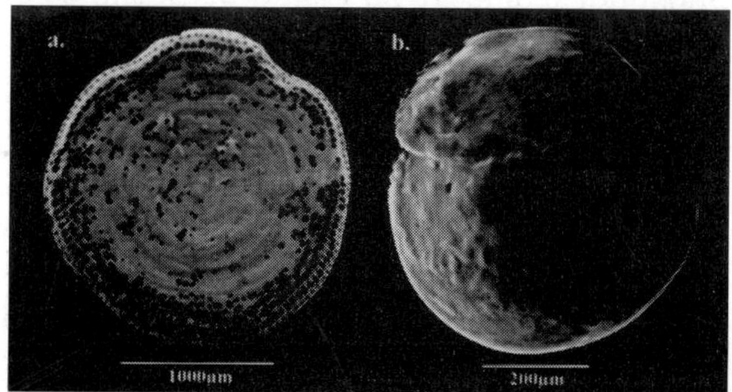

Fig.2. Photographs of the analysed species, *Marginopora kudakajimaensis* (a) and *Amphistegina lessonii* (b).

Amphistegina belong to the Order Rotaliida and *Marginopora* represents the Order Miliolida. *Amphistegina* is characterized by hyaline wall, low trochospiral and spirally coiled tests and small size (diameter ~ 1.5 mm). *Marginopora* is distinguished by its porcelaneous wall, planispirally coiled test, the initial embryonic chambers followed by cyclic chambers and large size (up to 10 mm in diameter). The life span of *A. lessonii* is about 4 months (Muller, 1977) and *M. kudakajimaensis*is lives for 12 months (Fujita *et al.*, 2000).

Two approaches were followed for isotope analysis of the foraminiferal tests. Bulk analysis was carried out on single specimens of both species. *Amphistegina* being small, the whole specimens were crushed for analysis. In *Marginopora*, due to its large size and cyclic growth, a cone was cut using a blade such that protoconch at the centre and the succeeding cyclic chambers were sub-sampled and crushed for isotope measurements. In addition, the outermost chambers of *Marginopora* were micro-sampled using 0.4 mm hand-held drill so that the carbonate secreted in most recent months of the growth was analyzed for isotope ratios. Considering a 12-month life span of *M. kudakajimaensis*, 0.4 mm growth in radius is estimated to correspond to calcification in 1 month. The $\delta^{18}O$ of *Marginopora* were measured at the Stable Isotope Laboratory, GNS Science (New Zealand) and *Amphistegina* were analysed at IISER, Kolkata (India).

The $\delta^{18}O$ and $\delta^{13}C$ values of seasonally collected *Marginopora* were measured by a Europa automated carbonate preparation system (CAPS) carbonate device attached to a Europa GEO 20-20 mass spectrometer (PDZ Europa Ltd., Crewe, UK) at the Stable Isotope Laboratory, GNS Science

(New Zealand). The CO_2 was obtained by reaction with phosphoric acid at 80°C. All measurements are normalised to VPDB using GNS Marble (internal standard), NBS-18 and NBS-19 (International Standards). Standard errors are ± 0.1‰. GNS Marble has an isotopic composition of $\delta^{13}C$ = 2.04‰, and $\delta^{18}O$ = -6.44‰ and we used a fractionation factor for calcite @ 80°C = 1.00844 (Sharma and Clayton, 1965). It may be mentioned that the analysis was carried out on whole specimens as well as hand-drilled outermost chambers of several seasonal specimens.

$\delta^{18}O$ value of *Amphistegina* was measured in MAT 253 with automated Kiel IV device at stable isotope laboratory of IISER, Kolkata. Powdered samples were reacted at 72°C. NBS-18 along with internal laboratory standard Z-Carrara marble was used to check the reproducibility (±0.1‰) and calibration.

RESULTS AND DISCUSSION

Table 1 provides $\delta^{18}O$ values and the corresponding seawater temperatures for both species. The $\delta^{18}O$ value of seawater at the location of the sampling is -0.3‰ (VSMOW) (Saraswati *et al.*, 2004). We do not expect large seasonal variation in this value as the sample location is reef-flat on an oceanic island. A small correction of -0.27‰ is applied to convert from VSMOW to VPDB and the corrected values are presented as $\delta^{18}O_w$ (see discussion by Bemis *et al.*, 1998). As discussed below, $\delta^{18}O$ values of the two species are adjusted for equilibrium calcite and the corrected values are presented as $\delta^{18}O_f$. The seawater temperatures represent average temperatures in the months of sample collection, except for the samples of July 1999 and March 2001 that were collected at the start of the months, and for which the average temperatures of the previous months are used for calibrations.

Not all organisms satisfy the assumption of equilibrium precipitation during calcification. The deviation from equilibrium calcite is especially noted in symbiont-bearing foraminifera due to photosymbiosis (Erez, 1978). Photosymbiosis has marked effect on carbon isotopic composition of LBF but the $\delta^{18}O$ values show only minor deviations from equilibrium calcite and it is mainly a function of surface-water temperatures (Wefer *et al.*, 1981; Luz *et al.*, 1983; Langer, 1995). Correction factors are suggested for several species of foraminifera to remove "vital effects" from isotopic values (Wefer and Berger, 1991; Katz *et al.*, 2003; Saraswati *et al.*, 2004, 2006b). In the present study, the $\delta^{18}O$ values are corrected by +0.3‰ for *M. kudakajimaensis* and -0.3‰ for *A. lessonii* (Saraswati *et al.*, 2004). The corrected values are denoted by $\delta^{18}O_f$. The temperature: $\delta^{18}O$ relationship is conventionally expressed in the following form of quadratic equation:

$$T\ (^{\circ}C) = a + b\ (\delta_c - \delta_w) + c\ (\delta_c - \delta_w)^2$$

where, δ_c is the $\delta^{18}O$ value of CO_2 evolved from the carbonate after reaction with phosphoric acid at 25°C and δ_w is the $\delta^{18}O$ value of CO_2 equilibrated isotopically at 25°C with the seawater from which the carbonate was precipitated, and both measured against the same working standard. It is found that linear equations, instead of quadratic, also give equally good fits at warm oceanic temperatures and accordingly later proposed paleotemperature equations are linear (Bemis *et al.*, 1998; Bouvier-Soumagnac and Duplessy, 1985; Spero *et al.*, 2003). We have also used linear calibration in this study.

Table 1. The dates of sample collections, measured average temperatures of seawater for the months of collection, 4-months average temperatures, salinity and statistical summary of $\delta^{18}O$ values of *Marginopora kudakajimaensis* and *Amphistegina lessonii* (*indicates temperature of the previous months for July 1999 and March 2001 collections).

Date of collection	Monthly Average Temp. (°C)	4 months Average Temp. (°C)	Salinity ‰	$\delta^{18}O$ *Amphistegina lessonii*		*Marginopora kudakajimaensis* (whole specimen)		*Marginopora kudakajimaensis* (outermost chamber)	
				Max	-1.21				
6 Jul 1999	26.2*	23.9	34.3	Min	-2.63				
				Mean	-1.75		–		–
				SD	0.44				
				(n=12)					
				Max	-1.79	Max	-1.25	Max	-1.38
10 Mar 2001	22.1*	23.5	34.6	Min	-2.46	Min	-2.04	Min	-1.87
				Mean	-2.08	Mean	-1.63	Mean	-1.61
				SD	0.24	SD	0.37	(n=3)	
				(n=12)		(n=4)			
				Max	-2.54	Max	-2.85	Max	-3.02
29 Sep 2003	28.7	27.8	34.4	Min	-3.57	Min	-3.24	Min	-3.18
				Mean	-3.02	Mean	-3.00	Mean	-3.10
				SD	0.33	SD	0.17	SD	0.07
				(n=11)		(n=4)		(n=5)	
				Max	-2.29	Max	-2.36	Max	-1.79
29 Nov 2003	25.2	27.5	34.6	Min	-3.01	Min	-2.42	Min	-2.53
				Mean	-2.60	Mean	-2.39	Mean	-2.31
				SD	0.22	SD	0.03	SD	0.27
				(n=11)		(n=4)		(n=6)	
				Max	-1.81	Max	-1.51	Max	-1.49
30 Jan 2004	22.0	24.3	34.8	Min	-2.86	Min	-1.91	Min	-1.95
				Mean	-2.30	Mean	-1.69	Mean	-1.70
				SD	0.31	(n=3)		SD	0.22
				(n=12)				(n=5)	
				Max	-1.84	Max	-1.41	Max	-1.12
18 Mar 2004	21.2	21.8	34.7	Min	-2.90	Min	-2.43	Min	-1.59
				Mean	-2.34	Mean	-1.84	Mean	-1.35
				SD	0.32	SD	0.45	SD	0.16
				(n=11)		(n=4)		(n=6)	
				Max	-1.59				
31 May 2004	25.1	22.6	34.6	Min	-2.38				
				Mean	-2.03	-	-		
				SD	0.25				
				(n=12)					

For temperature calibrations linear regressions are explored between ($\delta^{18}O_f - \delta^{18}O_w$) and (i) average monthly temperatures for single-specimens data of both *A. lessonii* and *M. kudakajimaensis*, (ii) average temperatures of 4-months in *A. lessonii* and, (iii) average monthly temperatures for the micro-drilled samples of *M. kudakajimaensis*.

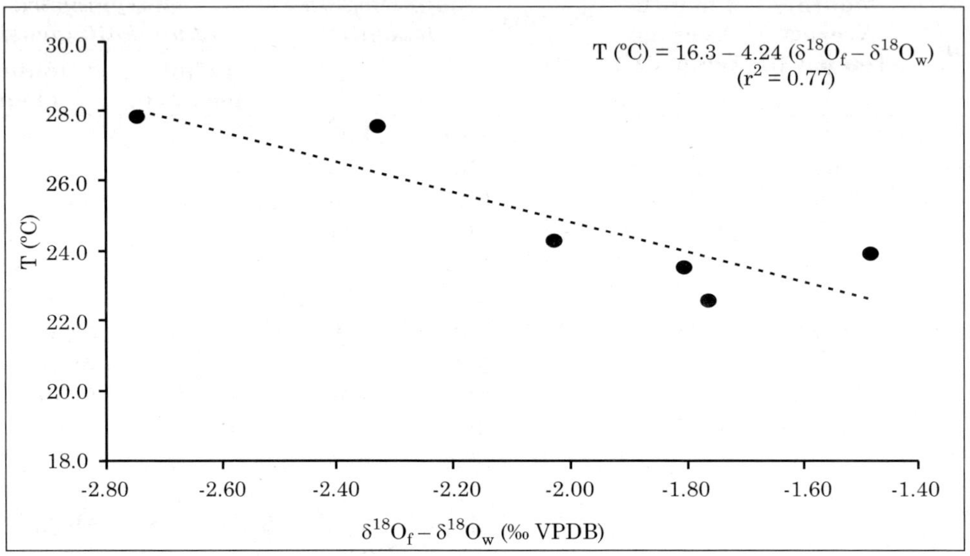

Fig.3A. Linear regression of mean values of ($\delta^{18}O_f - \delta^{18}O_w$) and 4-months average temperatures for *Amphistegina lessonii*. The $\delta^{18}O_f$ values are for the crushed samples of whole-specimens corrected for disequilibrium effect.

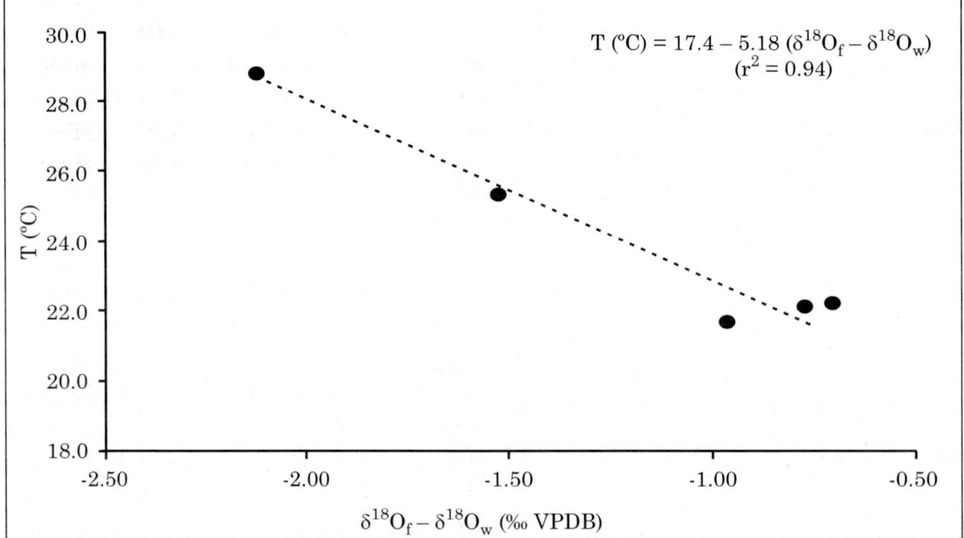

Fig.3B. Linear regression of mean values of ($\delta^{18}O_f - \delta^{18}O_w$) and average temperatures of the months of sample collection for *Marginopora kudakajimaensis*. The $\delta^{18}O_f$ values are for the crushed samples of whole-specimens corrected for disequilibrium effect.

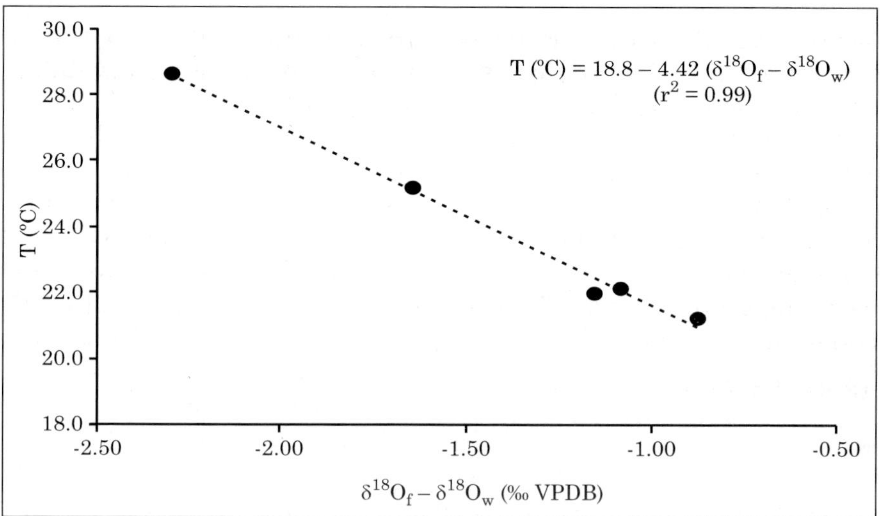

Fig.3C. Linear regression of mean values of ($\delta^{18}O_f - \delta^{18}O_w$) and average temperatures of the months of sample collection for *Marginopora kudakajimaensis*. The $\delta^{18}O_f$ values are for the dissected outermost cycles of chambers corrected for disequilibrium effect.

The isotope data are poorly correlated with temperatures in *A. lessonii* when regressed over the average monthly temperatures in the months of sampling. Assuming that the reported life span of *A. lessonii* is also valid for the Akajima populations, the regression remarkably improves over the mean temperatures of 4-months (Fig. 3A). The regression equation is as follows:

$$T\ (^{\circ}C) = 16.3 - 4.24(\delta^{18}O_f - \delta^{18}O_w) \quad \text{(Eq. 1)}$$

The data plots are still scattered and the r^2 value is not very high implying that the temperature: $\delta^{18}O$ value relationship is fairly good but not very strong. Improved correlation with 4-monthly temperatures signifies that for a robust calibration in whole-specimen data, the regression should be over the complete range of growth temperatures of the individuals. Although true for all foraminifera, it is particularly important for LBF that live longer (from a few months to one or two years) compared with smaller benthic and planktic foraminifera (having life span of a few weeks to few months). It is a major constraint in developing an exact temperature: $\delta^{18}O$ value relationship even in field-based calibrations. The isotope data of March 2004 is used to validate the equation, indicates a difference of about 3.3°C in measured and estimated temperatures. Presence of individuals of different growth stages and populations of overlapping generations may be a primary cause of scatter in whole-specimen data. The whole-specimen data of *M. kudakajimaensis*, in comparison to *Amphistegina lessonii*, is better correlated with monthly average temperatures (Fig. 3B). It is possibly because in tests with cyclic growth when a cone of sample is dissected the weighted average of more recently secreted carbonates contributes to a better correlation. A major constraint in developing an accurate scale based on whole-specimen data is thus compositional averaging of the isotopic values over the unknown biological age of the analysed individuals.

The uncertainty of whole-specimen analysis is minimized by analysing the most recently calcified chambers. The dissected outermost cycles of *M. kudakajimaensis* show high correlation between $\delta^{18}O$ values and the average monthly temperatures. A perfect correlation is noticed when the mean $\delta^{18}O$ value is regressed over temperature (Fig. 3C). The regression equation is as follows,

$$T\ (^{\circ}C) = 18.8 - 4.42(\delta^{18}O_f - \delta^{18}O_w) \quad \text{(Eq. 2)}$$

Notwithstanding fewer isotope data on LBF we attempt to validate the above calibration with the published $\delta^{18}O$ values of the miliolid LBF. The estimated and measured temperatures are comparable if species is the same. The different species of *Marginopora* and different genera of the miliolid LBF may or may not give comparable temperatures and the difference may be as high as 3°C (Table 2). The use of species-specific calibration is therefore highly recommended.

Table 2. A comparison of the measured and estimated seawater temperatures using the proposed calibration and $\delta^{18}O$ values of miliolid foraminifera reported in literature.

Species	Location	$\delta^{18}O$	Measured Temperature (°C)	Estimated Temperature (°C)	Remarks	Reference
Marginopora kudakajimaensis	Akajima	-2.49	26.2	26.0	$\delta^{18}O_{water}$ =-0.3	Saraswati et al. (2004)
		-2.45		25.8	$\Delta\delta^{18}O$ = -0.3	
Marginopora vertebralis	Philippines	-2.84	26-30	30.1	$\delta^{18}O_{water}$ = 0.0±0.2	Wefer et al. (1981)
		-3.44		32.8		
	Lakshadweep	-2.12	28-30	28.3	$\delta^{18}O_{water}$ = 0.3	Saraswati et al. (2006)
		-2.27		29.0		
		-2.24		28.8	$\Delta\delta^{18}O$ = -0.1	
		-2.12		28.3		
	Papua New Guinea	-3.9	26.7-32.3	34.8	$\delta^{18}O_{water}$ = 0.0	Langer (1995)
		-3.6		33.5	(Assumed)	
		-3.6		33.5		
		-3.6		33.5		
		-3.0		30.9		
Sorites orbiculus	Akajima	-2.60	26.2	30.0	$\Delta\delta^{18}O$ = 0.5	Saraswati et al. (2004)
		-2.08		27.7		
Amphisorus hemiprichii	Akajima	-2.29	26.2	26.8	$\Delta\delta^{18}O$ = -0.1	
Peneroplis planatus	Akajima	-0.79	26.2	23.3	$\Delta\delta^{18}O$ = 0.8	
		-0.86		23.6		

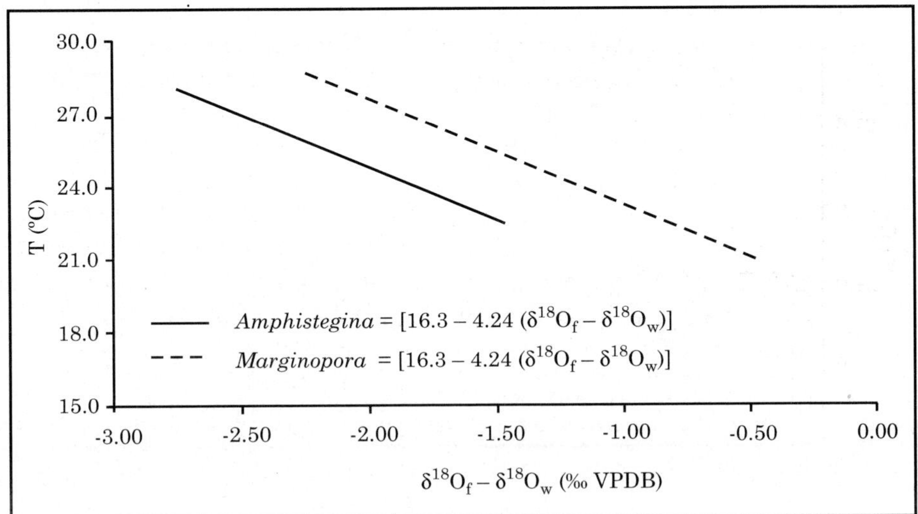

Fig.4. Comparison of the proposed $\delta^{18}O$: temperature calibrations of a miliolid foraminifer *Marginopora kudakajimaensis* and rotaliid foraminifer *Amphistegina lessonii*.

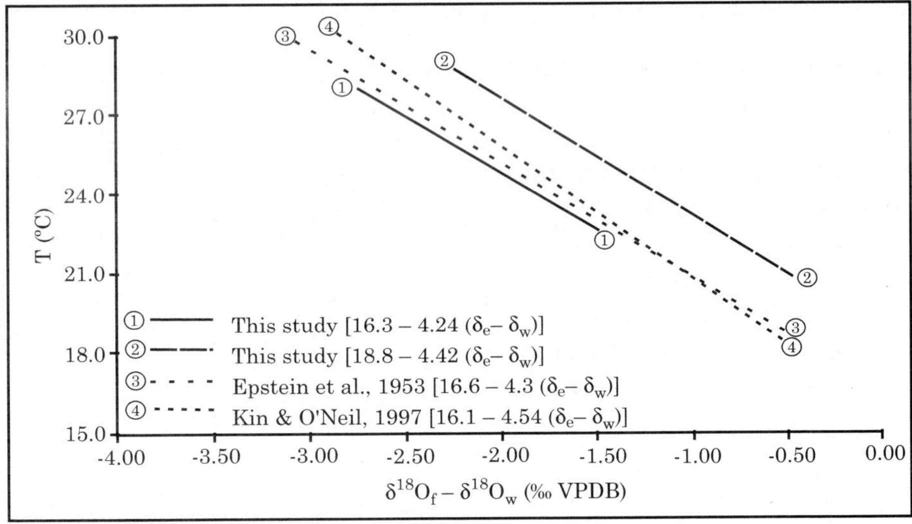

Fig.5. Comparison of the proposed $\delta^{18}O$: temperature calibrations with that of inorganic calcite (Kim and O'Neil, 1997) and mollusc (Epstein *et al.,* 1953).

There is a distinct shift in trend lines between the *A. lessonii* derived and *M. kudakajimaensis* derived equations (Fig. 4), and it may account for a difference of above ±2°C in estimated temperatures. The LBF derived equations are compared with the equation for inorganic-calcite (Kim and O'Neil, 1997) and the widely used mollusc derived equation of Epstein *et al.* (1953) (expressed as linear equation by Bemis *et al.*, 1998; Fig. 5). While *Marginopora* equation shows a distinct shift from both the inorganic calcite and mollusc derived equations, the *Amphistegina* derived equation is comparable with the mollusc derived equation.

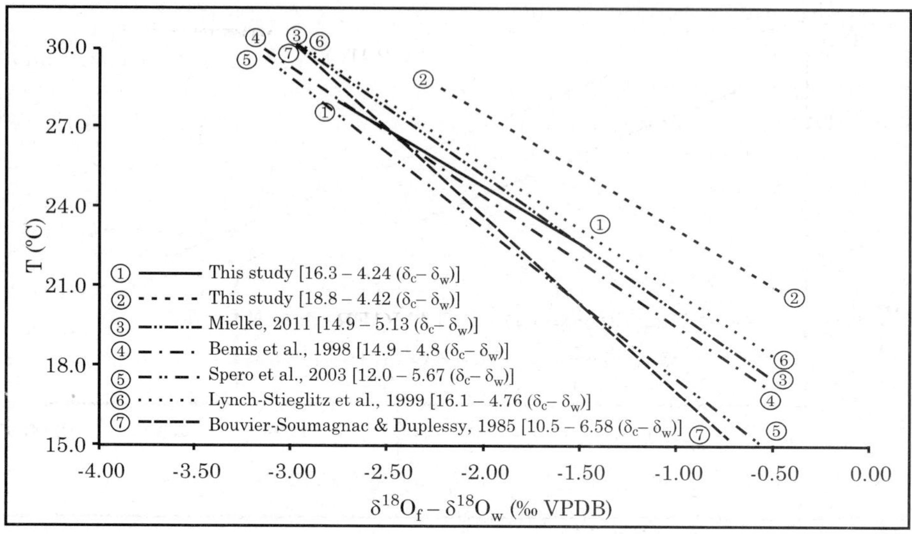

Fig.6. Comparison of the proposed $\delta^{18}O$: temperature calibrations with that of the other foraminifera-derived calibrations (Bemis *et al.*, 1998; Bouvier-Soumagnac and Duplessy, 1985; Lynch-Stieglitz *et al.*, 1999; Mielke, 2001; Spero *et al.*, 2003).

The benthic foraminiferal calibrations are mainly for the deep-water species and therefore calibrated for cool water temperatures. Marchitto *et al.* (2014) recently derived a quadratic equation for the combined *Cibicidoides* and *Planulina* indistinguishable from the inorganic calcite data of Kim and O'Neil (1997). We compared our equations with that of Lynch-Stieglitz *et al.* (1999) for *Cibicidoides* and *Planulina* that was calibrated over a wider range of 4°C to 26°C (Fig. 6). Both LBF calibrations are distinctly shifted from the *Cibicidoides* – *Planulina* equations. A comparison of calibrations between the symbiont-bearing LBF and symbiont-bearing planktic foraminifera further suggests that the LBF calibrations are not comparable with those of the planktic foraminifera (Fig. 6).

CONCLUSION

Longer life span and complex life cycles are the major constraints in LBF $\delta^{18}O$-temperature calibration although there is necessity of such calibration for shallow marine tropical to subtropical carbonates. In whole specimen analysis the exact temperature: $\delta^{18}O$ value relationship is masked by unknown biological age and presence of individuals of different growth stages and possibly of different generations in the sampled populations. It is eliminated to a reasonable extent by the analysis of dissected outermost chambers. The LBF derived calibrations differ distinctly from the other calibrations derived from small benthic and planktic foraminifera. The *Amphistegina* derived equation is indistinguishable from the original equation on mollusc but *Marginopora* derived equation is not comparable to any known foraminifer calibrations.

Acknowledgment: The work was initiated through funding by STA, Japan in 1999 and JSPS in 2000 to PKS. The University Grants Commission (India) is funding research fellowship to Asmita Singh.

REFERENCES

BEMIS, B.E., SPERO, H.J., BIJMA, J. and LEA., D.W. (1998) Reevaluation of the oxygen isotopic composition of planktonic foraminifera: Experimental results and revised palaeotemperature equations, Paleoceanography, v.13, pp.150-160.

BOUVIER-SOUMAGNAC, Y. and DUPLESSY, J.-C. (1985) Carbon and oxygen isotopic composition of planktonic foraminifera from laboratory culture, plankton tows and Recent sediment: implications for the reconstruction of paleoclimatic conditions and of the global carbon cycle, J. Foraminiferal Res., v.15, pp.302-320.

CRAIG, H. (1965) The measurement of oxygen isotope paleotemperatures. *In:* Tongiorgi, E. (Eds.), Stable isotopes in oceanographic studies and paleotemperatures, Spoleto 1965: Consiglio Nazionaledella Ricerche Laboratorio di Geologia Nucleare, Pisa. pp.161-182.

EPSTEIN, S., BUCHSBAUM, R., LOWENSTAM, H.A. and UREY, H.C. (1953) Revised carbonate – water isotopic temperature scale, Geol. Soc. Am. Bull., v.64, pp.315-1325.

EREZ, J. (1978) Vital effect on stable-isotope composition seen in foraminifera and coral skeletons. Nature, v.273, pp.199-202.

FUJITA, K., NISHI, H. and SAITO, T. (2000) Population dynamics of *Marginopora kudakajimaensis* Gudmundsson (Foraminifera: Soritidae) in the Ryukyu Islands, the subtropical northwest Pacific, Mar. Micropaleontol., v.38, pp.267-284.

HALLOCK, P. (1985) Why are larger foraminifera large? Paleobiology, v.11, pp.195-208.

KATZ, M.E., KATZ, D.R., WRIGHT, J.D., MILLER, K.G., PAK, D.K., SHACKLETON, N.J. and THOMAS, E. (2003) Early Cenozoic benthic foraminifera isotopes: species reliability and interspecies correction factors, Paleoceanography, v.18, doi:10.1029/2002PA000798.

KHANOLKAR, S. and SARASWATI, P.K. (2015) Ecological response of shallow marine foraminifera to early Eocene warming in equatorial India, J. Foram. Res., v.45, pp. 293-304.

KIM, S.-T. and O'NEIL, J.R. (1997) Equilibrium and nonequilibrium oxygen isotope effects in synthetic carbonates, Geochim. Cosmochim. Acta, v.61, pp.3461-3475.

LANGER, M. R. (1995) Oxygen and carbon isotopic composition of Recent larger and smaller foraminifera from the Madang lagoon (Papua New Guinea), Mar. Micropaleon., v.26, pp.215-221.

LUZ, B., REISS, Z. and SEGEV, T. Z. (1983) Seasonal variation of stable isotopes in *Amphisorus* from the Gulf of Aqaba (Elat), Red Sea. *In:* MEULENKAMP, J.E. (Eds.) Reconstruction of marine paleoenvironments, Utrecht Micropal. Bull., v.30, pp.229-237.

LYNCH-STIEGLITZ, J., CURRY, W.B. and SLOWEY, N. (1999) A geostrophic transport estimate for the Florida current from the oxygen isotopic composition of benthic foraminifera. Paleoceanography, v.14, pp.360-373.

MARCHITTO, T.M., CURRY, W.B., LYNCH-STIEGLITZ, J., BRYAN, S.P., COBB, K.M. and LUND, D.C. (2014) Improved oxygen isotope temperature calibrations for cosmopolitan benthic foraminifera, Geochim. Cosmochim. Acta, v.130, pp.1-11.

MIELKE, K.M. (2001) Reconstructing surface carbonate chemistry and temperature in paleoceans: geochemical results from laboratory experiments and the fossil record, M.S. Thesis, Univ. California, Davis.

MULLER, P.H. (1977) Some aspects of the ecology of several Large, Symbiont-Bearing Foraminifera and their contribution to warm, shallow-water biofacies, Ph.D. Thesis, Univ. Hawaii, Honolulu, pp.1-179.

ROHLING, E.J. and COOKE, S. (2003) Stable oxygen and carbon isotopes in foraminiferal carbonate shells, *In:* Sen Gupta, B.K. (Eds.) Modern Foraminifera, Kluwer, pp.239-258.

SARASWATI, P.K., SETO, K. and NOMURA, R. (2004) Oxygen and carbon isotopic variation in co-existing larger foraminifera from a reef flat at Akajima, Okinawa, Japan. Mar. Micropaleontol., v.50, pp.339-349.

SARASWATI, P.K., IWAO, K., ROGERS, K. and RAJA, R. (2006a) $\delta^{18}O$ of a shallow water benthic foraminifer *Marginopora* and palaeotemperature calibration, FORAMS 2006, Anuario do Instituto de Geociencias – UFRJ, Brazil, v.29, pp.627-628.

SARASWATI, P.K., ROGERS, K.M. and RAJA, R. (2006b) Disequilibrium effect in oxygen and carbon isotopic composition of modern foraminifera from Lakshadweep, India. J. Geol. Soc. India, v.68, pp.1003-1007.

SCHIEBNER, C. and SPEIJER, R.P. (2008) Late Paleocene-Early Eocene Tethyan carbonate platform evolution: A response to long- and short-term paleoclimatic change: Ear. Sci. Rev., v.90, pp.71-102.

SHARMA, T. and CLAYTON, R.N. (1965) Measurement of $^{18}O/^{16}O$ ratios of total oxygen in carbonates, Geochim. Cosmochim. Acta, v.29, pp.1347-1353.

SPERO, H.J., MIELKE, K.M., KALVE, E.M., LEA, D.W. and PAK., D.K. (2003) Multispecies approach to reconstructing eastern equatorial Pacific thermocline hydrogra phy during the past 360 kyr, Paleoceanography, v.18, doi:10.1029/2002PA000814.

WEFER, G. and BERGER, W.H. (1991) Isotope paleontology: growth and composition of extant calcareous species, Mar. Geol., v.100, pp.207-248.

WEFER, G., KILLINGLEY, J.S. and LUTZE, G.F. (1981) Stable isotopes in Recent larger foraminifera, Palaeogeogr. Palaeoclimatol. Palaeoecol., v.33, pp.253-270.

Micropaleontology and its Applications
Pradeep K. Kathal, Rajiv Nigam & Abu Talib (Editors)
Scientific Publishers (India), 2018, 71-81 pp.

Integrated Benthic Foraminiferal and Ammonite Biostratigraphy of Middle to Late Jurassic Sediments of Keera Dome, Kachchh, Western India

Abu Talib[1*], Sreepat Jain[2] and Roohi Irshad[1]

[1]Department of Geology, Aligarh Muslim University, Aligarh 202002, India
[2]Department of Applied Geology, Adama Science & Technology University, Adama, Oromia, Ethiopia
[*]**Email**: talib04@gmail.com

Abstract

Early Callovian to Middle Oxfordian foraminiferal assemblages are tagged with precise ammonite occurrences for the first time from the Jurassic sediments of Chari Formation exposed at Keera Dome, Kachchh, Western India, with precise dating and marking of the Callovo-Oxfordian boundary. Four ammonite zones and nine subzones are correlated with seven foraminiferal zones, enabling accurate and reliable regional biostratigraphic analysis. Such integrated work will lead to precise dating of the otherwise hard-to-date foraminiferal assemblages from Kachchh.

Keywords: Benthic foraminifera, Ammonites, Biostratigraphy, Keera Dome, Kachchh, Western India.

INTRODUCTION

Kachchh is well known for its prolific ammonite records (Waagen, 1873-75; Spath, 1924, 1927-33; Singh et al., 1982; Krishna and Westermann, 1985, 1987; Bhaumik et al., 1993; Krishna and Cariou, 1990, 1993; Callomon, 1993; Pandey and Callomon, 1995; Datta et al., 1996; Jain et al., 1996; Jain and Pandey, 1997, 2000;

Jain, 1997, 1998, 2002; Krishna and Ojha, 1996, 2000; Shome and Bardhan, 2005, 2007, 2009; Roy et al., 2007; Krishna et al., 2009a, b; Bardhan et al., 2010, 2011; Rai and Jain, 2012). However, rich foraminiferal occurrences have been recently recorded from the Kachchh Jurassics (Bhalla and Abbas, 1978; Bhalla and Talib, 1991; Pandey and Dave, 1993; Gaur and Talib, 2009; Alhussein, 2014; Talib et al., 2016; Bhat et al., 2016) but only a few biostratigraphic studies using foraminifera have been carried out so far. An integration of foraminiferal and ammonite data may provide more accurate and reliable results in biostratigraphic analysis of the Jurassic sediments of Kachchh.

Hence, it is imperative that an attempt be made to identify and establish marker Jurassic foraminiferal species (at least on a regional scale) and integrate the foraminiferal biozones with the available high resolution ammonite zonal data to make the former more useful for biostratigraphic applications. The present study is, thus, an attempt in this direction. A well exposed Kachchh Jurassic outcrop (Keera Dome) was selected for this purpose (Fig. 1). Another reason for this selection is the fact that this dome has also been extensively studied for its ammonite content and thus, has a well established and dated ammonite biozonation (Prasad, 1993, 1998; Krishna et al., 1998; Krishna and Ojha, 1996, 2000; SJ, personal observation).

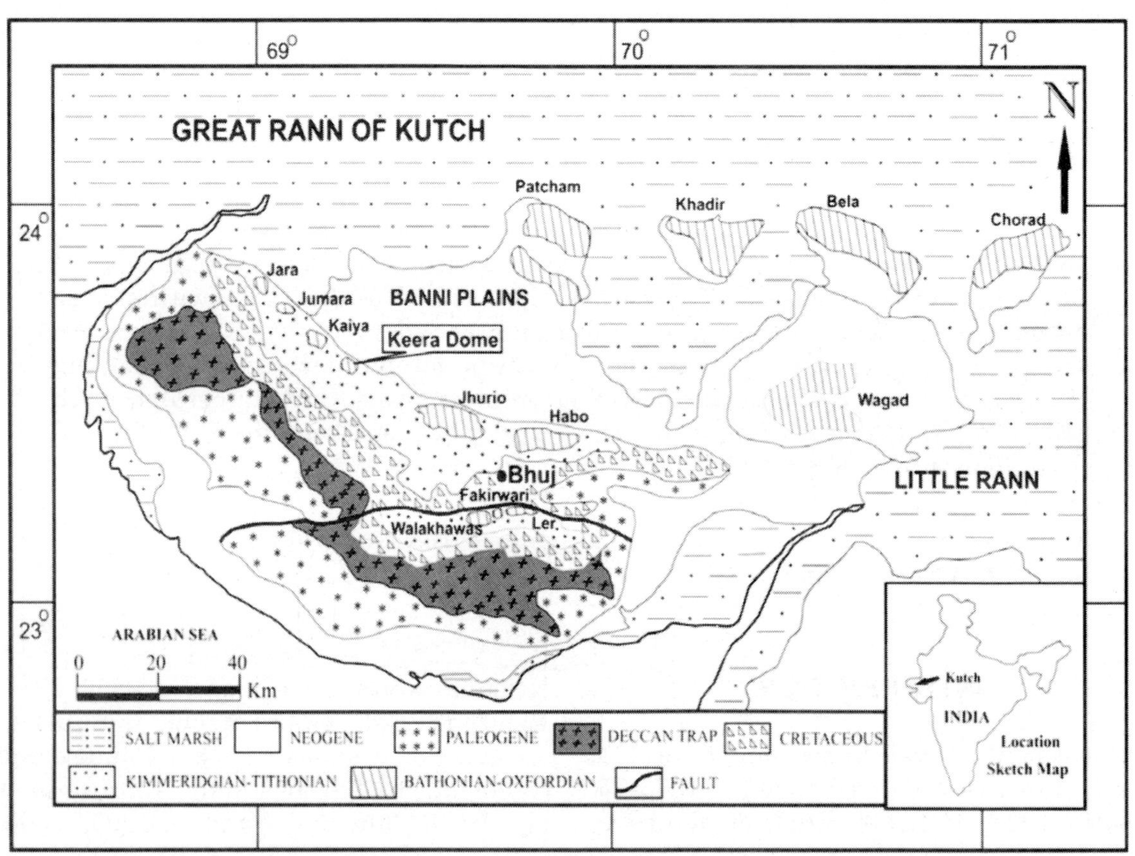

Fig.1. Geological Map of Kachchh showing the study area (after: Fürsich et al., 1991).

The Jurassic rocks of Kachchh are grouped into four formations, viz., Patcham, Chari, Katrol, and Umia, in ascending order (Waagen, 1873-75). The Keera Dome displays sediments from the Chari to Katrol formations. However, the presence of Patcham Formation at the base of the section is controversial (Prasad, 1998). Here, it must be mentioned that the basal few meters of sediments, viz., Friable Sandstones (see also Prasad, 1998) has not been sampled but have yielded fragmentary specimens of the Late Bathonian marker *Macrocephalites triangularis* Spath (Prasad, 1998), indicative of late Bathonian age.

The Chari Formation in this area, which is the subject of the present study, is grouped into four informal members, viz., A, B, C, and D. These are further divided into seven lithounits designated as lithounits Kr-1 to Kr-7 in ascending order (Fig. 2). The Katrol Formation is devoid of foraminifera and, therefore, not included in the present study.

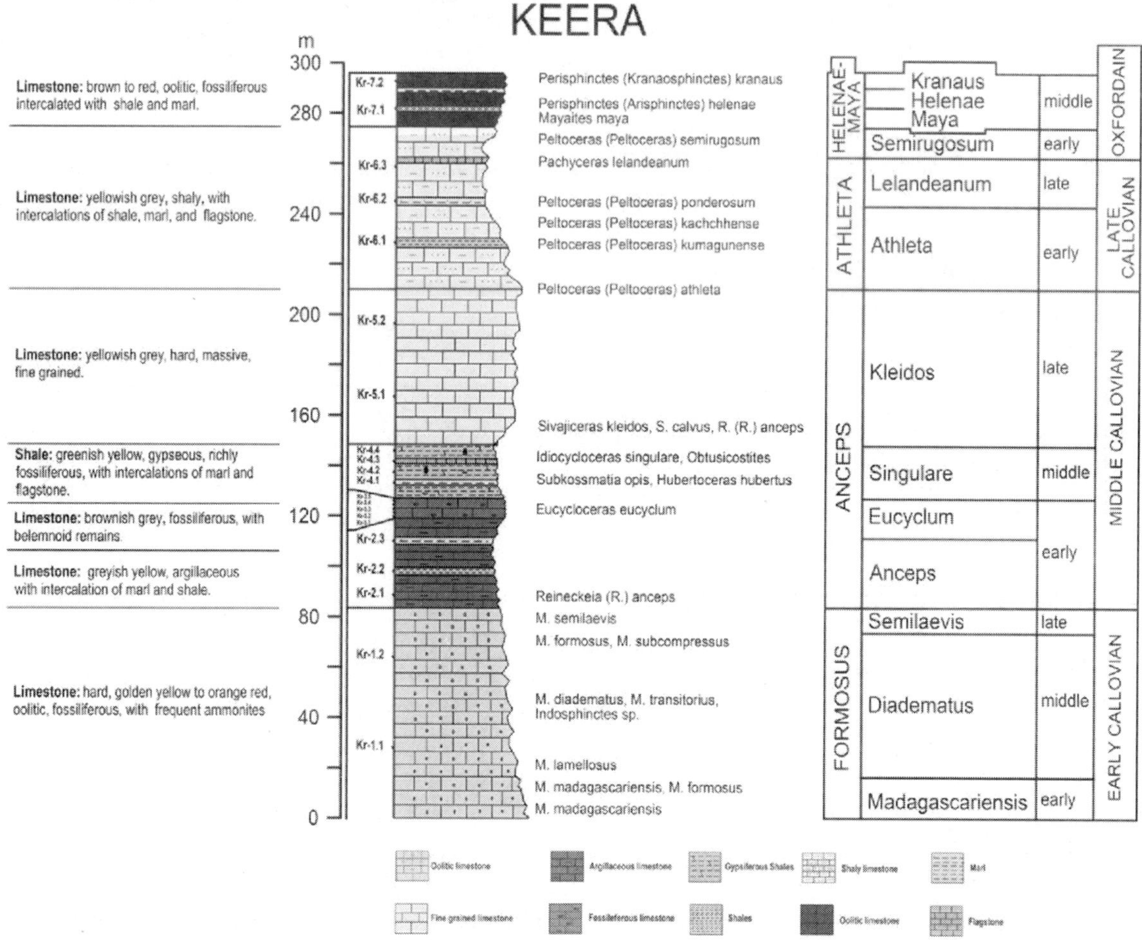

Fig.2. Litholog of the studied sequence at Keera Dome, Kachchh showing the ammonite occurrences and standard ammonite zones.

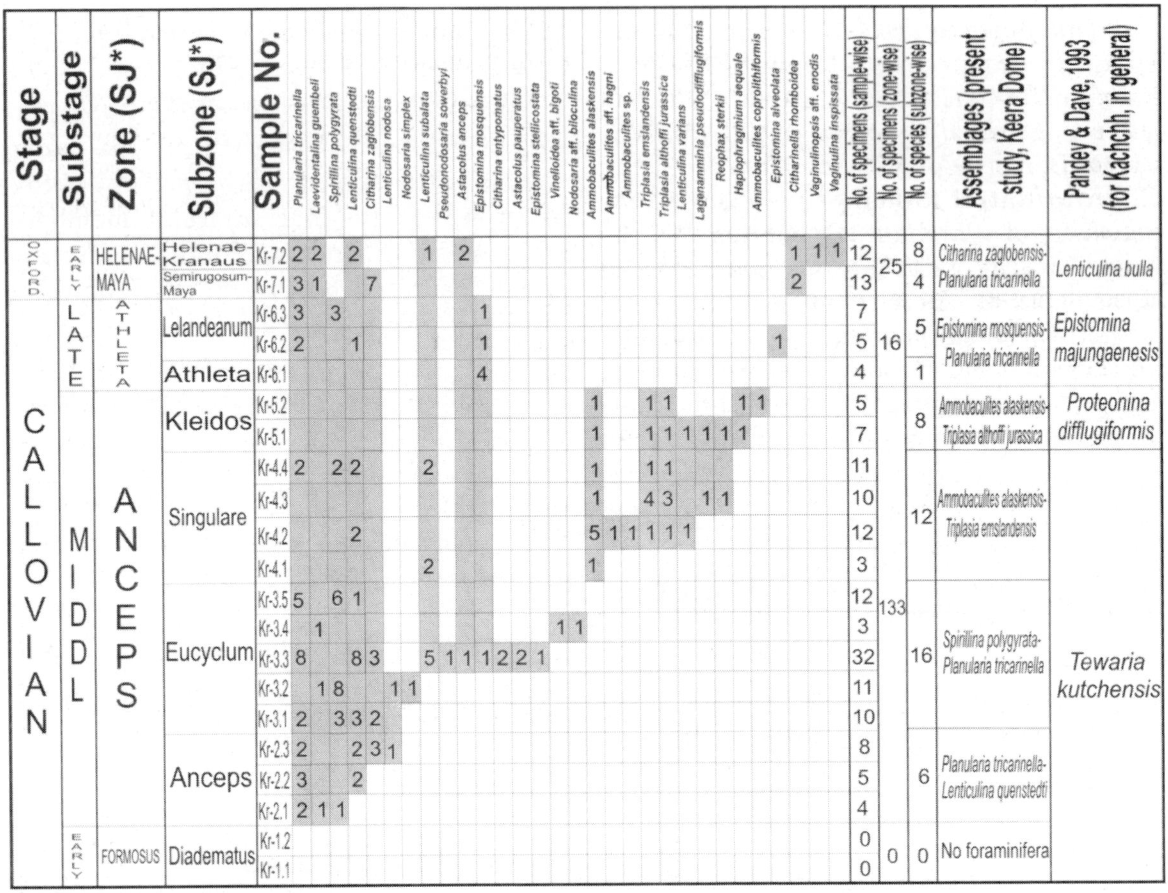

Stage	Substage	Zone (SJ*)	Subzone (SJ*)	Sample No.	*Planularia tricarinella*	*Laevidentalina guembeli*	*Spirillina polygyrata*	*Lenticulina quenstedti*	*Citharina zaglobensis*	*Lenticulina nodosa*	*Nodosaria simplex*	*Lenticulina subalata*	*Pseudonodosaria sowerbyi*	*Astacolus anceps*	*Epistomina mosquensis*	*Citharina entypomatus*	*Astacolus pauperatus*	*Epistomina stellicostata*	*Vinelloidea aff. bigoti*	*Nodosaria aff. biloculina*	*Ammobaculites alaskensis*	*Ammobaculites aff. hagni*	*Ammobaculites sp.*	*Triplasia emslandensis*	*Triplasia althoffi jurassica*	*Lenticulina varians*	*Lagenamminia pseudodifflugiformis*	*Reophax sterkii*	*Haplophragmium aequale*	*Ammobaculites coprolithiformis*	*Epistomina alveolata*	*Citharinella rhomboidea*	*Vaginulinopsis aff. enodis*	*Vaginulina inspissata*	No. of specimens (sample-wise)	No. of specimens (zone-wise)	No. of species (subzone-wise)	Assemblages (present study, Keera Dome)	Pandey & Dave, 1993 (for Kachchh, in general)
OXFORD.	EARLY	HELENAE-MAYA	Helenae-Kranaus	Kr-7.2	2	2		2				1		2																		1	1	1	12	25	8	*Citharina zaglobensis-Planularia tricarinella*	*Lenticulina bulla*
			Semirugosum-Maya	Kr-7.1	3	1			7																							2			13		4		
CALLOVIAN	LATE	ATHLETA	Lelandeanum	Kr-6.3	3		3								1																				7	16	5	*Epistomina mosquensis-Planularia tricarinella*	*Epistomina majungaenesis*
				Kr-6.2	2			1							1																1				5				
			Athleta	Kr-6.1											4																				4		1		
	MIDDLE	ANCEPS	Kleidos	Kr-5.2																	1			1	1				1	1					5	133	8	*Ammobaculites alaskensis-Triplasia althoffi jurassica*	*Proteonina difflugiformis*
				Kr-5.1																	1			1	1	1	1	1	1						7				
			Singulare	Kr-4.4	2		2	2				2									1			1	1										11		12	*Ammobaculites alaskensis-Triplasia emslandensis*	*Tewaria kutchensis*
				Kr-4.3																	1			4	3		1	1							10				
				Kr-4.2				2													5	1	1	1	1	1									12				
				Kr-4.1								2									1														3				
			Eucyclum	Kr-3.5	5		6	1																											12		16	*Spirillina polygyrata-Planularia tricarinella*	
				Kr-3.4		1													1	1															3				
				Kr-3.3	8			8	3			5	1	1	1	2	2	1																	32				
				Kr-3.2		1	8			1	1																								11				
				Kr-3.1	2		3	3	2																										10				
			Anceps	Kr-2.3	2			2	3	1																									8		6	*Planularia tricarinella-Lenticulina quenstedti*	
				Kr-2.2	3			2																											5				
				Kr-2.1	2	1	1																												4				
	EARLY	FORMOSUS	Diadematus	Kr-1.2																															0	0	0	No foraminifera	
				Kr-1.1																															0				

Fig.3. Frequency of foramininiferal species and correlation of foraminiferal and ammonite biozones, Keera Dome, Kachchh (SJ*: The ammonite biozones are based on the observations by one of us (SJ). The present biozonation also incorporates previous works by Prasad 1993, 1998; Krishna et al., 1988 and Krishna and Ojha 1996, 2000 with field inputs from Late J. H. Callomon, United Kingdom, to SJ).

FORAMINIFERAL BIOSTRATIGRAPHY

The foraminiferal assemblages recovered from Keera Dome comprises of thirty species (Fig. 3). All the species recovered are illustrated with a systematic account of the species reported for the first time from Indian region in a separate publication (Talib et al., 2012). However, most are long ranging and hence, unsuitable for precise dating (Fig. 4). In recent years some workers have attempted to establish marker foraminiferal species with biostratigraphic utility in the Indian region but with little success (Pandey and Dave, 1993; Talib and Bhalla, 2006; Talib et al., 2007, Talib and Gaur, 2008).

Interestingly, only one species of the present assemblages, *Vaginulina inspissata* appears to be globally restricted within two stages, i.e., Callovian and Oxfordian (Fig. 4) but a number of other species show short ranges within the Indian region, essentially confined to Callovian and Oxfordian stages (Fig. 5). Therefore, on the basis of these species, viz., *Ammobaculites* aff. *A. hagni, Triplasia emslandensis, Laevidentalina gümbeli, Nodosaria* aff. *N. biloculina, N. simplex, Lenticulina varians, Astacolus puperatus, Citharina entypomatus, Citharinella rhomboidea, Vaginulina inspissata,* and *Epistomina stellicostata,* a Callovian to Oxfordian age is suggested for the sequence at Keera Dome.

On the basis of a few short ranging species in the present assemblages restricted to a single stage as well as some species although long ranging but frequently reported from Callovian or Oxfordian strata from different parts of the world, an attempt is made here to mark the Callovo-Oxfordian boundary within the studied sequence.

Species identified as characteristic of Callovian in the present assemblages include *Pseudonodosaria sowerbyi* and *Epistomina mosquensis. Pseudonodosaria sowerbyi* is restricted to Callovian in the Indian region and occurs in Lithounit Kr-3. *Epistomina mosquensis* is known for its long range globally but in Kachchh Pandey and Dave (1993) and in Iran (Kalantari, 1969) considered it a representative Callovian species, its last appearance is in Lithounit Kr-6. In view of the above, it may be inferred that Lithounits Kr-1 to Kr-6 belong to Callovian.

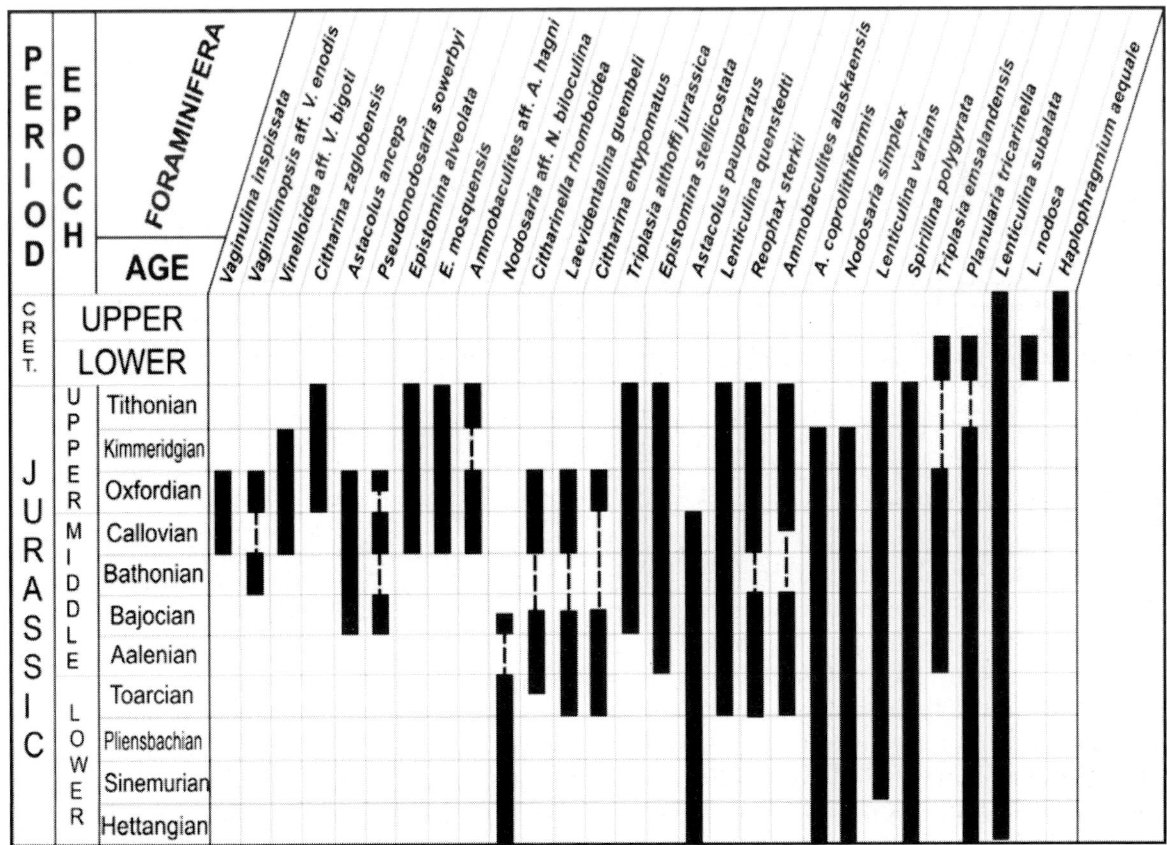

Fig.4. Global ranges of the foraminiferal species recovered from Keera Dome, Kachchh.

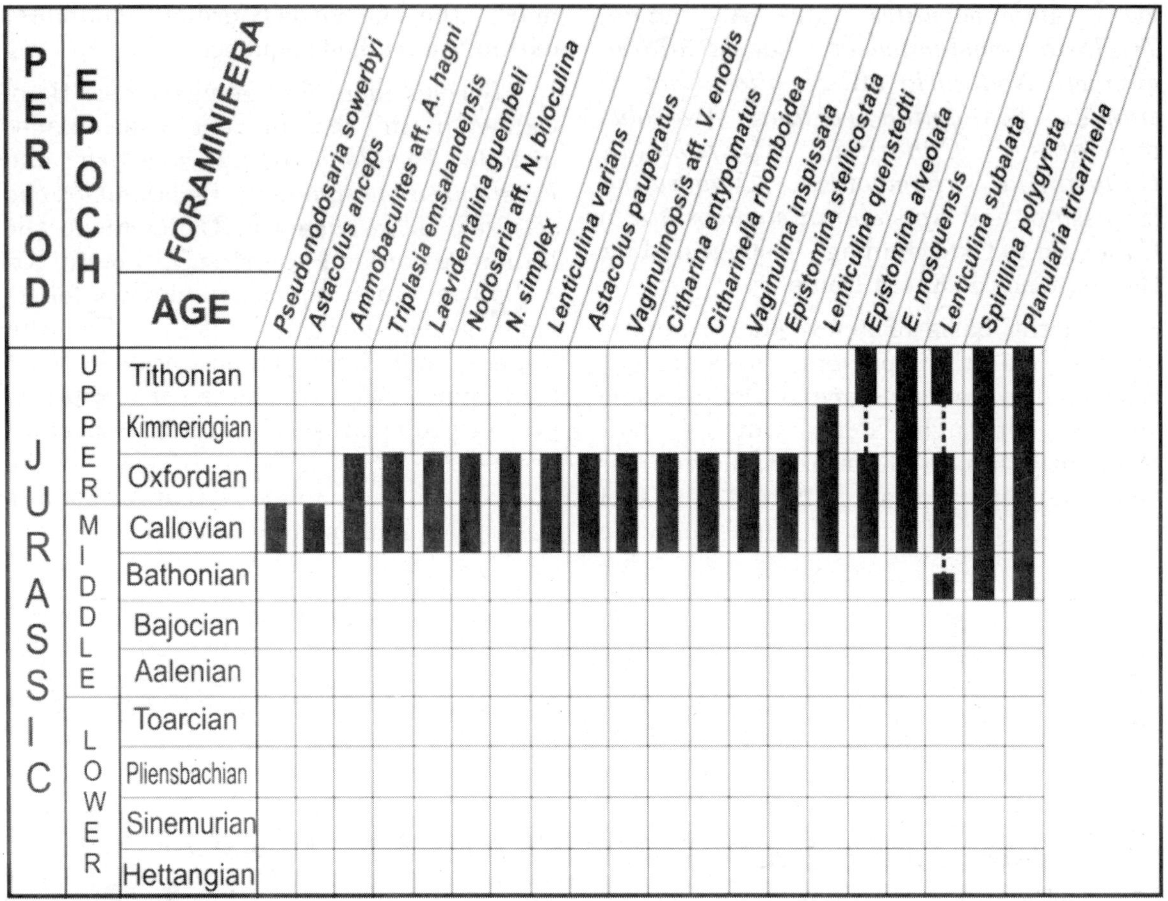

Fig.5. Indian ranges of the foraminiferal species recovered from Keera Dome, Kachchh.

Representative Oxfordian species recognized in the present assemblages is *Vaginulinopsis* aff. *V. enodis*. *Vaginulinopsis enodis* ranges from Callovian to Oxfordian in India but Kalantari (1969) regarded it as representative of Oxfordian in Iran. This species is, therefore, regarded as representative of Oxfordian in this region and on this basis Lithounit Kr-7 is considered as belonging to Oxfordian. The Callovian/Oxfordian boundary, consequently, lies between lithounits Kr-6 and Kr-7.

The broad foraminiferal biozones erected for the Kachchh Basin by Pandey and Dave (1993) are compared with the more refined and ammonite-constrained assemblage-based current benthic foraminiferal biozones (Fig. 3). These are:

1. Barren Zone - No foraminifers: This zone includes Lithounit Kr-1 and is devoid of foraminifera. This is equivalent to the Diadematus subzone of middle Early Callovian age.
2. *Planularia tricarinella-Lenticulina quenstedti* Assemblage Zone: Includes Lithounit Kr-2. Apart from zonal name species it includes *Laevidentalina guembeli, Spirillina polygyrata, Citharina zaglobensis* and *Lenticulina nodosa*. This Zone has an early

Middle Callovian age and is equivalent to Anceps subzone.

3. *Spirillina polygyrata-Planularia tricarinella* Assemblage Zone: Incorporates Lithounit Kr-3 and includes *Laevidentalina gümbeli, Lenticulina quenstedti, Citharina zaglobensis, Lenticulina nodosa, Nodosaria simplex, Lenticulina subalata, Pseudonodosaria sowerbyi, Astacolus anceps, Epistomina mosquensis, Citharina entypomatus, Astacolus pauperatus, Epistomina stellicostata, Vinelloidea* aff. *V. bigoti,* and *Nodosaria* aff. *N. biloculina*, in addition to zonal name species. This zone is equivalent to the Eucyclum subzone of upper early Middle Callovian age.
4. *Ammobaculites alaskensis-Triplasia emslandensis* Assemblage Zone: Includes Lithounit Kr-4. Along with the zonal name species, this zone includes *Planularia tricarinella, Spirillina polygyrata, Lenticulina quenstedti, Lenticulina subalata, Ammobaculites* aff. *A. hagni, Ammobaculites* sp., *Triplasia althoffi jurassica, Lenticulina varians, Lagenammina pseudodifflugiformis* and *Reophax sterkii.* This zone is assigned a middle Middle Callovian age as it is equivalent to Singulare subzone.
5. *Ammobaculites alaskensis-Triplasia althoffi jurassica* Assemblage Zone: This zone includes Lithounit Kr-5. In addition to zonal name species it includes *Triplasia emslandensis, Lenticulina varians, Lagenammina pseudodifflugiformis, Reophax sterkii, Haplophragmium aequale,* and *Ammobaculites coprolithiformis.* It is equivalent to Kleidos subzone of late Middle Callovian age.
6. *Epistomina mosquensis-Planularia tricarinella* Assemblage Zone: This zone encompasses Lithounit Kr-6 and apart from the zonal name species includes *Spirillina polygyrata, Lenticulina quenstedti,* and *Epistomina alveolata.* This zone is equivalent to Athleta Zone of Late Callovian age.
7. *Citharina zaglobensis-Planularia tricarinella* Assemblage Zone: Incorporates Lithounit Kr-7 and along with zonal name species includes *Laevidentalina gümbeli, Lenticulina quenstedti, L. subalata, Astacolus anceps, Citharinella rhomboidea, Vaginulinopsis* aff. *V. enodis,* and *Vaginulina inspissata.* This zone is correlated with Helenae-Maya ammonite Zone of Early to Middle Oxfordian age.

AMMONITE BIOSTRATIGRAPHY

The present ammonite biostratigraphy of the Keera Dome is based on data from Prasad (1993, 1998), Krishna et al. (1998) and Krishna and Ojha (1996, 2000) along with field observations by one of us (SJ) and Late J. H. Callomon, United Kingdom. (personal communication). Four ammonite zones and nine subzones are identified here spanning Early Callovian to Middle Oxfordian (Fig. 3). The oldest ammonite appearing at the bottom of the sequence in Lithounit Kr-1 is *Macrocephalites madagascariensis* indicating Early Callovian age while the last ammonite which occurs at the top of Lithounit Kr-7 is *Perisphinctes* (*Kranaosphinctes*) *kranaus* suggesting Middle Oxfordian age and indicating an Early Callovian to Middle Oxfordian interval for the studied sequence.

Thus, the age suggested by foraminiferal assemblages is less refined but does support the larger stage age inferred by ammonites (Callovian and Oxfordian) (Fig. 3).

The ammonite *Macrocephalites formosus* spans the entire Early Callovian duration and forms the first ammonite zone divisible into three subzones: Madagascariensis, Diadematus and Semilaevis dominated by *M. madagascariensis, M. diadematus* and *M. semilaevis,* respectively.

The presence of *Reineckeia* (*Reineckeia*) *anceps* marks the beginning of Middle Callovian and the Anceps Zone as such is divisible into four subzones: Anceps, Eucyclum, Singulare and Kleidos dominated by *Reineckeia* (*R.*) *anceps, Eucycloceras eucyclum, Idiocycloceras singulare,* and *Sivajiceras kleidos,* respectively.

The Late Callovian Athleta Zone is divisible into two subzones - Athelta (dominated by *Peltoceras athleta*) and Lelandeanum (*Pachyceras lelandeanum*).

The Oxfordian Helenae-Maya Zone is divisible into four subzones namely Semirugosum (*Peltoceras semirugosum*), Maya (*Mayaites maya*), Helenae [*Perisphinctes* (*Kranaosphinctes*) *helenae*] and Kranaus [*Perisphinctes* (*Kranaosphinctes*) *kranaus*]. The occurrence of *Pachyceras lelandeanum* in the topmost beds of Lithounit Kr-6 indicates a Late Callovian age and of *Peltoceras* (*Peltoceras*) *semirugosum* in the lowermost beds of the overlying Lithounit Kr-7, an Early Oxfordian age.

The Callovo-Oxfordian boundary, therefore, lies between lithounits Kr-6 and 7 for the present study.

CONCLUSION

The present study concludes:

1. The study helped in the identification of a number of relatively short ranging representative foraminiferal species that are confined to Callovian and Oxfordian stages. These include *Ammobaculites hagni, Triplasia emslandensis, Laevidentalina gümbeli, Nodosaria biloculina, N. simplex, Lenticulina varians, Astacolus pauperatus, Vaginulinopsis enodis, Citharina entypomatus, Citharinella rhomboidea, Vaginulina inspissata,* and *Epistomina stellicostata. Pseudonodosaria sowerbyi* and *Epistomina mosquensis* are considered as representative of Callovian and *Vaginulinopsis enodis* of Oxfordian in the Kachchh region. All these are likely to be useful for biostratigraphic studies in this region.
2. Based on ammonite data, the Jurassic sequence exposed at Keera Dome, Kachchh, Western India is dated from Early Callovian to Middle Oxfordian. The foraminiferal assemblages indicate a slightly less accurate but almost similar age range - from Callovian to Oxfordian. Both the ammonite and foraminiferal data suggest the placement of the Callovo-Oxfordian boundary between Litho-units Kr-6 and Kr-7.
3. Four ammonite Zones and nine subzones are correlated with seven foraminiferal Zones. This correlation has yielded a considerably more refined foraminiferal biozonation than any one proposed so far for the Kachchh Jurassics. The ammonite biostratigraphy, as put forward by Prasad (1998), well reflects field observations.
4. The correlation of foraminiferal and ammonite zones provides an accurate and refined resolution to foraminiferal biozones making them more useful for

biostratigraphic work. This integrated approach is the way forward for doing foraminiferal biostratigraphy in Kachchh Jurassics as well as globally.

Acknowledgements: We thank the Chairman, Department of Geology, Aligarh Muslim University, Aligarh, India, for providing various facilities. Logistic support during the field work provided by the district authorities of Kachchh, Gujarat and P. H. Bhatti is gratefully acknowledged.

REFERENCES

ALHUSSEIN, M. (2014) Taxonomy of the Middle Jurassic benthic foraminifera of the Kachchh basin, Western India. Beringeria, v.44, pp.51-97.

BARDHAN, S., DUTTA, R., CHANDA, P. and MALLICK, S. (2011) Systematic revision and sexual dimorphism in *Choffatia* (Ammonoidea: Perisphinctoidea) from the Callovian of Kutch, India. Palaeoworld, v.21, pp.29-49.

BARDHAN, S., JANA, S., and ROY, P. (2010) Sexual dimorphism and polymorphism in a Callovian *Phlycticeras* (Ammonoidea) assemblage in Kutch, India. Geobios, v.43, pp.269-281.

BHALLA, S.N. and ABBAS, S.M. (1978) Jurassic foraminifera from Kutch, India. Micropaleontology, v.24, pp.60-209.

BHALLA, S.N. and TALIB, A. (1991) Callovian-Oxfordian foraminifera from Jhurio hill, Kutch, Western India. Rev. Paléobiol., v.10, pp.85-114.

BHAT, B.A., TALIB, A. and WASIM, S.M. (2016) Systematics, age, paleoecology and paleobiogeography of Jurassic foraminifra from Fakirwari Dome, Kutch, Gujarat, India. Micropal. v.62, pp.171-194.

BHAUMIK, D., DATTA, K., JANA, S.K. and BARDHAN, S. (1993) Taxonomy and intraspecific variation of *Macrocephalites formosus* from the Jurassic Chari Formation, Kutch, Western India. J. Geol. Soc. India, v.42, pp.163-179.

BISWAS, S.K. (1991) Stratigraphy and sedimentary evolution of the Mesozoic Basin of Kutch, West India: *In*: TONDON, S. K., PANT, C. C. and CASSHYAP, S. M. (eds.), Sedimentary Basins of India: Tectonic context, pp.74-103, Gyanodaya Prakashan.

CALLOMON, J. H. (1993) On *Perisphinctes congener* Waagen, 1875, and the age of the Patcham Limestone in the Middle Jurassic of Jumara, Kutch, India. Geol. Blätt. Nord.-Bayer., v.43, pp.227-246.

DATTA, K., BHAUMIK, D., JANA, S.K. and BARDHAN, S. (1996). Age, ontogeny and dimorphism of *Macrocephalites triangularis* Spath- the oldest Macrocephalitid Ammonite from Kutch, India Jour. Geol. Soc. India, v.47, pp.447-458.

FÜRSICH, F.T., OCHMANN, W., JAITLY, A.K. and SINGH, I.B. (1991) Faunal response to transgressive-regressive cycles: examples from the Jurassic of western India. Palaeogeogr. Palaeoclimatol. Palaeoecol. v.85, pp.149-159.

GAUR, K.N. and TALIB, A. (2009) Middle-Upper Jurassic foraminifera from Jumara Hills, Kutch, India. Rev. Micropaleontol., v.52, pp. 227-248.

JAIN, S. (1997) On the earliest occurrence of Genus *Phlycticeras* Hyatt in Kachchh, Western India. J. Geol. Soc. India, v.49, pp.75-80.

JAIN, S. (1998) On some new discoveries of subfamily Bullatimorphatinae from the Lower Chari sediments of Kachchh, W. India. J. Geol. Soc. India, v.43, pp.107-118.

JAIN, S. (2002) Middle Jurassic Ammonite Biozonation in Western India: global implications. Geol. Soc. America Ann. Meet., Colorado, Paper No.141-10, p.34.

JAIN, S. and PANDEY, D.K. (1997). Revision of the age of heteromorph ammonite *Parapatoceras* from Kachchh, Western India. J. Pal. Soc. India, v.42, pp.133-139.

JAIN, S. and PANDEY, D.K. (2000) Middle Jurassic Ammonite Biozonation in Kachchh, western India. Bull. Ind. Geol. Assoc., v.33, pp.1-12.

JAIN, S., CALLOMON, J.H. and PANDEY, D.K. (1996) On the earliest known occurrence of the Middle Jurassic ammonite genus *Reineckeia* in the Late Bathonian of Jumara, Kachchh, Western India. Paläontol. Zeitsch., v.70, pp.129-143.

KALANTARI, A. (1969) Foraminifera from the Middle Jurassic-Cretaceous successions of Koppet-Dagh Region (N.E. Iran). Nat. Iran. Oil Co., Geol. Lab., Publ. No. 3, pp.1-298.

KRISHNA, J. and CARIOU. E. (1990) Ammoniod faunal exchanges during Early Callovian between the Indo-East African and Submediterranean Provinces: implication for the long distance east-west correlation. Newslet. Strat., v.23, pp.109-122.

KRISHNA J. and CARIOU, E. (1993) The Tethyan Macrocephalitinae: evolutionary environmental and dispersal strategies. Geobios, v.15, pp.217-226.

KRISHNA, J. and OJHA, J.R. (1996) The Callovian Ammonoid chronology in Kachchh, W. India. GeoRes. Forum, v.1-2, pp.151-165.

KRISHNA, J. and OJHA, J.R. (2000) The intrabasinal correlation in the Middle Jurassic Callovian Stage of Kachchh (Gujarat) and Ammonoids-Foraminifer integration. Geophytology, v.28, pp.101-120.

KRISHNA, J. and WESTERMANN, G.E.G. (1985) Progress report on the Middle Jurassic ammonite zones of Kachchh, W. India. Newslet. Strat., v.14, pp.1-11.

KRISHNA, J. and WESTERMANN, G.E.G. (1987) The faunal associations of the Middle Jurassic ammonite genus *Macrocephalites* in Kachchh, Western India. Canad. Jour. Earth Sci., v.24, pp.1570-1582.

KRISHNA, J., CARIOU, E. and ENAY, R. (1998) Succession of Macrocephalitinae assemblages as revealed at Keera in Kachchh, Western India. Proc. II Int. Symp. Jur. Strat., 1989, pp.383-394.

KRISHNA, J., PANDEY, B. and OJHA, J.R. (2009a) *Gregoryceras* in the Oxfordian of Kachchh (India): Diverse eventful implications. Geobios, v.42, pp.197-208.

KRISHNA, J., PANDEY, B. and PATHAK, D.B. (2009b) Characterization of *Dichotomoceras* in the Oxfordian of Kachchh. J. Geol. Soc. India, v.74, pp.469-479.

PANDEY, D.K. and CALLOMON, J.H. (1995) Contribution to the Jurassic of Kachchh, Western India. III. The Middle Bathonian ammonite families Clydoniceratidae and Perisphinctidae from Pachchham Island. *Beringeria*, v.16, pp.125-145.

PANDEY, J. and DAVE, A. (1993) Studies in Mesozoic foraminifera and chronostratigraphy of Western Kutch, Gujarat. Paleontographica Indica, no.1, pp.1-221.

PRASAD, S. (1993) A note on the Middle Jurassic stratigraphic succession of Keera, Kachchh District, Gujarat. J. Geol. Soc. India, v.41, pp.156-161.

PRASAD, S. (1998) Ammonite Biozonation of the Middle-Late Jurassic sediments with special reference to Keera and Jara, Kachchh district, Gujarat. J. Geol. Soc. India, v.52, pp.25-40.

RAI, J, and JAIN, S. (2012) Early Jurassic Gondwanaland Break up - A nannofossil Story. DST Sponsored Field Workshop and Brainstorming Session on 'Geology of Kachchh Basin, Western India: Present Status and Future Perspectives', 2012, Kachchh University, Bhuj, Kachchh (Abstract).

ROY, P., BARDHAN, S., MITRA, A. and JANA, S.K. (2007) New Bathonian (Middle Jurassic) ammonite assemblages from Kutch, India, J. Asian Earth Sci., v.30, pp.629-651.

SHOME, S. and BARDHAN, S. (2005) Record of a new species of *Erymnoceras* Hyatt, 1900 (Ammonoidea) from the Middle Jurassic of eastern Kutch and its stratigraphic and evolutionary significance. J. Asian Earth Sci., v.25, pp.679-685.

SHOME, S. and BARDHAN, S. (2007) Genus *Himalayites* (ammonoidea) from the Upper Tithonian of Spiti Himalaya - a systematic revision of Uhlig's (1910) material. J. Pal. Soc. India, v.52, pp.217-224.

SHOME, S. and BARDHAN, S. (2009) A New Late Tithonian ammonite assemblage from Kutch, Western India. J. Pal. Soc. India, v.54, pp.1-18.

SINGH, C.S.P., JAITLY, A.K. and PANDEY, D.K. (1982) First report of some Bajocian-Bathonian (Middle Jurassic) ammonoids, and age of the oldest sediments from Kachchh (Gujarat), India. Newslet. Strat., v.11, pp.37-40.

SPATH, L.F. (1924) On the Blacke collection of ammonites from Kutch, India. Palaeoontol. Indica, New Sr. 9, Mem.2.

SPATH, L.F. (1927-33) Revision of the Jurassic Cephalopod fauna of Kachh (Cutch): Mem. Geol. Surv. Ind. Palaeontol. Indica, New Ser. 9, Mem. 2, pp.1-945.

TALIB, A. and BHALLA, S.N. (2006) Composition and age of the foraminiferal assemblage from Chari Formation, Jhurio hill, Kutch. Ind. Jour. Pet. Geol., v.15, pp.67-81.

TALIB, A. and GAUR, K.N. (2008) Foraminiferal composition and age of the Chari Formation, Jumara Dome, Kutch. Curr. Sci., v.95: pp.367-373.

TALIB, A., GAUR, K.N. and BHALLA, S.N. (2007) Callovian-Oxfordian boundary in Kutch Mainland, India-A foraminiferal approach. Rev. Paléobiol., v.26, pp.625-630.

TALIB, A., GAUR, K.N., SISODIA, A.K., BHAT, B.A. and IRSHAD, R. (2012) Foraminifera from Jurassic sediments of Keera Dome, Kutch. J. Geol. Soc. India, v.80, pp.667-675.

TALIB, A., WASIM, S.M., SABEEHA AND ARKAN, M. (2016). Jurassic foraminifera from the Dharang Member, Habo Formation, Habo dome, Kutch, India: systematics, age, palaeoecology and palaeobiogeography. J. Syst. Palaeontol., pp.1-24 (online).

WAAGEN, W. (1873-75) Jurassic fauna of Kutch. The Cephalopoda: Mem. Geol. Surv. Ind., Palaeontologica Indica, New Series 914, pp.1- 247.

Micropaleontology and its Applications
Pradeep K. Kathal, Rajiv Nigam & Abu Talib (Editors)
Scientific Publishers (India), 2018, 83-87 pp.

Bloom of Zygnematalean Algae found from the Gechang Sandstone, Spiti Valley in the Tethys Himalaya, India

Gaurav Kumar Singh[1*], Valentin A. Krassilov[2], Naresh Chandra Mehrotra[3], Neerja Jha[3] and N. L. Chhabra[4]

[1]Dr. Harisingh Gour Vishwavidyalaya, Sagar-470003, M.P., India
[2]Paleontological Institute, Profsoyuznaya-117997, Moscow, Russia and The Institute of Evolution, University of Haifa, Israel
[3]Birbal Sahni Institute of Palaeobotany, Lucknow-226024, India
[4]University of Lucknow, Lucknow-26024, India
[*]**Email:** gsgeogenius3@gmail.com

Abstract

An algal bloom of Zygnematalean affinity is found in the ferruginous sandstone of Gechang Formation, which is considered to be of Asselian-Sakmarian age. Generally, such algal blooms are common around the Permian-Triassic boundary or other extinction boundaries, here it occurs well within the Mid-Permian, suggesting that the algal blooms are not confined only to the boundaries.

Key words: Zygnematalean algae, bloom, Gechang, Permian-Triassic boundary.

MEMORIAL

Dr. Valentin A. Krassilov did not live to see the publication of this paper. I remember him as a dedicated scientist, who, had a successful career in the field of palaeobotany and felt compelled to repay an obligation to his profession by enriching it with knowledge. A brief moment with him in BSIP and later on through e-mail encouraged me enough to complete the draft. This paper is a lasting tribute to his courage and devotion to palaeobotany. **GKS**

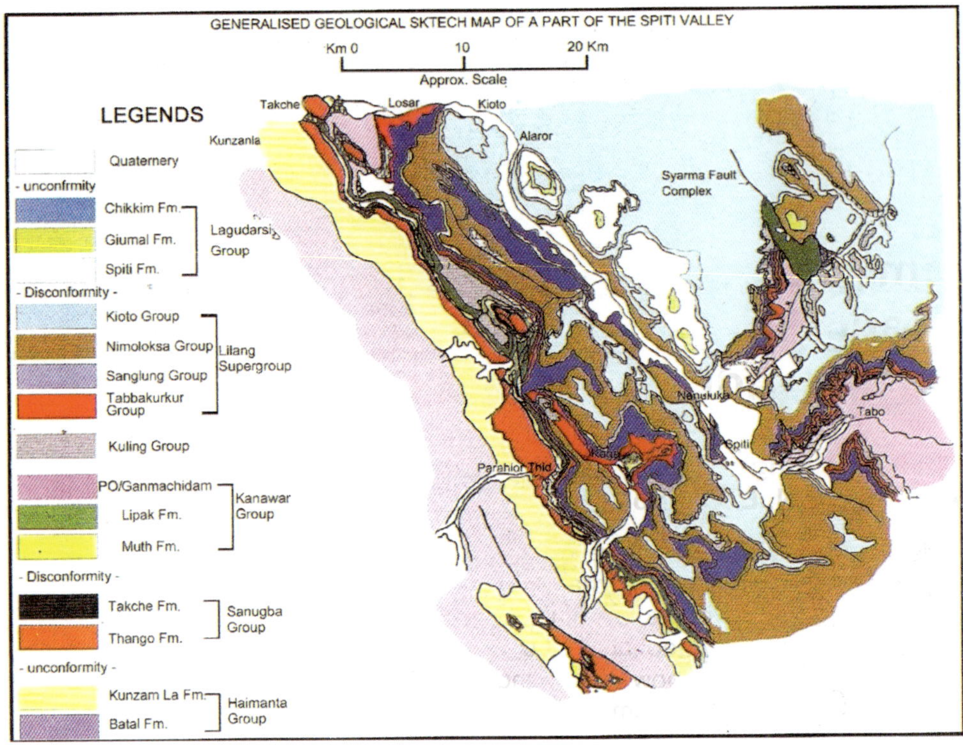

Fig.1. Generalised geological sketch map of a part of Spiti valley (after: Bhargava, 2008).

INTRODUCTION

An abnormal rise of Zygnematalean algae has been described near the Permian-Triassic boundary by Ouyang and Utting (1990) further complemented by the work of Krassilov et al. (1999) and Afonin et al. (2001). Foster et al. (2002) have suggested an algal affinity on the basis of systematics and geochemical analysis of the specimens of *Reduviasporonites* Wilson 1962. The same entity has been interpreted as a fungal spike earlier (Visscher et al., 1996; Foster et al., 1997; Elsik., 1999) and has been related to the mass extinction around the Permian-Triassic boundary (Visscher et al., 1996). These microfossils, however, are not only restricted to the Permian-Triassic boundary but also known from the Mid-Permian (Foster & Jones 1994; Foster et al., 1997). The microfossils described in this work show the following characters that are in the favour of its algal affinity:

i. The general morphology of cells;

ii. The presence of transverse septa;

iii. The shape and size of the filaments corresponding to Zygnematalean dimensions; and

iv. Cells twisted along their length and at an angle as well sometimes (Foster et al., 2002).

MATERIAL AND METHODS

The algal bloom described in this work has recently been found from the Gechang Formation of Kuling Group exposed at Mandaksa section near Losar village of Spiti Valley, India. The Gechang Formation rests unconformably over the Ganmachidam diamictite and commences with a clast supported conglomerate followed by a succession of ferruginous sandstones and siltstones (Fig.1). The compressed fossils were recovered from ferruginous sandstone by giving them hydrofluoric acid treatment. The released, dispersed cells were mounted in Canada balsam on glass slides. Light micrographs were taken with a Leica (Axioplane) microscope.

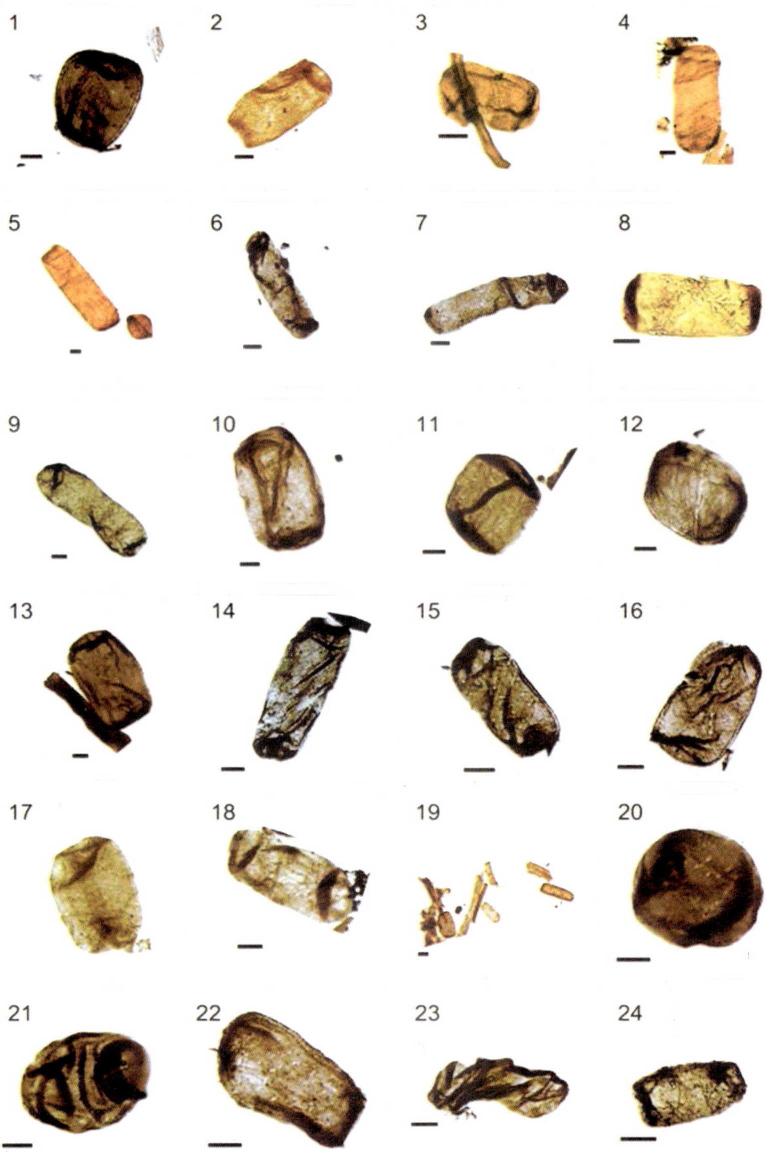

Fig.2. 1-24 Specimens of Zygnematalean algae showing nearly rectangular outline and folding of terminal rim of outer wall (scale bar =20μm).

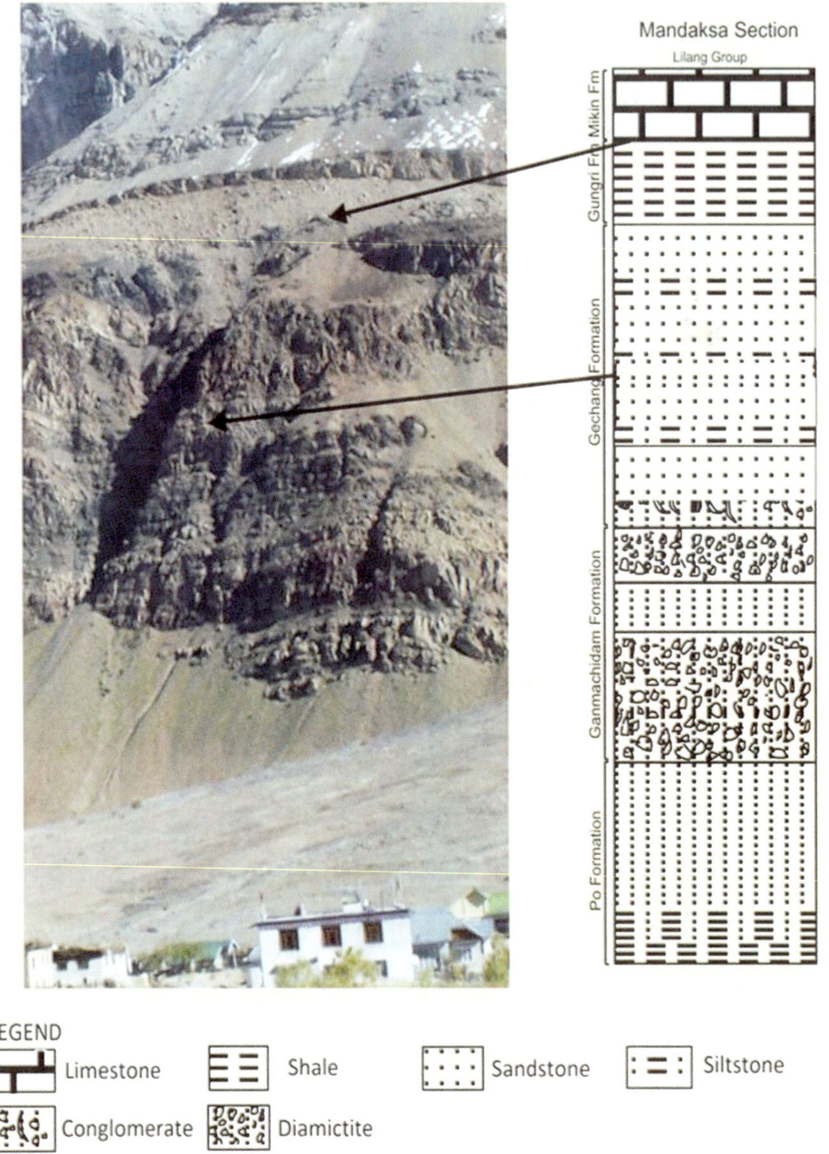

Fig.3. The stratigraphic succession, Mandaksa section, Spiti. The lower arrow shows Gechang Formation and the upper arrow marks the boundary between Gungri Formation and Mikin limestone.

RESULT

The specimens are present as single dispersed cells having almost a rectangular outline Fig. 2 (1-24). The cells are barrel shaped with 60-150 μm length and 34-75 μm width having smooth and perforated wall. Outer wall is slightly thickened at terminal rims, smooth, and folded. Dark material is also present in some of the specimens. Cells have some characteristic diagonal folding pattern caused by twisting of the cell body parallel to the long axis of the cell (Foster et al.,

2002). These microfossils are more closely allied to the Zygnematalean algae under order Cladophorales/Zygnematales.

CONCLUSION

Some global events in geological time scale like mass extinction for example, are attractive and at the same time; offer the occurrence of similar and correlatable events. The answers to such stratigraphic riddle is not as simple as it seems in the first sight and that they can be time transgressive sometimes; making correlations difficult. Zygnematalean algal bloom was initially considered to be a fungal spike and was interpreted as a consequence of mass extinction. Later on, the idea of its algal origin started gaining confidence, and now; some blooms are also being found outside the Permian-Triassic boundary.

The early interpretations, therefore, land in question. The Zygnematalean algae are found in running or stagnant fresh water and its identification as green algae rather than as fungi, complemented by its occurrence beyond the Permian-Triassic boundary indicates that its rise is unrelated to the mass extinction.

Acknowledgements: We are grateful to the Director, Birbal Sahni Institute of Palaeobotany (BSIP), Lucknow, India for his kind support and encouragement. This work is a part of the Ph.D. thesis of Gaurav Kumar Singh funded by the Birbal Sahni Research Scholarship (BSRS) of BSIP.

REFERENCES

AFONIN, S.A., BARINOVA, S.S. and KRASSILOV, V.A. (2001) A bloom of *Tympanicysta* Balme (green algae of zygnematalean affinities) at the Permian-Triassic boundary. Geodiversitas, v.23(4), pp.481-487.

BHARGAVA, O.N. (2008) An updated introduction to the Spiti geology. J. Pal. Soc. India, v.53(2), pp.113-129.

ELSIK, W.C. (1999) Reduviasporonites Wilson 1962: synonymy of the fungal organism involved in the late Permian crisis. Palynology, v.(23), pp.37-41.

FOSTER, C.B. and JONES, P.J. (1994) Correlation between Australia and type Tatarian, Russian Platform, evidence from palynology and conchostraca: a discussion. Permophiles, v.24, pp.36-42.

FOSTER, C.B., LOGAN, G.A., SUMMONS, R.E., GORTER, J.D. and EDWARDS, D.S. (1997) Carbon isotopes, kerogen types and the Permian-Triasssic boundary in Australia: implications for exploration. Australian Petrol. Produc. Expl. Assoc. J., pp.472-489.

FOSTER, C.B., STEPHENSON, M.H., MARSHALL, C., LOGAN, G.A. and GREENWOOD, P.F. (2002) A Revision of *Reduviasporonites* Wilson 1962: Description, Illustration, Comparison and Biological Affinities. Palynology, v.26, pp.35-58.

KRASSILOV, V.A., AFONIN, S.A., and LOZOVSKY, V.R. (1999) Floristic evidence of transitional Permian-Triassic deposits of the Volga-Dvina region. Permophiles, v.34, pp.12-14.

OUYANG, S. and UTTING, J., (1990) Palynology of Upper Permian and Lower Triassic rocks, Meishan, Changxing County, Zhejiang Province, China. Rev. Palaeobo. Paly., v.66, pp.65-103.

VISSCHER, H., BRINKHUIS, H., DILCHER, D.L., ELSIK, W.C., ESHET, Y., LOOY, C.V., RAMPINO, M.R. and TRAVERSE, A. (1996) The terminal Paleozoic fungal event: evidence of terrestrial ecosystem destabilization and collapse. Proc. Nat. Acad. Sci., v.93, pp.2155-2158.

Micropaleontology and its Applications
Pradeep K. Kathal, Rajiv Nigam & Abu Talib (Editors)
Scientific Publishers (India), 2018, 89-106 pp.

An Overview on the Comparative Morphology of the Recent Benthic Foraminifera from the East Coast of India and the Part of the Southern East Coast of Japan, Warm-water Indo-Pacific Province

P. K. Kathal[1]* and V. K. Singh[2]

[1]Department of Applied Geology, Dr Harisingh Gour University, Sagar- 470003, M. P., India
[2]Department of Geology, National Institute of Technology (NIT), Raipur, Chhattisgarh, India
***Email**: kathalpk@rediffmail.com

Abstract

Comparative morphology of the benthic foraminiferal fauna from the east coast of India and the southern part of the east coast of Japan have been discussed in the light of the view that the biogeographic connectivity was lost between the Pacific and Indian Oceans between a period from 10 to 4.4 Ma, due to the restriction of the Indonesian Throughflow (caused due to collision of Australia and Southeast Asia).

An overview on the intraspecific variations in the 41 Indo-Pacific benthic foraminiferal species of Recent foraminifera (out of the 122 common reported species) from the east coast of India and the southern part of the east coast of Japan has been presented together with the foramgeographical affinity between the regions, located in the two fairly distal ends of the warm-water Indo-Pacific province.

Keywords: Indo-Pacific, benthic foraminiferal morphology, east coast, India and SE coast of Japan.

INTRODUCTION

Many benthic foraminifera migrated from the Atlanto-Caribbean province and entered in the warm-water masses of the Indian Ocean under the influence of west-ward flowing 'Middle Miocene Equatorial Currents' (Glaessner, 1943; Kennett *et al.*, 1985; Crouch and Poag, 1987; Gibson, 1989; and Kathal and Bhalla, 1996 a and b and 1998).

The migration of Indo-Pacific species northward to Japan by the latest Pliocene (after 3 Ma) with the initiation of the modern nutrient deficient Kuroshio Current coincided with the intensification of the North Pacific Gyre caused by Northern Hemisphere ice sheet expansion (Gallagher *et al.,* 2009). Some could reach to as far as the northern Pacific end, close to Japan as the sublittoral fauna, from off Sendai, with warm 'Kuroshio currents' (Matoba 1976).

Comparison of Recent paralic benthic foraminiferal assemblages from the east coast of India with those of the East Malaya estuaries and Port Hacking, Australia shows some similarities (Rao and Rao, 1976). Eighty-five species of foraminifera (mainly benthic) were reported as common from the east coast of India and east coast of Japan (off Tiba Province, Kathal and Matoba, 2001). Report of 37 additional species (Singh, 2009; and Kathal and Singh, 2009 and 2010) and Singh and Kathal (2010) enhanced the total number of common species up to 122. Of these, morphological comparison of the common 41 benthic species from the littoral zone (covering a stretch of 2500km from 8° N to 21° N), east coast of India, and bottom-sediments (18m and 80m), off Tiba Province, (35° N, flanking the 'warm water core', terminating at Kuroshio Front, 37° N) east coast of Japan was made along with the systematics and SEM pictures (Singh and Kathal, 2012).

The present paper aims to review the morphological comparison of the common elements of benthic foraminifera of the two so distantly located water warm masses- the east coast of India and the part of the southern east coast of Japan in the 'Indo-Pacific region' along with the 'foramgeographic affinity' occurring.

The intraspecific morphological variations in the benthic foraminifera recovered from two regions show wide to narrow range of variation in size, shape and other morphological features (Table 1, 2 and 3, where the Indian forms are shown as 'a' while the Japanese as 'b'):

A. Species showing medium to wide range of variation

1. *Nonionellina* cf. *N. labradorica* (Dawson), (Tab. 1, Figs. 1a,b): The Indian forms show wide variation with Japanese forms in having basal aperture extending upwards, more inflated basal lobes in last chamber near umbilicus and subrounded periphery compared to interiomarginal equatorial aperture, less inflated basal lobes in the last chamber and rounded periphery.

2. *Spiroloculina angulata* Cushman (Tab. 1, Figs. 2a, b): Indian specimens differ widely from Japanese specimens in having shorter size, longer and flaring apertural neck, less projected basal chamber and smooth surface.

3. *Triloculina* cf. *T. rotunda* d'Orbigny (Tab. 1, Figs. 3a, b): Indian forms show wide variations with Japanese forms, which are smaller in size with more pointed periphery, less rounded aperture and larger simple tooth compared to small bifid tooth.

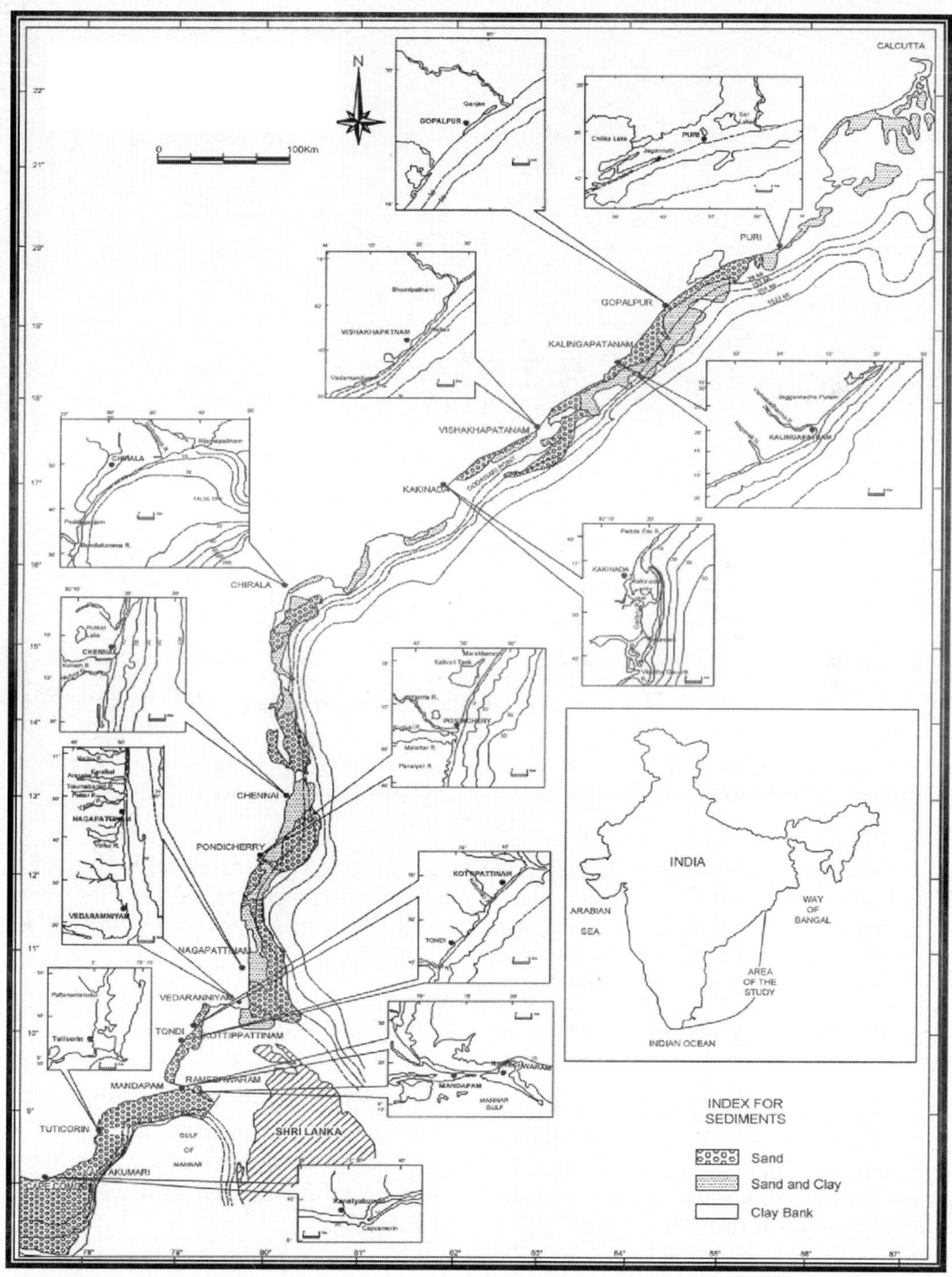

Fig.1. Location of samples (•) along the east coast of India.

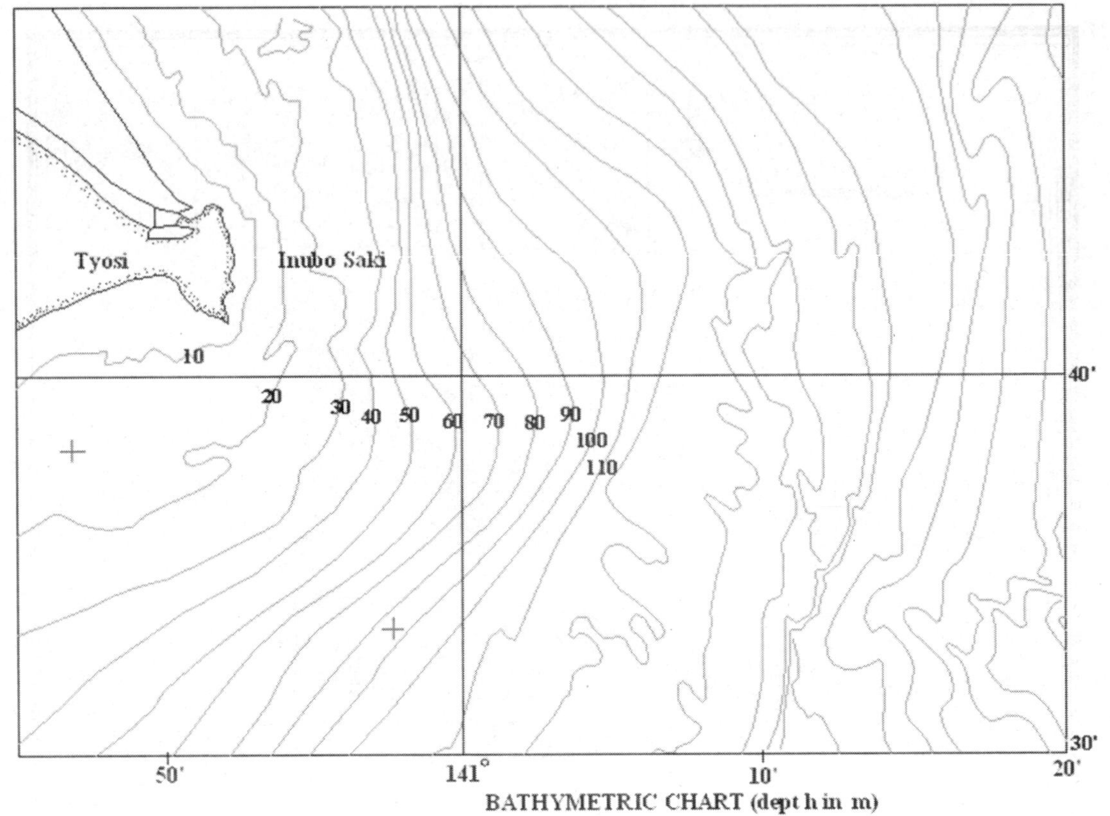

Fig.2. Location of samples (+) from off Tiba Province, east coast of Japan. (Source: Tokyo Maritime Safety Agency, 1994, Bathymetric Chat-6367).

4. *Textularia agglutinans* d'Orbigny (Tab.1, Figs. 4a, b): Indian forms show high to moderately variation with Japanese forms in size, number of chambers, shape of periphery and proximal end.

B. Species showing medium range of variation

1. *Bolivina semicostata* Cushman (Tab. 2, Figs. 1a, b): The Indian forms depict medium range of variation with the Japanese forms, which are larger in size, more elongated with thinner costae.

2. *Bolivina subreticulata* Parr (Tab. 2, Figs. 2a, b): Indian forms differ moderately from Japanese forms in having larger, elongated and compressed tests with more defined costae compared smaller, oval, less compressed tests with irregular costae.

3. *Elphidium advenum* (Cushman), (Tab. 2, Figs. 3a, b): The Indian forms depict moderate morphological variation with Japanese forms as they have smaller size, outward pointed last chamber, convex umbonal area and less prominent retral pores compared to larger test size, less convex umbilical area and well developed retral pores.

4. *Fissurina marginata* (Montagu), (Tab. 2, Figs. 4a, b): The Indian specimens differ moderately from Japanese speci-

mens, which are smaller and has more projected apertural ends.

5. *Glabratella patelliformis* (Brady), (Tab. 2, Figs. 5a, b): The Indian forms show medium range of variations with Japanese forms, which have less conical test and fine perforations. The former show fissures/radial rips and striations dorsally and cross striations/finger like structures ventrally.

6. *Loxostomum limbatum* (Brady), (Tab. 2, Figs. 6a, b): The Indian forms vary moderately from Japanese forms. The tests of latter are slightly curved and have projected aperture, finely perforated and ornamented test compared to straight tests with coarse perforations, non-projected aperture and smooth surface.

7. *Nonionoides grateloupi* (d'Orbigny), (Tab. 2, Figs. 7a,b): The Indian forms differ moderately from Japanese forms in having subrounded test, less flaring chambers, more granular secondary shell material in umbilical area and slightly angled periphery compared to elongated test, more flaring chambers, umbilical area filled with less material and subrounded periphery.

Table 1. Species showing moderate range of variations (after: Singh and Kathal, 2011)

S. No.	Name of Species	Variation Range	Indian Form 'a'	Japanese Form 'b'
1.	*Nonionellina cf. N. labradorica*	Indian form differs in having basal aperture extending upwards and inflated basal last chamber compared to interiomarginal-equatorial aperture and less inflated basal last chamber.		
2.	*Spiroloculina angulata*	Indian forms are shorter with longer and flaring apertural neck, less projected basal chamber and lack ornamentation.		
3.	*Triloculina cf. T. rotunda*	Indian forms are smaller with pointed periphery, less rounded aperture and with large simple tooth compared to small and bifid tooth.		
4.	*Textularia agglutinans*	Indian forms are smaller with less number of chambers, and shape of periphery and proximal end.		

(SEM pictures are not to scale as the Table shows only comparison between the Indian and Japanese forms)

Table 2. Species showing wide range of variations (after: Singh and Kathal, 2011)

S. No.	Name of Species	Variation Range	Indian Form 'a'	Japanese Form 'b'
1	*Bolivina semicostata*	Indian forms are smaller, less elongated with thicker costae.	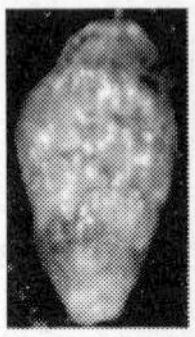	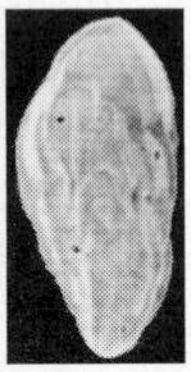
2	*Bolivina subreticulata*	Indian forms are larger, elongated, test compressed with more defined costae compared smaller, oval, less compressed tests with irregular costae.	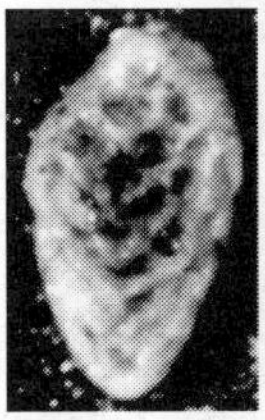	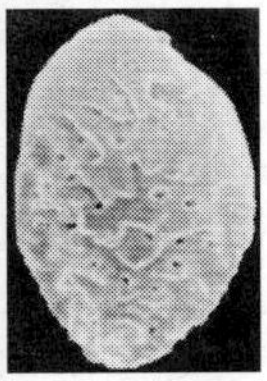
3	*Elphidium advenum*	Indian forms are smaller, outward pointed nature of last chamber more convex umbonal area and less prominent retral pores compared to larger average tests and well developed retral pores.		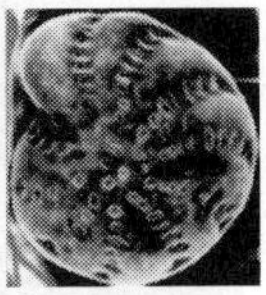
4	*Fissurina marginata*	Indian forms are larger with less projected apertural end.	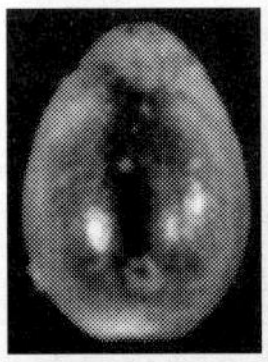	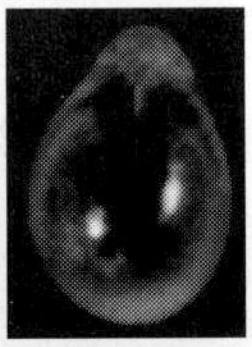
5	*Glabratella patelliformis*	Indian tests are more conical test, coarse perforation, also ornamented with fissures/ radial rips and striations on dorsal side and cross striations or finger like structures on ventral side.		
6	*Loxostomum limbatum*	Indian forms are slightly curved, slightly projected aperture, finely perforated and ornamented test in comparison to coarsely perforated straight test, non-projected aperture and smooth surface.	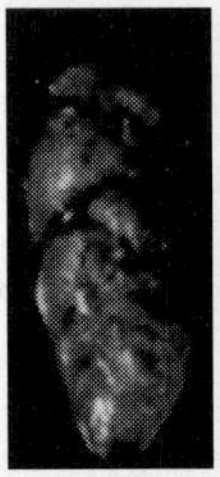	
7	*Nonionoides grateloupi*	Indian tests are subrounded, less flaring chambers, umbilical area with more granular secondary shell material and slightly angled periphery compared to elongated test, more flaring chambers and sub-rounded periphery.	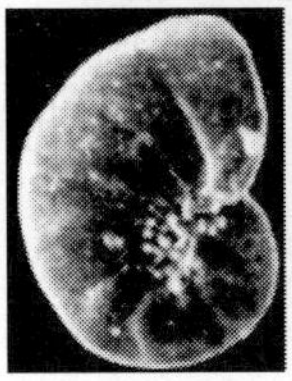	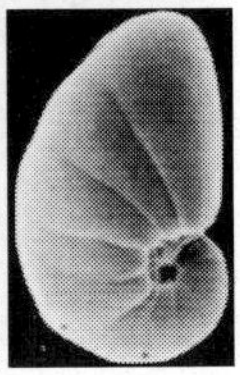

S. No.	Name of Species	Variation Range	Indian Form 'a'	Japanese Form 'b'
8	*Pararotalia nipponica*	Indian forms are less excavated with narrower sutures on ventral side, even umbilical plug and less excavated area around plug, narrower keel and less prominent spines.	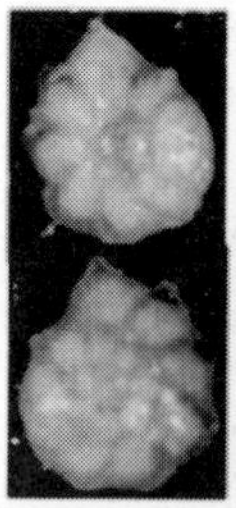	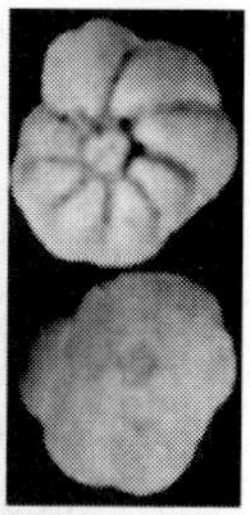
9	*Planorbulina variabilis*	Indian tests are smaller and slightly irregular, thicker wall imperforate attached side and finely perforated opposite side		
10	*Poroeponides lateralis*	Indian tests are smaller with regular in comparison to larger and irregular test (in latter stage) in Japanese forms		
11	*Pseudorotalia indopacificana*	Indian tests are smaller with raised dorsal sutures, protruding umbilical plug, excavated umbilical area, 4 whorls and, in not having flush dorsal suture, beaded ventral sutures and 5 whorls	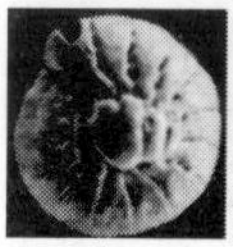	
12	*Quinqueloculina cf. Q. Venusta*	Indian forms are fusiform with subrounded aperture as compared to elongated test and slightly quadrangular aperture		
13	*Rosalina australis*	Indian forms are more rounded with; indistinct ventral sutures; highly-thickened dorsal sutures; entire and moderately keeled periphery; and small umbilical plug compared to radial ventral-suture (middle part hook-shaped), thicker dorsal-suture, slightly lobulate and keeled periphery	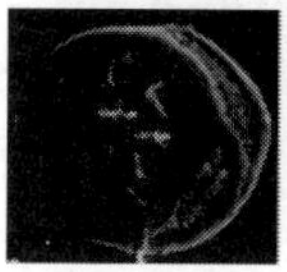	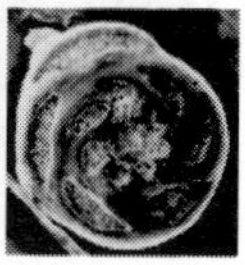
14	*Sigmavirgulina tortuosa*	Indian tests are more sigmoidal with pointed distal end, projected aperture and absence of costae	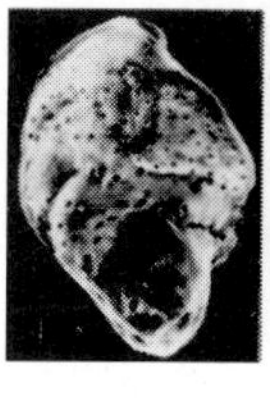	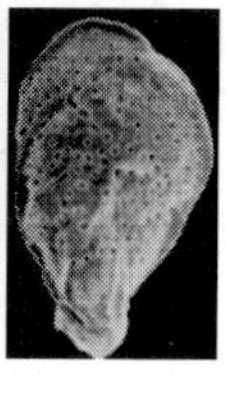

S. No.	Name of Species	Variation Range	Indian Form 'a'	Japanese Form 'b'
15	*Spiroloculina communis*	Indian tests are larger and vary in shape, base of last-chamber, central portion, neck, tooth and lip		
16	*Textularia foliacea*	Indian tests vary in shape of test, number of chambers, periphery and aperture		
17	*Triloculina tricarinata*	Indian forms show sharper periphery and have a simple tooth		

8. *Pararotalia nipponica* (Asano), (Tab. 2, Figs. 8a, b): The Indian forms depict moderately range of variations with Japanese forms, which have more excavated and broader sutures on ventral side, uneven umbilical plug and more excavated area around umbilical plug, broader keel and more prominent spines.

9. *Planorbulina variabilis* (d'Orbigny) differs (Tab. 2, Figs. 9a,b): The Indian material differs moderately from Japanese material having smaller size and slightly irregular test, thicker wall and imperforate attached side and finely perforated opposite side.

10. *Poroeponides lateralis* (Terquem), (Tab. 2, Figs. 10a, b): Our specimens show medium range of variation as they are smaller in size with regular test compared to the larger and irregular test (in latter stage) in Japanese specimens.

11. *Pseudorotalia indopacificana* (Thalmann), (Tab. 2, Figs. 11a, b): The Indian forms show moderate variation than Japanese forms, which are smaller in size with raised dorsal sutures, protruding umbilical plug, excavated umbilical area, 4 whorls and, in lacking flush dorsal suture, beaded ventral sutures and 5 whorls.

12. *Quinqueloculina* cf. *Q. venusta* Karrer (Tab. 2, Figs. 12a, b): The Indian forms show medium range of variation compared to the Japanes forms, which are fusiform in shape with subrounded aperture as compared to elongated test and slightly quadrangular aperture.

13. *Rosalina australis* (Parr), (Tab. 2, Figs. 13a, b): The Indian forms shows moderate variation with Japanese forms as the former are more rounded with indistinct ventral sutures; highly thickened dorsal sutures; moderately keeled periphery; and small umbilical plug compared to radial ventral-suture (middle part hook-shaped), thicker dorsal-suture, slightly lobulate and keeled periphery.

Table 3. Species showing narrow range of variations (after: Singh and Kathal, 2011)

S.No.	Name of Species	Variation Range	Indian Form 'a'	Japanese Form 'b'
1	*Adelosina laevigata*	**Narrow**	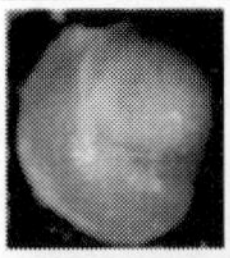	
2	*Amphistegina radiata*	**Narrow**		
3	*Asterorotalia pulchella*	**Narrow**	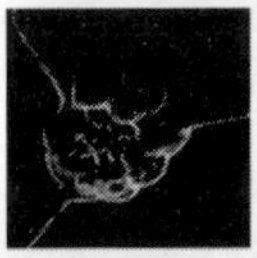	
4	*Elphidium jenseni*	**Narrow**	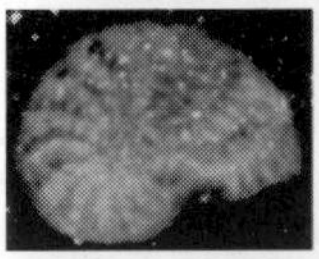	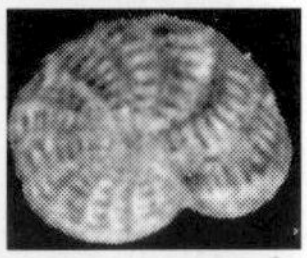
5	*Elphidium subgranulosum*	**Narrow**	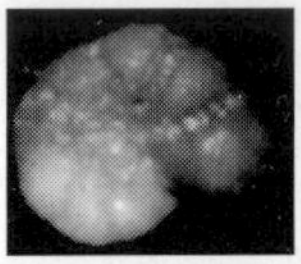	
6	*Hanzawaia nipponica*	**Narrow**	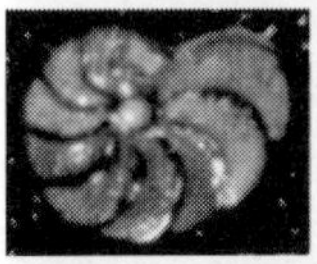	
7	*Lobatula lobatulus*	**Narrow**	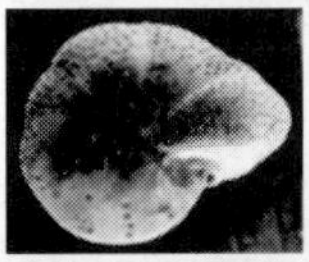	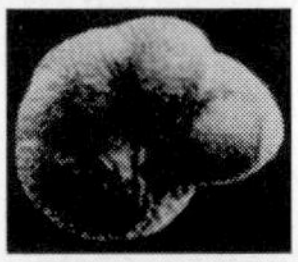
8	*Nonion commune*	**Narrow**	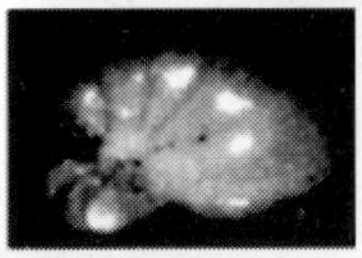	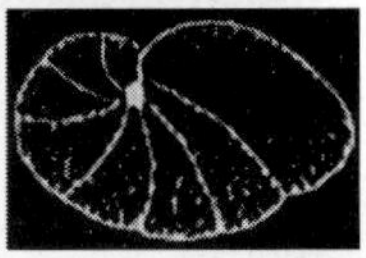
9	*Pararotalia? minuta*	**Narrow**	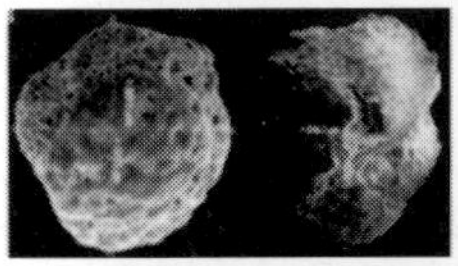	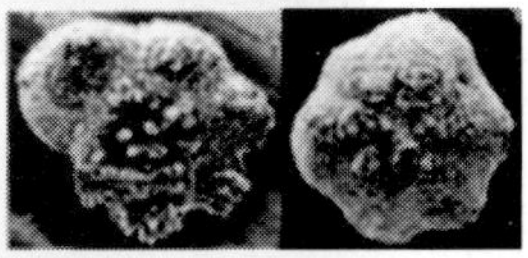
10	*Pseudononion japonicum*	**Narrow**	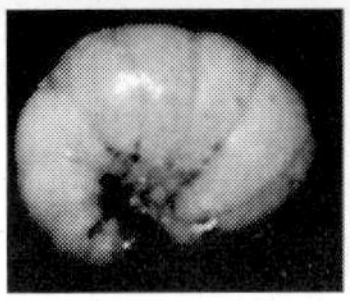	

S.No.	Name of Species	Variation Range	Indian Form 'a'	Japanese Form 'b'
11	*Quinqueloculina agglutinans*	**Narrow**	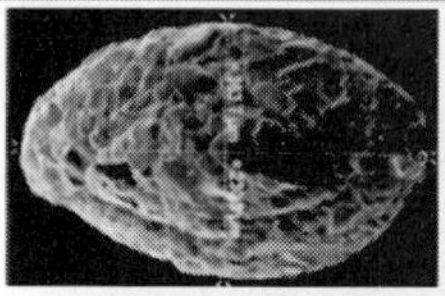	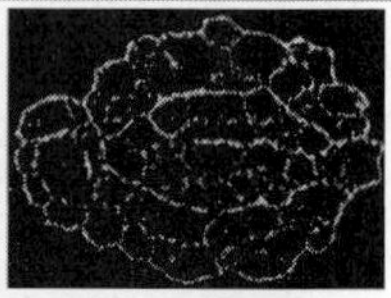
12	*Quinqueloculina parkeri*	**Narrow**	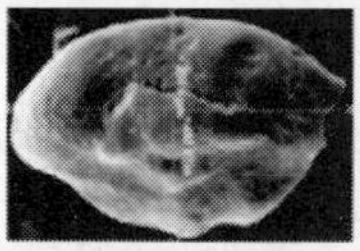	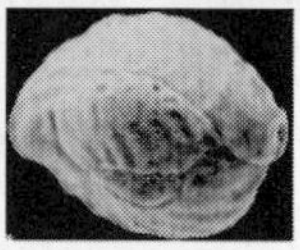
13	*Quinqueloculina pseudoreticulata*	**Narrow**	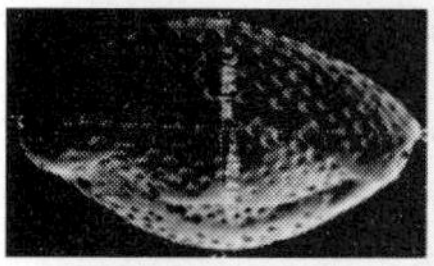	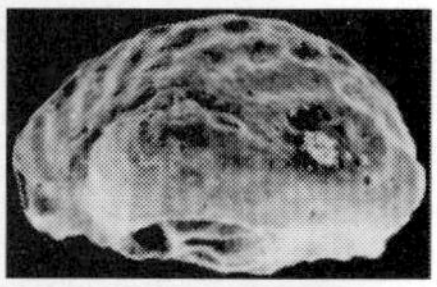
14	*Quinqueloculina subcuneata*	**Narrow**	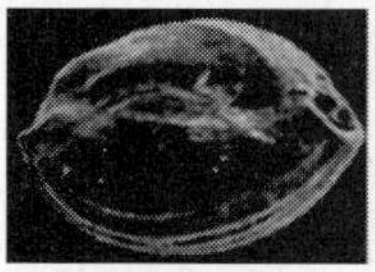	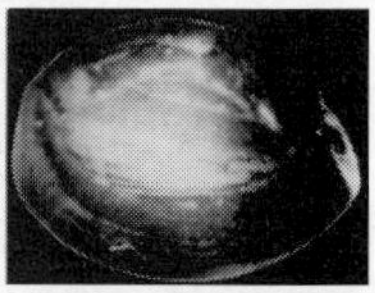
15	*Reussella haizumensis*	**Narrow**		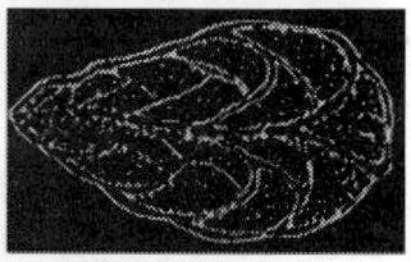
16	*Rosalina bradyii*	**Narrow**	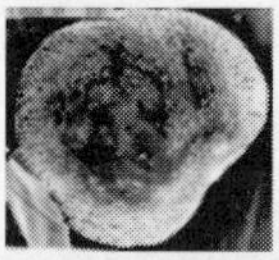	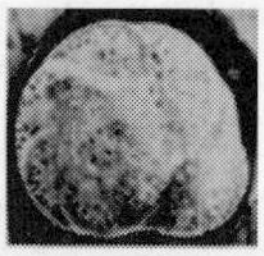
17	*Spiroloculina planulata*	**Narrow**	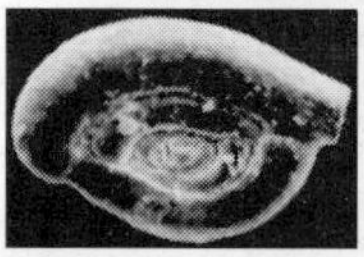	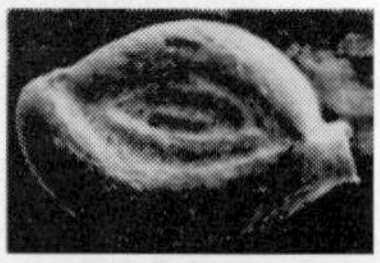
18	*Triloculina terquemiana*	**Narrow**	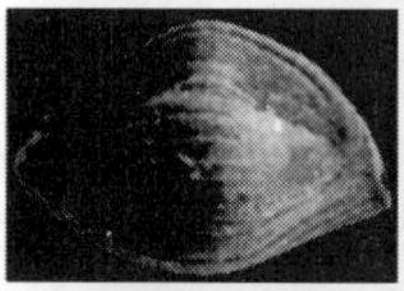	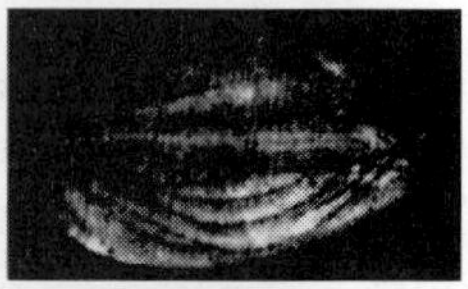
19	*Triloculina trigonula*	**Narrow**		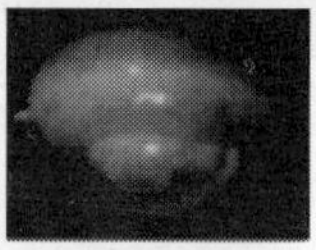
20	*Elphidium crispum*	**Very narrow**		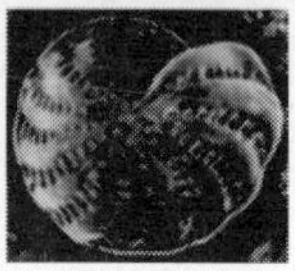

14. *Sigmavirgulina tortuosa* (Brady), (Tab. 2, Figs. 14a, b): The Indian forms differ moderately from Japanese forms in having more sigmoidal test, pointed distal end, projected aperture and absence of costae.

15. *Spiroloculina communis* Cushman and Todd (Tab. 2, Figs. 15a, b): Specimens from Indian waters differ moderately from Japanese waters as the former are larger with wide variations is test's shape, base of last-chamber, central portion, neck, tooth and lip.

16. *Textularia foliacea* Heron-Allen and Earland (Tab. 2, Figs. 16a, b): Our forms shows medium range of variations with Japanese forms in shape of test, number of chambers, wall agglutinants, periphery and aperture.

17. *Triloculina tricarinata sensu* Parker, Jones and Brady (Tab. 2, Figs. 17a, b): The Indian specimens show moderate range of variations with Japanese specimens in lacking very sharper periphery and having a simple tooth.

C. *Species showing narrow range of variations*

1. *Adelosina leavigata* d'Orbigny (Tab. 3, Figs. 1a, b): The Indian forms varies slightly from the Japanese forms in having more sub-rounded shape, smoother surface with sharper periphery, more rounded aperture and more rounded basal end of last chamber compared to elongated undulatory test, slightly produced subrounded aperture and rough surface.

2. *Amphistegina radiata* (Fichtel and Moll), (Tab. 3, Figs. 2a, b): The Indian forms slightly differ from their Japanese counterparts in having more rounded test, more convex ventral side, raised sutures with parallel beads and papillae, highly convex umbonal boss on ventral side and more developed keel.

3. *Asterorotalia pulchella* (d'Orbigny), (Tab. 3, Figs. 3a, b): Indian forms differ slightly from Japanese forms in having larger size and longer and narrower spines.

4. *Elphidium jenseni* (Cushman), (Tab. 3, Figs. 4a,b): The Indian forms, compared to the Japanese are smaller in size and have subrounded to rounded test, less prominent retral processes and less prominent keel.

5. *Elphidium subgranulosum* Asano, (Tab. 3, Figs. 5a, b): Indian forms vary slightly with Japanese forms in having smaller size, more inflated final chamber, less depressed sutures and depressed umbilical area.

6. *Hanzawaia nipponica* Asano (Tab. 3, Figs. 6a, b): Indian forms are sinstrally coiled while the Japanese forms are dextrally coiled. The former differs slightly in having smaller size, more raised and narrow sutures, normal final chamber, broader aperture without lip, more prominent umbonal boss, and coarser perforations.

7. *Lobatula lobatulus* (Walker and Jacob), (Tab. Figs. 3, 7a, b): Indian forms differ slightly with their Japanese counterparts in having less number of chambers with less inflated final chamber, larger aperture and more prominent lip and coarser perforations.

8. *Nonion commune* (d'Orbigny), (Tab. 3, Figs. 8a, b): Indian forms differ slightly from the Japanese forms in having broader umbilical area (filled with granular secondary shell material) and entire to slightly lobulate periphery.

9. *Pararotalia? minuta* (Takayanagi), (Tab. 3, Figs. 9a,b): Indian specimens

differ slightly from their Japanese specimens in lacking hirsute test and in having ventral chambers more narrowing near umbilicus and larger and more depressed umbilical area.

10. *Pseudononion japonicum* Asano (Tab. 3, Figs. 10a, b): Indian forms differ slightly from the Japanese forms in having gradually enlarging chambers, slightly lobulate periphery in side view and subrounded periphery in the apertural view.

11. *Quinqueloculina agglutinans* d'Orbigny (Tab. 3, Figs. 11a, b): Indian forms depict narrow range of variation by having smaller size and lacking projected apertural end as well as the base of last chamber than Japanese forms.

12. The Indian forms differ from the holotype of *Q. parkeri* (Brady), (Tab. 3, Figs. 12a,b). (Barker, 1960, pl. 8 figs. 3a-c): The Indian forms have rounded periphery in apertural view, less projected neck, less developed lip, ridges and in having thick undulations. While the Japanese forms show narrow range of variations as they are more rounded with projected lip and fine undulations.

13. *Quinqueloculina pseudoreticulata* Parr (Tab. 3, Figs. 13a, b): Indian forms differ slightly from Japanese forms in having larger size, smaller, deeper and denser pits on surface.

14. *Quinqueloculina subcuneata* Cushman (Tab. 3, Figs. 14a, b): The Indian forms differ slightly from Japanese forms in aperture, middle chamber and ornamentation.

15. *Reussella haizumensis* Asano (Tab. 3, Figs. 15a, b): The Indian forms show very small range of variations in having smaller size, pistules on septa and thick keel in comparison elongated test, pits on septa and sharper keel in the Japanese forms.

16. *Rosalina bradyii* (Cushman), (Tab. 3, Figs. 16a, b): The Indian forms differ slightly from the Japanese forms in having flush dorsal sutures and thinly keeled periphery, in early part compared to slightly depressed dorsal sutures and thick keel.

17. *Spiroloculina planulata* Cushman (Tab. 3, Figs. 17a, b): The Indian forms differ slightly from Japanese forms in dimensions and the ratio of tests, shape, apertural end, nature of lip, shape of basal chamber, number of chambers and test-surface.

18. *Triloculina terquemiana* (Brady), (Tab. 3, Figs. 18a, b): The Indian forms differ slightly from Japanese forms in having more elongated test and smaller tooth and narrower costae.

19. *Triloculina trigonula* (Lamarck), (Tab. 3, Figs. 19a, b): The Indian forms differ slightly from Japanese forms in having ovate to slightly elongate test and rough/pitted surface.

20. *Elphidium crispum* (Linné), (Tab. 3, Figs. 20a, b): Our forms differ very slightly from Japanese forms in having smaller size, more prominent keel and more developed retral pores.

CONCLUSION

The benthic formainiferal assemblages from the east coast of India and the southern part of the east coast of Japan (from 18m and 80m depths, at 37°N) show faunal affinity as their 122 species are common to the two regions despite of their locations in the two ends of the Indo-Pacific Province. Forty-one of their common species show morphological variation as shown in Plate I, II & III and Table 1, 2 and 3:

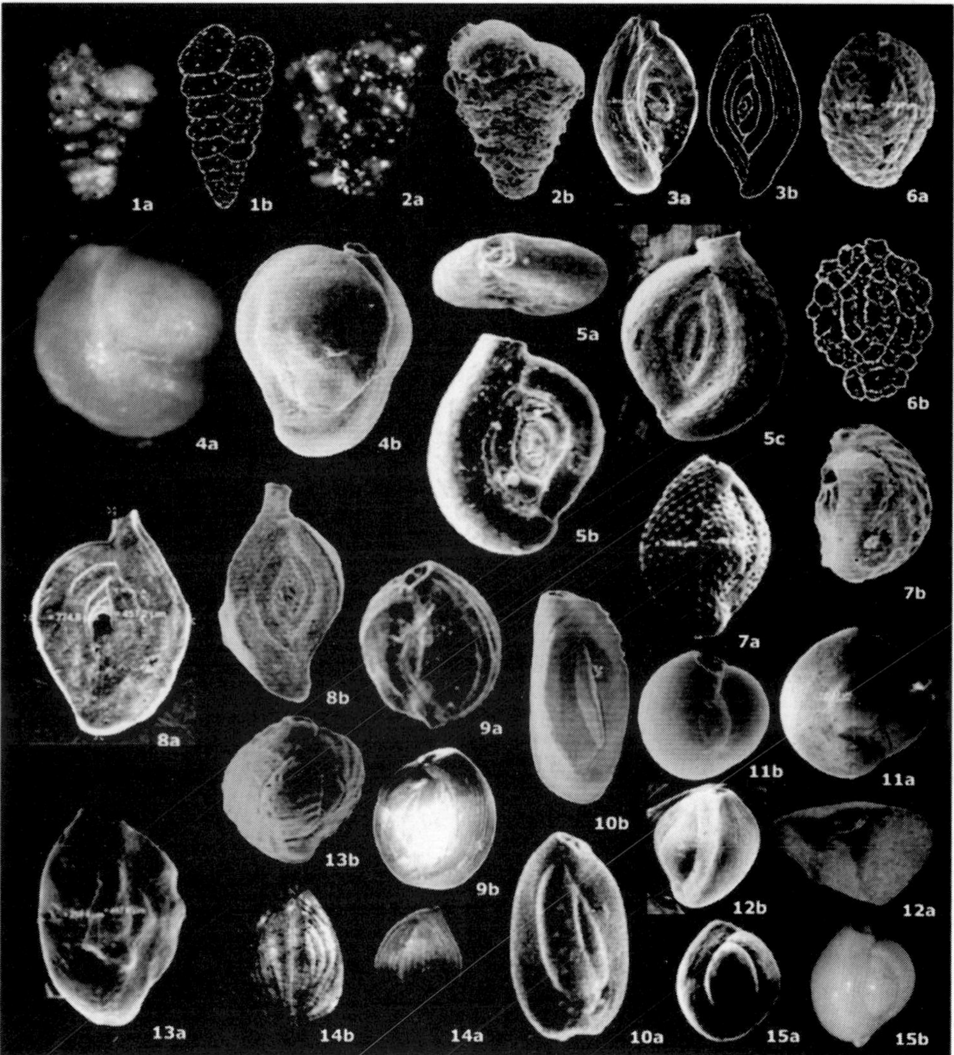

Plate 1. 1. *Textularia agglutinans* d'Orbigny, **a.** Indian form, (X 84), **b.** Japanese form, (X 26), (Asano, 1950); **2.** *T. foliacea* Heron-Allen and Earland, **a.** Indian form, (X 70), **b.** Japanese form, (X 117), (Akimoto *et al.*, 2002); **3.** *Spiroloculina angulata* Cushman, **a.** Indian form, (X 74), **b.** Japanese form, (X 38), (Asano, 1951); **4.** *Adelosina laevigata* d'Orbigny, **a.** Indian form, (X 127), **b.** Japanese form, (X 200), (Akimoto *et al.*, 2002); **5.** *Spiroloculina planulata* Cushman, **a.** Aperture view, (X 90), **b.** Side view, (X 97), Indian form, **c.** Japanese form, (X 67); **6.** *Quinqueloculina agglutinans* d'Orbigny, **a.** Indian form, (X 83), **b.** Japanese form, (X 40), (Asano, 1951); **7.** *Q. pseudoreticulata* Parr, **a.** Indian form, (X 47), **b.** Japanese form, (X 71), (Akimoto *et al.*, 2002); **8.** *Spiroloculina communis* Cushman and Todd, **a.** Indian form, (X 55), **b.** Japanese form, (X 67), (Akimoto *et al.*, 2002); **9.** *Quinqueloculina subcuneata* Cushman, **a.** Indian form, (X 76), **b.** Japanese form, (X 66); **10.** *Q.* cf. *Q. venusta* Karrer, **a.** Indian form, (X 61), **b.** Japanese form, (X 105), (Akimoto *et al.*, 2002); **11.** *Triloculina* cf. *T. rotunda* d'Orbigny, **a.** Indian form, (X 63), **b.** Japanese form, (X 64), (Akimoto *et al.*, 2002); **12.** *T. tricarinata sensu* Parker, Jones and Brady, **a.** Indian form, (X 104) **b.** Japanese form, (X 56); **13.** *Quinqueloculina parkeri* (Brady), **a.** Indian form, (X 92), **b.** Japanese form, (X 60), (Akimoto *et al.*, 2002); *T. terquemiana* (Brady) **a.** Indian form, (X 40), **b.** Japanese form, (X 64), (Matoba, 1970); **15.** *T. ulina trigonula* (Lamarck), **a.** Indian form, (X 42), **b.** Japanese form, (X 42).

Plate 2. **1.** *Fissurina marginata* (Montagu), **a.** Indian form, (X 112), **b.** Japanese form, (X 135); **2.***Bolivina semicostata* Cushman, **a.** Indian form, (X 91), **b.** Japanese form, (X 75), (Akimoto *et al.*, 2002); **3.** *B. subreticulata* Parr, **a.** Indian form, (X 61), **b.** Japanese form, (X170), (Akimoto *et al.*, 2002); **4.** *Loxostomum limbatum* (Brady), **a.** Indian form, (X 76), **b.** Japanese form, (X 111); **5.** *Reussella haizumensis* Asano, **a.** Indian form, (X 139), **b.** Japanese form, (X 50), (Asano, 1951-52); **6.** *Sigmavirgulina tortuosa* (Brady), **a.** Indian form, (X 170), **b.** Japanese form, (X 164), (Akimoto *et al.*, 2002); **7.** *Poroeponides lateralis* (Terquem), **a.** Indian form, (X 51), **b.** Japanese form, (X 39), (Matoba, 1970); **8.** *Rosalina australis* (Parr), **a.** Indian form, (X 106), **b.** Japanese form, (X 65), (Asano, 1951-52); **9.** *Rosalina bradyii* (Cushman), **a.** Indian form, (X 124), **b.** Japanese form, (X 70); **10.** *Glabratella patelliformis* (Brady), **a.** Indian form, (X 88), **b.** Japanese form, (X 82), (Matoba, 1970); **11.** *Lobatula lobatulus* (Walker and Jacob), **a.** Indian form, (X 90), **b.** Japanese form, (X 34); **12.** *Nonion commune* (d'Orbigny, **a.** Indian form, (X 122) **b.** Japanese form, (X 53), (Asano, 1951-52); **13.** *Amphistegina radiata* (Fichtel and Moll), **a.** Indian form, (X 60), **b.** Japanese form, (X 135), (Akimoto *et al.*, 2002); **14.***Nonionellina* cf. *N. labradorica* (Dawson), **a.** Indian form, (X 96), **b.** Japanese form, (X 37), (Matoba, 1967); **15.** *Planorbulina variabilis* (d'Orbigny), **a.** Indian form, (X 100), **b.** Japanese form, (X 100), (Akimoto *et al.*, 2002).

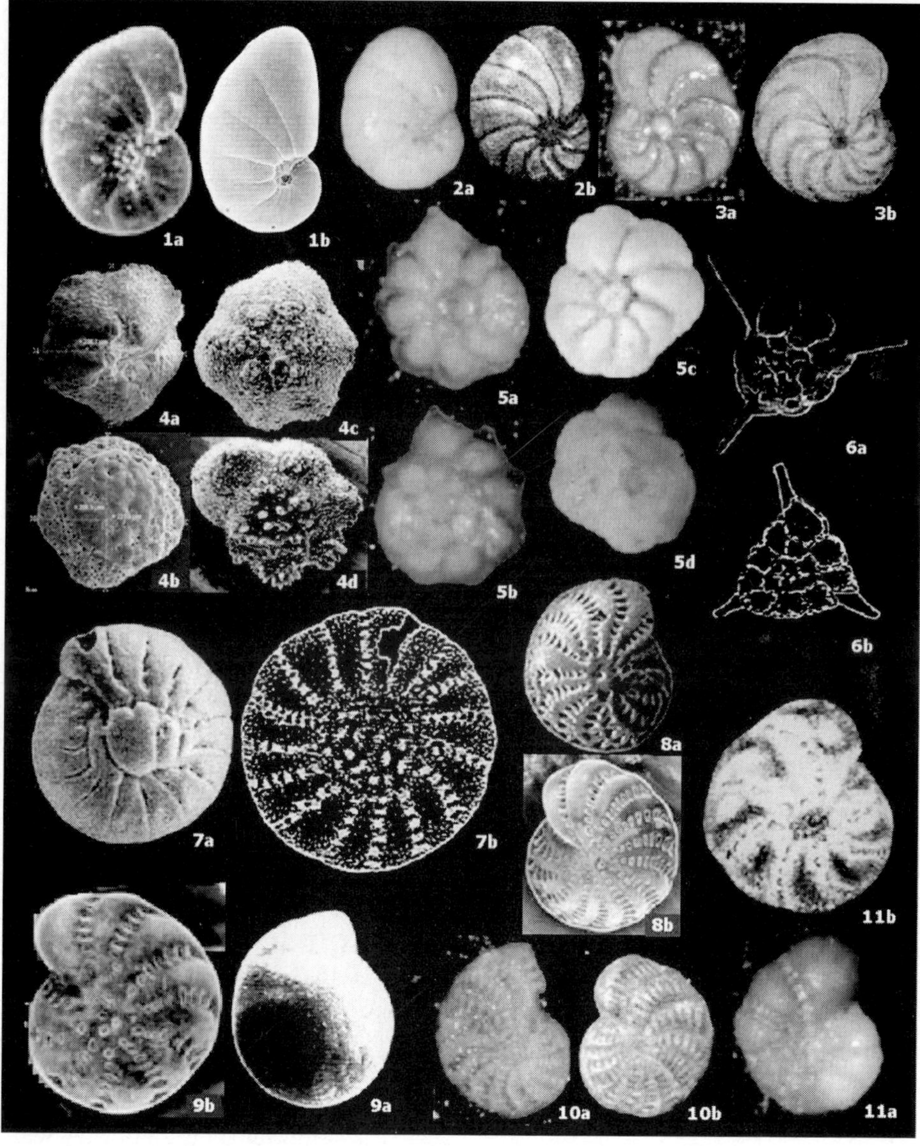

Plate 3. 1. *Nonionoides grateloupi* (d'Orbigny), **a.** Indian form, (X 165), **b.** Japanese form, (X 94) (Akimoto *et al.,* 2002); **2.** *Pseudononion japonicum* Asano, **a.** Indian form, (X 59), **b.** Japanese form, (X 66), (Matoba, 1970); **3.** *Hanzawaia nipponica* Asano, **a.** Indian form, (X 85), **b.** Japanese form, (X 43), (Matoba, 1967); **4.** *Pararotalia? minuta* (Takayanagi), **a.** Ventral view, (X 117), **b.** Dorsal view, (X 117), Indian form, **c.** Ventral view, (X 148), **d.** Dorsal view, (X 148), Japanese form; **5.** *P. nipponica* (Asano), **a.** Ventral view, (X 107), **b.** Dorsal view, (X 108), Indian form, **c.** Ventral view, (X 66), **d.** Dorsal view, (X 63), Japanese form; **6.** *Asterorotalia pulchella* (d'Orbigny), **a.** Indian form, (X 31), **b.** Japanese form, (X 66), (Asano, 1951); **7.** *Pseudorotalia indopacificana* (Thalmann), **a.** Indian form, (X 35), **b.** Japanese form, (X 23), (Asano, 1951); **8.** *Elphidium crispum* (Linné), **a.** Indian form, (X 78), **b.** Japanese form, (X 66); **9.** *E. advenum* (Cushman), **a.** Indian form, (X 96), **b.** Japanese form, (X 105); **10.** *E. jenseni* (Cushman), **a.** Indian form, (X 72), **b.** Japanese form, (X 56) (Matoba, 1970); **11.** *E. subgranulosum* Asano, **a.** Indian form, (X 119), **b.** Japanese form, (X 93) (Matoba, 1970).

1. Nineteen species show narrow range of variations (*Adelosina laevigata, Amphistegina radiata, Asterorotalia pulchella, Elphidium jenseni, E. subgranulosum, Hanzawaia nipponica, Lobatula lobatulus, Nonion commune, Pararotalia? minuta, Pseudononion japonicum, Quinqueloculina agglutinans, Q. parkeri, Q. pseudoreticulata, Q. subcuneata, Reussella haizumensis, Rosalina bradyii, Spiroloculina planulata, Triloculina terquemiana,* and *T. trigonula*);
2. Seventeen show moderate rage of variations (*Bolivina semicostata, B. subreticulata, Elphidium advenum, Fissurina marginata, Glabratella patell iformis, Loxostomum limbatum, Nonionoides grateloupi, Pararotalia nipponica, Planorbulina variabilis, Poroeponides lateralis, Pseudorotalia indopacificana, Quinqueloculina* cf. *Q. Venusta, Rosalina australis, Sigmavirgulina tortuosa, Spiroloculina communis, Textularia foliacea* and *Triloculina tricarinata*);
3. Three species show wide ranges of variations in their morphological parameters (*Nonionellina cf. N. Labradorica, Spiroloculina angulata, Triloculina cf. T. rotunda*); and
4. *Textularia agglutinans* shows wide to medium ranges of variations and *Elphidium crispum* shows very narrow ranges of morphological variations.

The present study shows that two so distantly located regions- the 'east coast of India and the part of the southern east coast of Japan' of the water warm 'Indo-Pacific region show remarkable level of 'foramgeographic affinity'.

Acknowledgement: PKK thanks S. Hasegawa, Kumamoto University, Kumamoto Japan for fruitful discussions, the Japanese Society for Promotion of Science (JSPS) for sponsoring the work in Akita University, Akita and Kumamoto University, Kumamoto, Japan and DFG, Germany for sponsoring work in Tubingen University, Tubingen, Germany. Thanks are due to Head, Department of Applied Geology, Dr. Harisingh Gour University, Sagar for providing infrastructural facilities.

REFERENCES

AKIMOTO, K., CHISATO, M., SHIMOKAWA, A. and FURUKAWA, K. (2002) Atlas of Holocene, benthic foraminifers of Shimabara Bay, Kyushu, southwest Japan. Kagoshima Univ. Mus., Monogr., v.2, 34p., pls.1-70.

ASANO, K. (1950-52) Illustrated catalogue of Japanese Tertiary smaller foraminifera. Hosokawa Printing Company, Tokyo Japan, (compiled & Ed. LEO, W. STACH), parts 1-15, supplement 1, 183p.

CROUCH, R.W. and POAG, C.W. (1987) Benthic foraminifera from the Panamanian Province: distribution and origin. Jour. Foram. Res., v.17, pp. 153-176.

GALLAGHER, S.J., WALLACE, M.W., Li, CHUNG LEONG, KINNA, BELINDA, BYE, John T., AKIMOTO, KAZUMI and TORII, M. (2009) Neogene history of the West Pacific Warm Pool, Kuroshio and Leeuwin currents, Paleooceanogr., v.24(1) http://agupubs.onlinelibrary.wiley.com/ /jour.,10.1002 (ISSN)1944-9186).

GANAPATI, P.N. and SATYAVATI, P. (1958) Report on the foraminifera in bottom sediments in the Bay of Bengal off the east of India. Andhra Univ. Mem. Oceanogr., v.62(2), pp.100-127, pls. 1-6.

GANDHI, M.S., RAJAMANICKAM, G.V.M. and NIGAM, R. (2002) Taxonomy and distribution of benthic foraminifera from sediments off Palk Strait, TN, east coast of India. J.Pal.Soc.India, v.47, pp.47-64.

GIBSON, T.G. (1989) Miocene evolution of Tubulogenerina in the Indo-Pacific and African areas. J. Foram. Res., v.19, pp.126-145.

GLAESSNER, M. F. (1943) Problems of stratigraphic correlation in the Indo-Pacific region. *In*: Proc. Royal Soc. Victoria, v.55(II), new series,41-80.

GRAHAM, J.J., MILITANTE, P.J. (1959) Recent foraminifera from Puerto Galera area northern Mindoro, Philippines. Stanford Univ. Pub., Geol. Sciences, v.6(2), pp.1-171.

HASEGAWA, S., SPROVIERI, R. and POLUZZI, A. (1990) Quantitative analysis of Benthic foraminiferal assemblages from Plio-Pleistocene sequences in the Tyrrhenian sea, ODP LEG 107. *In*: Proc. ODP Sci. Results, (Eds. KASTENS K. A., MASCLE, J.), v.107, pp.461-478, pls.1-5.

HERON-ALLEN, E. and EARLAND, A. (1915) The Foraminifera of the Kerimba Archipelago (Portugese East Africa). Part II, Transactions of the Zoological Society of London, v.20(17), pp.543-794.

HOFKER, J. (1951) The foraminifera of the Siboga Expedition. Part III. Siboga-Expeditie, Monographie IVa. Leiden: E. J. Brill, 1-513p.

HUANG, T. Y. (1964) "Rotalia" Group from the Upper Cenozoic of Taiwan. Micropal. v.10(1), pp.49-62, pls. 1-32.

ISHIZAKI, K. (1941) A new foraminifera, *Streblus trispinosus*, from Neogene formation of Taiwan (Formosa).v. 12(4), pp. 63-66, pl. 6.

JAIN, S.P. and BHATIA, S.P. (1978) Recent foraminifera from Mandvi, Kutch, *In*: Proc. VII Ind. Colloqu. Micropal. Stratigr., 153-174, pls. 1-2.

JONES, R. W. (1994) The Challenger foraminifera. Natural History Museum and Oxford Univ. Press, Oxford. 151p., 123pls, 22text figs.

KARRER, F. (1868) Die Miocene Foraminiferen fauna von Kostej im Banat. Sitzungsberichte der Kaiserlichen Akad. Wissenschaften Wien, Math.-Nat. Klasse. v.58(1), pp. 121-193.

KATHAL, P.K. (1998) Taxonomy and palaeolatitudinal significance of Rotorboides granulosum- a less known foraminifera from tropical waters. J. Geol. Soc. Ind., v.51(6), pp.799-802.

KATHAL, P.K. (2002a) Taxonomy, distribution pattern and ecology of the Recent littoral foraminifera of the east coast of India. N. Jb. Jah. Geol. Palä., Abh., v.224(1), pp.115-160.

KATHAL, P.K. (2002b) Distribution and ecology of Recent foraminifera from littoral sediments of eastern India. J. Geol. Soc. India, v.60, pp. 429-454.

KATHAL, P.K. and BHALLA, S.N. (1996a) Intraspecific variation and Palaeolatitudinal significance of *Rotorboides granulosum*- less known recent foraminifera of tropical waters. Rev. de Paleobio., v.15(10), pp.79-85.

KATHAL, P.K. and BHALLA, S.N. (1996b). On the first report from the Indo-Pacific region, migratory trends and paleolatitudinal significance of *Rotorboides granulosum*- a tropical water foraminifer since Middle Pliocene (Late Miocene?), *In*: Contri. XV Ind. Colloq. Micropal. Stratigr. (Eds. PANDEY, J., AZMI, R. J., BHANDARI, A., DAVE, A.), pp.317-320.

KATHAL, P.K. and MATOBA, Y. (2001) Morphological variations in the common elements of Recent 'Indo-Pacific' and 'Atlanto-Caribbean' foraminifera, off the eastern Japanese and Indian coasts. *In*: Neogene Climate, (Ed. GUPTA, A.K.), Ind. Inst. Tech., Kharagpur, India, Ext. Abs., pp. 20-21.

KATHAL, P.K. and SINGH, V.K. (2009) Migration of the additional common Indo-Pacific Recent foraminifera of the Indian and Japanese water since Middle Miocene. *In*: Nat. Sem. Geody. Sedi. Biotic Res.

Context of India-Asia Collision, Mizoram Univ., Extended Abs., pp. 212-213.

KATHAL, P.K. and SINGH, V.K. (2010) First report of some Recent benthic foraminifera from the east coast of India. J. Geol. Soc. India, v.76(1), pp. 69-74.

KENNETT, J.P., KELLER, G. and SRINIVASAN, M.S. (1985) Miocene planktonic foraminiferal biostratigraphy and palaeooceanographic development of the Indo-Pacific region. Mem. Geol. Soc. America, v.163, pp.196-236.

MATOBA, Y. (1967) Younger Cenozoic foraminiferal assemblages from the Choshi district, Chiba prefecture. Science Reports Tohoku Univ., 2nd series (Geol.), v.38(2), pp.221-263, pls.25-30.

MATOBA, Y. (1970) Distribution of Recent shallow water foraminifera of Matsushima Bay, Miyagi prefecture, northeast Japan. Sci.Rep. Tohoku Univ., 2nd series (Geol.), v.42(1), pp.1-85, pls.1-8.

RAO, M.S., RAO, T.V. (1976) Ecology of foraminifera from certain marginal marine environments from east coast of India. Ind. J. Marine Sci., v.5, pp.77-82.

SINGH, V.K. (2009) Study of comparative morphology of common Recent benthic foraminifera from the east coast of India and off Tiba Province, east coast of Japan. Ph.D. thesis, Dr. Harisingh Gour Univ., Sagar, India (Unpub.), 173p.

SINGH, V.K. and KATHAL, P.K. (2010) First report of thirteen Recent benthic foraminifera from the Indian waters. Gond. Geol. Mag, v.10, pp.59-266.

SINGH, V.K. and KATHAL, P.K., (2011) Morphological variations in common recent benthic foraminifera from the east coast of India and the southern east coast of Japan, J. Pal. Soc. India, v.56(1), pp.65-81.

Micropaleontology and its Applications
Pradeep K. Kathal, Rajiv Nigam & Abu Talib (Editors)
Scientific Publishers (India), 2018, 107-118 pp.

Late Cretaceous Pelagic Red Shales in the Cauvery Basin, Southeast India, its Biostratigraphy and Paleoceanographic Implications

Abiraman Govindan

Asian Biostratigraphic Services, H-53, Central Avenue, Korattur, Chennai - 600080, India
Email: abiramangovindan@gmail.com

Abstract

The Cretaceous Period witnessed a succession of contrasting global changes marked by sediments of early Cretaceous organic carbon rich black shales (OAE's) and organic carbon depleted late Cretaceous red shales (CORB). These events are identified in the Cauvery basin. This contribution focuses on Cretaceous pelagic red shales occurrence and their biostratigraphic position. The presence of red shales in a bore hole close to Kunnam in Ariyalur area between 9m and 45m of subsurface underlain by light to dark grey shales and argillaceous sandstones, was recorded earlier. Later in Ramnad sub basin 3 deep exploratory wells revealed the presence of pelagic red shales of nearly 25m thickness at 2000m depth. The red shales in these subsurface sections revealed planktic and benthic foraminiferal fauna correlatable to *Dicarinella asymetrica* planktonic zone of Santonian age. This age assignment is in accordance with the various reports of the occurrences of Cretaceous Pelagic "Red beds" in the Tethyan belt. In this region, the presence of oxygenated ocean bottom water moving to the shelf and slope in the warming period gave rise to sedimentation of red shales in Santonian. Dysoxic bottom water circulation in intervening adverse cool environment favoured the deposition of dark grey black shales correlatable to Global Cretaceous Anoxic Event (OAE1b, 1d and 2). Oxidation and retention of organic carbon in the Cretaceous sedimentary succession in the basin could be due to local tectonics, also influenced by paleoceanographic as well as paleoclimatic conditions.

Keywords: Pelagic red beds, Late Cretaceous, Cauvery Basin, Biostratigraphy, oxic environment paleoceanography, eastern Tethys.

INTRODUCTION

The Cretaceous Period has witnessed isochronous perturbations of the global carbon cycle resulting in the episodic deposition of organic carbon rich black shales in the Early Cretaceous. These anoxic events have been defined as Oceanic Anoxic Events (OAE's) by Schlanger and Jenkyns (1976). Appropriate climatic and oceanographic conditions would have prevailed then inducing the prevalence of OAE. The Cretaceous period has also witnessed transition from anoxic to highly oxic environmental conditions at a few stratigraphic levels in the Late Cretaceous. These Cretaceous oxygenated red facies "CORB" are wide spread in the North Atlantic and in the Tethyan belt (Hu et al., 2005).

The Cauvery Basin, a divergent marginal basin in the southeast India, having a thickness of 3 - 3.5km Cretaceous sedim-ents in the deeper part of the basin. Fossiliferrous Cretaceous rocks are well exposed in the western margin of the basin adjoining the Peninsular craton. The basin in this part of Tethys, offers opportunities for a detailed foraminiferal biostratigraphy, sea level changes, paleoclimatic conditions, paleoceanographic implications and input of terrigenous clastics from the adjoining craton in the Cretaceous sea.

Wide spread deposition of organic rich dark grey to black shales in the deeper part has been met in the basin. The total organic carbon contents up to 2.5 to 4% suggests the sequestrations of organic carbon in the possible marine source rocks. These source prone sediment depositional events in the basin are comparable with the global Cretaceous oceanic anoxic events OAE1b (Early Albian), 1c (Late Albian) and 2 (Late Cenomanian). These events coincide globally with the sea level rise and it is also evident in this basin (Govindan, 1982, 1993; Govindan and Ramesh, 1996). Agglutinated foraminifera are most prominent at these stratigraphic levels. Radiolarian silica is also present in the black shales associated with the OAE in some wells in the basin (Kalyanasunder, et al., 1996). Agglutinated foraminifera and the presence of radiolarians at these levels in the basin can be ascribed to increase in productivity, organic carbon burial and bio-siliceous accumulations related to dyserobic conditions during OAE's.

The Cauvery Basin has also witnessed oxic sedimentation with deposition of pelagic red shales during Late Cretaceous. The basin provides geological evidence indicating the transition from an anoxic to an oxic environmental condition. The anoxic/oxic events indicate rhythmic transition in this basin.

The intent of this contribution is to focus on the occurrence of pelagic red shales in this basin and discuss its stratigraphic position and paleoenvironmental conditions.

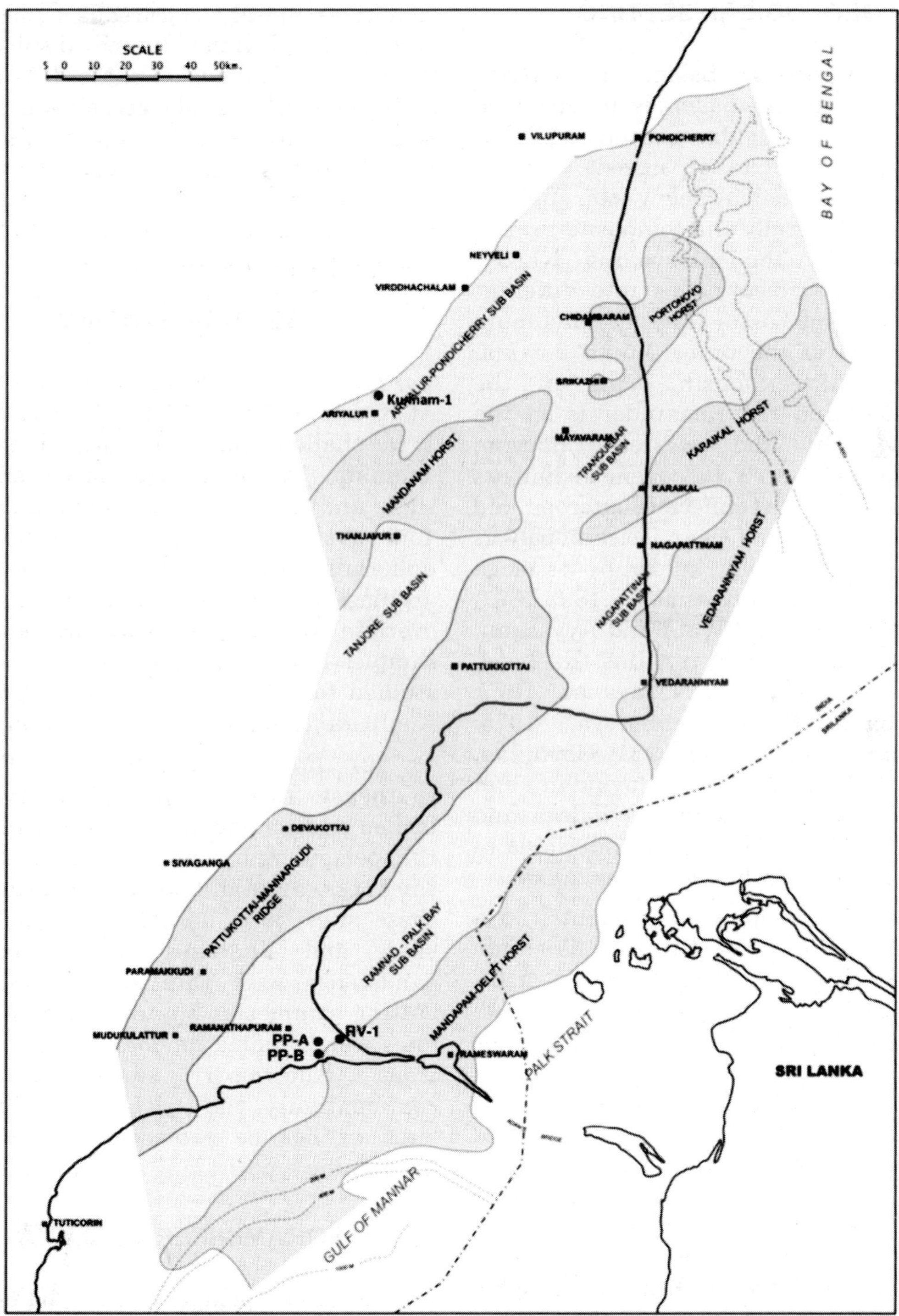

Fig.1. Tectonic frame work of Cauvery Basin, southeast India showing different sub-basins and ridges. The location of studied sites are shown.

GEOLOGICAL SETTING

The Cauvery Basin, a passive continental margin basin, is bounded on the west by Pennisular Craton, while on the east by Sri Lankan massif with its subsurface extension below the Bay of Bengal. Tectonically the composite graben is divided by the intervening NE-SW trending subsurface ridges into different sub-basins and horsts (Fig. 1). Maximum basin fill is of the order 4.5 to 5.5 km, sediments three fourth being in the Cretaceous and the remainder is in the Tertiary. On the western margin, Cretaceous and early Paleogene sediments are exposed as disconnected outcrops and the one outcropping near Trichinopoly is the largest and widely known by its mega (Stoliczka, 1868; Kossmat, 1895,1897; Ayyasami, 1990; and Wani and Ayyasami, 2004) and microfaunal remains (Rasheed, 1963; Rao et al., 1968; Narayanan, 1977; Narayanan, and Scheibnerova, 1975; Rasheed and Ravindran 1981; Govindan, 1972, 1977 and 1978; Govindan and Ramesh, 1996; Govindan, Ravindran and Rangaraju, 1996; Govindan and Chidambaram, 2000; and Chidambaram, 2000).

The outcropping sediments are broadly grouped into Uttatur, Trichinopoly, Ariyalur and Niniyur in the ascending order of superposition (Blanford, 1862). These well defined groups and their members have been brought in a subsurface correlatable lifthostratigraphic scheme, taking into account microfauna, lithology and E.log and seismic markers for a better understanding of the subsurface sedimentary pile (Venktarangan et al., 1993). Exposed groups and their equivalent subsurface lithostratigraphic units of the basin are shown in the Fig 2.

Pelagic red shales occur in the Trichinopoly Group in the Anaipadi formation. It comprises mainly calcareous sandstones, siltstones and shales. Its coeval subsurface litho unit is Kudavasal shale. It is chiefly calcareous silty shale occasionally interstratified with thin calcareous sandstone. Pelagic red shales occur in some deep wells in this unit. The well sections, having red shale investigated in detail for this report are shown in Fig 4.

SECTION STUDIED

(1) *Ariyalur - Pondicherry sub-basin* - At Kunnam (79°02' 15"E, 11°12'22"N; Fig. 1) a shallow bore hole drilled close to Kunnam - Ariyalur road encountered red shale and thin silt stone from 9m down to 45m subsurface. The pelagic red shales are underlain by light to dark grey shale and argillaceous sandstones down to 120m and overlain by Recent alluvium. Fourteen samples from every 3m core interval were studied for foraminiferal record from the Anaipadi formation of Uttatur Group.

(2) *Ramnad sub-basin* - Further southeast, three deep exploratory wells drilled in Ramnad sub-basin has revealed the pelagic red shales of about 25m in thickness around 2000m subsurface. At these sites, it is underlain by dark grey shale and siltstones and overlain by sandstones with thin shales. About 30 cutting samples at 5m interval from these sites were studied for foraminiferal fauna from Kudavasal shale. Due to confidentiality, the exact location of these sites cannot be provided. However, its geographical position is shown in Fig. 2.

FORAMINIFERAL FAUNA

In the pelagic red shales the preservation of foraminifera is moderately good with well diversified planktonic and benthic assemblage. The planktonic

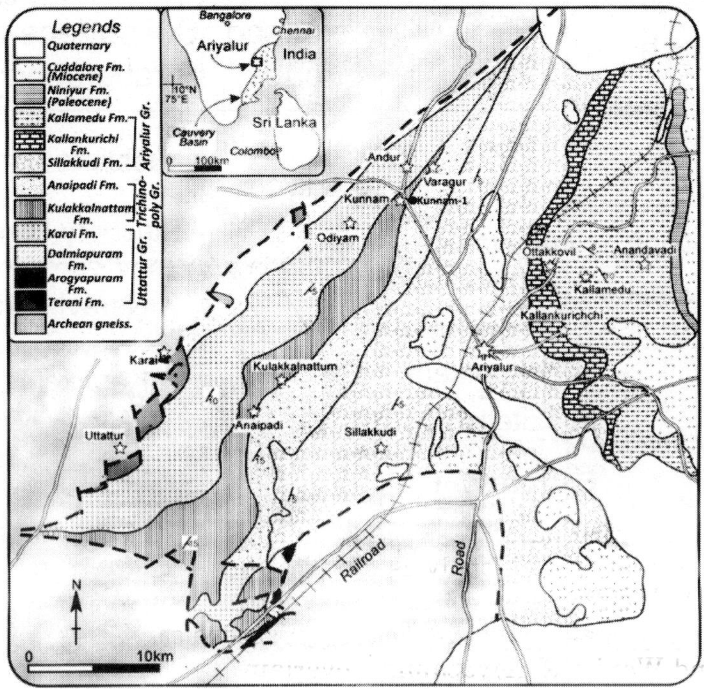

Fig.2. Geological Map of Ariyalur Area, Cauvery Basin, Southeast India, Source: Wani and Ayyasami (2004).

portion is moderately high in the Kunnam well in contrast to other studied sites in the Ramnad sub-basin. The benthic assemblage in the red shales is diversified with representatives of gavelinellids, laginids and trochaminids in good numbers and other forms are few in number.

(1) *Ariyalur- Pondicherry sub-basin section and age assignment-* In the pelagic red shales at Kunnam site the planktonic assemblage is dominated by *Marginotruncana pseudolinneiana, M. marginata, M. coronata, M. schneegansi, Dicarinella asymetrica, Contusotruncana fornicata, Costellagerina bulbosa, Heterohelix reussi* and *Hedbergella delrioensis.* The species *Falsotruncana* cf. *loeblichae,* though less common, has been recognized. The basal section is impoverished in foraminifera at this site. Distribution of planktonic foraminiferal species in the red shales and in the underlying shales at this site is shown in Fig. 5. In the red shales the benthic foraminiferal assemblages are dominated by Gavelinellids represented by *Gavelinella monterelensis, G. excolata, G. clementiana convexa,* besides other forms such as *Gyroidinoides depressa, Globorotalites michelinianus, Globorotalites subconicus, Cibicides subcarinatus, Lenticulina macrodisca, Lenticulina muensteri, Bolivinoides strigillatus, Neoflabellina projecta* and *Allomorphina cretacea* (Plate-1).

The presence of zonal taxa *Dicarinella asymetrica* in the red shales at this site indicate Santonian age (Caron, 1985). Besides the record of benthics as *Globorotalites subconicus, Gavelinella excolata, Neoflabellina projecta, Bolivinoides strigillatus* also support the

Period	Epoch	Stage	Exposed Group	Subsurface Litho Stratigraphic Units
NEO GENE	Post Miocene - Miocene	Tortonian - Aquitanian	Cuddalore	Cuddalore Sandstone Titachery Formation
PALEOGENE	Oligocene	Chattian - Rupelian		Neravi Formation
	Eocene L	Priabonian		Karaikal Shale Fm.
	Eocene M	Lutetian		Kamalapuram Fm.
	Eocene E	Ypresian		
	Paleocene L	Thanetian	Niniyur	
	Paleocene E	Danian		
CRETAECOUS	Late	Maestrichtian	Ariyalur	Porto Nova Shale Fm.
		Campanian L M E		Nannilam Formation
		Santonian	Trichinopoly	Kudavasal Shale Fm.
		Coniacian		
		Turonian	Uttatur	Bhuvanagiri Formation
		Cenomanian		Sattapadi Shale Fm.
	Early	Albian		Andimadam Formation / Palk Bay Fm.
		Aptian		
		Pre Aptian	Pre Uttatur	

Fig.3. Relationship between exposed groups in the Ariyalur-Pondicherry sub-basin and their subsurface lithologic unis in the Cauvery Basin, after: Venkatrangan et al. 1993).

? Late Turonian	Coniacian	Santonian	Stage	Kunnam-1
D. primitiva M. sigali	*Dicarinella concavata*	*Dicarinella asymetrica*	Zone	

mbs: 120, 115, 110, 105, 100, 95, 90, 85, 80, 75, 70, 65, 60, 55, 50, 45, 40, 35, 30, 25, 20, 15, 10

Planktonic species:
Archaeoglobigerina bosquensis
Archaeoglobigerina cretacea
Archaeoglobigerina portsdownensis
Contusotruncana fornicata
Costellagerina bulbosa
Dicarinella asymetrica
Dicarinella canaliculata
Dicarinella concavata
Dicarinella primitiva
Falsotruncana cf. loeblichae
Globigerinelloides bollii
Globigerinelloides ultramicrus
Globotruncana bulloides
Hedbergella delrioensis
Hedbergella flandrini
Hedbergella hoelzi
Hedbergella holmdelensis
Hedbergella planispira
Heterohelix. globulosa
Heterohelix reussi
Marginotruncana coronata
Marginotruncana linneiana
Marginotruncana marginata
Marginotruncana pseudolinneiana
Marginotruncana renzi
Marginotruncana schneegansi
Marginotruncana sigali

Fig.4. Planktonic forminiferal species distribution and biozonation in the studied site Kunnam-1, Ariyalur-Pondicherry sub-basin. The occurrence of red shales are indicated by shaded background in the chart.

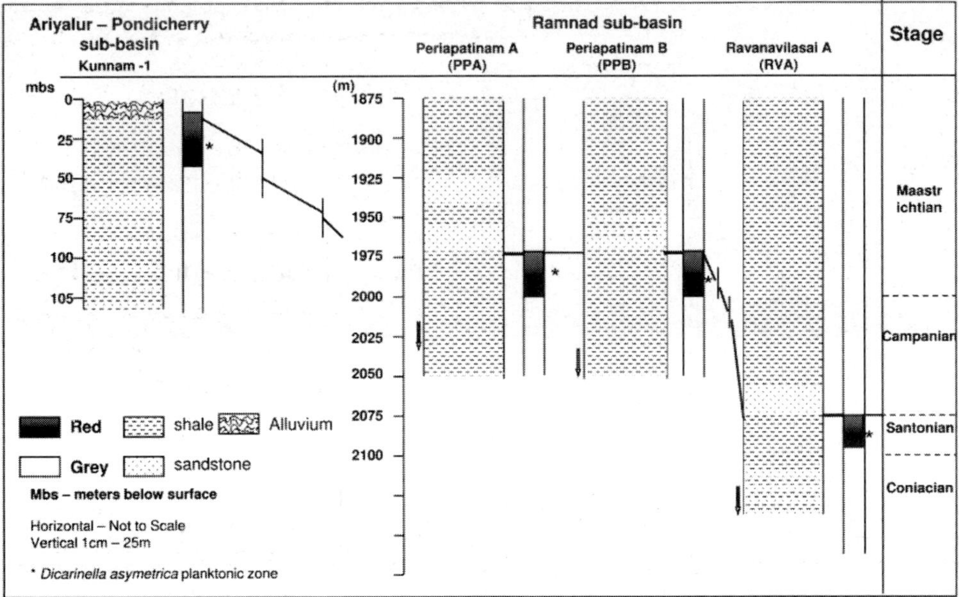

Fig.5. Stratigraphic correlation of red shales between Kunnam-1 in the Ariyalur-Pondicherry sub-basin and three other deep exploratory sites further southeast in the Ramnad sub-basin.

age assignment. These benthics are widely reported at this level elsewhere (Bolli et al., 1994). The species *Bolivinoides strigillatus*is often observed from late Santonian from many well sites in the Cauvery Basin (Govindan 1977, Chidambaram, 2000).

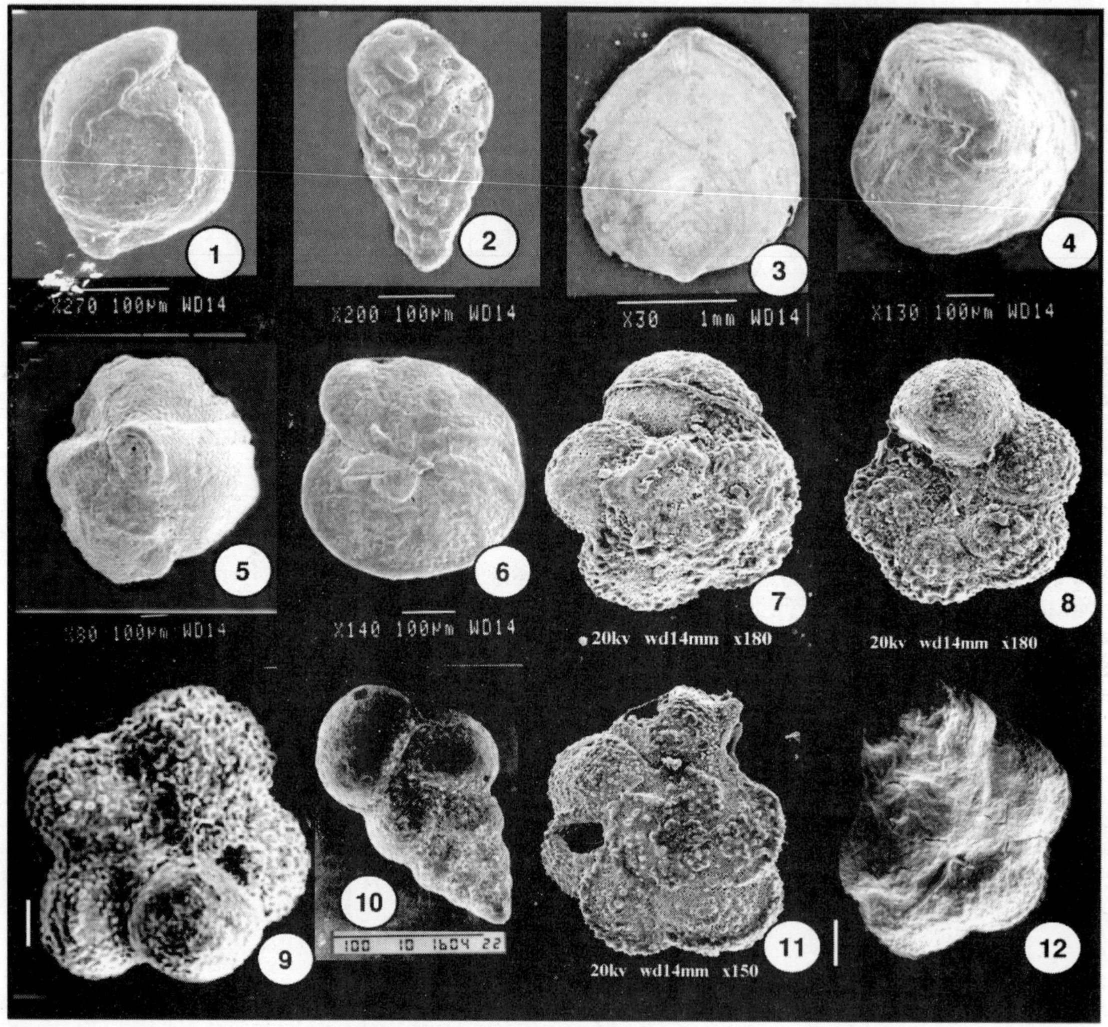

Plate 1. 1. *Praebulimina reussi* (Morrow) Kunnam-1, cc- 3. depth 12.85m; **2.** *Bolivinoides strigillatus* (Chapman) Kunnam-1, cc-3; depth 12.85m; **3.** *Neoflabelina projecta* (Carsey) Kunnam -1, cc-17; depth 37m; **4.** *Globorotalites michelinianus* (d'Orbigny), Kunnam-1, cc-3; depth 12.85m; **5.** *Globorotalites subconicus* (Morrow) Kunnam-1. cc-3, depth 12.85m; **6.** *Gavelinella monterelensis* (Marie) Kunnam-1,cc-2; depth 12m; **7-8.** *Archeoglobigerina bosquensis* Pessagno, Kunnam-1, cc-3; depth-12.85m; **9.** *Archeoglobigerina portsdownensis* (Williams and Mitchell) Kunnam-1, cc-4, depth-17m; **10.** *Heterohelix reussi* (Cushman) Kunnam-1, cc-4, depth-17m; **11-12.** *Dicarinella asymetrica* (Sigal) Kunnam-1, cc-5; depth-18m, Spiral and umbilical views; raised umbilical shoulder not well portrayed due to extraneous material cover (scale bar = 100μ).

(2) *Ramnad sub-basin section and age assignment-* Further southeast of Kunnam site, three deep exploratory wells drilled in the Ramnad sub-basin penetrated pelagic red shales of about 25m thickness at 2000m subsurface. At these sites the calcareous red shales interbedded with thin siltstones. The red shales are underlain by calcareous to feebly calcareous dark grey shales, siltstones and sandstones and overlain by sandstones with interbedded grey shales. The planktonic foraminiferal species recorded in the red shale at these sites include *Dicarinella asymetrica, Marginotruncana schneegansi, M. pseudolinneiana, M. coronata,* and *Whiteinella inornata.* The record of index taxon *Dicerinella asymetrica* indicates Santonian age as in Kunnam site. The benthic forms represented at these sites include *Lenticulina muensteri, Globorotalities subconicus, Gavelinella clemntiana convexa, G. stelligera, Haplophragmoides* cf. *H. walteri, Trochamina deformis, Hyperammina indivisa* and *Vernulinoides* cf. *tailleuri.*

The record of pelagic red shales of Santonian age at these sites has revealed wide spread red shale sedimentation in the basin under oxic environmental condition.

DISCUSSION AND CONCLUSION

The present report of pelagic red shales from Cauvery Basin, southeast India in *Dicarinella asymetrica* planktonic zone of Santonian age is significant. At this stratigrpahic level, the occurrence of pelagic red shales have been reported in many basins in the Tethyan belt (Hu, *et al.*, 2005). The report of red shales from Chuangde section, southern Tibet, Himalayan region from *D. asymetrica* zone and *Globotruncanita elevata* planktonic zones of Santonian and Early Campanian confirms further its widespread distribution in the eastern Tethyan region (Wang, et al., 2005; Scott *et al.*, 2005).

After the dismemberment of the Indian continent from Gondwana in the Early Cretaceous (Larsson and Lad, 1973 and Markl, 1974) drifted northward in isolation and was positioned in southern high paleolatitudes at 45°S in warm tropical belt in Santonian, 85Ma. The Late Cretaceous foraminiferal assemblage of Cauvery Basin is distinctively of tropical Tethyan affinity (Govindan, 1980 and 1981). The assemblages are represented by the greater diversity of planktonic species all the single and double keeled Globotruncanids are represented as well as full range of Heterohelicidae including species of *Pseudotextularia, Racemiguembelina, Planoglobulina, Guembelitria, Gublerina, Hetrohelix* besides Rugoglobigerinids. During the deposition of the oldest Cretaceous sediments (Apitan - Early Albian) the climate in the east coast Indian basin was cold as reflected by the simple morpho types among the planktonics. This assemblage bears close affinity to the cold Austral Realm fauna reported from the southern high paleolatitudes (Govindan and Narayanan, 1980). Paleotemperature derived from isotopic data of foraminifera indicates, globally a cool climate in middle-late Aptian and Albian followed thereafter by a warm trend (Huber et al., 1995). The initiation of warming trend from middle Albian onwards in this basin is evident by the paleotemperature of 23°C derived from *Belemnites* from the exposed basal Uttatur Group (Bowen, 1961). Gupta et al. (2007) reported the maximum temperature of sea surface water to be 29°C estimated from Cauvery Basin planktonic foraminiferal isotope composition and bottom water temperature as indicated by benthic foraminifera varies between 21°C and 23°C in the Albian-Turonian times. These data when viewed in conjunction with the other published information on isotopic paleotemperatures clearly indicates the warmth of the Cretaceous climate in Turonian - Coniacian - Santonian times and suggest higher than modern day ocean water temperature in high latitudes (Bice et al., 2003). The presence of oxygenated ocean bottom waters in the shelves and slopes giving rise to the sedimentation of red shales in Santonian in the basin is well evident. The record of organic carbon rich dark grey/ black shales, correlatable to global Cretaceous anoxic events OAE1b, 1c and 2 envisages dysoxic bottom water circulation in the intervening adverse cool environmental conditions. Oxic bottom water environmental conditions resulting in the deposition of pelagic red shales in the basin would have influenced by local tectonics, high south paleolatitudinal position, paleoceanographic and warmer paleoclimatic condition.

Acknowledgement: The author gratefully acknowledges L. Chidambaram, ONGC (retd.) for sparing foraminiferal assemblage slides of Kunnam well and sharing faunal data of other sites of Cauvery Basin. He also thanks the project leaders of IGCP 463 and 494 for their support in the author's participation and presenting this study in the workshop at Beijing in 2007. The author also thanks Ewa Malata, Krakow, Poland for comments on early draft of text.

References

AYYASAMY, K. (1990) Cretaceous heteromorph Ammonoid biostratrigraphy of Southern India, Stratigraphy. v.22, pp.111-117.

BICE, K.L., HUBER, B.T. and NORRIS, R.D. (2003) Extreme polar warmth during the Cretaceous greenhouse? Paradox of the late Turonian $\delta^{13}O$ recorded at Deep Sea Drilling Project Site 511. Paleoceanography, v.8, doi: 10.1029/2002PA000848.

BLANFORD, H.F. (1862) On the Cretaceous and other rocks of the South Arcot and Trichinopoly district. Geol. Sur. India, Mem. no.4, pp.1-217.

BOLLI, H.M., BECKMANN, J.-P. and SAUNDERS, J.B. (1994) Benthic foraminiferal biostratigraphy of the south Caribbean region: Cambridge University Press, 408p.

BOWEN, R. (1961) Oxygen isotope paleotemperature measurement of Cretaceous Belemnoidea from Europe, India and Japan. J. Palaeo., v.35, pp.1071-1084.

CARON, M. (1985) Cretaceous planktic foraminifera. *In*: Plankton Stratigraphy, BOLLI, H.M., SAUNDERS, J.B., and PERCH-NIELSEN, K. (Eds). pp. 17-86.

CARON, M. (1985) Cretaceous planktonic foraminifera. *In*: BOLLI, H.M. et al. (Eds.), Foraminiferal Stratigraphy. Cambridge University Press. pp.17-86.

CHIDAMBARAM, L. (2000) Middle Turonian-Santonian Foraminferal biostratigraphy of Trichinopoly Group, South India. *In*: GOVINDAN, A. (ed.). Geol. Soc. India, Mem. no.46, pp.173-204.

GOVINDAN, A. (1977) Late Cretaceous and Lower Tertiary foraminiferal genus *Bolivinoides* from the Cauvery Basin, south India and its paleobiogeographical significance. J. Geol. Soc. India, v.18, pp.160-193.

GOVINDAN, A. (1980) Cretaceous foraminifera from the eastern part of Indian peninsula and their paleoclimatological significance. Bull. Oil Natural Gas Comm., v.17, pp.1-9.

GOVINDAN, A. (1981) Selected Upper Cretaceous benthonic foraminifera from the Cauvery Basin, south India and paleobathymetric interpretation. *In*: RASHEED, D.A.(ed.), *Proc. VII Indian Colloq. Micropal. Strat.*, pp.40-59.

GOVINDAN, A. (1982) Imprint of global "Cretaceous Anoxic Events" in East Coast India and their implications. Bull. Oil Natural Gas Comm., v.19, pp.257-270.

GOVINDAN, A. (1993) Cretaceous anoxic events, sea level changes and microfauna in Cauvery Basin, India. *In* BISWAS, S.K., et al. (eds.), In: *Proc. II Sem. Petrol. Basins India,* v.1, pp.161-176.

GOVINDAN, A. and NARAYANAN, V. (1980) Affinities of Cretaceous foraminifera of the east coast Indian basins and the drifting of Indian Shield. J. Geol. Soc. Iraq, v.13, pp.269-280.

GOVINDAN, A. and RAMESH, P. (1996) Cretaceous anoxic events in Cauvery basin and its significance for hydrocarbon exploration. Indian Petrol. Geol., v.14, pp.1-15.

GOVINDAN, A. and CHIDAMBARAM, L. (2000) *Lenticulina kossmatina* a new foraminiferal species from the Turonian of Cauvery basin, south India. J. Geol. Soc. India., Mem. no.46. pp.205-211.

GOVINDAN, A., RAVINDRAN, C.N. and RANGARAJU, M.K. (1996) Cretaceous Stratigraphy and Planktonic Foraminiferal Zonation of Cauvery Basin, South India. Geol. Soc. India, Mem. no.37, pp.155-187.

GUPTA, K.D., SARASWATI, P.K., KRAMAR, U., RAVINDRAN, C.N., STÜBEN, D. and BERNER, Z. (2007) Oxygen isotopic composition of Albian-Turonian Foraminifera from Cauvery Basin, India: evidence of warm sea-surface temperature. J. Geol. Soc. India, v.69, pp.390-396.

HU, X.M., JANSA, L., WANG, C., SARTI, M., BAK, K., WAGERIECH, M., MICHALIK, J. and SOTÁK, J. (2005) Upper Cretaceous oceanic red beds (CORBs) in the Tethys: occurrences, lithofacies, age, and environments. Cretaceous Res., v.26, pp.3-20.

HUBER, B.T., HOWELL, D.A. and HAMILTON, C.P. (1995) Middle-Late Cretaceous climate of the southern high latitudes. Palaeo. Palaeo. Palaeo., v.92, pp.325-320.

KOSSMAT, F. (1895) Untersuchungen über die Südindische Kreideformation. Beiträge zur Paläontologie Österreich-Ungarns und des Orients. v.9, pp.97-203.

KOSSMAT, F. (1897) Untersuchungen über die Südindische Kreideformation. Beiträge Paläontologie und Geologie Österreich-Ungarns und des Orients. v.11, pp.1-46.

KALYANSUNDAR, R., RAVINDRAN, C.N. and GOVINDAN, A. (1996) Cretaceous bio and lithostratigraphy of Kovil Kallappal area, Cauvery Basin. *In:* PANDEY, J., AZMI, A. R., BHANDARI, A., DAVE, A. (eds), Contr. XV Indian Colloq. Micropal. Strat., pp.7-18.

LARSSON, R.L. and LAD, J.W. (1973) Evidence for opening of South Atlantic in the Early Cretaceous and Late Jurassic. Nature, v.246, pp.233-254.

MARKL, R.G. (1974) Evidence for the breakup of eastern Gondwanaland by the Early Cretaceous. Nature, v.251, pp.196-200.

NARAYANAN, V. and SCHEIBNEROVA, V. (1975) *Lingulogavelinella* and *Orithostella* (Foraminifera) from the Utatur Group of the Trichionopoly Cretaceous, South (peninsular) India, Revista Esp Micropaleont, v.71, pp.25-36

NARAYANAN, V. (1977) Biozonation of the Uttatur Group, Trichinopoly, Cauvery Basin. Jour. Geol. Soc. India, v.18(8), pp.415-428.

RAO, B.R.J., MAMGAIN, V.D. and SASTRY, M.V.A. (1968) *Globotruncana* in Ariyalur group of Trichinopoly Cretaceous. South India, Geol. Soc. India, Mem. no.2, pp.18-29.

RASHEED, D.A. (1963). Some calcareous Foraminifera belonging to the families Rotaliidae, Globigerinidae, Globorotaliidae and Anomalinidae from the Cullygoody Limestone, Trichinopoly Cretaceous of South

India. Part III. Jour. Madras Univ., v.33(3), pp.231-248

RASHEED, D.A. and RAVINDRAN, C.N. (1981) Foraminiferal biostratigraphic studies of the Ariyalur Group of Tiruchirapalli Cretaceous rocks in Tamilnadu State. *In* Proc: VII Indian Colloq. Micropal. Strat., pp.321-336, figs.

SCHLANGER, S.O. and JENKYNS, H.C. (1976) Cretaceous Oceanic Anoxic Events: Causes and Consequences. Geologie en Mijnbouw, v.55, pp.179-184.

SCOTT, R.W., WANG, C. and HU, X.M. (2005) Cretaceous Oceanic Red Beds (CORB), response to paleoclimatic/paleoceanographic global changes and regional tectonics- Workshop of IGCP 463 & 494: Episodes, v.20, pp.1-3.

STOLICZKA, F. (1868) Cretaceous fauna of southern India II. The Gastropoda. Paleo. Indica, ser.5 v.2, 498p.

VENKATARANGAN, R., PRABHAKAR, K.N., SINGH, D.N., AWASTHI, A.K., REDDY, P.K., ROY, S.K. and PALAKSHI, K. (1993) Lithostratigraphy of Indian Petroliferous basins – Cauvery basin. ONGC Document no. VII.

WANG, C., HU, X.M., SARTI, M., SCOTT, R.W. and LI., X. (2005) Upper Cretaceous oceanic red beds in southern Tibet: a major change from anoxic to oxic, deep-sea environments, Cretaceous Res., v.26, pp.21-32.

WANI, R. and AYYASMI, K. (2004) A review of the Cretaceous biostratigraphy in the Ariyalur area, southern India. Bull. Mikasa City Museum, v.8, pp.1-15

Micropaleontology and its Applications
Pradeep K. Kathal, Rajiv Nigam & Abu Talib (Editors)
Scientific Publishers (India), 2018, 119-133 pp.

Foraminiferal Palaeoecology and Depositional Environment of Middle to Late Jurassic Sequence, Kaiya Dome, Kutch, India

Abu Talib[1*], K. N. Gaur[2], Y. P. Singh[3], Shabana Anjum[4] and S. M. Abbas[5]

[1]Department of Geology, Aligarh Muslim University, Aligarh - 202 002, India.
[2]K-69, Gyansarover, Aligarh- 202002, India
[3]Department of Geology, Dharam Samaj College, Aligarh - 202 002, India
[4]Palaeontology Lab, K.D.M. Institute of Petroleum Exploration, Oil and Natural Gas Corporation Limited, 9, Kaulagarh Road, Dehradun - 248195, India
[5]Retd. Senior Geologist, KDMIP, ONGC, Dehra Dun - 248 195, India
***Email**: talib04@gmail.com

Abstract

The present study attempts to interpret detailed palaeoecology and depositional environment employing foraminiferal assemblages recovered from a well exposed Middle to Late Jurassic sequence belonging to Chari Formation at Kaiya Dome, Kutch, India. The foraminiferal assemblages comprise forty two species dominated by the family Vaginulinidae. Several contemporary techniques are employed to draw a fairly accurate picture of the palaeobathymetry and palaeoenvironment of the Chari sequence exposed in the studied area which indicate that the deposition of the present sequence took place in an open ocean, shallow marine shelf environment ranging from mid to outer shelf with normal to slightly reduced salinity and high to slightly low level of dissolved oxygen. In order to detect finer and subtle palaeoenvironmental changes during the deposition of the studied sequence, these techniques were applied to divide the sequence into three palaeoecological units. The foraminiferal analysis suggests that the deposition commenced in a transgressing sea in the mid shelf region which gradually became deeper, attaining its maximum depth at the termination of the Chari deposition in the studied area.

Keywords: Foraminifera, Jurassic, Kutch, palaeoecology, depositional environment.

INTRODUCTION

The Jurassic foraminifers of the Indian region are not well investigated as compared to their Palaeogene counterpart. The only detailed studies available on Indian Jurassic foraminifera are by Bhalla and Abbas (1978), Bhalla and Talib (1991), Pandey and Dave (1993), Gaur and Talib (2009), Alhussein (2014), Talib et al. (2016), and Bhatt et al. (2016). All these studies are devoted mainly to taxonomy and biostratigraphy and no remarkable detailed palaeoecological investigation using Jurassic foraminifera has been attempted so far in this region, although some palaeoenvironmental studies have been carried out using mega-invertebrates (Fürsich et al., 1991, 1994, 2004; Pandey and Fürsich, 2001).

In India, Jurassic rocks are well developed in the Kutch region and are globally famous for varied and well preserved megafossils, especially ammonites which have been amply studied. However, microfossils entombed within these sediments, although varied and well-preserved, have not been adequately investigated and applied for various interpretations including palaeoenvironmental analysis.

Foraminifera are well established as reliable palaeoecological indicators and are widely used world over for interpretation of palaeoecology including palaeobathymetry and depositional environment of the enclosing sediments. The present study for the first time attempts a relatively detailed palaeoenviromental analysis of Jurassic rocks of Kutch utilizing and synthesizing various currently available techniques for palaeoenvironmental analysis employing foraminifera, in order to interpret a fairly accurate depositional environment of these sediments. A well developed sequence of Middle to Late Jurassic rocks exposed at Kaiya Dome, near Nara village (69^0 07' 30" : 23^0 38' 30"), about 75 km northwest of Bhuj, the district headquarter of Kutch region in Gujarat state of India is selected for this purpose (Fig. 1).

GEOLOGY AND STRATIGRAPHY

Kutch Basin is a peri-continental rift basin situated at the western margin of the Indian peninsula. The basin has a thick sediment in-fill ranging from Middle Jurassic to Holocene. Jurassic rocks occupy a prominent place in this thick pile of sediments and are exposed in three parallel east-west trending anticlinal ridges and an isolated rock mass in the east near Wagad. Of these, the middle ridge is the most prominent and is broken up into several domed outcrops. Most of the well exposed sequences of Jurassic rocks are situated in this ridge and are in the form of structural domes with their northern limbs having steeper dips. Following a widely used classification of Sastry and Mamgain (1971) modified by Kumar (1985), the Jurassic rocks of Kutch are grouped into four formations, viz., Patcham, Chari, Katrol and Umia in ascending order, ranging in age from Bajocian to Tithonian (Table 1). Kaiya Dome is one of the well exposed domes of the middle ridge in which only Chari and Katrol formations are exposed, the latter is devoid of foraminifera and hence excluded from the present study. The Chari Formation of Kaiya Dome is divided into seven lithounits from bottom to top, viz., Ky-1 to Ky-7 (Fig. 2).

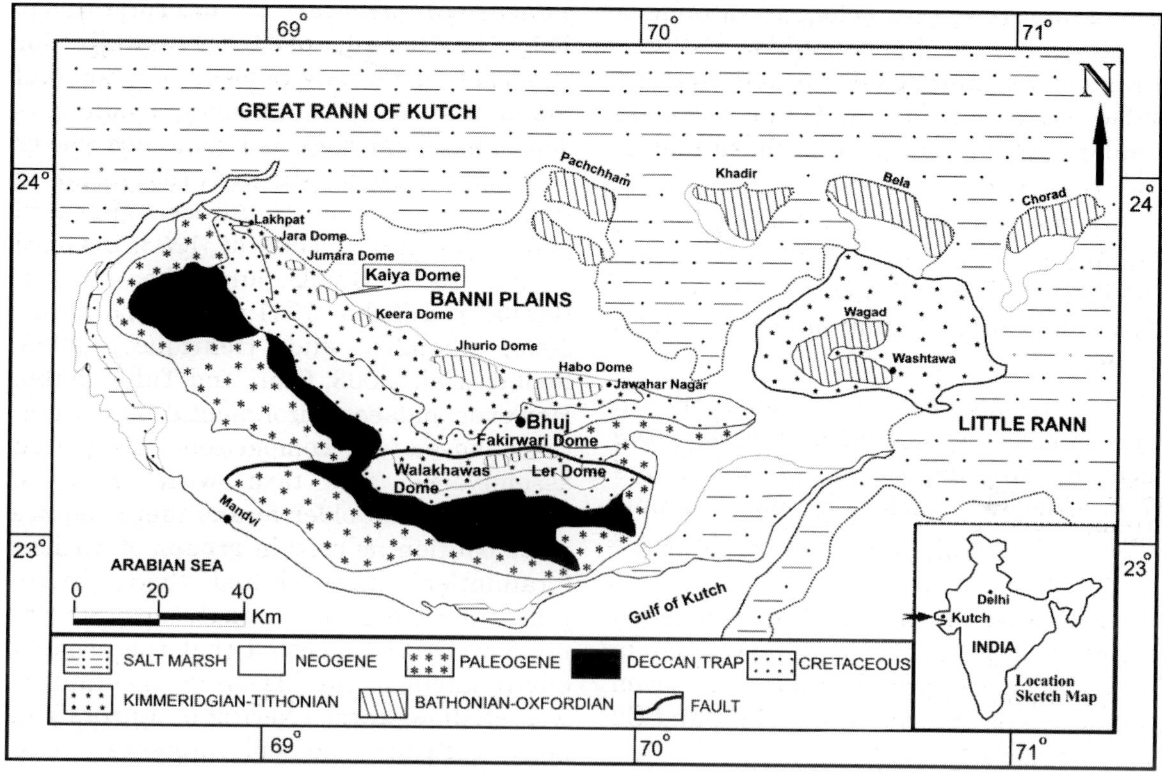

Fig.1. Geological map of Kutch showing the study area (after Fürsich et al., 1991).

MATERIALS AND METHODS

Twenty seven samples were collected from a well exposed section of the Chari Formation of Kaiya Dome, mainly on the basis of lithological variation. About 500gm of dried rock sample was crushed and boiled with sodium carbonate and the disintegrated material was washed and screened through a set of standard sieves of 30, 60 and 120 mesh and dried in oven. 10 gm material from 60 and 120 mesh fractions were subjected to picking under a stereozoom binocular microscope and all the foraminiferal tests were picked out. Some specimens were treated in an ultrasonic cleaner for further cleaning. One well preserved specimen of each species was photographed with the help of SEM make Zeiss model EVO-40 at Wadia Institute of Himalayan Geology, Dehradun, India.

FORAMINIFERAL COMPOSITION

Foraminiferal assemblages comprising forty two species belonging to eight families are recovered from the Chari Formation exposed at Kaiya Dome (Fig. 3). The assemblages are characterized by both calcareous and agglutinated species, the calcareous component constitutes 76.19% of the total foraminiferal species and represented by only hyaline forms. The ratio of agglutinated to calcareous forms is 1 : 3.2. The foraminiferal assemblages are dominated by the species belonging to family Vaginulinidae constituting 52.37% of the total species and represented by 22 species belonging to nine genera. Of the remaining, six species belong to Lituolidae (14.29%), five to Epistominidae (11.9%), three each to Hormosinidae and Nodosariidae (7.15% each), and one each to families Ammobaculinidae, Spirillinidae, and Lagenidae (2.38% each).

Of the 27 samples analyzed 24 yielded foraminifera whereas three are barren. Calcareous species are dominant in 21samples and agglutinated in three samples.

Genus *Lenticulina* is dominant in nine samples, *Citharina* and *Epistomina* each in two samples, and *Reophax* and *Vaginulinopsis* each in one sample whereas in nine samples no single genus is dominant.

Following the morphogroup scheme of Reolid et al. (2008) (Table 2), morphogroup K and J2 each are dominant in seven samples, G in two samples, and C1 and C2 each in only one sample.

According to life habit group (after Reolid et al., 2008), potentially deep infauna is dominant in nine samples, shallow infauna in eight samples, and epifauna in two samples whereas five samples do not show dominance of any life habit group.

Following the feeding strategy group (after Reolid et al., 2008), omnivores are dominant in 16 samples, detrivores and bacterial scavengers in three samples, and active herbivores and phyto-detrivores in two samples whereas in three samples dominance of any feeding strategy group is not observed.

PALAEOECOLOGY AND DEPOSITIONAL ENVIRONMENT

Voluminous literature on foraminiferal ecology and palaeoecology attests to the reliability of these organisms for palaeoenvironmental interpretations. They have been used successfully for interpreting various palaeoenvironmental conditions of the ancient sediments deposited as far back as Cretaceous (Sliter and Baker, 1972; Murray, 1991) and in some cases even up to Jurassic (Barnard and Shipp, 1981; Gebhardt, 1998; Barnard et al., 1981; Bhalla and Abbas, 1984; Bhalla and Talib, 1991; Talib and Gaur, 2005; Canales and Henriques, 2008; Reolid et al., 2008; Gaur and Talib, 2009). However, palaeoenvironmental interpretations based on comparison of modern assemblages and taxa with those of Cretaceous and older foraminifera require some caution as certain groups of ancient foraminifera have shifted their environmental preferences through time. Scheibnerova (1972) suggested that the original environment of the Mesozoic foraminifera was shallow-water (Neritic or Littoral) in both Tethyan and epicontinental seas. However, several workers on ecology and palaeoecology of foraminifera (Natland, 1957; Skolnick, 1958, Phleger, 1960; Burnaby, 1962; Ager, 1963) noted that some groups of foraminifera have changed their habitat since the Palaeogene.

Few workers (Said, 1950; Wall, 1960) advocated the use of individual foraminiferal species rather than genera or families in palaeoecological studies as different species of a genus may thrive in different environments, making them unreliable for palaeoecological interpretations. However, it is also true that palaeoecological interpretations based on individual species may be unreliable, especially in the pre-Palaeogene sediments, as many foraminiferal species occurring in older rocks may show entirely different environmental preferences from their modern representatives (Phleger, 1960). A number of researchers favoured the use of taxa higher than species for interpretations of palaeoenvironments of the Pre-Palaeogene rocks because most of the foraminiferal species of this age do not survive up to Recent (Bhalla and Abbas, 1978; Shipp and Murray, 1981; Gebhardt, 1998; Bhalla and Talib, 1991; Nagy and Seidenkrantz, 2003).

Table 1. Classification of Jurassic rocks of Kutch (after: Kumar, 1985).

Formation	Subdivisions	Characteristic fossils
	Deccan Traps	
	--------unconformity---------	
BHUJ	Umia Plant Beds	*Ptilophyllum* flora
	Ukra Beds (calcareous shales)	*Australiceras* sp.
	Sandstones and shales	Unfossiliferous
UMIA	*Trigonia* Beds	*Trigonia*
	Umia Ammonite Beds	*Virgatosphinctes* sp.
	Upper Katrol Shale	*Hildoglochiceras*
	Upper Katrol Sandstone	Mainly unfossiliferous
	Middle Katrol Sandstone	*Torquatisphinctes* sp. and *Katroliceras*
KATROL		
	Lower Katrol Shale	Ammonites
	Belemnites Marls of Jurun	*Belemnites*
	Kantkote Sandstone	*Euaspidoceras*, *Taramelliceras* sp.
	Dhosa Oolite	*Mayaites*, *Epimayaites*
	Athleta Beds	*Metapeltoceras*, *Peltoceras*, *Reineckeia*
CHARI	Anceps Beds	*Kinkeliniceras*, *Hubertoceras* *Indosphinctes*
	Rehmanni Beds	*Reineckeia tyranniformis*, *R. rehmanni*
	Macrocephalus Beds	*Macrocephalites*, *Dolichocephalites*
	Coral Beds	*Macrocephalites*, *Sivajiceras*, *Procerites*
PATCHAM	Shelly Limestone	*Macrocephalites*
	Kuar Bet Beds	*Corbula lyrata*, *Protocardia pseudotrapezium*
	---------unconformity----------	
Precambrian Basement (not exposed)		

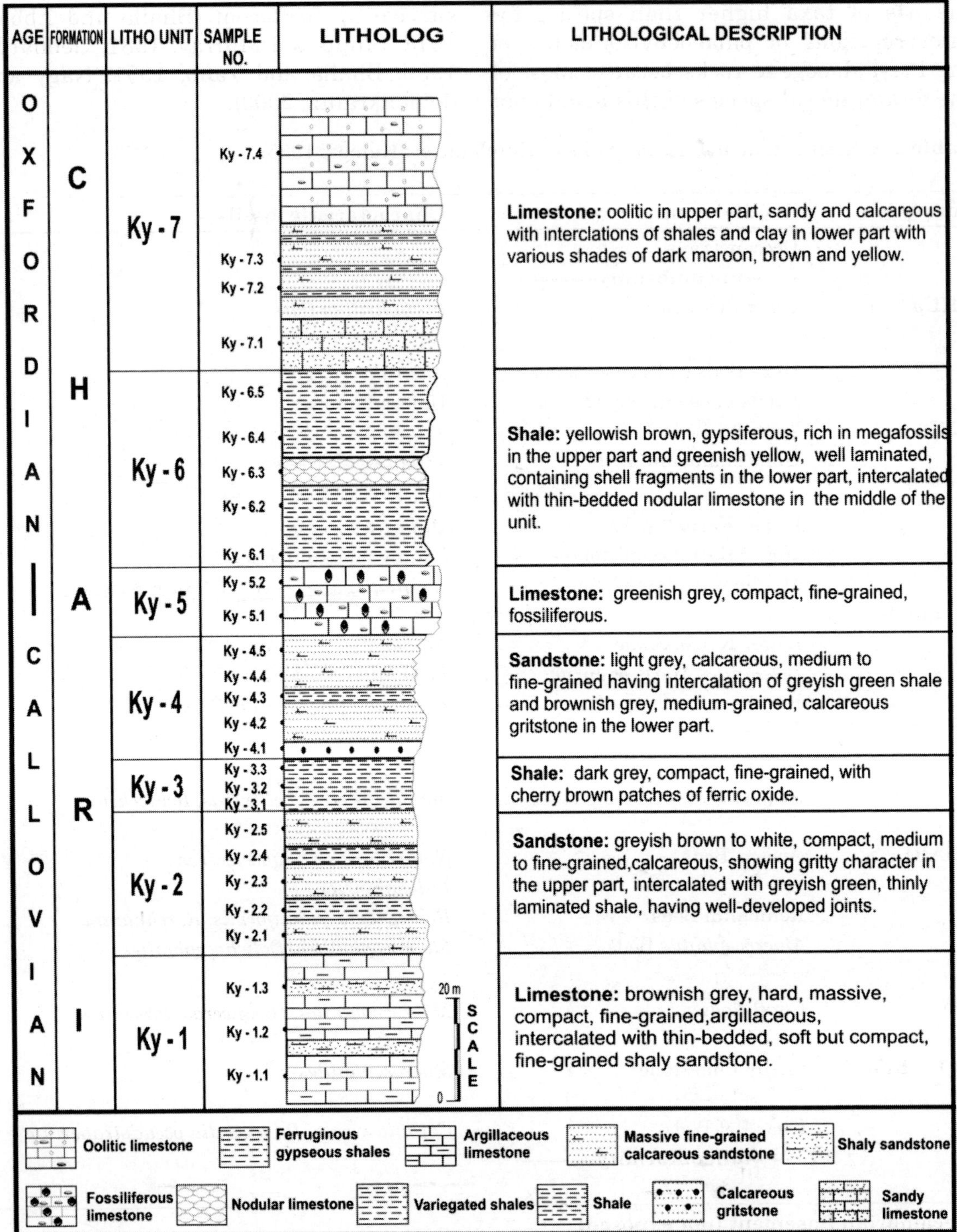

Fig.2. Litholog of Chari Formation, Kaiya Dome, Kutch.

FORAMINIFERA

STRATIGRAPHIC SEQUENCE

Age	Formation	Litho unit	Sample No.	Reophax aff. R. [illegible]	R. subgoodlandensis	R. aff. R. variabilis	Ammobaculites coprolithiformis	A. fisheri	A. suprajurassicum	Flabellammina bharatica	Triplasia emslandensis	T. kingakensis	Bulbobaculites aff. B. willowensis	Spirillina polygyrata	Dentalina aff. D. undulosa	Frondicularia kutchensis	Frondicularia cf. F. pseudoconcinna
OXFORDIAN	CHARI	Ky-7	Ky - 7.4												1		
			Ky - 7.3											3		1	1
			Ky - 7.2														
			Ky - 7.1														
		Ky-6	Ky - 6.5*														
			Ky - 6.4											2			
			Ky - 6.3*														
			Ky - 6.2														1
			Ky - 6.1														
		Ky-5	Ky - 5.2	1	1					1		1					
			Ky - 5.1	1	1					7		1	1				
		Ky-4	Ky - 4.5					1	1	2	5						
			Ky - 4.4							1		1		1			
			Ky - 4.3			1	1			1		1					
			Ky - 4.2								8						
CALLOVIAN			Ky - 4.1*														
		Ky-3	Ky - 3.3														
			Ky - 3.2											2			
			Ky - 3.1														
		Ky-2	Ky - 2.5											5			
			Ky - 2.4											1			
			Ky - 2.3											1			
			Ky - 2.2											6			
			Ky - 2.1											1			
		Ky-1	Ky - 1.3											1			
			Ky - 1.2														
			Ky - 1.1											1			

Sample No.	Lenticulina bulla	Lenticulina cf. L. humilis	L. muensteri	L. quenstedti	L. subalata	L. varians	Saracenaria cornucopiae	Astacolus anceps	A. aphrastus	A. filosa	A. pauperatus	A. stilla	Marginulina bhatiai	M. jumarensis
Ky - 7.4	2	1		2		1	1	5						
Ky - 7.3	1			3									1	
Ky - 7.2				1										
Ky - 7.1				1										
Ky - 6.5*														
Ky - 6.4	3	1	1	1	2									1
Ky - 6.3*														
Ky - 6.2					2									
Ky - 6.1														
Ky - 5.2									1					
Ky - 5.1			2	1										
Ky - 4.5											1	1		1
Ky - 4.4					1									
Ky - 4.3										1				
Ky - 4.2	4		1	2										
Ky - 4.1*														
Ky - 3.3	2													
Ky - 3.2				5	2									
Ky - 3.1														
Ky - 2.5				8	8			3			1			
Ky - 2.4					1									
Ky - 2.3				2	2									
Ky - 2.2				1	1	1								
Ky - 2.1				4	5									
Ky - 1.3	1				2									
Ky - 1.2	1													
Ky - 1.1	1			1	1									

Sample No.	Vaginulinopsis mirarensis	Vaginulinopsis aff. V. bartensteini	Citharina clathrata	C. rudocostata	Citharinella rhomboides	Citharinella aff. C. compara	Planularia tricarinella	Vaginulina woodi	Lagena sulcata	Epistomina cretosa	E. mosquensis	E. omninoreticulata	E. regularis	E. stellicostata	Total No. of Specimens
Ky - 7.4		1					2	1				1	1	2	21
Ky - 7.3															10
Ky - 7.2												4	7		12
Ky - 7.1									1	2					04
Ky - 6.5*															00
Ky - 6.4							1								12
Ky - 6.3*															00
Ky - 6.2												1	1		05
Ky - 6.1	1														01
Ky - 5.2															05
Ky - 5.1															14
Ky - 4.5			2	1	2	1					1				19
Ky - 4.4								1							05
Ky - 4.3															05
Ky - 4.2															15
Ky - 4.1*															00
Ky - 3.3											8				10
Ky - 3.2							2								11
Ky - 3.1			5	3	1										09
Ky - 2.5							5								30
Ky - 2.4			4	2			1								09
Ky - 2.3							1								06
Ky - 2.2							1				1				11
Ky - 2.1							9								19
Ky - 1.3							1								05
Ky - 1.2							1								02
Ky - 1.1			1	1			1								07

• Foraminifera absent

Fig.3. Frequency distribution of foraminifera in the Chari Formation, Kaiya Dome, Kutch.

Jurassic shallow water foraminiferal assemblages are nearly always dominated by vaginulinids and nodosariids which have evidently migrated from shallow to deeper waters since the Mesozoic (Barnard, 1948; Bhalla and Abbas, 1978; Coleman, 1981; Bhalla and Talib, 1991). Problems of taxonomic determination due to extreme inter- and intra-specific variation and poor preservation are other constraints in drawing accurate palaeoecological interpretations using Jurassic foraminifera. It is a well known fact that the Jurassic foraminiferal faunas show wide range of morphological variation which further creates hindrance in their accurate identification and this has resulted in the creation of a host of new species and long synonymies for Jurassic foraminiferal species.

Nevertheless, in spite of all these constraints, recently several advance techniques are available which may be used to draw fairly accurate palaeoecological interpretations based on foraminiferal assemblages. An excellent account of most of these techniques is provided by Valchev (2003). Most widely used methods for palaeoenvironmental interpretations using foraminifera include diversity indices (commonly Fisher α Index), triangular plot of test composition, planktic/benthic ratio (Tau Index), environmentally significant taxa, and relatively recently developed morphogroup concept.

Table 2. Morphogroups (after Reolid et al., 2008) in the foraminiferal assemblages of Chari Formation, Kaiya Dome, Kutch.

MORPHO- GROUP			TEST FORM	LIFE-STYLE	FEEDING STRATEGY	EXAMPLES
AGGLUTI-NATED	A		Tubular and unilocular	Epifaunal	Suspension-feeders	NILL
	B		Plano-convex irregular	Sessile epifaunal	Passive herbivores (suspension feeders?)	NILL
	C	C1	Elongated uniserial	Shallow to Deep infaunal	Detritivores bacterial scavengers	*Reophax*
		C2	Elongated uniserial, initial coiled phase	Shallow infaunal	Detritivores bacterial scavengers	*Ammobaculites, Flabellammina, Bulbobaculites*
		C3	Elongated biserial, triserial and high trochospiral	Shallow to Deep infaunal	Detritivores bacterial scavengers	*Triplasia*
	D		Rounded, globular and plano-convex planispiral to low trochospiral	Epifaunal	Active herbivores, omnivores, detritivores and bacterivores	NILL
	E		Discoidal coiled (unilocular)	Epifaunal (phytal)	Active herbivores, detritivores	NILL
CALCA-REOUS	F		Plano-convex and meandering, initial phase coiled	Sessile Epifaunal	Passive herbivores (suspension feeders?)	NILL
	G		Plano-convex trochospiral	Epifaunal	Primary weed fauna grazing herbivores	*Epistomina*
	H		Discoidal flattened (planispiral) and plano-convex (trochospiral)	Epifaunal	Primary weed fauna grazing herbivores/ phytodetrivores	*Spirillina*
	I		Discoidal flattened spiral, elongated	Epifaunal	Active deposit-feeders herbivores, detritivores	NILL
		J1	Elongated uniserial	Shallow infaunal	Active deposit-feeders, herbivores, bacterial scavengers	*Dentalina, Marginulina, Lagena*

MORPHO- GROUP	TEST FORM	LIFE-STYLE	FEEDING STRATEGY	EXAMPLES
J	Elongated flattened	Shallow infaunal	Active deposit-feeders, grazing omnivores	*Frondicularia, Saracenaria,*
J2				*Astacolus, Vaginulinopsis,*
				Citharina, Citharinella, Planularia, Vaginulina
K	Biconvex (lenticular) planispiral	Epifaunal to deep infaunal	Active deposit-feeders, grazing omnivores	*Lenticulina*

Out of the above mentioned palaeoecological methods, agglutinated/ calcareous ratio, foraminiferal morpogroups, including life habit groups and feeding strategy groups, have been used in the present palaeoecological analysis. Fisher's index could not be applied due to paucity of specimens in the studied samples. In addition to these, some available ecological data on individual foraminiferal genera and families have also been employed. Triangular plots of the three types of test composition could not be used here as the present foraminiferal assemblages include only two types of test compositions, i.e., agglutinated and calcareous hyaline. However, ratio of calcareous and agglutinated forms have been determined and interpreted. Furthermore, planktic/benthic ratio and Tau index could not be used as there is no planktic form present in the Kaiya Dome foraminiferal assemblages. Morphogroup analysis (after Reolid et al, 2008) has been carried out. Representative foraminiferal species belonging to various morphogroups occurring in the present assemblages are depicted in Plate 1. A synthesis of interpretations drawn on the basis of these methods and evidences facilitated in drawing a reasonably coherent and fairly accurate picture of the palaeoecology and depositional environment of the Chari sequence exposed at Kaiya Dome, Kutch.

The foraminiferal assemblages recovered from Kaiya Dome are dominated by the family Vaginulinidae. The study of relevant literature reveals that the depth distribution of vaginulinids is controversial. Some authors (Norton, 1930; Natland, 1933) inferred a moderately deep marine environment for this group while others (Barnard, 1948; Brouwer, 1969) suggested a shallow marine environment for the family. Nevertheless, it appears that they have shifted from near shore shallow marine waters of the Mesozoic to deep waters from Palaeogene onwards but always preferred an open marine environment with normal salinity condition (Bhalla and Abbas, 1978, 1984; Bhalla and Talib, 1991). The dominance of vaginulinids in the present foraminiferal assemblages suggests that it belongs to shallow water, open marine environment, preferably the deeper parts of the shelf. The presence of only calcareous hyaline and agglutinated forms and absence of porcelaneous forms rules out proximity of the present assemblages to the shallower inner shelf region (Valchev, 2003) because dominant porcelaneous tests are typical for inner shelf and in view of this the present assemblages, in all probability, belong to the middle and outer shelf regions.

The above mentioned palaeoecological and palaeobathymetric techniques were employed to identify a number of sub-assemblages, representing changes in palaeobathymetric and palaeoecological conditions during the deposition of the studied sequence. The foraminiferal analysis helped in dividing the studied sequence into three palaeoecological units (Fig. 4), the depositional environment of each is discussed below in considerable detail:

Palaeoecological Unit I (Lithounit Ky-1 to Ky-3, Samples Ky-1.1 to Ky-3.3)

This palaeoecolgical unit includes the Lithounit 1 to 3 of the exposed Chari sequence of the Kaiya Dome.

Presence of only hyaline taxa supports normal salinity as environments dominated by hyaline tests indicate normal salinity (Valchev, 2003).

Normal oxygen condition is suggested by the abundance of unornamented forms belonging to genera such as *Planularia, Astacolus,* and *Vaginulina* (Bernhard, 1986). Abundance of these forms (>50%) in this unit also supports the presence of high oxygenated realm during the deposition of this unit. Dominance of genus *Lenticulina* and morphogroup K suggests high level of dissolved oxygen. According to life habit, this palaeoecological unit is dominated by potentially deep infauna which suggests deposition in mid shelf environment (Reolid et al., 2008)

The Palaeoecological Unit I appears to have deposited in relatively shallow mid shelf environment, having normal salinity and well-oxygenated waters.

Palaeoecological Unit II (Lithounits Ky-4 & 5, Samples Ky-4.1 to Ky-5.2)

This Palaeoecological unit comprises of Lithounits 4 and 5 and shows considerable number of agglutinated forms. Agglutinated and calcareous groups are present in equal proportion. Although no genus is dominant, presence of considerable number of large coarse-grained species of *Reophax* and *Ammobaculites* as well as dominance of morphogroup C suggests deposition in the deeper shelf setting having reduced salinity and oxygen levels (Reolid et al., 2008; Kaminski et al., 1995). Dominance of potentially deep infauna indicates mid shelf environment whereas dominance of detrivores and bacterial scavengers suggests outer shelf condition (Reolid et al., 2008).

The Palaeoecological unit II appears to have deposited in the deepest part of the middle shelf near the boundary of the mid and outer shelf with slightly reduced salinity and low level of dissolved oxygen.

Palaeoecological Unit III (Lithounits Ky-6 & 7, Samples Ky-6.1 to Ky-7.4)

Palaeoecological unit III displays a dominance of calcareous hyaline forms suggesting normal salinity.

The dominant morphogroup of this unit is J2 followed by G. Dominance of J2 with unornamented taxa suggests normal oxygen level whereas morphogroup G is indicative of deeper water conditions in the outer shelf region as this morphogroup is represented by *Epistomina* in this palaeoecological unit. *Epistomina* is known to be associated with maximum flooding surfaces (Henderson and Hart, 2000; Oxford et al., 2000; Oloriz et al., 2003).

In view of the above, Paleoecological Unit III is inferred to have deposited in outer shelf environment with normal salinity and oxygen levels.

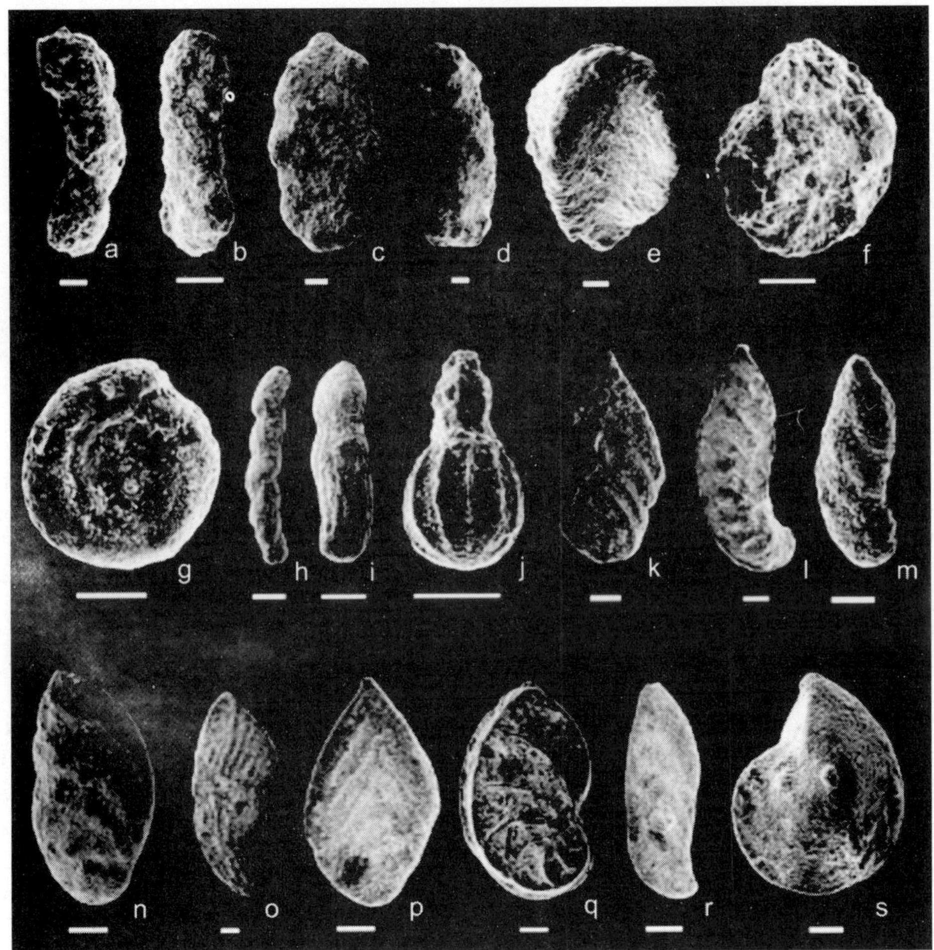

Plate 1. Representative foraminiferal species of various morphogroups (after Reolid et al. 2008) occurring in the Jurassic sequence, Chari Formation, Kaiya Dome, Kutch (Bar scale = 100 µm). Morphogroup C1: **a.** *Reophax* aff. *R. acosta* Bermudez 1949; Morphogroup C2: **b.** *Ammobaculites fisheri* Crespin 1953; **c.** *Flabellammina bharatica* Gaur and Talib 2009; **d.** *Bulbobaculites* aff. *B. willowensis* Hedinger 1993; Morphogroup C3: **e.** *Triplasia kingakensis* Loeblich and Tappan 1952; Morphogroup G: **f.** *Epistomina mosquensis* Uhlig 1883; Morphogroup H: **g.** *Spirillina polygyrata* Gümbel 1862; Morphogroup J1: **h.** *Dentalina* aff. *D. undulosa* Terquem 1870; **i.** *Marginulina jumaraensis* Bhalla and Gaur 1989; **j.** *Lagena sulcata* (Walker and Jacob 1798); Morphogroup J2: **k.** *Frondicularia* cf. *F. pseudoconcinna* Bartenstein and Brand 1951; **l.** *Saracenaria cornucopiae* (Schwager 1865); **m.** *Vaginulinopsis misrensis* Said and Barakat 1958; **n.** *Astacolus pauperatus* (Jones and Parker 1860); **o.** *Citharina clathrata* (Terquem 1864); **p .** *Citharinella rhomboidea* Loeblich and Tappan 1950; **q.** *Planularia tricarinella* (Reuss 1863); **r.** *Vaginulina woodi* (Bhalla and Abbas 1978); Morphogroup K: **s.** *Lenticulina bulla* (Lalicker 1950).

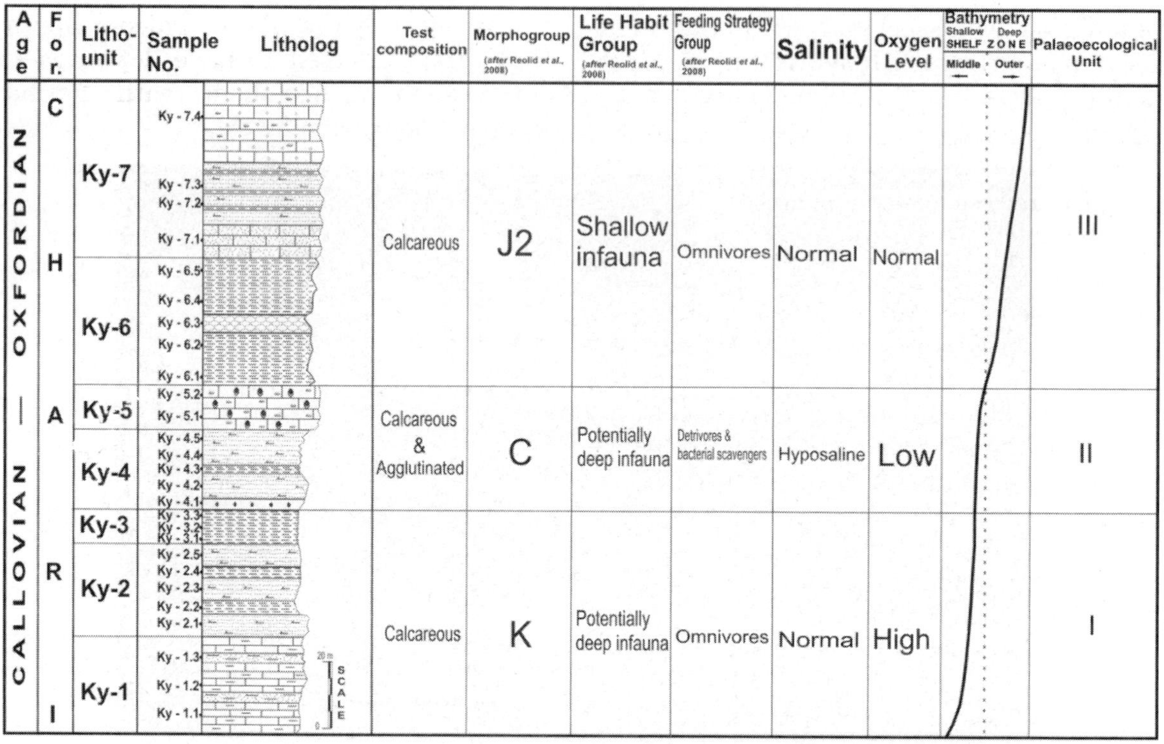

Age	For.	Litho-unit	Sample No.	Litholog	Test composition	Morphogroup (after Reolid et al., 2008)	Life Habit Group (after Reolid et al., 2008)	Feeding Strategy Group (after Reolid et al., 2008)	Salinity	Oxygen Level	Bathymetry Shallow Deep SHELF ZONE Middle · Outer	Palaeoecological Unit
OXFORDIAN	CHARI	Ky-7	Ky-7.4, Ky-7.3, Ky-7.2, Ky-7.1		Calcareous	J2	Shallow infauna	Omnivores	Normal	Normal		III
		Ky-6	Ky-6.5, Ky-6.4, Ky-6.3, Ky-6.2, Ky-6.1									
CALLOVIAN		Ky-5	Ky-5.2, Ky-5.1		Calcareous & Agglutinated	C	Potentially deep infauna	Detrivores & bacterial scavengers	Hyposaline	Low		II
		Ky-4	Ky-4.5, Ky-4.4, Ky-4.3, Ky-4.2, Ky-4.1									
		Ky-3	Ky-3.3, Ky-3.2, Ky-3.1		Calcareous	K	Potentially deep infauna	Omnivores	Normal	High		I
		Ky-2	Ky-2.5, Ky-2.4, Ky-2.3, Ky-2.2, Ky-2.1									
		Ky-1	Ky-1.3, Ky-1.2, Ky-1.1	20 m SCALE 0								

Fig.4. Main faunal parameters and inferred palaeoenvironment of Chari Formation, Kaiya Dome, Kutch.

Therefore, on the basis of foraminiferal assemblages, it may be inferred that the sediments of the Chari Formation exposed at Kaiya Dome, Kutch were deposited in shallow water, open marine conditions within middle to outer shelf environment with normal to slightly reduced salinity conditions and well to slightly lowered oxygen levels. The depositional environment shifted from middle to outer shelf regions along with the changes of ecological parameters, especially salinity and oxygen level.

The accumulation of sediments begins with a transgressive phase in the mid shelf region with normal salinity and well oxygenated waters (Palaeoecological Unit I). The site of deposition then shifted to deeper part of the mid shelf environment near the boundary of the mid and outer shelf (Palaeoecological Unit II) with slightly reduced salinity and low oxygen level. Thereafter, outer shelf conditions prevailed with further deeper water environment having normal salinity and oxygen levels, where the sea reached at its maximum flooding surface near the culmination of the Chari sedimentation in the studied area (Palaeoecological Unit III).

CONCLUSION

The Palaeoecological analysis employing foraminiferal assemblages recovered from the Middle to Late Jurassic Chari Formation exposed at Kaiya Dome, Kutch facilitated in drawing a fairly accurate interpretation regarding the depositional environment of the studied sequence. The foraminiferal assemblages suggest that the overall deposition of the present sequence took place in an open ocean, shallow marine shelf environment in the mid to outer shelf having normal to slightly reduced salinity and well oxygenated to slightly oxygen depleted waters.

Various palaeoecological techniques were used to detect finer and subtle palaeoenvironmental changes during the deposition of the studied sequence which helped in dividing the sequence into three palaeoecological units. The sedimentation in the area commenced in a transgressive phase in the mid shelf region with normal salinity and well oxygenated waters. The site of deposition then experienced deeper water conditions in the mid shelf environment but close to the outer shelf region with slightly reduced salinity and oxygen level. Outer shelf condition then prevailed with deeper shelf environment having normal salinity and oxygen levels, when the sea attained its maximum depth.

Acknowledgements: We thank the Chairman, Department of Geology, A.M.U., Aligarh and Head, Department of Geology, D. S. College, Aligarh for providing necessary facilities. Thanks are also due to Director, Wadia Institute of Himalayan Geology, Dehradun, India for SEM photography. Help received from N.K. Juiyal, Wadia Institute of Himalyan Geology, Dehradun, during SEM photography is gratefully acknowledged.

References

AGER, D.V. (1963) Principles of palaeoecology. McGraw-Hill, New York, 371p.

ALHUSSEIN, M. (2014) Taxonomy of the Middle Jurassic benthic foraminifera of the Kachchh basin, western India, Western India. Beringeria, v.44, pp.51-97.

BARNARD, T. (1948) The uses of foraminifera in Lower Jurassic stratigraphy. Rept.18th Sess., IGC, Great Britain, 15, pp.3-10.

BARNARD, T., CORDEY, W.G. and SHIPP, D.J. (1981) Foraminifera from the Oxford Clay (Callovian-Oxfordian) of England. Rev. Espanol Micropal., v.13, pp.383-462.

BARNARD, T. and SHIPP, D.J. (1981) Kimmeridgian foraminifera from the Boulonnais. Rev. Micropaléontol., v.24, pp.3-26.

BERNHARD, J.M. (1986) Characteristic assemblages and morphologies from anoxic organic rich deposits: Jurassic through Holocene. Jour. Foram. Res., v.16, pp.207-215

BHALLA, S.N. and ABBAS, S.M. (1978) Jurassic foraminifera from Kutch, India. Micropal. v.24, pp.160-209.

BHALLA S.N. and ABBAS, S.M. (1984) Depositional environment of the Jurassic rocks of Habo hills, Kutch, India. In: Benthos '83- 2nd Inter. Sym. Ben. Foram., Pau, France, pp.55- 60.

BHALLA, S.N. and TALIB, A. (1991) Callovian-Oxfordian foraminifera from Jhurio hill, Kutch, Western India. Rev. Paleobiol., v.10, pp.85-114.

BHAT, B.A., TALIB, A. and WASIM, S.M. (2016) Systematics, age, palaeoecology and palaeobiogeography f Jurassic fpraminifra from Fakirwari Dome, Kutch, Gujarat, India. Micropal., v.62, pp.171-194.

BROUWER, J. (1969) Foraminiferal assemblages from the Lias of northwestern Europe. K. Neder. Akad. Weten. Verh, Afd. Nat., v.25, pp.1-64

BURNABY, T.P. (1962) The palaeoecology of the foraminifera of the Chalk Marl. Paleontol., v.4, pp.599-608.

CANALES, M.L. and HENRIQUES, M.H. (2008) Foraminifera from the Aalenian and the Bajocian GSSP (Middle Jurassic) of Murtinheira section (Cabo Mondego, West Portugal): Biostratigraphy and paleoenvironmental implications. Mar. Micropaleontol. v.67, pp.155-179.

COLEMAN, B. (1981) The Bajocien to Callovian. *In*: JENKINS, D.G. and MURRAY, J.W. (eds.) Stratigraphic atlas of fossil foraminifera. Ellis-Horwood Ltd. Publ., London, pp.106-124.

FÜRSICH, F.T., OSCHMANN, W., PANDEY, D.K., JAITLY, A.K., SINGH, I.B. and LIU, C. (2004) Paleoecology of Middle to Lower Upper Jurassic Macrofaunas of the Kachchh Basin, Western India - An overview. J. Pal. Soc. India, v.49, pp.1-26.

FÜRSICH, F.T., OSCHMANN, W., SINGH, I.B. and JAITLY, A. K. (1991) Faunal response to transgressive regressive cycles: example from the Jurassic of western India. Palaeogeogr. Palaeoclimatol. Palaeoecol., v.85, pp.149-159.

FÜRSICH, F.T., PANDEY, D.K., OSCHMANN, W., JAITLY, A.K. and SINGH, I.B., (1994) Ecology and adaptive strategies of corals in unfavourable environments: examples from the Middle Jurassic of Kachchh Basin, Western India. N.Jb.Geol. Paläo. Abh., v.194, pp.269-303.

GAUR, K.N. and TALB, A. (2009) Middle-Upper Jurassic Foraminifera from Jumara Hills, Kutch, India. Rev. Micropaléontol., v.52, pp.227-248.

GEBHARDT, H. (1998) Benthic foraminifera from the Maastrichian Lower Mamu Formation near Leru (Southern Nigeria): Paleoecology and paleogeographic significance. J. Foram. Res, v.20, pp.76-89.

HENDERSON, A.S. and HART, M.B. (2000) The distribution of foraminiferida in the Oxfordian sequences of north Dorset, UK. GeoRes Forum, v.6, pp.311-320.

KAMINSKI, M.A., BOERSMA, A., TYSZKA, J. and HOLBOURN, A.E.L. (1995) Response of deep water agglutinated foraminifera to dysoxic conditions in the California Borderland Basins. Grzybowski Found. Publ., v.3, pp.131-140.

KUMAR, R. (1985) Fundamentals of Historical Geology and Stratigraphy of India. Wiley Eastern Limited, New Delhi, 254 p.

MURRAY, J. (1991) Ecology and Paleoecology of Benthic Foraminifera. Longmans, London.

NAGY, J. and SEIDENKRANTZ, M-S. (2003) New foraminiferal taxa and revised biostratigraphy of Jurassic marginal marine deposit on Anholt, Denmark. Micropaleont., v.49, pp.27-46.

NATLAND, M.L. (1933) The temperature and depth distribution of some Recent and fossil foraminifera in the southern California region. Bull. Calif. Univ. Scripps Inst. Oceanogr., Tech. Ser., v.10, pp.225-230 .

NATLAND, M.L. (1957) Paleoecology of West Coast Tertiary sediments. Mem. Geol. Soc. Am, v.67, pp.543-572.

NORTON, R.D. (1930) Ecologic relations of some foraminifera. Bull. Calif. Univ. Scripps Inst. Oceanogr., Tech. Ser., v.2, pp.331-388.

OLORIZ, F., REOLID, M. and RODRIGUEZ-TOVAR, F.J. (2003) Palaeogeographic and stratigraphic distribution of Mid-Late Oxfordian foraminiferal assemblages in the Prebetic Zone (Betic Cordillera, southern Spain). Geobios, v.36, pp.733-747.

OXFORD, M. J., HART, M. B. and WATKINSON, M.P. (2000) Micropalaeon-tological investigations of the Oxford Clay-Corallian succession of Dorset coast. Geosci.,v.10, pp.9-13.

PANDEY, J. and DAVE, A. (1993) Studies in Mesozoic foraminifera and chronostratigraphy of Western Kutch, Gujarat. Paleontogr. Indica, v.1, pp.1-221.

PANDEY, D.K. and FÜRSICH, F.T. (2001) Environmental distribution of scleractinian corals in the Jurassic of Kachchh, Western India. J. Geol. Soc. India, v.56, pp.479-495

PHLEGER, F. B. (1960) Ecology and distribution of Recent foraminifera. John Hopkins Press, 297 p.

REOLID, M., RODRIGUEZ-TOVAR, F.J., NAGY, J. and OLORIZ, F. (2008) Benthic

Foraminiferal morphogroups of mid to outer shelf environments of the Late Jurassic (Prebetic Zone, Southern Spain): Characterization of biofacies and environmental significance. Palaeogeogr. Palaeoclimatol. Palaeoecol., v. 261, pp. 280-299.

SAID, R. (1950) The distribution of foraminifera in the northern Red Sea. Contr. Cushman Found. Foram. Res., v.1, pp.9-29.

SASTRY, M.V.A. and MAMGAIN, V.D. (1971) The marine Mesozoic formations of India. Rec. Geol. Surv. India, v.101, pp.162-177.

SHEIBNEROVA, V. (1972) Foraminifera and their Mesozoic biogeoprovinces. 24th IGC, Sec. 7, pp.331-338.

SHIPP, D. and MURRAY, J.W. (1981) The Callovian to Portlandian. In: Jenkins DG and Murray (eds) Stratigraphic atlas of fossil foraminifera. Ellis-Horwood Ltd. Publ., London, pp.125-144.

SKOLNIÇK, H. (1958) Lower Cretaceous foraminifera of the Black hills area. Jour. Pal., v.32, pp.275-285.

SLITER, W.V. and BAKER, R.A. (1972) Cretaceous bathymetric distribution of benthic foraminifers. J. Foram. Res., v.2, pp.167-183.

TALIB, A. and GAUR, K.N. (2005) Foraminiferal palaeoecology, microfacies and palaeoenvironment of Middle-Upper Jurassic sequence, Jumara Hills, Kutch. Ind. J. Pet. Geol., v.14, pp.9-21.

TALIB, A., WASIM, S.M., SABEEHA and ARKAN, M. (2016). Jurassic foraminifera from the Dharang Member, Habo Formation, Habo Dome, Kutch, India : systematics, age, palaeoecology and palaeo-biogeography. J. Syst. Pal., pp.1-24 (online).

VALCHEV, B. (2003) On the potential of small benthic foraminifera as Paleoecological indicators: Recent advance. 50yrs Univ Min Geo "St. Ivan Rilski", Geol. Geophys., Sofia, v.46, pp.189-194.

WALL, J.H. (1960) Jurassic microfauna from Saskatchewan. Dept. Min. Res, Pet. Nat. Gas Br., Geol. Div., Saskatchewan, v.53, pp.1-299.

Micropaleontology and its Applications
Pradeep K. Kathal, Rajiv Nigam & Abu Talib (Editors)
Scientific Publishers (India), 2018, 135-162 pp.

An Overview of Ostracoda Studies from the Freshwater, Marginal Marine and Marine Ecosystems of Andaman and Nicobar islands and the Coasts of India

S.M. Hussain

Department of Geology, University of Madras, Chennai- 600 025, India
Email: smhussain7@hotmail.com

Abstract

It is evident from the analysis of literature that Recent freshwater, marginal marine water (brackish water such as estuaries, lagoons, salt marshes, mangroves etc.) and marine ostracods are not fully known from the Indian subcontinent, from such a long coast line of India, stretching about 7500 km in length and with a highly diversified terrestrial environments. Fresh water habitat such as rivers, streams, lakes, ponds and coastal habitats *viz.*, estuaries, creeks and bays, have been degraded to varying degrees, depending on the fluxes of pollutants and flushing characteristics of the area. Bioindicators, like ostracods can provide evidences for degradation and have proven useful in monitoring and assessing the coastal ecosystems. The research work carried out on the tiny Crustaceans, Ostracoda from the fresh water, marginal marine and marine waters of Andaman and Nicobar Islands, as well as east and west coasts of India has been reviewed and presented in this paper. The review shows that majority of the research papers deal with the taxonomy/biodiversity of Ostracoda and few papers with ecological implications and very few articles deal with the applied aspects such as pollution, biogeography, tsunami deposits and other related topics of micropaleontology. It is observed that the ostracod abundance and diversity is less along the east coast than the west coast of India, may be due to the monsoonal variability in the Arabian Sea. Among the 9 coastal states of India, the ostracod research work is comparatively more from the coast of Tamil Nadu. As a sequel, it is opined that ostracod research study pertaining to a very vast Indian coastal margin with many estuaries, intertidal, lagoonal, marshy zones and mangrove environments, provides a lot of scope to work on siltation, major, trace and isotope geochemistry of ostracod shells, paleoclimate, progradation/ degradation of the deltaic regions, and delineation of paleotsunami deposits etc. and future directions for further studies are given in this paper.

Keywords: Ostracoda, research studies, Indian coast, review, future directions.

INTRODUCTION

Ostracods, the tiny, calcareous bivalved Crustaceans, live in all types of aquatic environments, and found in the entire Phanerozoic Eon. They live in an environment in which the controlling factors are temperature, bottom topography, depth, salinity, dissolved oxygen, substrate, food supply and organic matter (Puri, 1966; Yassini and Jones, 1995). They are considered one of the important microfossil groups in interpreting and reconstructing the paleoclimate and/or paleoenvironment. The literature shows that the studies on Recent marine, marginal marine water (brackish water habitat such as estuaries, lagoons, salt marshes, mangroves etc) and freshwater ostracods are still in infancy and ostracods are not fully known from the Indian subcontinent. India has about 7500 km long coastal margin in addition to the estuarine, intertidal, lagoonal and marshy zones (Desai et al., 2000), which provides a lot of scope to work on ostracods. There are few studies on the taxonomy, distribution and ecology of Ostracoda from marine and marginal marine environments of the Indian sub-continental region. The review further reveals that out of the 9 coastal states of India, the research work is comparatively more from the coast of Tamil Nadu, and, therefore, an overview of the same is presented habitat-wise and theme-wise, in this paper.

Honnappa (1983) gave a brief account on the work done on Quaternary and Recent ostracods, east and west coasts of India; Bhatia (1984) reviewed the reports on the ostracods of the Lower Paleozoic to Quaternary sediments of Indian subcontinent, which included the Recent ostracods from beaches regimes and discussed their paleozoogeographic and paleoecologic implications; and Hussain and Rajeshwara Rao (1996) presented gave an account with a checklist of Recent Indian marine ostracods and discussed their zoogeographic distribution and faunal affinity. Thereafter, much work has been carried out on ostracods during the past two decades. The existing information on Recent ostracod taxonomy and ecology from this part of the world is scanty. So far, only sporadic reports of research work that have been carried out to bring out an understanding of the faunal diversity and relationship of the ostracod fauna to the substrate as well as other ecological parameters that are congenial for the population abundance in the shallow inner shelf regions. The present reviews deals with the taxonomy, ecology and other related applied studies of ostracods reported from different locations of Indian coast and Andaman Islands (Fig. 1), prior and post to 1996. A total of 209 ostracod species reported from the marine and marginal environs of India (Table 1) among which 62 are from east coast and 56 are from the west coast of India exclusively and the remaining 90 species are common to both the coasts in their occurrence. Since, many marine and marginal marine areas are studied when compared to west coasts higher number of taxa is recorded from the east coast of India. Otherwise, the maximum biodiversity along the west coast than east coast of India is observed comparatively, which may be due to the monsoonal variability in the Arabian Sea and other congenial environmental parameters.

MARINE OSTRACODA

Studies on taxonomy and ecology marine ostracods of east coast of India

High species diversity, a useful indicator of maturity in communities and of ecological trends shows that areas supporting diverse species have, generally, better life conditions. Brady (1886) studied the dredged samples from Calpentyn, Gulf of Mannar, and reported 35 species of Ostracoda; Gramman (1975) gave a casual reference to some Indian ostracods; and Misra and Shrivastava (1979) recorded 25 species of Ostracoda from the bottom sediments collected off the coast of Tuticorin.

Hussain (1998) for the first time gave a detailed systematic account of 52 ostracod species from the east coast of India, along with their ecological studies, from the sediments of Gulf of Mannar, off Tuticorin, based on a season-wise study of environmental parameters and ostracod population size and analysis of the distribution of the fauna in relation to ecological factors (Hussain et al., 1996a and 1996b) and discussed the faunal diversity from the Gulf of Mannar. Hussain et al. (1998) reported two new species (*Triebelina tutitcorensis* and *Hemicytheridea khoslai*) from off Tuticorin.

Naidu et al. (1997) presented the diversity and distribution of Recent Ostracoda from the shelf sediments off Pentakota and Kalingapatnam, near Visakhapatnam; Sridhar et al. (1998) studied the responses of Recent benthic Ostracoda, inner shelf sediments, Palk Bay, off Rameswaram, east coast of India; Mohan et al. (2001) identified 51 species and discussed the ecology of ostracoda, Bay of Bengal, off Karikattukuppam and also reported four new species (*Hemitrachyleberis siddiquii, Puricythereis whatleyi, Neocythromorpha reticulata* and *Pterigocythereis chennaienis*) and noted that salinity, water depth and calcium carbonate content control the overall abundance and distribution of ostracod populations in this region; and Mohan et al. (2002) observed that the living/total population sizes are more during April than July, attributable to the high temperature, salinity, and dissolved oxygen of the bottom waters. Sridhar et al. (2002) illustrated 48 species, Palk Bay, off Rameswaram and observed thatsiltysand and sand substrates are favourable for species, sediments of Palk Bay, off Rameswaram and deduced a relatively faster rate of sedimentation in the inner-shelf off Rameswaram based on the occurrence of carapace and valve ratio along with recording species, namely-*Kotoracythere inconspicua, Neocytheretta snellii* and *Paradoxstoma subtile* for the first time in this region. Twenty-one species belonging to 18 genera have been reported from 22 surface sediment samples collected through SSK-35 from the coastal region along the northwestern part of the Bay of Bengal and reported *Echinocythereis* sp., *Cytherelloidea leroyi* and *Keijia demissa* occurring abundantly (Mohammed Nishath et al., 2015).

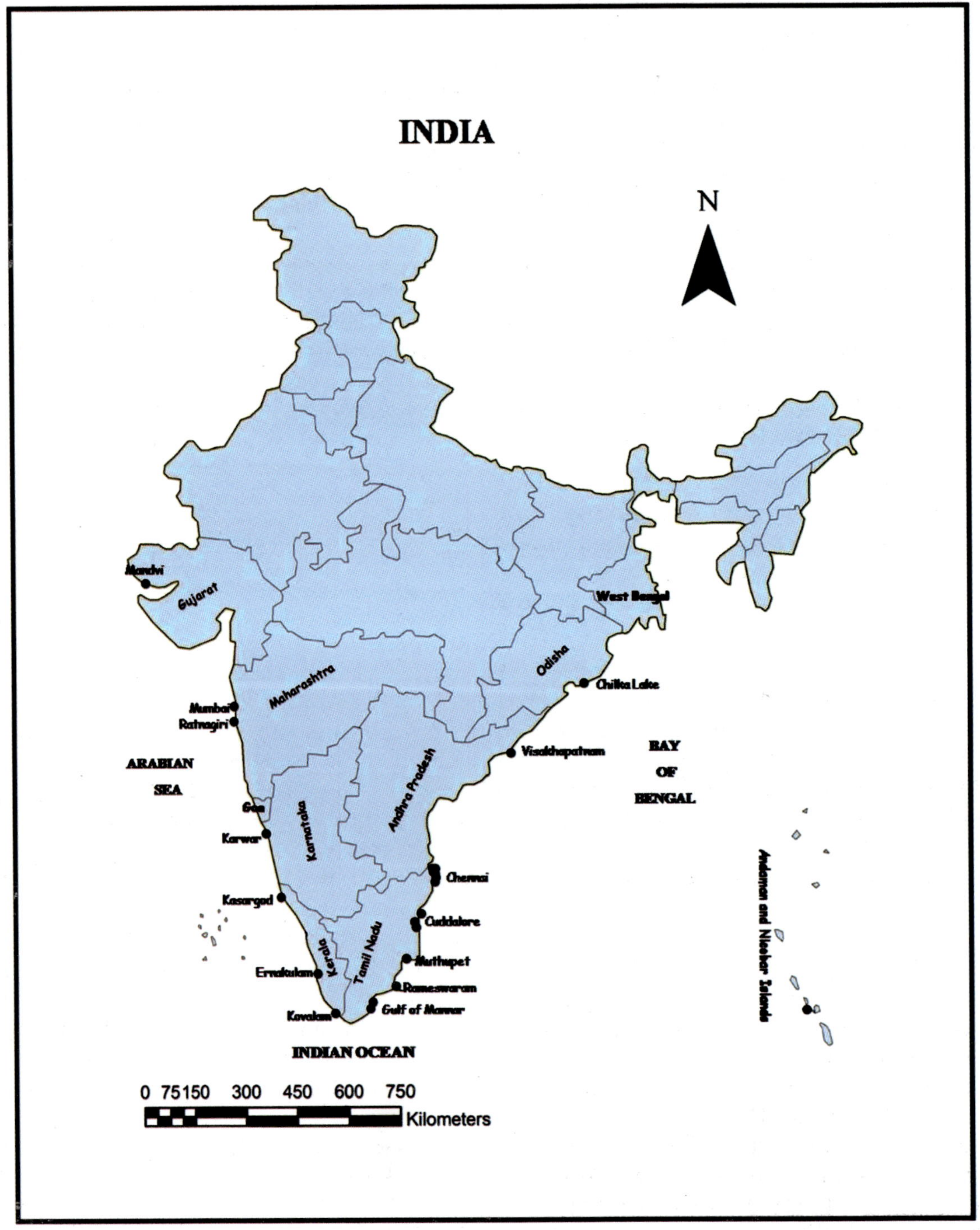

Fig.1. Locations along the east coast of India including Andaman Islands, reviewed in the study.

Table 1. ARE- Adyar River Estuary. **ACT**- Andhra coast. **CHL**-Chilka lagoon. **ENE**- Ennore creek. **GOM**-Gulf of Mannar, **IPP**-Inner shelf off Pulicat & Palar, **MLM**-Mullipallam lagoon, near Muthupet, **CRE**- off Cuddalore, **KPM**- off Karikkattukuppam, **PBR**- Palk Bay, off Rameshwaram, **PMS**- Pitchavaram mangroves, **TME**- Tamirabarani estuary, **KET**- Kerala coast, GUT- Gujarat coast, and **KAT**- Karnataka coast.

S.No.	Name of Ostracoda	East coast of India												West coast of India		
		ARE	ACT	CHL	ENE	GOM	IPP	MLM	CRE	KPM	PBR	PMS	TME	KET	GUT	KAT
1	*Actinocythereis roesmanni*													P		
2	*A. scutigera*					P	P		P	P				P	P	P
3	*Alocopocythere kendengensis*													P		
4	*A. bhatiai*															P
5	*A. reticulata indoaustralica*						P				P			P	P	P
6	*Anchistrocheles bradyi*	P														
7	*A.* sp.cf.*A.mcquadei*					P										
8	*Arculacythereis indoarabica*						P							P		
9	*Atjehella semiplicata*													P	P	P
10	*Bairdia faveolata*	P												P		
11	*Bairdoppilata (B.) alcyonicola*	P				P	P		P	P	P		P	P		P
12	*B. carinata*													P		
13	*B. cushamani*													P		
14	*B.(B) paraalcyonicola*													P		P
15	*Baltraella banerjii*															P
16	*Basslerites liebaui*	P	P		P			P							P	P
17	*Bradleya andamanae*	P														P
18	*B. dictyon*													P		
19	*B. (Quasibradleya) plicocarinata*	P														
20	*Bythoceratina alatum*															P
21	*B. mandviensis*					P				P			P	P	P	
22	*B. oertlii*															P
23	*B. siddiquii*															P
24	*Callistocythere flavidofusca intricatoides*	P			P		P	P		P			P	P	P	
25	*C. undulatifacialis*								P							
26	*Carinocythereis (C) batei*														P	
27	*C. (C) hamata*					P								P	P	P
28	*C. (Tandonella)* cf. *indica*					P								P	P	P
29	*C. simsoni*	P														
30	*Caudites javana*	P			P	P	P	P	P	P	P		P	P	P	P
31	*C. rectangularis*		P													
32	*C. rosaliensis*										P					
33	*C. sublevis*					P					P			P		P
34	*C. keeni*															P

S.No.	Name of Ostracoda	East coast of India												West coast of India		
		ARE	ACT	CHL	ENE	GOM	IPP	MLM	CRE	KPM	PBR	PMS	TME	KET	GUT	KAT
35	*Chrysocythere keiji*	P	P		P	P	P			P	P		P	P	P	P
36	*Copytus coramandalensis*		P													
37	*C.* sp. cf. *rara*													P	P	
38	*Costa quadricostatum*	P														
39	*Cuneocythere hornelli*	P														
40	*Cushmanidea guhai*					P				P	P			P	P	
41	*Cyclocypris laevis*											P				
42	*Cyprideis mandviensis*				P								P		P	
43	*Cypridopsis obesa*		P					P				P				
44	*C.vidua*	P														
45	*Cyprinotus (Heterocypris) incongruensis*													P		
46	*C. salinus*	P											P			
47	*Cypris subglobosa*													P		
48	*Cythere* sp.cf. *cribriformis*													P		
49	*C.reticulata indoaustralica*															P
50	*C.subcuneata*	P														
51	*C.thompsoni*	P														
52	*Cytherella dictyon*				P	P					P					
53	*C. hemipunctata*									P						P
54	*C.pulchra*	P														
55	*C.cf.punctata*													P		
56	*C. semitalis*						P							P	P	P
57	*Cytherelloidea javana*													P		
58	*C. leroyi*	P	P		P	P	P	P		P	P		P	P		P
59	*C.panagaensis*													P		
60	*C. praecipua*					P			P							
61	*C.* sp. aff. *C. trifurcata*					P										
62	*C. subreticulata*		P													
63	*C. whatleyi*															P
64	*Cytheretta maddocksae*														P	
65	*C. subradiosa*													P		
66	*C. trifurcata*									P				P		
67	*Cyprinotus salinus*	P														
68	*Cytheromorpha knoxi*	P														
69	*Cytheropteron alatum*	P														
70	*C. rhombiformis*													P		P
71	*Cytherura halyi*	P														
72	*C. subradiosa*													P		
73	*C. sumatraensis*		P													
74	*Darwinula stevensoni*												P			

S.No.	Name of Ostracoda	East coast of India												West coast of India		
		ARE	ACT	CHL	ENE	GOM	IPP	MLM	CRE	KPM	PBR	PMS	TME	KET	GUT	KAT
75	*Echinocythere ruperti*	P														
76	*Eopaijenborchella subcaudatum*	P														
77	*Falsocythere maccagnoi*									P				P		P
78	*Hemicythridea keyseri*														P	
79	*H. bhatiai*		P		P		P	P				P	P			
80	*H. khoslai*					P	P				P		P			
81	*H. jaini*															P
82	*H. paiki*		P		P		P	P				P	P		P	P
83	*H. reticulata*	P				P		P			P		P	P		
84	*Hemicytherua subulata*					P							P			
85	*H. truncatula*	P														
86	*Hemikrithe orientalis*									P						
87	*H. peterseni*				P	P	P							P	P	
88	*Hemitrachylebris nealei*															P
89	*H. siddiquii*						P			P						
90	*Hentryhowella (Neohenryhowella) hartmanni*												P	P	P	P
91	*Ilyocypris bradyi*		P					P								
92	*I. gibba*		P					P					P			
93	*Jankeijcythere mckenzeii*		P		P		P	P				P	P	P		P
94	*Kalingella mckenzii*		P	P	P			P				P	P			
95	*Kangarina abyssicola*				P						P				P	
96	*Keijcyoidea praecipua*				P						P			P		
97	*Keijella darwini*		P				P							P		
98	*K. karwarensis*					P	P	P		P	P			P		P
99	*K. neali*													P	P	P
100	*K. oertlii*		P	P												
101	*K. papuensis*		P													
102	*K.reticulata*	P			P	P	P	P		P	P		P			P
103	*K. whatleyi*				P								P	P		P
104	*Keijia. demissa*		P			P	P			P	P		P			
105	*Kotoracythere inconspicua*										P					
106	*Krithe bartonensis*	P														
107	*K, kroemmelbeini*													P	P	
108	*Lankacythere coralloides*				P	P	P				P			P		P
109	*L. reticulata*				P											P
110	*Leguminocythereis donnani*	P														
111	*L. papuensis*													P	P	
112	*Leptocythere pulchra*													P		
113	*Limnocythere imthurni*	E														

S.No.	Name of Ostracoda	East coast of India											West coast of India			
		ARE	ACT	CHL	ENE	GOM	IPP	MLM	CRE	KPM	PBR	PMS	TME	KET	GUT	KAT
114	*Loxoconcha cericinata*									P						
115	*L. gibbera*	P														
116	*L. gruendeli*							P		P			P	P	P	P
117	*L. guhai*		P													
118	*L. liljeborgii*				P	P					P	P		P		P
119	*L. mandviensis*							P	P	P	P		P	P	P	P
120	*L. megapora indica*				P		P	P					P	P	P	P
121	*L. tamarindoidea*	P														
122	*L. tekkaliensis*		P										P	P		
123	*L. tewarii*			P												
124	*Loxoconchella anomala*									P	P			P	P	P
125	*L. dorsobullata okhaensis*													P		
126	*Macrocyprina decora*					P				P	P		P			
127	*Microceratina punctata*									P						
128	*Miocyprideis spinulosa*					P	P	P		P						
129	*M. okhaensis*													P		
130	*Moosella cochinensis*													P		P
131	*M.paiki*															P
132	*Mutilus pentoekensis*	P			P	P		P		P	P		P	P		P
133	*M.khoslai*															P
134	*Neocaudites purii*													P		
135	*Neocytheromorpha goguleruensis**		P										P			
136	*N. reticulata*	P								P			P	P		
137	*N.* sp.cf. *N.indoarabica*					P							P			
138	*Neocytheretta murilineata*	P			P	P	P	P		P	P	P		P		
139	*N. snellii*						P			P						
140	*Neonesidea guildertonensis*														P	
141	*N.woodwardiana*													P		
142	*N.* sp.aff. *N.cracenticlavula*					P										
143	*Neomonoceratina columbiformis*	P														
144	*N. delicata*		P					P						P	P	
145	*N. halyii*													P		
146	*N. indica*	P														
147	*N. iniqua*	P	P		P	P	P	P	P	P	P	P	P	P		P
148	*N. jaini*	P	P		P	P	P	P		P	P		P			
149	*N. microreticulata*			P										P		
150	*N. porocostata*				P	P	P	P	P	P	P	P	P	P		
151	*N. spinosa*	P			P	P							P			
152	*Neonesidea antonbrunna*	P														

S.No.	Name of Ostracoda	East coast of India												West coast of India		
		ARE	ACT	CHL	ENE	GOM	IPP	MLM	CRE	KPM	PBR	PMS	TME	KET	GUT	KAT
153	*N. elegans*															P
154	*N. guildertonensis*													P		
155	*N. hartmanni*															P
156	*N. schulzi*													P		
157	*N. tenera*	P														
158	*N. woodwardiana*															P
159	*Neosinocythere dekrooni*		P		P		P	P			P	P				
160	*N. indica*													P		
161	*N. mckenziei*		P													
162	*N. murilineata*													P		
163	*Ornatoleberis morkhoveni*				P	P	P				P			P		
164	*O. quiloniensis*					P										
165	*Paijenborchella iocosa*	P														
166	*P. keiji*		P										P			
167	*P. kingmai*															P
168	*P. malaiensis*		P													
169	*Paijenborchellina caudata*		P													P
170	*P.indoarabica*					P	P						P	P		P
171	*P. prona*	P			P					P	P			P		
172	*P. reticulata*		P													
173	*Paracypris polita*	P														
174	*Paracytheridea psuedoremanei*									P				P		P
175	*P. longicaudata*													P		
176	*P.*sp.cf. *aqabaensis*					P										
177	*Paracytheroma ventrosinuosa*					P				P			P	P		P
178	*Paradoxostoma bhatiai*		P		P			P		P			P			
179	*P.* sp.cf. *P. subtile*										P					
180	*Paranesidea fracticorallicola*					P				P				P		
181	*Phlyctenophora bhatiai*			P												
182	*P. orientalis*	P	P		P		P	P		P	P		P	P		P
183	*P. karwarensis*															P
184	*P. zealandica*					P			P					P		
185	*Propontocypris (P.)crocata*					P	P			P	P		P			
186	*P. (S.)bengalensis*		P		P	P		P		P	P		P	P		P
187	*P. chennaiensis*									P						
188	*P. maddocksii*															P
189	*P. mannikerii*															P
190	*P. whatleyi*									P				P		
191	*Puricythereis indica*															P
192	*Quadracythere truncula*	P														

S.No.	Name of Ostracoda	East coast of India												West coast of India		
		ARE	ACT	CHL	ENE	GOM	IPP	MLM	CRE	KPM	PBR	PMS	TME	KET	GUT	KAT
193	*Q. willisi*	P														
194	*Quasibradleya* sp. cf. *Q. pliocarinata*					P										
195	*Radimella* sp. cf. *packardi*													P		
196	*Ruggeria darwini*													P		P
197	*R. indoiranica*													P		P
198	*R. indopacifica*															P
199	*R. zhaoi*															P
200	*Semicytherura contraria*					P	P			P	P					
201	*Spinoceratina spinosa*									P			P			
202	*Stigmatocythere indica*	P	P		P		P	P		P	P	P	P	P		
203	*S. kingmai*				P		P			P	P		P	P		
204	*Tanella gracilis*	P	P	P	P	P	P	P		P	P	P	P	P		P
205	*Triebelina sertata*										P			P		
206	*T. tuticorensis*					P										
207	*Xestoleberis curta*	P														
208	*X. malaysiana*												P	P		
209	*X. variegata*				P	P	P	P		P	P		P			P

Studies on Taxonomy and Ecology Marine Ostracods of West Coast of India

Guha (1968a) recorded 32 taxa from the Archipelago Series, exposed in Interview and Guitar Islands of the Andamans; James (1973a, b) recorded 4 species of pelagic Ostracoda from off the Kerala coast; and Maddocks (1969a, b) presented a report on the ostracods from the Bengal coast, Cochin, Mangalore coast and Andaman Island studied during the Indian Ocean Expedition and also recorded a new species- *Propontocypris (Schedopontocypris) bengalensis*, which occurs commonly all along the coasts of the Bay of Bengal and Arabian Sea.

Honnappa (1975) presented the morphology, taxonomy and a statistical account of *Actinocythereis tumefacentis* (Lubimova and Guha) from the Mangalore harbour area. Jain (1978) described and illustrated 56 taxa from the beach sands of Mandvi, Kutch, Gujarat and reported one new genus- *Vijaiella*, one new subgenus- *Carinocythereis* (*Tandonella*) and 16 new species and later (1981) documented 34 species from the beach sands of Kerala; and Honnappa and Venkatachalapathy (1978) dealt with pyritisation and colour variation observed in the ostracod carapaces recovered from Mangalore beach and in the subsequent year (1979), they made a mention of predation noticed on the shells of *Leguminocythereis*.

Bhatia and Kumar (1979) recorded 13 ostracod species, including a new genus, *Jainella*, and one new subgenus, *Neohenryhowella*, from the vicinity of Karwar area (Anjadiv Island); Guha (1980) listed 18 Recent species, belonging to 4 families from the sea bed samples of Bombay and Ratnagiri; Tapaswi (1980) recorded 12 species from the beach sediments of Nalgaon and Adgarn (Maharashtra); and Honnappa and Abrar (1983) reported 4 species of family Bairdiidae and 21 species (including six new forms) from the coastal sediments of Bhatkal and Karwar areas but a thorough taxonomic and morphologic revision is essential.

Honnappa et al. (1984) reported a new genus *Neomangaloria*, from the Mangalore to Karwar coastal sediments as a junior synonym of *Phlyctenophora*. *Actinocythereis tumefacentis* recorded by Honnappa (1975) from the same area and has been considered conspecific with *Henryhowella* (*Neohenryhowella*) *hartmanni*, by Bhatia and Kumar (1979). Vaidya and Mannikeri (1994) reported a bulk assemblage of 80 species from around Karwar and discussed their faunal affinity and zoogeography and Vaidya *et al.* (1995) studied the relationship between the bottom sediments and recent benthic Ostracoda from Karwar. Rajesh Reghunath et al. (1999) reported 21 species belonging to 17 genera from the inner shelf off Kasargod, west coast of India. A good assemblage of 61 inner shelf species from Malabar coast, Kerala were reported inferring a slow rate sedimentation based C/V ratio (Gopalakrishna et al., 2007). Twenty tropical, brackish to epineritic ostracods were reported from the backwaters of Ernakulum, Kerala (Hussain et al., 2013) including a few shallow marine species drifted into the backwaters due to tidal influence. Of the 16 species of Ostracoda belonging to 14 genera encountered in beach sands of Kovalam, Thiruvananthapuram, Kerala. Most of the species are either moderately calcified, pitted or highly ornate forms, and also ostracod carapaces are broken, abraded indicating the high surf action of beach, adopting themselves under a turbulent/agitated environmental condition (Hussain et al., 2016).

MARGINAL MARINE OSTRACODA

Studies on Taxonomy and Ecology of Brackish Water Ostracods of India

When compared to shallow marine ostracoda, the brackish water species are still less explored from the Indian region. Ratnam and Rao (1965) recorded 3 genera (*Perissocytheridea, Candona* and *Limnocythere*) from the Chilka Lake; Jain (1976) illustrated and described 6 species from the Chilka Lake, Orissa including 4 new species (*Kalingella mckenzieii, Loxoconcha tewarii, Phlyctenophora bhatiai* and *Keijella oertlii*); and Annapurna and Rama Sarma (1979 a, b; 1981) and Annapurna (1981) studied taxonomic and ecological parameters of brackish water along with 2 new species recorded (*Tanella indica* and *Bradleya ganapatii*), which appear to be conspecific with *T. gracilis* and *Chrysocythere keiji*, respectively, from the Bimili-Balacheruvu tidal stream.

Though Khosla et al. (1982) reported 58 Ostracoda species from the sediments of Miani lagoon, however many species were kept under open nomenclature. Annapurna and Rama Sarma (1982) correlated the distribution of Ostracoda with sedimentological parameters, like-sand-silt-clay percentage, organic matter content and physical environmental parameters, like- bottom water temperature, dissolved oxygen and salinity in the Balacheruvu tidal stream and Bimili backwaters. From these areas, they later (1986) recorded 3 new species (*Leptocythere andhraensis, Tanella estuarii* and *T. kingmai*). Subsequently, Varma et al. (1993) revised and opined that these species are conspecific to *Hemicytheridea andhraensis* and *T. gracilis*, respectively. Annapurna and Rama Sarma (1987) identified 40 more species belonging to 27 genera from these marginal marine environments. Bhatia (1991) pointed out many inconsistencies and incorrect interpretations made in the taxonomic descriptions of some Ostracoda reported by Sreenivas et al. (1991) from Pulicat Lake. Varma et al. (1993) recorded 25 species from the Tekkali Creek, east coast of India including 3 new species (*Hemicytheridea bhatiai, Loxoconcha tekkaliensis* and *Neomonoceratina jaini*) and recorded *Neosinocythere dekrooni and Paijenborchella (Eopaijenborchella) malaiensis* for the first time from the Indian waters. Shyam Sunder et al. (1995) reported 33 species from the Goguleru Creek and adjacent beaches of the east coast of India including 5 new species (*Copytus coromandalensis, Loxoconcha guhai, Neocytheromorpha goguleruensis, Paijenborchella keiji* and *Paradoxostoma bhatiai*) and also recorded *Keijia demissa* for the first time from Indian waters. Kumar and Hussain (1997) reported for the first time 10 species from Pitchavaram mangroves, Tamil Nadu.

Hussain and Mohan (2000) reported 26 ostracod species from the Adyar River estuary and later also inferred fast rate of sedimentation in the estuary based on 'carapace valve ratio'(Hussain and Mohan, 2001); Arul et al. (2003) identified and illustrated 29 species with ecological conditions from the sediments of Pitchavaram mangroves; Hussain et al. (2005) recorded 22 benthic specie from Tamiraparani estuary, Punnakayal, Southeast coast of India; and Bhandari and Singh (2006) recorded 24 species belonging to 14 genera from the Krishna river estuary and Gautami-Godavari estuary and discussed upon the frequency and diversity in the upper, lower estuaries. Environmental implications of ostracod distribution in the Ennore creek and Tamiraparanai estuary were presented by Elumalai et al., 2010 and Ganesan and Hussain, 2010, respectively.

FRESHWATER OSTRACODA

Studies on Taxonomy and Ecology of Freshwater Ostracods from the Southern India

Despite many limnological investigations, the freshwater ostracods have remained, to quote Kesling (1956), "neglected little Crustaceans in India". Work on the freshwater ostracods of India has been mainly and rarely taxonomic and a review of earlier work on these microcrustaceans shows a paucity of ecological information. This is because of meager and scattered literature, earlier descriptions are less adequate and a key to the species is still lacking.

Literature on freshwater ostracods of India shows that majority of the publications deal with systematics and ecology. Victor and Michael (1975) described 9 new species and Michael and Victor (1975) redescribed 2 little known ostracods from Madurai, Tamil Nadu. Victor and Fernando (1979) in a monographic work revised and brought-up-to-date of earlier described species, mainly from Tamil Nadu, Kerala and some parts Andhra Pradesh and emphasized the need for further research on freshwater ostracods of India. Though it is a pioneering work needs revision in the light of modern work and new literature. Mannikeri and Vaidya (1987, 1990) and Mannikeri et al. (1989) described freshwater ostracods from perennial water bodies in and around Dharwad, in Northern Karnataka. Besides, Vaidya (1996) discussed in brief the zoogeographic distribution of freshwater ostracods of Dharwad, north Karnataka. It is observed that in India, which is highly diverse geographically and ecologically, many localities are there to explore on these aspects. Our present knowledge of freshwater ostracods is confined to lakes, ponds, and rice fields only. Sunny George et al. (1993) described a new species from temporary habitats of Kerala. Ravi et al. (2007) have reported nine species belong to six genera of freshwater taxa from Perumal Eri lake and discussed the zoogeography and faunal affinity.

Studies on Taxonomy and Ecology of Freshwater Ostracods from the Northern India

Deb (1972) described *Stenocypris biswasi* and *S. krisnakantai* from Rajasthan and Maharastra, respectively. The research work of freshwater ostracods of Northern India gained considerable interest as evidence by a series of papers published by different workers (Battish, 1977, 1981, 1982; Bhatia and Singh, 1970, 1971, 1977; Deb, 1973a, 1973b; Bhatia, 1982; and Bhatia and Mannikeri, 1977). Baird (1859) described 3 species, namely- *Cypris subglobosa, C. cylindrica* var. *major* and *C. marginatodentata* from Nagapur, and Klie (1927) reported 2 new species, *Stenocypris swelli* and *Cypridopsis horai* from Punjab, Darjeeling and Cheerapunji, and *Cypris javana* from Kerala. Habibnia and Mannikeri (1988), while reporting additional freshwater ostracods from north of Jaisalmer town, discussed their zoogeographical affinities and provinces. Malik and Amarjit Singh (1994) discussed the ecology of ostracods from Mansar Lake, Jammu.

Distribution of Freshwater Ostracods of India and their Zoogeographic Implications

Zoogeography, the study of the geographical distribution of the fauna, of Ostracoda with well established worldwide distribution patterns, which ideally include a fossil record, are used to analyze and reconstruct possible centers of speciation, lineage, patterns of migration and dispersion. The entire world is divided into six faunal realms, viz.- Palaearctic, Ethiopian, Oriental, Australian, Nearctic and Neotropical zoogeographical regions, based upon their distribution.

The Indian region falls under Oriental Province, bounded by Himalayas in the north but without any physical boundary in the southeast corner and includes Indian, Sri Lankan, Indo-Chinese and Indo-Malayan sub-regions. It is known for its varied physical features as it represents tropical climate compared to the temperate climate of the remaining northern part of India. Its fauna exhibits considerable resemblance with that of Ethiopian region, so much so that few zoogeographers prefer to club these two regions under a single realm, the Palaeotropical region. Ethiopian region consists of Africa south of Sahara, Madagascar and southern Arabia. As nearly the whole of the Ethiopian region lies between the tropics, the conditions here for the existence of life are more or less uniform. Since Sahara desert forms an effective barriers between Palaearctic and Ethiopian regions, the fauna in the north and south are quite distinct. The fauna of this region is rich, varied and well marked. Some of the forms are endemic to this region with no living representatives outside the region. Moreover, a few forms are strongly similar with Oriental fauna, due to similar tropical climate. The freshwater forms are also called limno-biotic fauna. The terrestrial waters in lakes and rivers contain very limited fauna due to their freshness and continuous flow. The fauna living in standing water bodies like pond, lake or swamp are called lentic. *Stenocypris major* and *S. hislopi* are known from Oriental Province. *Ilyocypris gibba* and *I. bradyi* are cosmopolitan species, known to occur from Recent lakes and Pleistocene deposits of different parts of the Oriental zoogeographical province.

STATISTICAL ASPECTS OF OSTRACODA

Studies on Sedimentation Rate (Siltation) Based on Ostracod Studies

The application of statistical data on Ostracoda, such as juveniles and adults; closed and isolated (open) valves; males and females; right and left valves; smooth and ornamented forms, etc., besides colour variation, pyritisation and predation, to interpret the environment of deposition and rate of deposition of sediments (siltation) has attained importance, during the last five decades (Kathal, 2012).

In an environment where rate of sedimentation is low, the carapaces are likely to open up by bacterial action. While if it is rapid, the carapaces will sink into the soft bottom and will be quickly covered by sediments. Thus, the carapaces will have less chances of opening up, after the destruction of muscles and ligaments. After the death of animal, the adductor muscles of carapace disintegrate and the valve tend to open up during the transportation. Further due to intense bacterial activity, the ostracod carapace gets open up into valves. However, if they get buried earlier than opening of their valves, the percentage of closed carapaces will be more than open valves. More is the number of closed carapace higher will be the rate of sedimentation. The usage of carapace-valve ratio yield palaeoecological information was pioneered by Pokorny (1965). Oertli (1971) reviewed Pokorny's work and related the carapace-valve ratio to potential for the information of hydrocarbons. He summarized that when the ratio is high, sedimentation is rapid, which minimizes disarticulation of carapaces into separate valves. With sufficiently rapid burial, organic matter is not absorbed by mineral particles and so retains potential for conversion into hydrocarbons. Faster rate of deposition of the sediments favours conversion of organic matter into hydrocarbons, in reduced aquatic conditions, under optimum temperature and pressure.

Honnappa and Venkatachalapathy (1978) studied the carapace-valve ratio to interpret the rate of deposition of sediments in the Mangalore harbour area, southwest coast of India. They found that the occurrence of open valves is much more in number than the closed one (ratio being 24:1) According to them, this is an indicative of a slow rate of seidimentation in more agitating waters. While comparing Eocene/Oligocene ostracods from southeastern Australia and India for petroleum potential indicators, McKenzie and Guha (1987) inferred a rapid rate of sedimentation through the presence of high percentage of carapaces. Sreenivas et al. (1991) found rapid rate of sedimentation in Pulicat lake estuary on the basis of occurrence of more number of closed carapaces. Sridhar et al. (2002) from the Palk Bay, off Rameswaram observed the carapace to valve ratio to 5:1 indicative of the fairly faster rate of sedimentation.

The ratio between the carapaces and open valves has been taken into consideration for determining the relative rate of sedimentation. Hussain and Rajeshwara Rao (1996) observed more number of closed carapaces than open valves along inner shelf sediments, the east coast of India. While the number is much less from the sediments off the west coast of India. From this observation, they inferred that the rate of sedimentation is rapid on the east coast whereas it is slow on the west coast of India, which is attributed to more number of streams/rivers flowing and debouching the sediments into the Bay of Bengal. Hussain et al. (2002) studied carapace and valve ratio and observed faster rate of sedimentation in the inner shelf of Gulf of Mannar, off Tuticorin, southeast coast of India. They counted 4 occurrence of carapaces to open valves, which was attributed to the inflow of sediments through Tamirabarani River. Hussain et al. (2004) observed a slow rate of sedimentation from off Karikkattukuppam, near Chennai which they attributed that there is no major river to discharge the sediments into the bay. Ganesan and Hussain (2010) observed a faster rate of sedimentation in Tamiraparni Estuary, Punnaikayal, near Tuticorin, Tamil Nadu. Hussain et al. (2013) have inferred comparatively a fast rate of deposition of sediments in the Ennore creek, near Chennai, due to flux of sediments from the Kosisthalaiar, Kortalaiar and Araniar rivers, which pass through the brackish waters of the creek before meeting the Bay of Bengal. Hussain and Kalaiyarasi (2010) have also noticed a very high rate of sedimentation in the Mullipallam creek, near Muthupet through the carapace and valve ratio of Ostracoda. A relatively faster rate of sedimentation is observed in the Pulicat lagoon area by Mahalakshmi and Hussain (2015). Since the siltation is more in the lagoon, periodically desiltation and dredging activities are being carried out at the mouth and middle portion of the lagoon.

Studies on Ostracod Predation

Predation can be stated as an interaction between two organisms which results in negative effects on the growth and survival of one of the populations (Odum, 1971). It can also be defined as a relationship between animals wherein one species eats another species. Predators include coelenterates and many gastropods (muricids and naticids). Muricids drill oblique holes in a deep-water environment, whereas naticids drill vertical holes in a shallow marine environment. There are active predators of many pelecypods, gastropods, scaphopods, ostracods and barnacles. Gastropod predators exhibit different patterns of predatory behaviour and drill holes of different shapes and sizes. Predators generally confine their activity to the upper sedimentary layers. Predation is a more common phenomenon in a community of organisms of benthic habit of shallow water environment. It is one of the limiting factors that affects the abundance, distribution and individuals of species (Clarke, 1954). According to Hecker (1965), predation may be considered as a "palaeopathological phenomenon" which occurs during the life of an organism. Accordingly, these holes are made on the shells of different organisms lying on the sedimentary layers. Such organisms are characteristic of certain facies, particularly those of shallow seas, not far from shorelines and, therefore, serve as good indicators of the regions. They can also be used successfully for determining the upper and lower surface of the beds.

Sliter (1971) proposed that evidence of predation provides ecologic and paleoecologic information on living habits and habitats, and community structures of the organisms. This also gives an indirect clue in the interpretation of rate of sedimentation. He further found that predation is more common in benthic fauna of shallow water environment. Reyment (1971) while studying the predation of ostracods from Cretaceous, Tertiary and Recent sediments of Nigeria, stated that predation is one of the important biological factors, which involve interaction between two or more species for limited food supply or, sometimes, one species is a part of the food for the other. From Texas region, Maddocks (1988) observed moderate predation in Cenomanian ostracods, noteworthy increase in predation of ostracod shells of Late Campanian, Cretaceous-Tertiary sediments and fewer rates of predatory shells in Holocene assemblages. Accordingly, ostracods are one of the food sources for juvenile naticids and smooth forms are more susceptible to predation and the large size of the carapace and maximum population also favour predation. The position, shape and dimension of the predatory drills found on ostracod carapaces can be utilized to interpret the environment of deposition and ecological implications. Murray (1973) found that the population of bored shells varies from 3% in shallow estuarine bodies to around 16% at a depth of 2000 m, in the Gulf of Mexico. Hussain et al. (2002) while discussing the predatory aspects of ostracod assemblage of the Gulf of Mannar, noticed only 4.9% of drilled shells. The drilled ostracod shells constitute 0.9% of the entire ostracod specimens recovered from the Pulicat lagoon, which indicate a shallow brackish water lagoon. Multiple predatory holes are also seen in the brackish water environments.

Sediment Texture and Surface Ornamentation of Ostracods

Environmental changes have been frequent in the geological past and are observed even today. New habitats have developed as a result of the changes in the ecosystem due to these environmental changes. Although at times, the physical environment remains comparatively constant, the organic environment is not. Ostracods are very sensitive microorganisms and react rapidly to any change in the ecosystem. In order to survive any change in the environment, they have a tendency to adopt different morphological characters. A direct relationship between the particle size of the sediment and the surface ornamentation of the carapace can be observed. A few species have smooth surface, and devoid of any sculpture, whereas in many others, the surfaces have simple to complex surface ornamentation, which serve as direct evidences for ecological interpretations (Annapurna and Rama Sarma, 1982; Vaidya et al. 1995; Sridhar et al. 1998 and Hussain et al. 2002). Such studies and interpretations may be made based on Scanning Electron Microscope (SEM) micrographs of the specimens.

The substrate sediment texture has a control on the kind of ostracod fauna that can colonise a particular sediment type (Armstrong and Brasier, 2005). The texture stability of sediment comprising the substrate exerts a strong influence on marine ostracods. Smooth forms are predominant in fine-grained muds, whereas more ornamented forms are being found in coarser or in more calcareous sediment (Benson, 1961; and Armstrong and Brasier, 2005). The surface sculptures of ostracod carapaces have a direct relationship with the substrate type. Puri (1966) and Malz and Lord (1988) opinioned that more ornate and rather heavily calcified forms commonly show shallow water, high-energy environments and inhabit sandy substrates in modern seas. In the Bimili backwaters and Balacheruvu tidal stream, Annapurna and Rama Sarma (1982) noticed that the genus *Phlyctenophora* occurs in sand dominated areas and not in muddy areas and the moderately ornamented forms like *Tanella, Loxoconcha, Paijenborchellina* and *Kalingella* occur in considerable numbers in sandy areas. Vaidya et al. (1995) observed good faunal content in the substrates consisting of medium- to fine-grained sand, whereas poor occurrence was noticed in the clean, coarse-grained sand. Sridhar et al. (1998) found more number of species in siltysands and sandy substrates and less in fine materials and forms like *Macrocyprina decora, Paracytheroma ventrosinuosa, Phlyctenophora orientalis* and *Keijella karwarensis* occur in considerable numbers in the samples with a mixture of clayeysand and clayey-silt, thus indicating that smooth forms occupy fine-grained sediments. Similar observations have also been made Tamiraparani estuary and adjoining shelf area by Ganesan and Hussain (2010). Their study also infers that the smooth, less ornamented and feebly calcified carapaces reflect the fine grained nature of the sediment (Plate 1).

Studies on Shell (Valve) Chemistry of Ostracods

Trace element geochemistry of ostracod shell, a recently developed method is increasingly used to infer past environmental conditions in diverse water bodies (Chivas et al., 1983, 1986; Correge and De Deckker, 1997; Xia et al., 1997; and Danielopol et al., 2002). The other possible application of trace elements in Ostracod shells would be to compare the shell chemistry of benthic and planktic species from the same ocean. This could help detect temperature and chemical gradients in the water column. In addition, analyses of valves belonging to the various juvenile growth stages and of adults from within a single sample might reveal seasonal temperature as well as chemical changes. Trace-element contents recorded in ostracod shells may be related to environmental changes that occur at a very short time scale seasonally. They secrete low-Mg calcite valves, and moult up to nine times before attaining the adulthood. During the moulting of the valves as the ostracod grows, the chemical elements of the ostracod carapace come directly from the ambient water, and can be used as hydrochemical indicators. Chivas et al. (1983, 1986) and De Deckker et al. (1999) studied and determined the geochemical relationships between the uptake of trace metal (Mg, Sr) substitution of Ca in the CaCO3 by ostracods and the ambient water. Chivas et al. (1983, 1986) found that for Mg there is a strong temperature effect on Mg incorporation in the ostracod calcite, but that this relationship depends on the Mg/Ca of the host water and the Sr uptake is controlled by Sr/Ca of the host water. Baskar et al. (2014) and Sridhar et al. (2015) discussed the ostracod shell chemistry from off Rameswaram using SEM-EDAX. Accordingly, the elements that constitute the shell might have come either from water or sediment or from both. But, most of the elements are from water, whereas Al might have come from sediment, as it is not found in water Sridhar et al. (2015).

Studies on Pollution Implications of Ostracods

Pollution ecology of the fauna in the Ennore creek, north of Chennai was studied based on ostracod live/dead ratio, size, colour, morphology and ecdysis (Hussain et al., 2013). It was observed that the presence of more dead forms and less occurrence of living forms in the creek. The low count of living forms is due to the effect of effluents being discharged from the surrounding industries viz., Kothari Fertilizers, EID Parry Fertilizers, Ennore Foundries, Ennore Thermal Power Station, Ashok Leyland, Chennai Rubber Factory, Fertilizer Unit- I and II, TAMIN, Oil Mill and Chennai Petrochemicals. The carapaces/valves are pale yellow to white in colour, finely preserved show that the creek is not highly polluted except a few species exhibit only one or two juvenile stages pointing to partial pollution/ deterioration in the environment.

Studies on (Paleo) Tsunami Sediment Through Ostrtacods

Different features of ostracods such as- variability of the local assemblages, population density, species diversity, age population structure (like percentages of adults and juvenile stages) and taphonomical signatures suggest that these microorganisms may be included amongst the most promising tracers of these high-energy events in marshes, lakes, lagoons or shallow marine areas (Ruiz et al., 2009).

A study conducted immediately after December, 2004 tsunami on the marginal marine environments of Andaman Islands yielded benthic forms of neritic and upper bathyl zones (Hussain et al., 2006) brought from the inner to outer shelf region by the tsunami waves. It is also observed that this is the first report of ostracods from the sediments of these islands. Similar studies, carried out along the coast of Tamil Nadu, SE coast of India by opening few trenches and pits (Hussain et al., 2010 and Elakkiya et al., 2013) to document ostracod distribution in the coastal sediments in stratigraphic context (geochronology through ^{14}C and OSL dating) to study paleo-tsunami (catastrophic) events.

Distribution of Brackish and Marine Ostracods of India- Their Faunal Affinity and Zoogeographic Implications

It is broad in its concept which seeks knowledge not only of the present day geography but also of the patterns and timings of changes in the location of land-masses during the earth's development. It is a common feature to note that some of the forms migrate or disperse inter-oceanically and inter-continentally due to climatic variations through active as well as passive migration. A few other forms are, however, endemic and confined to a locality within the geographical limit. Such migration of the fauna can be correlated with the observed environmental parameters from different localities. The ostracod marine fauna of Bay of Bengal and Arabian Sea has got affinity to the Indo-Pacific faunal province and their dispersal/migration is mainly due to south-equatorial currents and also passive dispersal through ships and floating algae (Titterton and Whatley, 1988; Witte, 1993; Vaidya and Mannikeri, 1994; Hussain and Rajeshwara Rao, 1996, Mohan et al., 2001; Sridhar et al., 2002; and Elumalai et al., 2010), as also concluded based on the Recent Foraminifera (Kathal et al., 2000).

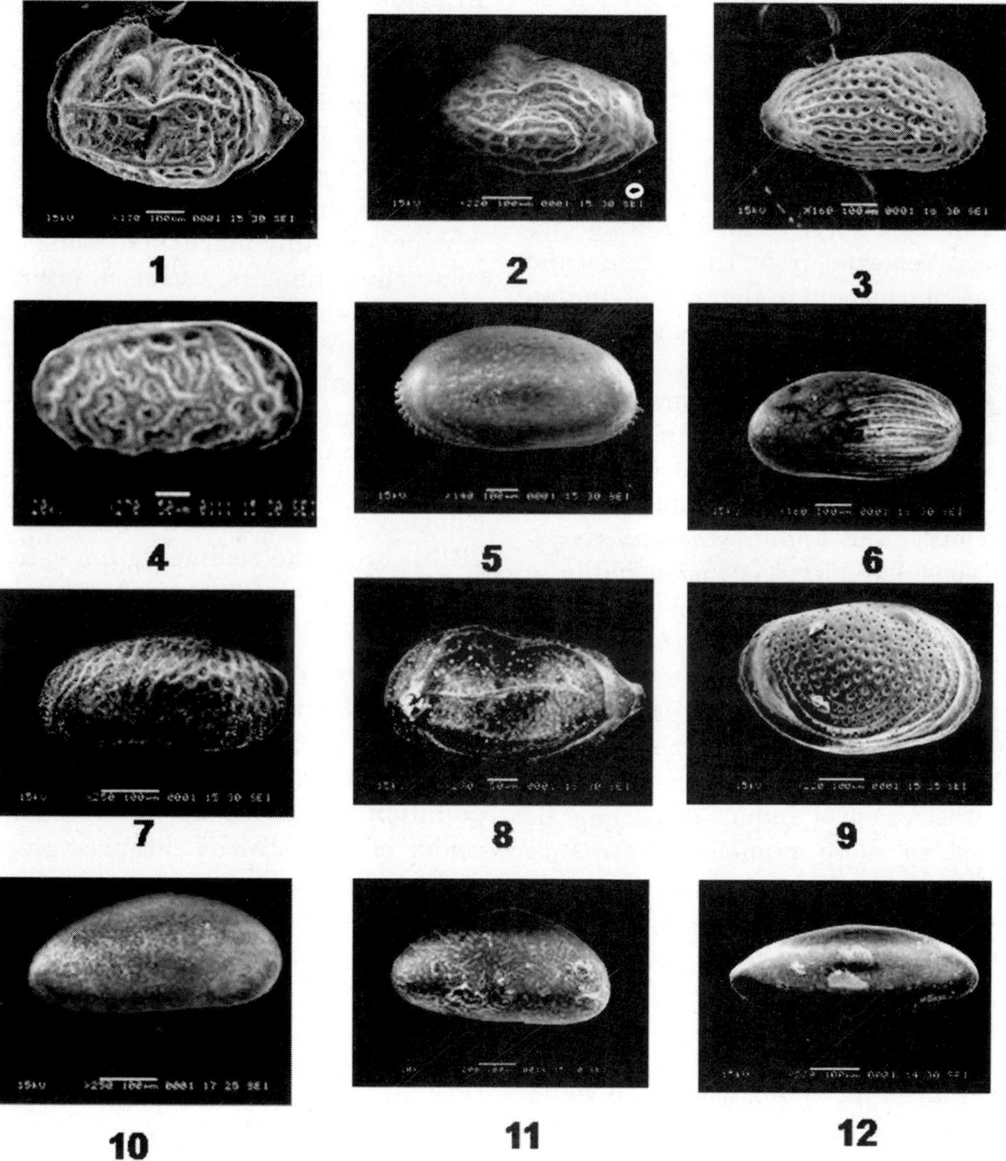

Plate 1 Highly calcified and ornate forms: 1 . *Neomonoceratina delicate,* Carapace, left valve external view; 2. *Neomonoceratina iniqua,* Carapace, left valve external view; 3. *Keijell, reticulate,* Carapace, right valve external view; 4. *Callistocythere flavidofusca intricatoides*, Carapace, right valve external view; 5. *Miocyprideis spinulosa,* Carapace, left valve external view; **Moderately ornamented Forms:** 6. *Basslerites liebaui*, Carapace, right valve external view; 7. *Tanella gracilis,* Carapace, right valve external view; 8. *Neomonoceratina porocostata,* Carapace, left valve external view; 9. *Loxoconcha gruendeli,* Carapace, right valve external view; **Smooth and less ornamented Forms:** 10. *Phlyctenophora orientalis,* Carapace, right valve external view; 11. *Propontocypris (Schedopontocypris) bengalensis*, Carapace, right valve external view; 12. *Paradoxostoma bhatiai*; Carapace, left valve external view; (Bar scale 100μm, unless specified).

CONCLUSION

The marine ostracod species reported in the up-to-date literature from the coasts of India is reviewed in this paper and is observed that the faunal assemblage is characteristic of shallow water, neritic, tropical in nature. A total of 209 species have been reported from the marine and marginal marine environments of Indian coast; 63 from the east coast and 56 from the west coast of India. Ostracod abundance and diversity is comparatively less along the east coast than the west coast of India, due to monsoonal variability in the Arabian Sea. *Tanella gracilis, Kejia demissa* and *Kotoracythere inconspicua* are considered to be of cosmopolitan in distribution and occurrence while *Basslerites liebaui, Jankeijcythere mckenziei, Kalingella mckenziei* and *Neomonoceratina jaini* are endemic to Indian waters only. The rate of sedimentation is rapid on the east coast whereas it is slow on the west coast of India, which may be attributed to more number of streams/ rivers flowing and debouching the sediments into the Bay of Bengal than the Arabian Sea. The ostracods along the east coast of India show zoogeography of affinity to the Indo-Pacific faunal assemblage as also shown by Recent foraminifera (Kathal et al, 2000). The fresh-water taxa is strongly similar having affinity with Oriental fauna, due to similar tropical climate, whereas few forms exhibit affinity with the Ethiopian zoogeographic province as the conditions here for the existence of life are more or less uniform.

Ostracoda distribution in the sediments of the lagoons helps to demarcate the zones such as freshwater, brackish water and marine environment. In modern marginal-marine settings, it is known that variations in salinity determine the composition and diversity of the ostracoda and there are well-defined assemblages, each dominated by just two or three species, which represent oligohaline, freshwater to brackish, brackish to marine and fully marine environments. In some environments, ostracod assemblages are dominated by a single taxon, which is often the case in biologically 'stressed' environments such as hypersaline water bodies, intertidal settings or reduced-oxygen conditions as observed in the Mullipallam creek, near Muthuper, Cauvery delta (Hussain and Kalaiyarasi, 2013). From the sedimentation rate in the creek, it is also observed that it is a prograding delta of the Cauvery River.

The presence of more dead ostracods and less occurrence of living forms; the colour of ostracod carapaces and valves and their taphonomical features reflect the pollution ecology of the environment. The colour of the ostracod carapace gives information on the oxic and anoxic conditions of the study area. A study undertaken at different areas along the coast of Tamil Nadu indicate that if the open valves and carapaces are light yellow and white in colour reflect on oxic conditions whereas if they are in dark grey, black, brown or green colour supports the fact that the sediments were deposited under anoxic environment. The ostracod valve chemistry infers salinity, temperature and water level changes in an environment. Based on the principle Uniformitarianism, the baseline data generated on these micro-Crustaceans, may be utilized to infer the paleoenvironmental interpretations of the older sediments of a sedimentary basin.

Acknowledgements: I am thankful to the DST-PURSE programme for financial support and the Head of the Department of Geology for his encouragement and also for providing facilities to carry out this work. I thank S. G. D. Sridhar for his collaborative research work and thanks are also due to A. Raj Kumar, Mohd. Noohu Nazeer and D. Sunitha for their help in the preparation of the map and table for this manuscript. I profusely thank the editors for having given an opportunity to submit a paper on Ostracoda, particularly P.K. Kathal for sparing his time to make few corrections in the paper.

References

ANIL BHANDARI and SINGH. S.D. (2006) Recent Ostracoda from the deltaic subenvironment and shallow inner shelf, east coast of India. J. Geol. Soc. India. v.67, pp.663-672.

ANNAPURNA, C. (1981) Seasonal quantitative variations in benthic ostracods in the tidal stream of Balacheruvu, east coast of India. Proc. Ind. Acad. Sci., (Anim. Sci.). v.90 (5), pp.519-527.

ANNAPURNA, C. and RAMA SARMA, D.V. (1982) Sediment-ostracod relationship in the Bimili backwaters and Balacheruvu tidal stream. Proc. Ind. Acad. Sci. (Anim. Sci.). v.91 (3), pp.297-303.

ANNAPURNA, C. and RAMA SARMA, D.V. (1979a) Occurrence of a new podocopan ostracod *Tanella vasishta* in the Vasishta-Godavari estuary, east coast of India. Curr. Sci., v.48 (1), pp.42-43.

ANNAPURNA, C. and RAMA SARMA, D.V. (1979b) Occurrence of a podocopan ostracod *Tanella indica* sp. nov. in marginal water bodies, east coast of India. Ind. J. Mar. Sci., v.8 (2), pp.117-118.

ANNAPURNA, C. and RAMA SARMA, D.V. (1981) Occurrence of a podocopan ostracod *Bradleya ganapatii* sp. novo in the marginal water bodies, east coast of India. J. Geol. Soc. India, v.22 (1), pp.51-53.

ANNAPURNA, C. and RAMA SARMA, D.V. (1986) Taxonomic studies on the marine Ostracoda from India. Family Leptocytheridae Hanai, 1957. J. Bombay Nat. Hist. Soc., v.83 (3), pp.618-622.

ANNAPURNA, C. and RAMA SARMA, D.V. (1987) Taxonomic studies on the marine Ostracoda from the east coast of India. J. Bombay Nat. Hist. Soc., v.84 (1), pp.177-180.

ARMSTRONG, H.A. and BRASIER, M.D. (2005) Microfossils, Blackwell Publishing pvt. Ltd., Australia, 296p.

ARUL, B., SRIDHAR, S.G.D., HUSSAIN, S.M., DARWIN FELIX, A. and PERIAKALI, P. (2003) Distribution of recent benthic Ostracoda from the sediments of Pitchavaram mangroves, Tamil Nadu, Southeast coast of India. Bull. Pure Appl. Sci. v.22, pp.55-73.

BAIRD, W. (1859) Description of some new entomostracan from Nagpur, Collected by Rev. S. Hislop. Proc. Zool. Soc. Lond., pp.389-234.

BASKAR, K., SRIDHAR, S.G.D., MANIYARASAN, S., HUSSAIN, S.M. and BALAKUMAR, S. (2014) Shell chemistry of Loxoconchidae family, Recent benthic Ostracoda, off Rameswaram, Tamil Nadu, Palk Bay, southeast coast of India. Inter. J. Innovative Res. Sci. Engg. (IJIRSE), ISSN 2347-3207, pp.1-8.

BASKAR, K., SRIDHAR, S.G.D., SIVAKUMAR, T., HUSSAIN, S.M. and MANIYARASAN, S. (2015) Temporal variation of physico-chemical parameters and Ostracoda population, off Rameswaram, Gulf of Mannar, southeast coast of Tamil Nadu. India. J. Geol. Soc. India, v.86, pp.663-670.

BASKAR, K., SRIDHAR, S.G.D., SIVA KUMAR, T. and HUSSAIN, S.M. (2013) Seasonal Comparison of Recent Benthic Ostracoda Response to Sediment Type in

Palk Bay, off Rameswaram Island, Tamil Nadu, Southeast coast of India. Inter. J Earth Sci. Engg. ISSN 0974-5904, v.6(5), pp.1047-1059.

BASKAR, K, SRIDHAR, S.G.D., HUSSAIN, S.M., SOLAI, A. and KALAIVANAN, R. (2013) Distribution of Recent benthic Ostracoda off Rameswaram, Palk strait, Tamil Nadu, south east coast of India. Sp. Pub., Geol. Soc. India, no.1, pp.195-212.

BATTISH, S.K. (1977) Record of genus *Pseudocypris* Dady (1908) (Crustacoda: ostrocoda) with the description of new species from India. Zool. Jour. Linean Soc. v.60, pp.363-366.

BATTISH, S.K. (1981) Freshwater ostracoda of the Family Cyprinotinae from Punjab, India, with the description of eight new species. J. Nat. Hist. v.15, pp.645-669.

BATTISH, S.K. (1982) Freshwater ostrocoda of the Family Cypridopisidae from Punjab, India, with the description of two new species. J. Nat. Hist., v.16, pp.83-94.

BENSON, R.H.(1961) Ecology of ostracod assemblages. *In*: Treatise on Invertebrate Paleontology, Part Q, Arthropoda 3, Ostracoda, MOORE, R.C. (Ed.), Geol. Soc. Amer., pp.Q56-Q63.

BHATIA, S.B. and KUMAR, S. (1976) Recent benthonic foraminifera from the inner shelf area around Anjidiv Island, off Binge, west coast of India, Maritime Sediments, spl. publn., v.1, pp. 239 249.

BHATIA, S.B and MANNIKERI, M. S. (1977) Recent freshwater ostracodes from Rajasthan. Proc. IV Colloq. Indian Micropal. Strat., Dehra Dun, pp.81-86.

BHATIA, S.B. (1982) Quaternary ostracoda of the Indo-Gangetic Plain and their Paleozoogeographic implications. *In*: MADDOCKS, R.F. (Ed.), Applications of ostracoda. *In* Proc. VIII Int. Symp. on ostracoda, Houston, Texas, pp.442-458.

BHATIA, S.B. (1984) Ostracod fauna of the Indian subcontinent- their paleozoogeographic and paleoecologic implications. J. Pal. Soc. India, v.20, pp.1-8.

BHATIA, S.B. (1991) Comment 3: Ostracoda in the estuarine sediments, Pulicat Lake estuary, east coast of India. J. Geol. Soc. India, v.38 (5), pp.541-542.

BHATIA, S.B. and KUMAR, S. (1979) Recent Ostracoda from off Karwar, west coast of India. *In* Proc.VIII Internatl. Symp. Ostracods, Beograd, pp.173-178.

BHATIA, S.B. and SINGH. D. (1970) A note on some ostrocodes from Dal Lake, Kashmi. Res. Bull. (N.S), Punjab Univ., v. 21, pp.257-259.

BHATIA, S.B. and SINGH. D. (1971) Ecology and distribution of some Recent ostracodes of the Vale of Kashmir, India . Micropal., v.17, pp.214-220.

BHATIA, S.B. and SINGH. D. (1977) Some Late Pleistocene and Recent ostracoda from parts of Punjab and the Union Territory of Chandigarh, India. Rec. Res. in Geo., Hindustan Pub., v.3, pp.399-414.

BRADY, G.S. (1886) Notes on Entomostraca collected by Mr. A. HALY in Ceylon. J. Linn. Soc. London, v.19, pp.293-317.

CHIVAS, A.R., DE DECKKER, P. and SHELLEY, J.M.G. (1983) Magnesium, strontium and barium partitioning in nonmarine astrict shells and their use in paleoenvironmental reconstructions: A preliminary study. *In* MADDOCKS, R. F. (ed.), Applications of Ostracoda. Geosci. Univ., Houston, pp.238-249.

CHIVAS, A.R., DE DECKKER, P. and SHELLEY, J.M.G. (1986) Magnesium and strontium in nonmarine ostracode shells: A new paleosalinometer and paleothermometer. Palaeogeog. Palaeocli. Palaeoecol. v.54, pp. 43-61.

CLARKE, G.L. (1954) Elements of Ecology, Wiley, New York, 534 p.

CORREGE, T. and DE DECKKER (1997) Faunal and geochemical evidence for changes of intermediate water temperature and salinity in the coral sea (N.E. Austalia) During the late Quaternary.

Palaeogeogr. Palaeoclimatol. Palaeoecol. v.131, pp.183-205.

DANIELOPOL, D.L., ITO, E., WANSARD, G., KAMIYA, T., CRONIN, T.M., and BALTANAS, A. (2002) Techniques for collection and study of Ostracoda. *In* the Ostracoda: Application in Quarternary Research. (HOLMES, J.A. and CHIVAS, A.R. (Eds). American Geophy. Union, Washington D.C., pp.65-97.

DE DECKKER, P., CHIVAS, A.R., SHELLEY, J.M.G. (1999) Uptake of Mg and Sr in the euryhaline ostracod Cyprideis determined from in vitro experiments. Palaeogeogr. Palaeoclimatol. Palaeoecol., v.148, pp.105-116.

DEB, M. (1924) First record of the Genus Scleocypris Sars, (Crustacea: Ostracoda), from India, with description of two new species (1973a). J. Zool. Surv. Ind., v.23, pp.47-52.

DEB, M. (1972) Fauna of Rajasthan (India) Crustacea Ostracoda. Rec. Zool. Surv. Ind., v.67, pp.233-259.

DEB, M. (1973) Notes on a collection of ostracoda from Delhi. Rec. Zool. Surv. Ind., v.67, pp.309-313.

DESAI, P.S., GOWDA, H.H., and KASTURIRANGAN, K. (2000) Ocean research in India: Perspective from space. Curr. Sci., v.78, p.268-278.

ELAKKIYA, P., HUSSAIN, S. M. and ELUMALAI, K. (2013) Distribution of Foraminifera and Ostracoda in the Kameshwaram coast, Nagapattinam, South India: Implications for recognition of over wash/extreme wave events deposits. (Chapter 7) In: On a sustainable future of the earth's natural resources". *In* RAM KUMAR, M.U. (Ed.), Springer Earth Sys. Sci., DOI 10.1007/978-3-642-32917-3_7, Springer-Verlag Berlin, Heidelberg, Germany, pp.139-148.

ELUMALAI, K., HUSSAIN, S.M. and SCOTT, IMMANUEL DHAS.C. (2010) Recent benthic Ostracoda from the sediments of Ennore creek, Chennai, Tamil Nadu. J. Pal. Soc. India, v.55(1), pp.11-22.

GANESAN, P. and HUSSAIN, S.M. (2010) Distribution of Recent benthic Ostracoda in Tamiraparni Estuary, Punnaikayal, near Tuticorin, Southeast Coast of India-Implications on microenvironment. Gond. Geol. Mag., Sp. Issue App. Micropal. KUNDAL, P.K. and HUMNAE, (Eds.), v.25(1), pp.103-114.

GOPALAKRISHNA, K., HUSSAIN, S.M., BILWA, MAHESH and AYISHA, V.A. (2007) Recent benthic Ostracoda from the inner shelf off Malabar coast, Kerala, southwest coast of India. Jour. Pal. Soc. India, v.52(1), pp.59-68.

GRAMMAN, F. (1975) Ostracoda from Tertiary sediments of Burma with reference to living species. Geol. Jb., v.14, pp.1-46.

GUHA, D.K. (1968) On the Ostracoda from Neogene of Andaman Islands. J. Geol. Soc. India, v.9 (1), pp.58-66.

GUHA, D.K. (1980) On some Recent Ostracoda from the west coast of India. Geosci. J., v.1(2), pp.41-50.

HABIBNIA, B, A. and MANNIKERI, M. S. (1988) Additional Recent ostracodes from Rajasthan. Curr. Sci., v.57, pp.33-34.

HECKER, R.T. (1965) Introduction to Paleoecology, Elsevier, New York.

HONNAPPA (1975) Taxonomy, morphology, ecology and statistical interpretation on *Actinocythereis tumefacentis* (Lubimova and Guha), Ostracoda from the Recent sediments of Mangalore harbour area, west coast of India, Rev. Espanola Micropal., v.7, pp.417-449.

HONNAPPA (1983) Review of Quaternary ostracods from the west and east coast of India. Geoviews, v.10 (6), pp.223-240.

HONNAPPA and ABRAR, S. (1983) Bairdiidae (Ostracoda) from the Recent coastal sediments of Bhatkal area (Karnataka State), west coast of India. Curr. Sci., v.52 (12), pp.588-591.

HONNAPPA and ABRAR, S. (1984) Ostracoda from the coastal sediments of Karwar and Bhatkal areas, Karnataka state, west coast of India. *In* Proc. X Ind. Colloq. Micropal., Strat., pp.483-500.

HONNAPPA and VENKATACHALAPATHY, V. (1978) Paleoecological and ecological interpretations of sediments of Mangalore harbour area, west coast of India, on the basis of colour variation of ostracod shells. Curr. Sci., v.47 (20), pp.772-773.

HONNAPPA and VENKATACHALAPATHY, V. (1979) Predation and ecology of Ostracoda from the sediments of Mangalore harbour area, west coast of India. *In* Rec. Res. Geology, v.4, pp.351-355.

HONNAPPA, VENKATACHALAPATHY, V. and ABRAR, S. (1984) Occurrence of a new genus *Neomangaloria* gen. nov. Ostracoda in the Mangalore and Karwar sediments, west coast of India. Bull. Indian Geol. Assoc., v.17 (2), pp.167-173.

HUSSAIN, S.M. (1998) Recent benthic Ostracoda from the Gulf of Mannar, off Tuticorin, southeast coast of India. J. Pal. Soc. India, v.43, pp.1-22.

HUSSAIN, S. M., ELAKKIYA, P., ELUMALAI, K., JONATHAN, M. P. and NAGARAJAN, R (2013) Evidences for extreme wave events in Velanganni coast, southeast coast of India. Bol. Soc. Geológica Mexicana, v.65(1), pp.201-205.

HUSSAIN, S.M., GANESAN, P. and MOHAN, S.P. (2005) Recent Benthic response to environment in Tambraparni estuary, Punnaikayal, Tuticorin, Southeast coast of India. Gond. Geol. Magz. v.20, pp.25-30.

HUSSAIN, S.M., GANESAN, P., RAVI, G., MOHAN, S.P. and SRIDHAR, S.G.D. (2007) Distribution of Ostracoda in marine and marginal marine habitats off Tamil Nadu and adjoining areas, southern east coast of India and Andaman Islands: Environmental implications. Indian J. Mar. Sci., v.36(4), pp.369-377.

HUSSAIN, S.M. and KALAIYARASI, A. (2010) Distribution of *Neomonoceratina iniqua* in the sediments of Mullipallam creek, near Muthupet, Tamil Nadu, southeast coast of India: Ecological implications. International Journal of Earth Sciences and Engineering, (Special Issue), v.3 (5), pp.23-35.

HUSSAIN, S.M. and KALAIYARASI, A. (2013) Distribution of Ostracoda in the Mullipallam lagoon, near Muthupet, Tamil Nadu, southeast coast of India – Implications on microenvironment. (Chapter 11) *In*: Climate Change and coastal and Island Vulnerability (Eds. SUNDARESAN, J., SREEKESH, S., RAMANATHAN, A.L., SONNENSCHEIN, L. AND BOOJH, R.), Capital Publishing Co., (India), (ISBN: 978-93-81891-02-5), pp.166-176.

HUSSAIN, S.M., KULEEN, E. C., JISHA, K., ELUMALAI, K. and RAVI, G. (2013) Distribution of benthic Ostracoda in surface and subsurface backwater sediments of Ernakulam, Kerala, Southwest coast of India: microenvironmental implications. spl. publ. Geol. Soc. India, no.1, pp.213-224.

HUSSAIN, S.M., KRISHNAMURTHY, R., SURESH GANDHI, M., ILAYARAJA, P., GANESAN, P. and MOHAN, S.P. (2006) Micropaleontological investigations of tsunamigemnic sediments of Andaman Islands. Curr. Sci., v.91, pp.1655-1667.

HUSSAIN, S. M., MANIVANNAN, V. and RAGOTHAMAN, V. (1997) Sediment-Ostracod relationship in the Gulf of Mannar, off Tuticorin, east coast of India. J. Nepal Geol.Soc., v.15, pp.33-37.

HUSSAIN, S. M., MERIN MARIA JOY, RAJKUMAR, A. MOHAMMED NISHATH, N. and SHUBHANGI T FULMALI (2016) Distribution of calcareous microfauna (Foraminifera and Ostracoda) from the beach sands of Kovalam, Thiruvananthapuram, Kerala, southwest coast of India. J. Pal. Soc. India, v.61(2), pp.267-272.

HUSSAIN, S.M. and MOHAN, S.P. (2000) Recent Ostracoda from Adyar river estuary, Chennai, Tamil Nadu. J. Pal. Soc. India, v.45, pp. 25-31.

HUSSAIN, S.M. and MOHAN, S.P. (2001) Distribution of recent benthic Ostracoda in Adyar river estuary, east coast of India. Ind. J. Mar. Sci, v.30, pp.53-56.

HUSSAIN, S.M., MOHAN, S.P. and JONATHAN, M.P. (2010) Ostracoda as an aid in identifying 2004 tsunami sediments: An example from SE coast of India. Natural Hazards, v.55, (2), pp.513-522.

HUSSAIN, S.M., MOHAN, S.P. and MANIVANNAN, V. (1998) Recent cytherellids, bairdids and cytherids (Ostracoda) from the Gulf of Mannar, off Tuticorin, southeast coast of India. Indian J. Earth Sci., v.25(1), 1-4, pp.17-28.

HUSSAIN, S.M., MOHAN, S.P. and MANIVANNAN, V. (2001) Distribution, faunal affinity and zoogeography of Recent benthic Ostracoda from the Gulf of Mannar, off Tuticorin, Southeast coast of India. Oil Nat. Gas. Corp. (ONGC), Bull., v.37(1), pp.93-98.

HUSSAIN, S.M., MOHAN, S.P. and MANIVANNAN, V. (2002) Microenvironmental inferences of Recent benthic Ostracoda, from Gulf of Mannar, off Tuticorin, east coast of India. *In* Proc. Nat. Sem. Manag. Nat. Resources, SANKARA PITCHAICH, P. (ed), Nagarjuna University, pp.23-43.

HUSSAIN, S.M., MOHAN., S.P. and MANIVANNAN, V. (2003) Ecology and distribution of Recent benthic Ostracoda in the Gulf of Mannar, off Tuticorin southeast coast of India. Gondwana Geol. Mag., Sp. Vol. 6, pp.129-138.

HUSSAIN, S.M., RAGOTHAMAN, V. and MANIVANNAN, V. (1996a) Distribution of Ostracoda in waters off Tuticorin, southeast coast of India. Ind. J. Mar. Sci., v.25, pp.78-80.

HUSSAIN, S. M., RAGOTHAMAN, V. and MANIVANNAN, V. (1996b) Distribution and species diversity of Recent benthic Ostracoda from the Gulf of Mannar, off Tuticorin, Tamil Nadu. J. Pal. Soc. India, v.4l, pp.17-20.

HUSSAIN, S.M. and RAJESHWARA RAO, N. (1996) Faunal affinity, zoogeographic distribution and review on Recent Ostracoda from the east and west coasts of India. Bull. Pure Appl. Sci., v.15(1), pp.37-50.

HUSSAIN, S.M., RAJESHAWARA RAO, N. and MAHALAKSHMI, P. (2012) A Comparative Study on the distribution of Foraminifera and Ostracoda in the Inner shelf off Pulicat and Palar, Tamil Nadu, Southeast Coast of India. Gond. Geol. Mag., sp. vol. 13, pp.153-162.

HUSSAIN, S.M., RAVI, G., MOHAN, S.P. and RAJESHWARA RAO, N. (2004) Recent benthic Ostracoda from the Bay of Bengal, off Karikkattukuppam (near Chennai), southeast coast of India – implication on microenvironments. Inter. J. Environ. Micropal. Microbio. Meioben. v.1, pp.105-121.

HUSSAIN, S.M., RAVI, G., MOHAN, S.P. and RAJESHWARA RAO, N. (2009) Distribution of Recent Ostracoda in the Bay of Bengal, off Karikkattukuppam (near Chennai), southeast coast of India - implication for biodiversity. Gond. Geol. Mag., v.24 (1), pp.35-39.

JAIN, S.P. (1976) Holocene Ostracoda from the Chilka Lake, Orissa. *In* Proc. VI Indian Colloq. Micropal. Strat., pp.126-134.

JAIN, S.P. (1978) Recent Ostracoda from Mandvi Beach, west coast of India, Bull. Ind. Geol. Assoc. v.11 (2), pp.89-139.

JAIN, S.P. (1981) Recent Ostracoda from the south-west Kerala Coast, India Bull. Ind. Geol. Assoc., v.14(2), pp.107-120.

JAMES, C.M. (1973a) New distributional records of species of the Subfamily Asteropinae (Ostracoda) from Indian Seas. J. Mar. Biol. Assoc. India, v.15, pp.604-615.

JAMES, C.M. (1973b) *Bathyconchoecia lacunosa* (Muller) a rare Halocypris ostracod from the Arabian Sea. J. Mar. Biol. Assoc. India, v.15, pp.433-437.

KATHAL, P.K. (2012) Applied Geological Micropaleontology, Scientific Publishers, New Delhi, (ISBN 978-81-7233-753-7), 230p.

KATHAL, P.K., BHALLA, S.N. and NIGAM, R. (2000) Foramgeographical affinities of the west and east coasts of India: an approach through cluster analysis and comparison of taxonomical, environmental and ecological parameters of Recent foraminiferal thanatotopes. Bull. Oil & Nat. Gas Corp., India, v.37(2), pp.65-75.

KHOSLA, S.C., MATHUR, A.K. and PANT, P.C. (1982) Ecology and distribution of Recent ostracods in the Miani lagoon, Saurashtra coast. *In* Proc. *In* Rec. Res. Geol. Sr.), Sem. Quaternary Environ, Baroda MEHR, V. (Ed.), Nat. Hindustan Publishing Corp., v.9, pp.361-371.

KLIE, W. (1927). Zur Kenntnis der mikrofauna von British Indian. I. Ostracoda. Rec. Indian Mus., v.29, pp.157-165.

KUMAR, V. and HUSSAIN, S.M. (1997) A Report on Recent Ostracoda from Pitchavaram mangroves, Tamil Nadu. Geosci. J. v.18 (2), pp.131-139.

MADDOCKS, R.F. (1969a) Recent ostracods of the Family Pontocyprididae chiefly from the Indian Ocean, Smith. Inst., Contr., Zool., v.7, pp.1-56.

MADDOCKS, R.F. (1969b) Revision of Recent Bairdiidae (Ostracoda), Bull.U.S. Nat.Mus.,v.295,pp.1-26.

MADDOCKS, R.F. (1988) One hundred million years of predation on ostracods: the fossil record in Texas, In: Evolutionary Biology of Ostracoda, its Fundamentals and Applications. *In* Proc. IX Inter. Sym. Ostracoda, HANNAI, T., NORIYUKI, I. and ISHIZAKI, K. (eds.), pp.637-649.

MAHALAKSHMI, P. and HUSSAIN, S.M. (2015) Distribution of Ostracods from Pazhaverkadu (Pulicat Lagoon), Tamil Nadu, India. (Chapter- 16). *In*: lakes and wetlands, VASUDEVEN, S. et al. (eds.) Partridge India, pp.268-287.

MALIK, M. A. and AMARJIT SINGH (1994). Distribution and ecology of Recent ostracodes from Mansar Lake, Jammu. J. Pal. Soc. India, v.39, pp.5-11.

MALZ, H. and LORD, A. (1988) Recent ornate bairdiid ostracoda: origin and Distribution. *In* HANNAI, T., NORIYUKI, I. and ISHIZAKI, K. (eds.), evolutionary bioloigy of Ostracods, its fundamentals and applications. Dev. Pal. Stratigr., Elsevier, Kodansha, Tokyo, pp.63-74, pls.1-3.

MANNIKERI, M. S. and VAIDYA, A. S. (1987). Ecology and Distribution of Recent freshwater ostracodes from Dharwad, Karnataka. *In* Proc. Role of Earth Sciences in environment, Bombay: 209-305.

MANNIKERI, M. S. and VAIDYA, A. S. (1990). New freshwater ostracodes from Dharwad, Karnadaka. J. Pal. Soc. India, v.35, pp.17-24.

MANNIKERI, M. S., VAIDYA, A. S. and HABIBNIA, B. A. (1989) Additional freshwater Ostracods from Dharwad, Karnataka. Geo. Sci. Jour., v.10, pp.165-174.

McKENZIE, K.G. and GUHA, D.K (1987) A comparative analysis of Eocene/Oligocene boundary Ostracoda from south-eastern Australia and India with respect to their usefulness as indicators of petroleum potential. Trans. Roy. Soc., S. Australia, v.III, pt.l, pp.15-23.

MICHAEL, R.C. and VICTOR, R. (1975) Redescription of two little known Indian freshwater ostracodes *Oncocypris pustulose* Gurney, 1916 and *Cypris malthoaii* Arora, 1931. J. Nat. Hist., v.9, pp.509-512.

MISRA, R.S. and SHRIVASTAVA, P.C. (1979) Recent foraminifera and Ostracoda of Tuticorin, Gulf of Mannar. Geol. Surv. India, misc. pub. v.45, pp.289-309.

MOHAMMED NISHATH, N., HUSSAIN, S.M. and RAJKUMAR, A. (2015) Distribution of Ostracoda in the sediments of northwestern part of Bay of Bengal, India-implications for microenvironment. J. Pal. Soc. India, v.60 (2), pp.27-33.

MOHAN, S.P., RAVI, G. and HUSSAIN, S.M. (2002) Distribution of recent benthic Ostracoda off Karikkattukuppam (Near Chennai), southeast coast of India. Ind. Jour. Mar. Sci, v.31, pp.315-320.

MOHAN, S.P., RAVI, G., HUSSAIN, S.M. and RAJESHWARA, RAO. (2001) Recent Ostracoda from Bay of Bengal, off Karikkattukupam (Near Chennai), SE coast, India. J. Pal.Soc.India, v.46, pp.1-14.

MURRAY, J.W. (1973) Distribution and ecology of living benthic foraminiferids, Heinemann Education Books, London, 274p.

NAIDU, T.Y., VARMA, K.U. and SHYAM SUNDER, V.V. (1997) Diversity and distribution of Recent Ostracoda in the continental shelf sediments off Pentakota and Kalingapatnam, central east coast of India. J. Geol. Soc. India, v.50, pp.727-738.

ODUM, E.P. (1971) Fundamentals of Ecology, III Edition, W.B. Saunders, London Toppan Co. Ltd., Japan, 557p.

OERTLI, H.J. (1971) The aspects of ostracod faunas - a possible new tool in petroleum sedimentology, Bull. Centre. Rech. Pau-SNPA, 5, Suppl., pp.137-151.

POKORNY, V. (1965) Principles of Zoological Micropaleontology. 2, Pergamon Press, London, 465p.

PURI, H.S. (1966) Ecologic and distribution of Recent Ostracoda, Proc. Symp. Crustacea, Pt.I. Mar. Biol. Assoc. India, Mandapam, pp.457-495.

RAJESH RAGHUNATH, SREEDAHRA MURTHY, T.R. and HUSSAIN, S.M. (1999) Distribution of ostracodes in innershelf off Kasargod, SW coast of India. Indian J. Mar. Sci., v.28, pp.302-306

RATNAM, K. V. and RAO, M.P. (1965) Study of foraminifera and Ostracoda in the coarse fractions of the sediments of Chilka Lake. *In* Proc. Ind. Sci. Congr., 3, pp.193-194.

RAVI, G., HUSSAIN, S.M., VAIDYA, A.S., and MOHAN, S.P. (2007) Recent freshwater Ostracoda from Perumal Eri (lake), Cuddalore district, Tamil Nadu- zoogeographic implications. J. Pal. Soc. India, vol.52(2), pp.121-127.

REYMENT, R.A. (1971) Introduction to Quantitative Palaeoecology. Elsevier, Amsterdam, 213p.

RUIZ, F. (2009) Manuel Abad, Luís Miguel Cáceres, Joaquín Rodríguez Vidal, María Isabel Carretero, Manuel Pozo, María Luz González-Regalado, Ostracods as tsunami tracers in Holocene sequences, Quat. Res., doi:10.1016/j.yqres.2009.08.005.

SARASWATI, P.K. and SRINIVASAN, M.S. (2016) Microplaeontology- principles and applications. Springer Inter. Pub., Switzerland, 224 p.

SCOTT, A. (1905) Report on the Ostracoda collected by Professor Herdman at Ceylon in 1902. *In*: Report to the Government of Ceylon on the Pearl Oyster Fisheries of the Gulf of Mannar, pt.3(22), pp.365-384.

SHYAM SUNDER, V.V., VARMA, K.U. and NAIDU, T.Y. (1995) Recent Ostracoda of the Goguleru creek, east coast of India. J. Geol. Soc. India, v.45 (4), pp.471-481.

SLITER, W.V. (1971) Upper Cretaceous foraminifera from southern California and north western Baja California, Mexico, Univ. Kansas. Palaeont. Inst., pp.1-14.

SREENIVAS, K., RAJU, B.N. HONAPPA and REDDI, K.R. (1991) Ostracoda in the estiuraine sediments, Pulicat Lake esturay, east coast of India. J. Geol. Soc. India, v.37(5), pp.492-499.

SRIDHAR, S.G.D., BASKAR, K., MANIYARASAN, S. and HUSSAIN, S.M. (2015a) Sediment geochemistry with population of Recent benthic Ostracoda in Palk Bay, south-east coast of India. Geosci. J., DOI 10.1007/ s12303-015-0035-0.

SRIDHAR, S.G.D., BASKAR, K., MANIYARASAN, S. and HUSSAIN, S.M. (2015b) Shell chemistry of Cytherellidae, Bythocyprididae and Bairdiidae Family, Recent benthic Ostracoda, off Rameswaram, Tamil Nadu, Palk Bay, southeast coast of India. J. Acad. Ind. Res. (JAIR), v.3(9), pp.427-432.

SRIDHAR, S.G.D., HUSSAIN, S.M., KUMAR, V. and PERIAKALI, P. (1998) Benthic ostracod responses to sediments in the Palk Bay, off Rameswaram, south-east coast of India. J. Indian Assoc. Sedimentol., v.17 (2), pp.187-195.

SRIDHAR, S. G. D., HUSSAIN, S. M., KUMAR, V. and PERIAKALI P. (2002) Recent Ostracoda from Palk Bay, off Rameswaram, southeast coast of India. J. Pal. Soc. India. v.47, pp.17-39.

SRIDHAR, S.G.D., HUSSAIN, S.M., PERIAKALI, P. and KUMAR, V. (2007) Ostracod *Kotoracythere inconspicua* (Brady) from the Palk Bay, off Rameswaram, southeast coast of India: its zoogeography and ecology. J. Geol. Soc. India, v.70, no.6, pp.981-986.

SUNNY GEORGE, MARTENS, K. and NAYAR, C.K.G. (1993) Two new species of freshwater Ostracoda of the genus Parastenocypris Hartmann, 1964 from Kerala, India. Hydrobiologia, v.254, pp.183-193.

TAPASWI (1980) Recent ostracods from offshore sediments of Nalgon and Adgaru areas, west coast of India. *In*: Review of Quaternary ostracods from west and east coast of India. Geoviews, 1983, Honnappa (Ed.), v.10(6), 230 p.

TITTERTON, R and WHATLEY, R.C. (1988b) The provincial distribution of shallow water Indo-Pacific marine Ostracoda: origins, antiquity, dispersal routes and mechanism. *In* Proc. IX Internat. Sym. Ostracoda, pp.759-786.

UMA MAHESHWARI, HEMA ACHYUTHAN, GANDHI, S and HUSSAIN, S.M (2009) Foraminifera and Ostracods: signatures for middle Holocene environmental change, Muttukadu, Chennai, India. J. Pal. Soc. India, v.54(1), pp.19-26.

VAIDYA, A.S. (1996) A note on the zoogeographic distribution of Recent freshwater ostracoda from Dharwad, north Karnataka. Contr. *In* Proc. XV ICMS, Dehra Dun, pp.439-445.

VAIDYA, A.S. and MANNIKERI, M.S. (1994) Faunal affinity and zoogeography of Recent marine Ostracoda from Karwar, west coast of India. Curr. Sci., v.67 (9 & 10, 10 & 25), pp.735-738.

VAIDYA, A.S., MANNIKERI, M.S. and CHAVADI, V.C. (1995) Some relationship between the bottom sediments and Recent Ostracoda: A case study. J. Ind. Assoc. Sedimentol., v.14, pp.83-88.

VARMA, K.U., SHYAM SUNDER, V.V. and NAIDU, T.Y. (1993) Recent Ostracoda of the Tekkali Creek, east coast of India. J. Geol. Soc. India, v.41(6), pp.551-560.

VICTOR, R. and FERNANDO, C. H. (1979) The freshwater ostracods (Crustacea: Ostracoda) of India. Rec. Zool. Surv. India. v.74, pp.147-242.

VICTOR, R. and MICHAEL, R.G. (1975) Nine new species of freshwater ostracode from Madurai area in Southern India. J. Nal. Hist., v.9, pp.361-376.

WITTE, L. (1993) Taxonomy and biogeography of West African beach ostracods, Proc. Koninkl. Nederl. Akad. Wetensch., v.39, pp.13-105.

XIA, J., ITO, E. and ENGSTROM, D.R. (1997) Geochemistry of ostracode calcite: Part 1. The experimental determination of oxygen isotope fractionation. Geochemica et Cosmochemica Acta. v.61(2), pp.377-382.

YASSINI, I. and JONES, B. G. (1995) Recent Foraminifera and Ostracoda from Estuarine and Shelf Environments on the Southeastern Coast of Australia, University of Wollongo.

Micropaleontology and its Applications
Pradeep K. Kathal, Rajiv Nigam & Abu Talib (Editors)
Scientific Publishers (India), 2018, 163-189 pp.

Siliceous Microfossils of the Neogene Sediments from Northern Indian Ocean

Arindam Chakraborty and Amit K. Ghosh*

Birbal Sahni Institute of Palaeosciences, 53 University Road, Lucknow - 226 007, India
***E-mail:** akghosh_in@yahoo.com; amitbsip@gmail.com

Abstract

The continental margins of Indo-Pacific realm are predominated by Neogene marine siliceous rocks. Silica present in these sediments is essentially biogenic in origin that is represented by the abundance of several siliceous microfossils like diatoms, radiolarians, silicoflagellates and sponge spicules. Recent developments in the micropalaeontological research using microfossils help us towards global correlation of different outcrops and the sediment cores. Earlier, several works have been carried out in this region but still there are some gaps in correlation. Biostratigraphic and palaeoecological studies pertaining to the siliceous microfossils in deep sea sediments from the northern Indian Ocean are remarkably few in comparison to the information available from the equatorial Pacific and Atlantic oceans. Abundance and diversity of siliceous microfossils are very significant indicators of past climate as evident from the studies carried out in the equatorial Pacific. The climatic condition was cooler at the higher latitudes during the Neogene (except Middle Miocene Climate Optimum and early Pliocene) that was accompanied by the migration towards equator of the cold water climatic zones. During the early Miocene the climatic condition at the eastern and western sectors were completely different in the northern Indian Ocean but gradually in the late Miocene this disappeared and the latitudinal zonality became more prominent. Studies from the deep sea cores of Indian Ocean show that there were two major cycles of biogenic silica, once during the middle Eocene to early Oligocene and the other during the late Miocene. Overall diatom productivity dwindled during the late Oligocene to middle Miocene and moreover, there is a hiatus during the Oligo-Miocene boundary. The opaline silica value was moderate during the Oligocene through middle Miocene that increased in the latest Miocene, which is more or less equivalent to the present value. When investigation was carried out by us on the outcrop samples of Andaman and Nicobar Basin these events were also evident. The siliceous microfossils of Andaman and Nicobar are represented by diatoms, radiolarians, silicoflagellates, sponge spicules etc. As observed, in most of the samples, the silicified microfossils are present in association with the calcareous nannofossils and foraminifers. The siliceous microfossils from the early to middle Miocene outcrops in Havelock Island are poorly preserved and most of them are broken and dissolute. The assemblage of diatoms is represented by a marker diatom taxa *viz., Rosseilla paleacea* that broadly indicate the geological age from early to Middle Miocene. The age has also been reconfirmed by marker radiolarian *Stichocorys wolfii* (RN 4 Zone). Abundance and diversity of well preserved siliceous microfossils remarkably increased in the late Miocene sediments as observed in the outcrop samples of Neil Island. The diatom assemblage is represented by *Actinocyclus ellipticus*, *A. ellipticus* var. *javanica*, *Actinoptychus senarius*, *Paralia sulcata* and *Thalassionema nitzschioides* indicate a late Miocene age though *Actinoptychus senarius*, *Paralia sulcata* and *Thalassionema nitzschioides* are still extant. During microscopic observation we also have identified the marker radiolarians of RN 8 and RN 9 zones. In the early Pliocene, the biogenic silica content became very low and the organisms were probably dissolute and very less silicified microfossils were found as observed from the study of outcrop samples in Car Nicobar Island. The significant diatom taxa are represented by *Azpeitia nodulifera*, *Coscinodiscus radiatus*, *Stictodiscus nankoorensis* and *Triceratium favus*. The sponge spicules and the radiolarians were more dominant in comparison to other silicified microfossils. Occurrence of radiolarian taxon *Didyomocyrtis avita* (RN 10 Zone) in the studied section is significant as it is a marker of Zanclean sediments from latitude lower than 40°.

Keywords: Siliceous microfossils, Neogene, Northern Indian Ocean, Andaman and Nicobar Basin.

INTRODUCTION

Silica is an important component of the ocean water. Generally several organisms *viz.,* diatoms, radiolarians, sponges and silicoflagellates utilize silica from the surface water for their growth. These organisms mainly depend on the availability of silicic acid for constructing their cell walls. If the supply of this component is exhausted, the organisms collapse which led large fraction of them to sink at the ocean bottom. The sinking of these dead materials dissolves the silica from the cell wall with the sediments and thus increases the percentage of silica in the deep waters. To understand the Cenozoic evolution of global weather, productivity and carbon accumulation the history of oceanic silica deposition must be studied in detail. Mainly silica is released by weathering of continental and oceanic crusts (Tréguer et al., 1995). However, dissolved silica in oceans has a short residence time (Heath, 1974), for this marine opal deposition can be used as a proxy for the intensity of silicate weathering and cycling over the geological time. In the present day oceans, the residence time of silica are short because most of it is rapidly removed from the water and sequestered as opaline sedimentary silica by marine silicified organisms (Tréguer and Pondaven, 2000 and Renaudie, 2016). About half of the mass of biogenic sediments accumulating on the deep-sea floor are dominated by siliceous microfossils (Broecker and Peng, 1982). Mostly among them, diatoms and radiolarians are the dominant players. Diatoms contribute at least 20% of the annual primary productivity and also play an important role in the biogeochemical cycles. Diatoms are eukaryotic oxygenic photo-autotrophs characterized by opaline silica frustules that can be preserved in the fossil record. Marine diatoms have diversified dramatically over the Cenozoic (Lazarus et al., 2014) that initially originated from primarily coastal, benthic ancestors in the late Cretaceous (Fenner, 1985) whereas, the molecular clock estimates their origination to be as early as 165-240 Ma ago (Finkel et al., 2005). Over the latter half of the Cenozoic era, the rise of marine diatom diversity (Finkel et al., 2005) is mirrored by a simultaneous decrease in the weight of radiolarian tests. Sometimes rates of continental weathering and erosion could also increase the concentration of Orthosilicic acid (H_4SiO_4) in the oceanic surface waters over the geological time. This may also be responsible for the rise of diatoms that in turn may enhance the primary productivity. In the present-day oceans, radiolarians are also one of the most significant components of the silicified plankton community other than the diatoms. Like diatoms, radiolarians also build their tests of amorphous silica by precipitating the acid from seawater. Radiolaria are siliceous unicellular microprotozoans inhabiting in exclusively marine environment. Although, these organisms inhabit in almost all depths, but their abundance is more in the photic region of the water column (Kling, 1979) and in the deeper part of the ocean their abundance decreases drastically. They are also among the chief microfossils in the siliceous oozes of the world oceans and constitute major part of the preserved silica in fossil record. They have considerable influence on the oceanic silica cycle. This old group of protozoans have been recorded from the Precambrian strata (Campbell, 1954), however, this report is disputed. From the early Cambrian black cherts of Yangtze Platform, China (Braun et al., 2007) record of radiolarians is confirmed. Based on molecular data it is evident that radiolarians originated approximately 1 billion years ago (Knoll, 1992; Hedges et al., 2001). There is lack of any fossil evidence in support of these molecular data. Radiolarians have been utilized as a very useful tool all over the world for biostratigraphic, biogeographic, palaeoclimatic and palaeoceanographic investigations. Indian Ocean is poorly studied in comparison to the Pacific and Atlantic oceans. Qualitative studies on radiolaria from the surface sediments of Indian Ocean was done by Nigrini (1967), Petrushevskaya (1967) and Johnson and Nigrini (1980, 1982), whereas, Lozano and Hays (1976), Dow (1978) and Morley (1989) have made some quantitative estimates of radiolarian in the surface sediment samples from southern Indian Ocean. Morley (1989) has used transfer functions while studying down core variations in radiolarian fauna and reconstructed the palaeoceanography of the Indian Ocean region.

The intense climatic changes, from the 'greenhouse' climate of early to Middle Cenozoic to the 'icehouse' climate of late Cenozoic occurred in the Neogene period due to several factors like atmospheric CO_2, oceanic heat transport etc. (Fauquette et al., 2007). This was a time of great climatic, floral and faunal changes across the globe. The episodes of future climate change due to anthropological activities may be predicted by understanding the global climate during the Neogene (Salzmann et al., 2009). Observing the climatic evolution is currently a major scientific focus for the trend in the present day global warming. As a matter of fact, the focus of international scientific community is on several aspects of climate research. Geosciences play a major role in deducing the past climate change. Palaeoecological and palaeoenvironmental studies may help to understand these changes through time. Studies of past climate provide useful information that may predict future climate changes. Different instruments provide climate records for the last few centuries, but not on a geological scale (Plaut et al., 1995). To have an idea about these changes micro-fossils provide valuable evidences of that time. In respect the silicified microfossils may provide an important clue for minor climatic changes in the regional and global perspective. The present study is mainly concentrated on the northern Indian Ocean and more specifically in the Andaman and Nicobar Basin. Certain workers have designated region as northern Indian Ocean (Srinivasan and Azmi, 1979).

GEOLOGICAL SETTING

The Andaman-Nicobar Basin evolved due to a complex tectonic activity associated with convergent plate boundaries between Indian and west Burmese plates. The initiation of this basin occurred by the northern drift of Indian plate with respect to Asia. At about 59 Ma, the soft collision of the Indian plate with the Burmese plate took place (Klootwijk et al., 1992). This continental collision slowed down the spreading of sea floor in Indian Ocean and the subduction velocity along the Sunda-Arc system that gave rise to the depression between the main ridge and the volcanic arc. At the end of the Eocene and early Oligocene, the main ridge emerged and at that time major upliftment of the islands of Andaman and Nicobar Group initiated (Curray et al., 1979). However, according to Rodolfo (1969 a,b), this basin originated in the late Miocene and that was followed by another diastrophism in the late Pliocene. These islands are part of a continuous ridge belonging to Arakan-Yoma Range of western Burma to the festoon of islands, south and west of Sumatra. The islands can be separated into two concentric arcs: outer sedimentary arc (major islands of Andaman) and the Nicobars. The Andaman and Nicobar Islands in the Bay of Bengal represent a thick deposition of marine sediments ranging from late Mesozoic to Quaternary and is divisible into five lithostratigraphic units i.e., Porlob, Serpentine, Baratang, Port Blair and Archipelago groups in ascending sequence (Karunakaran et al., 1968 and Srinivasan, 1986). Neogene sediments are well exposed in different islands of Ritchie's Archipelago and Nicobar Group of islands. The present analysis is mainly focused on the Havelock and Neil islands of Ritchie's Archipelago and Car Nicobar Island of the Nicobar Group.

Havelock Island

The outcrops in Havelock Island consist of sediments dominated by calcareous mudstone and sandstone. The age of the sediments range from early to late Miocene. It has very good exposures of the fossiliferous rocks at several areas. Two lithological successions *viz.,* Inglis and Long formations are exposed in this island. Srinivasan and Azmi (1976b) suggested an early to late Miocene age on the planktonic foraminiferal zones (*Catapsydrax stainforthi* and *Neogloboquadrina acostaensis*) in Havelock Island. The Kalapathar Section is of Inglis Formation and the lithological composition of the section is soft creamish nanno-foram chalk with interbedded calcareous silty bands. Lacam Point Section consists of moderately hard, creamish white nanno-foram chalk that represents Inglis Formation. The outcrop is situated towards the northwest point of the island near the lighthouse as a small cliff. The South Point Cliff Section is situated in the southern most point of the island. The lithology of this section is represented by siltstone, chalk, sandy limestone and silty mudstone which are in alternate sequence.

Neil Island

Late Miocene to Pleistocene outcrops are distributed all along the east and west coasts of Neil Island. Two lithostratigraphic formations are recognised in this island i.e., Sawai Bay Formation and Neil West Coast Formation (Srinivasan and Azmi, 1976a). The Sawai Bay Formation is exposed on the eastern and central part of the island. Lithologically, it is characterised by light grey, calcareous mudstones. The type locality of Neil West Coast Formation is exposed on the west coast of the island with two units *viz.,* a lower silty mudstone and an upper limestone member. Lithologically it is comprised of fine to medium, compact to porous, buff, foraminiferal-algal biocalcarenite with moderately sorted, broken fragments of foraminifera, corals, algae and polycrystalline quartz grain. Based on planktonic foraminiferal biostratigraphy (*Globorotalia tosaensis* Zone to *Globorotalia truncatulinoides* Zone) (Sharma and Srinivasan, 2007), late Pliocene to Pleistocene was assigned for the Neil West Coast Formation.

Car Nicobar Island

The Car Nicobar Island belonging to Nicobar Group possesses three major lithostratigraphic formations *viz.,* Sawai Bay Formation (Srinivasan and Azmi, 1976b), Guitar Formation (Srinivasan and Azmi, 1976b) and Malacca Limestone Formation (Srinivasan and Sharma, 1973). According to Sharma and Srinivasan (2007), the "arenaceous limestone" (Jacob and Sastri, 1951), Sawai Bay limestone Member (Srinivasan and Sharma, 1973) and the Kakana Formation (Chandra and Saxena, 1998; Chandra et al., 1999; Ghosh et al., 2004) corresponds to the Guitar Formation. The fourth lithostratigraphic unit, Long Formation has recently been visualised near Sawai Bay locality after the 2004 tsunami (Sharma and Bajpai, 2011). The Sawai Bay Formation is the lowermost unit characterized by light grey to bluish grey, moderately hard, calcareous mudstone. Srinivasan and Sharma (1973) designated Sawai Bay of Car Nicobar as the type locality of the formation. The Guitar Formation is dominated by hard buff coloured limestones having abundant algal and forminiferal remains. Based on planktonic foraminifera, the Sawai Bay Formation has been considered as late Miocene to early Pliocene by Srinivasan and Azmi (1976b) in having the presence of *Globorotalia plesiotumida* to *Globigerina margaritae* zones and the Guitar Formation spans from middle to late Pliocene *Neogloboquadrina acostaensis* to *Globigerinoides fistulosus* zones.

SILICIFIED MICROFOSSILS FROM NORTHERN INDIAN OCEAN

Diatoms

The record of fossil diatoms from marine sediments of Northern Indian Ocean are very scanty though the first report of marine fossil diatoms was made by Ehrenberg in 1851 from the Miocene rocks of Nancowry Island. These microfossils from Nancowry were later worked out by Grunow (1867), Van Heurck (1880-1885) and many others. In 1870, Ball reported diatoms from Trinket Island, Andaman-Nicobar Group. The comprehensive report of fossil diatoms from the sediment cores of the Indian Ocean region were first described by Kolbe (1957). He described the Quaternary marine diatoms from seven geographical regions of Indian Ocean (23 selected sediment cores) under the Swedish Deep-Sea Expedition during 1947-48. He also depicted that the assemblage is more or less similar in all the major oceans but the relative frequencies differ in several regions. This was followed by the contributions of Schrader (1974). He described the Cenozoic diatom from ODP Leg 22, 23 and 24. He divided the sediments into twenty one tropical Indian Ocean diatom zones. At Site 236 (Leg 24) rare diatom valves were present. In those sediments the dominance of the sponge spicules indicates strong dissolution of biogenic opal. Moderate to well preserved diatoms were common to abundant in Site 238 (Leg 24). In Site 213 (Leg 22) and Site 215 (Leg 23) the diatom and radiolarian ooze suggests onset of high productivity with only siliceous components preserved below the Carbon Compensation Depth (CCD) during the late Miocene. Later on, Desikachary et al. (1987) and Desikachary (1989) worked on several DSDP samples of the Indian Ocean region and published an 'Atlas on diatoms'. Gowthaman and Desikachary (1989) have reported some new and interesting fossil diatoms from these cores of DSDP. A new genus *Quadrodiscus* was introduced by Prema and Desikachary (1989) from the DSDP samples of the Indian Ocean. A small number of publications on fossil diatoms appeared from Andaman and Nicobar Group of Islands during the last four decades. Ghosh and Maitra (1947) reported 27 forms belonging to 14 genera of diatoms from the Miocene of Nicobar Group of islands. Boileau (1950) and Jacob and Shrivastava (1952) identified the presence of diatoms in the Miocene rocks of Colebrook Island and in the islands of Ritchie's Archipelago without any detailed description and taxonomic identification. This gap was filled by the study of Desikachary and Maheshwari (1958) where they identified the species from the rocks of Colebrook Island and also recorded some new fossil diatoms. In addition to this they listed 133 species of diatoms known till then from the sediments of Nancowry Island. Srinivasan and Srivastava (1975) illustrated the stereoscan of *Annellus californicus* from the middle Miocene of Nancowry Island. Singh and Vimal (1973) listed 7 diatom species from the early Pliocene of Neil Island. Desikachary (1984) reported the occurrence of *Rossiella* (the marker of early Miocene) from the sediments of Nancowry Island. Mathur (1973) reported 15 diatom forms from the Round Formation and Strait Sandstone Formation (early Miocene) of Havelock Island, Ritchie's Archipelago. Singh et al. (1978) have done a detailed study of the fossil diatoms from the Sawai Bay Mudstone Formation, Neil Island and described 30 species of diatoms belonging to 13 genera with 3 new species. Further in 1979, Singh reported some additional diatom taxa. In 1981, Mathur described 10 species of diatoms from the Strait Sandstone Formation exposed at South Point in Outram Island of Ritchie's Archipelago and correlated with the middle Miocene *Craspedodiscus coscinodiscus* diatom zone. Subsequently, Mathur (1985) described 27 diatom species belonging to 18 genera including 1 new species of *Liostephania ovalis* from the middle Miocene sediments of Nicholson Island, Ritchie's Archipelago and correlated with the *Cestodiscus peplum* Zone of early middle Miocene. After a long time gap, Singh and Sharma (1996) worked on the Neil East coast Section and divided the sequence into ten diatom zones. In recent years, Chakraborty and Ghosh (2016) published an account of diatoms from the late Miocene outcrop of Neil Island that includes 82 planktonic and benthic diatoms with an interpretation on the prevalence of monsoonal activity during that time. Diatoms reported from different islands of Andaman and Nicobar Group are represented in Table 1.

Table 1. Contributions on diatoms from different islands of Andaman and Nicobar Group.

Author (Year)	Name of the Island	Age of the region
Ehrenberg (1851)	Nancowry Island	Miocene
Grunow (1867)	Nancowry Island	Age not specifically mentioned
Van Heurck (1880-1885)	Nancowry Island	Age not specifically mentioned
Ball (1870)	Trinket Island	Age not specifically mentioned
Ghosh and Maitra (1947)	Nicobar Island	Miocene
Boileau (1950)	Colebrook Island	Miocene
Jacob and Shrivastava (1952)	Colebrook & other islands, Ritchie's Archipelago	Miocene
Desikachary and Maheswari (1958)	Colebrook Island	Miocene
Srinivasan and Srivastava (1975)	Nancowry Island	Middle Miocene
Singh and Vimal (1973)	Neil Island	early Pliocene
Mathur (1973)	Havelock Island	early Miocene
Singh et al. (1978)	Neil Island	late Miocene-early Pliocene
Singh (1979)	Neil Island	late Miocene-early Pliocene
Mathur (1981)	Outram Island	Middle Miocene
Mathur (1985)	Nicholson Island	Middle Miocene
Singh and Sharma (1996)	Neil Island	late Miocene-early Pliocene
Chakraborty and Ghosh (2016)	Neil Island	late Miocene

Radiolarians

In 1851, Ehrenberg for the first time used the term related to Radiolaria i.e., 'Polycystine rocks'. In 1854 he reported radiolarian species *Rhopalastrum lagena* and *Stylosphaera hispida* from rocks of Kamorta and Nancowry islands respectively. Till the mid of twentieth century the true potential of radiolarians were not recognized in the Indian context.

In 1952, Jacob and Shrivastava reported a fairly rich assemblage of radiolarians from the clays of Colebrook Island and the islands of Ritchie's Archipelago. The radiolarian group Nassellaria are more dominant and Acantharia and Spumellaria are poorly represented in their record. They were not much confirmed about the age of the sediments but a probable age of Miocene was assigned to it. Singh and Vimal (1973) identified some radiolarians from the grey fossiliferous mudstone of East Coast Section, Neil Island. Singh (1981) reported an assemblage of radiolarians from the upper part of the Round Formation exposed in Neil Island. The assemblage shows affinity with the species reported from the stratotype of Zanclean sediments of Java, Philippines and Mariana trench. Further, Srinivasan et al., (1983) gave the first detailed documentation of radiolaria from the early Miocene sequence exposed at Piu Bay, Colebrook Island. Fifty species of radiolaria identified and the assemblage is preferable to the *Stichocorys wolffii* Zone of Reidel and Sanfilippo (1978) which indicates early Miocene age. Sharma and Sharma (1988) illustrated 104 species of radiolarians from late Miocene-early Pliocene sediments of Neil Island. They also compared the assemblage with the modern radiolarian species of Indian Ocean and opined that the fossil assemblage is dominated by more species. Some of the species are characteristic of cold water and upwelling indicators in modern oceans. Sharma and Sharma (1989) identified species from the East Coast Section of Neil Island. They identified two zones as *Didymocyrtis penultima* and *Stichocorys peregina*. They provided an integrated radiolarian and planktonic foramininferal biostratigraphy where the Miocene/Pliocene boundary has been recognised closely. Gupta and Srinivasan (1992) described the late Miocene radiolarians from the Sawai Bay Formation, Neil Island. The cluster analysis of the taxonomic groups divides it into two major clusters which signify the warm water species in one cluster and the colder water species in the other. The presence and absence of deep and intermediate water dwelling radiolarians shallowing of the basin during the late Miocene. Sharma and Singh (1993) studied the radiolarians from the early Pliocene deposits of Sawai Bay Formation exposed at Car Nicobar Island. Three low latitude radiolarian zones were identified and they were also correlated with the central Indian Ocean Basin. An integrated biostragraphic scheme was also presented when correlated with the planktonic foraminiferal zone in the study area. Sharma et al. (1993) recorded 95 species from North Passage Island and the assemblage indicates that the sediments are of late early Miocene sequence. Mahapatra and Sharma (1994) identified 35 trissocyclid radiolarian species from late early Miocene outcrops of several islands of Andaman and Nicobar Group. They also recognized 9 radiolarian events (all First Occurences). Based on 42 deep sea surface sediments, Gupta and Fernandes (1995) carried out a Q-mode factor analysis of the percentage data of 25 broadly defined supra-generic radiolarian groups. The analysis revealed four statistically significant and geographically distinct faunal assemblages and estimated the precipitation (PPT) during South West Monsoon (July-August) and sea surface temperature (SST) during spring (May-July). The changes in SST and PPT were correlated with the late Miocene palaeoceanographic interpretations based on the radiolarian assemblage of the Sawai Bay Formation, Neil Island, Andamans. Sharma and Singh (1997) identified 11 radiolarian events in the late Miocene-early Pliocene sequences of Andaman and Nicobar Islands. Stratigraphic ranges of the important taxa were discussed and the events were compared to that of central Indian Ocean and the other ODP sites of Indian Ocean and Pacific Ocean. In 1998, Sharma and Daneshian reported 133 radiolarian taxa from Nicholson and John Lawrence islands. Based on DSDP and ODP data on various radiolarian events of eastern Pacific region by Shackleton et al. (1995), they estimated an late early Miocene to early middle Miocene age (16.9 – 15.9 Ma). Sharma et al. (1999) provided a detailed database of early middle Miocene radiolarian taxa from land-based sections of Andaman Nicobar Islands that is helpful in the comparative studies between the tropical radiolarian faunas of the Pacific and Indian oceans. They illustrated 120 species belonging to the *Dorcadospyris alata* Zone (approximately 15-13 Ma). Singh et al. (2000) studied the radiolarians with other microfossils from the early Neogene sequence of John Lawrence Island of Andaman in order to compare the stratigraphic ranges for achieving more precise biostratigraphic resolution. They opined that the age connotation based on planktonic foraminifera are more precise than other microfossils. Sharma and Daneshian (2003) used the radiolarian assemblage from seven land based sections in Andaman-Nicobar Islands to understand the palaeoceanographic changes during early Neogene. The late early Miocene assemblage defines warmer surface water condition while in the later phase i.e. early middle Miocene the assemblage denotes cooler condition due to the expansion of cool surface water in the northward part. This cooling phase has also been correlated with the results of planktonic foraminiferal isotopic study. Sharma and Ram (2003) reported radiolarians from 6 sections in Andaman Islands. The assemblages range from early to middle Miocene where 10 radiolarian events were identified. The radiolarian events are comparable to low latitude tropical zonations. Sharma and Bhagyapati (2007a) proposed new radiolarian zonations in early to middle Miocene since the low latitude index taxa of that time cannot be applied satisfactorily. These zones are also useful for the correlation of tropical Indian Ocean. Sharma and Bhagyapati (2007b) studied the radiolarians from the land based sections of Andaman and Nicobar those are indicative of cooling and warming phases in late early Miocene to middle Miocene. They briefly described the Neogene oceanographic and climatic changes in the northern Indian Ocean with the help of radiolarians. Sharma et al., 2011 detected two intervals of faunal turnover i.e. first during late early Miocene and the second during middle Miocene. During the early Miocene cold water species began to increase and at middle Miocene a major radiolarian assemblage was replaced the abundances of taxa, number of species and diversity. Chakraborty and Ghosh (2015) reported

Acrobotrys disolenia from the late Miocene sediments of northern Indian Ocean (Sawai Bay Formation Neil Island) for the first time. This species was previously reported from the equatorial Pacific and South China Sea. Biogeographic studies on planktonic foraminifera and isotopic analysis postulated that the closure of the Indonesian Gateway took place during the early Pliocene (Srinivasan and Sinha, 1998). Chakraborty and Ghosh (2015) interpreted that *A. disolenia* could have migrated from the Pacific Ocean through the Indonesian Gateway with the help of South Equatorial Current. Radiolarians reported from different islands of Andaman and Nicobar Group are represented in Table 2.

Table 2. Contributions on radiolarians from different islands of Andaman and Nicobar Group.

Author (Year)	Name of the Island	Age of the region
Ehrenberg (1854)	Kamorta, Nancowry	Miocene
Jacob and Srivastava (1952)	Colebrook, Neil, Havelock, John Lawrence, Henry Lawrence, Outram islands	Miocene (?)
Singh and Vimal (1973)	Neil Island	early Pliocene
Singh (1981)	Neil Island	
Srinivasan et al. (1983)	Colebrook Island	early Pliocene
Sharma and Sharma (1988)	Neil Island	late Miocene to early Pliocene
Sharma and Sharma (1989)	Neil Island	Miocene/Pliocene boundary
Gupta and Srinivasan (1992)	Neil Island	late Miocene
Sharma and Singh (1993)	Car Nicobar Island	early Pliocene
Sharma et al. (1993)	North Passage	late early Miocene
Mahapatra and Sharma (1994)	Colebrook, North Passage, Great Nicobar Islands	late early Miocene
Sharma and Singh (1997)	Andaman and Nicobar Island	late Miocene to early Pliocene
Sharma and Daneshian (1998)	Nicholson, John Lawrence Island	late early to early middle Miocene
Sharma et al. (1999)	Andaman and Nicobar Island	early middle Miocene
Singh et al. (2000)	John Lawrence Island	early Neogene
Sharma and Daneshian (2003)	Colebrook, Nancowry, Nicholson, Kamorta Islands	early Neogene
Sharma and Ram (2003)	Strait, Colebrook, Nicholson, Havelock, Henry Lawrence Islands	early to middle Miocene
Sharma and Bhagyapati (2007a)	Andaman and Nicobar Islands	early to middle Miocene
Sharma and Bhagyapati (2007b)	Andaman and Nicobar Islands	early to middle Miocene
Sharma et al. (2011)	Colebrook, Nancowry, Nicholson, Kamorta Islands	early to middle Miocene
Chakraborty and Ghosh (2015)	Neil Island	late Miocene

Fig.1. Map showing different islands of Andaman-Nicobar Basin (studied islands are marked with red circles).

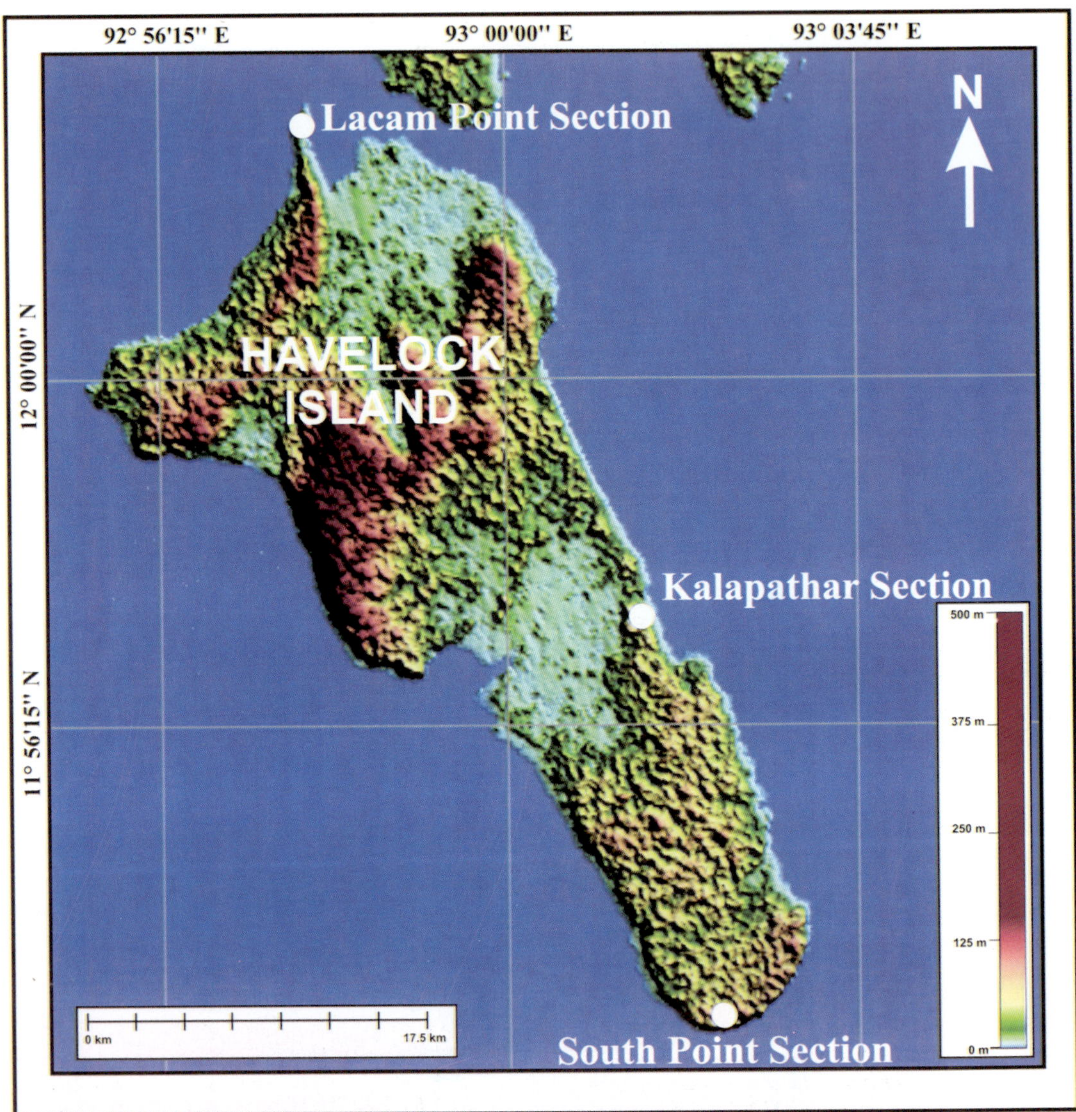

Fig.2. Map of Havelock Island, Andaman-Nicobar Island showing the studied outcrops.

PRESENT OBSERVATION ON SILICEOUS MICROFOSSILS FROM THE NEOGENE SEQUENCE OF ANDAMAN AND NICOBAR BASIN

During the last couple of years we have generated substantial datasets on the siliceous microfossils from the Neogene sequence exposed in different islands (Havelock, Neil and Car Nicobar) of Andaman and Nicobar Basin (Fig. 1). Out of these Havelock and Neil island belong to Ritchie's Archipelago and the Car Nicobar belongs to the Nicobar Group of islands. The siliceous microfossils ranging in age from early Miocene to early Pliocene are represented by diatoms, radiolarians, sponge spicules and silicoflagellates (Plate 1, Figs. A-Z).

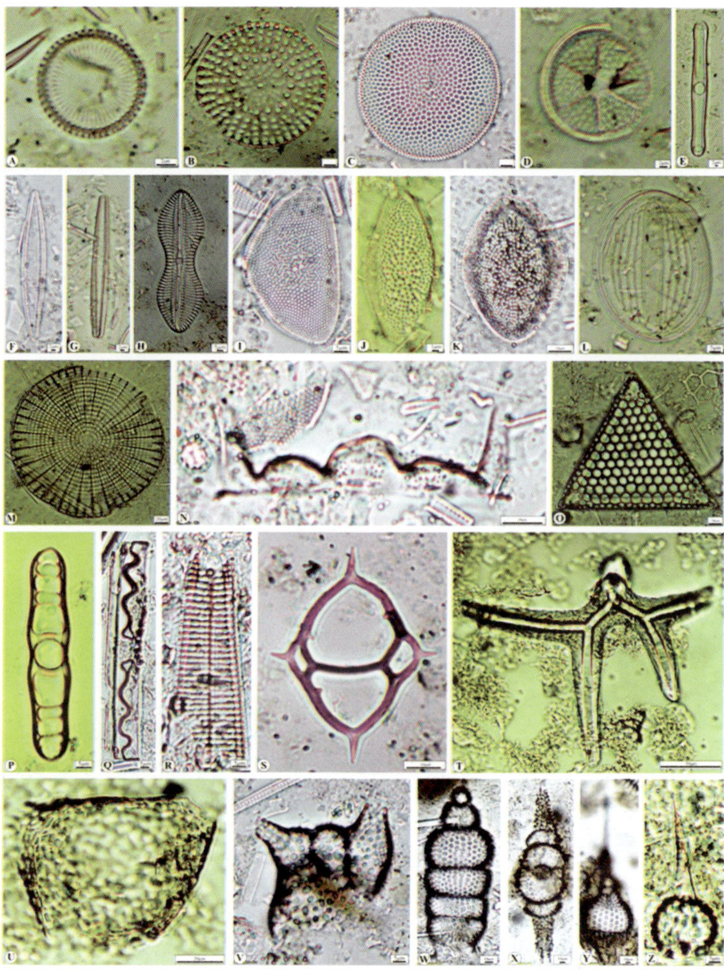

Plate 1. Diatoms (A-R): A. *Paralia sulcata* (Cave Point Section, Neil Island) **B.** *Stictodiscus nankoorensis* (Cave Point Section, Neil Island) **C.** *Azpeitia nodulifera* (East Coast Section, Neil Island) **D.** *Actinoptychus senarius* (Cave Point Section, Neil Island) E. *Grammatophora stricta* (South Point Section, Havelock Island) **F.** *Nitzschia reinholdii* (East Coast Section, Neil Island) **G.** *Thalassionema nitzschioides* (Cave Point Section, Neil Island) **H.** *Diploneis crabro* (East Coast Section, Neil Island) **I.** *Hemidiscus cuneiformis* (Cave Point Section, Neil Island) **J.** *Actinoptychus ellipticus* (Cave Point Section, Neil Island) **K.** *Actinoptychus ellipticus* var. *javanica* (Cave Point Section, Neil Island) **L.** *Cocconeis pellucida* (Cave Point Section, Neil Island) **M.** *Arachnoidiscus ornatus* (Cave Point Section, Neil Island) **N.** *Biddulphia tridens* (Cave Point Section, Neil Island) **O.** *Triceratium favus* (Mus Jetty Section, Car Nicobar Island) **P.** *Grammatophora serpentina* (South Point Section, Havelock Island) **Q.** *Grammatophora longissima* (South Point Section, Havelock Island) **R.** *Gephyria media* var. *ornata* (South Point Section, Havelock Island) **S.** Silicoflagellate - *Dictyocha fibula ausonia* (Mus Jetty Section, Car Nicobar Island) **T.** Sponge spicule - Orthodichotriaenes (Mus Jetty Section, Car Nicobar Island) **Radiolarians (U-Z): U.** *Solenosphaera polymorpha* (Mus Jetty Section, Car Nicobar Island) **V.** *Acrobotrys disolenia* (East Coast Section, Neil Island) **W.** *Stichocorys delmontensis* (Cave Point Section, Neil Island) **X.** *Didyomocyrtis penultima* (East Coast Section, Neil Island) **Y.** *Calocycletta costata* (Kalapathar Section, Havelock Island) **Z.** *Artophormis* sp. (Mus Jetty Section, Car Nicobar Island).

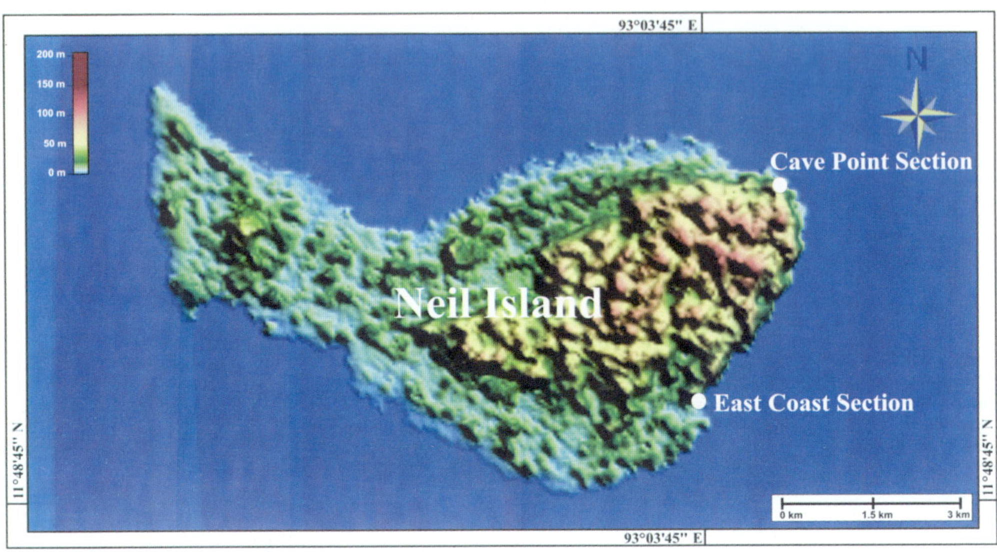

Fig.3. Map of Neil Island, Andaman-Nicobar Island showing the location of outcrops.

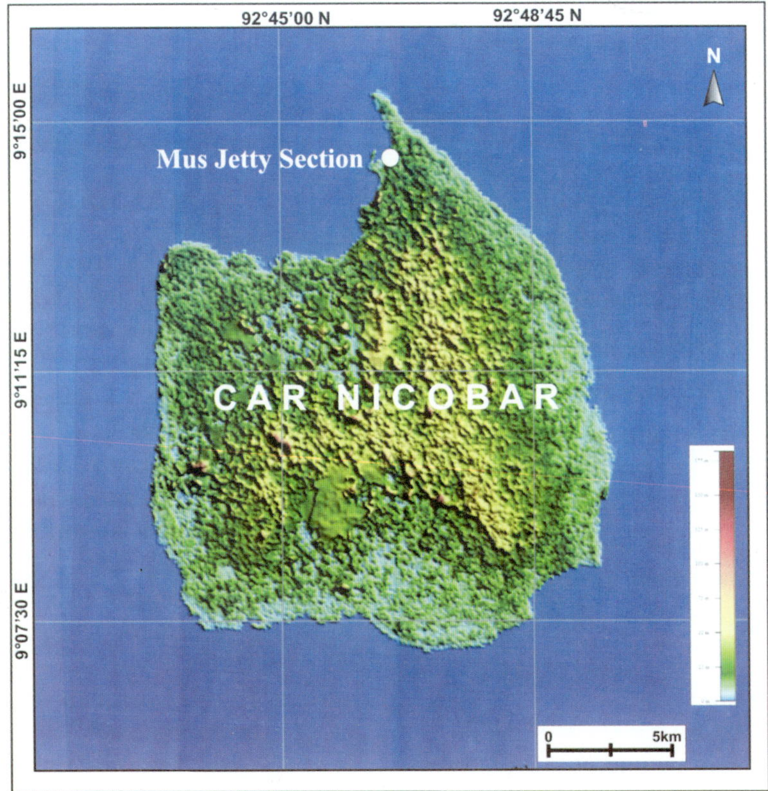

Fig.4. Map of Car Nicobar Island, Andaman-Nicobar Island showing the location of outcrop.

Early to Middle Miocene

We have studied the diversity of silicified organisms in three outcrops (Kalpathar, Lacam Point and South Point sections) representing early to middle Miocene sequence in Havelock Island, Ritchie's Archipelago (Fig. 2). In Kalpathar Section diatom valves are predominantly broken. The diversity is also very less, which is represented by 17 species belonging to 14 genera. The commonly occurring species are *Cestodiscus* sp. cf. *kugleri, Coscinodiscus rhombicus, Syndera* sp., *Nitzschia* sp. and *Thalassiothrix longissima*. The frequency of planktonic and benthic diatoms varies throughout the outcrop. Mostly the planktonic diatoms are useful for palaeo-environmental interpretations. Most of the planktonic diatoms recorded from the studied Section are representatives of warm water (Table 3).

Table 3. Preferred environment of reco-rded planktonic diatoms (early to middle Miocene, Havelock Island).

Species of Diatoms	Shape	Environment
Cestodiscus sp. cf. *kugleri*	Centric	Warm water
Coscinodiscus radiatus	Centric	Warm to Cool water
Cestodiscus peplum	Centric	Warm water
Rossiella paleacea	Pennate	Warm water
Thalassionema nitzschioides var. *obtusa*	Pennate	Warm water
Thalassiothrix longissima	Pennate	Cool water

Rarely complete valves of diatoms are preserved and only those diatoms have been taken into consideration for biozone demarcation. Three assemblage biozones have been recognised and these are as follows:

Assemblage Biozone III: *Rossiella paleacea - Synedra* sp.

Assemblage Biozone II: *Synedra* sp. - *Thalassiothrix longissima*

Assemblage Biozone I: *Actinocyclus octonarius - Nitzschia* sp.

The age of the studied outcrop has been confirmed by marker radiolarians of RN 3 - RN 4 zones of Sanfilipo and Nigrini, 1998. These are namely *Calocycletta costata* and *Stichocorys wolffi*. Planktonic/benthic ratio of the diatoms indicates the evidence of sea level fluctuations. Diatom assemblage of early to middle Miocene outcrop of Lacam Point Section in Havelock Island is dominated by *Amphora* sp., *Azpeitia* sp., *Coscinodiscus rhombicus, Cymatosira* sp., *Diploneis vacillans, Diadesmis* sp., *Stauronensis* sp. and *Biddulphia pulchella*. It should be mentioned here that the size of the diatom valves in this outcrop is substantially less in dimension, may be due to dissolution of silica and deficiency of nutrients. Early to middle Miocene diatoms are comparatively better preserved in the outcrop samples of South Point, Havelock Island. The assemblage is represented by *Stictodiscus nankoorensis, Stictodiscus californicus, Arachnoidiscus ornatus, Mastogloia ovata, Aulacodiscus patulus, Coscinodiscus radiatus, Paralia sulcata, Trigonium tabellarium, Triceratium favus, Biddulphia pulchella, Gephyria media* var. *ornata, Actinocyclus ellipticus, Cocconeis pellucida* var. *nankoorensis, Coscinodiscus minor, Rhizosolenia* sp., *Diploneis crabro, Denticulopsis seminae, Grammatophora angulosa, G. stricta, G. undulata, Omphal-*

opsis australis, *Grammatophora oceanica* var. *macilenta*. From the above analysis it has been observed that there are generic as well as species variation within early to middle Miocene in different outcrops of Havelock Island. However, the age of all the outcrops has been confirmed with the radiolarian assemblage as well as calcareous nannofossil assemblage (NN4-NN5 zone of Martini, 1971). This may be due to variations of facies and lithological features.

Late Miocene

Late Miocene outcrops containing well preserved siliceous microfossils are very well exposed in the Neil Island of Ritchie's Archipelago, Andaman and Nicobar Basin. Our studies on two outcrops of Neil Island (Cave Point and East Coast sections, Fig 3), reveal the occurrences of *Actinocyclus ellipticus*, *A. ellipticus* var. *javanica*, *Actinoptychus senarius*, *Paralia sulcata* and *Thalassionema nitzschioides*. The diatom assemblage indicates a late Miocene age (Kanaya, 1971; Barron, 1973; Mikkelsen, 1990; Gladenkov, 2013) though, some of the diatom taxa *viz.*, *Actinoptychus senarius*, *Paralia sulcata* and *Thalassionema nitzschioides* are extant. In the East Coast Section presence of *Nitzschia reinholdii* indicates that the sediments are also of uppermost Miocene that corresponds to Schrader's (1973) North Pacific Diatom Zone X established from DSDP leg 18 (Site 173) off North California and also *Nitzschia reinholdii* Zone of southern California (Barron 1976). The dominance of this particular species indicates the sediments of *Nitzschia reinholdii* Zone and definitely younger than 7 Ma (Clemens et al. 2016). *Thalassiothrix longissima* is consistently present in both the studied sections, however, its frequency remarkably increased in the upper part. It has been opined by Schrader (1974) that *Thalassiothrix longissima* is consistently present during the Miocene through Recent in the tropical Indian Ocean. Samples from the East Coast Section yielded 61 species of well preserved diatoms belonging to 23 genera represented by 36 planktonic and 25 benthic species.

On the other hand samples from the Cave Point Section yielded 82 species of fossil diatoms represented by 18 centric and 17 pennate genera, out of which 29 species are planktonic and 53 species are benthic (Chakraborty and Ghosh, 2016). The preservation potential of most of the diatom species is very good. Deformity of the diatom valves has not been noticed in any of the analysed samples. The dominance of planktonic diatoms throughout the studied sections indicates deeper waters. This also corresponds to the interpretations of Srinivasan and Azmi (1976b) and Srinivasan and Chatterjee (1981) based on foraminiferal fauna of that time slice, which is represented by more than 90% planktonic forms containing abundant deep water species. According to Srinivasan and Azmi (1976b) the deposition took place at abyssal depths. Based on different microfossils Singh and Vimal (1973) also opined that the sediments were deposited in deep waters. On the other hand, Roy (1983) and Pal et al. (2003) interpreted shallow marine conditions for the late Miocene sediments of Ritchie's Archipelago. Our findings from the late Miocene sequences of Neil Island belonging to Ritchie's Archipelago indicate that the deposition predominantly took place in deeper water with an alteration of a marginally shallow marine regime. Curray (2005) also opined the same view while he interpreted the tectonics and history of the Andaman Sea region. The base of the studied sections is demarcated by the top of *Diartus hughesi* (datum/ event), whereas, the secondary datum/ event is represented by the base of *Didymocyrtis penultima* and *Stichocorys delmontensis* belonging to the low latitude radiolarian zone RN8. On the other hand the RN9 zone is characterised by the top of *Stichocorys delmontensis* and the base of *Stichocorys peregina*. The planktonic/ benthic diatom ratio suggests that benthic habitats were not prevalent which may be due to reduced light intensity that prevented the growth and survival of benthic diatom taxa. The dominance of planktonic diatom taxa was responsible for high primary productivity that in turn resulted diatom bloom owing to river runoff. This may be due to heavy rainfall i.e., intense monsoonal activity. The planktonic/ benthic ratio of diatoms also indicates that the sea level was high throughout however, minor fluctuations of sea level changes also have been demarcated (Chakraborty and Ghosh, 2016). Abundance of siliceous microfossils specifically diatoms are indicative of a nutrient-rich condition. It has been interpreted that during the late Miocene there would have been a high amount of terrigenous input in the oceanic water, responsible for the nutrient-rich condition (Chakraborty and Ghosh, 2016). It has been opined by Curray (2005) that the sediments of the Archipelago might have been derived from the deltaic sediments of the Irrawady River.

Early Pliocene

We have conducted our studies on the siliceous microfossils from the type section of Sawai Bay Formation, Car Nicobar Island of early Pliocene (Zanclean) age (Fig. 4). The abundance and diversity of siliceous microfossils significantly decreased in the early Pliocene of Car Nicobar Island. It is evident from our observation on the late Miocene sequence of Andaman and Nicobar Basin exposed in the islands of Ritchie's Archipelago (Chakraborty and Ghosh, 2015, 2016), in which the biogenic bloom was well developed. The diversity of the siliceous organisms during Zanclean was substantially decreased where only 16 genera of diatoms and 14 genera of radiolarians have been identified in the studied samples. In addition, to this other siliceous microfossils *viz.*, sponge spicules and silicoflagellates have also been recovered. Amongst the diatoms, 13 species are planktonic and 10 are benthic. Some of the commonly occurring diatom species are *Azpeitia nodulifera*, *Coscinodiscus radiatus*, *Stictodiscus nankoorensis* and *Triceratium favus*. In having meagre and sporadic occurrence of diatoms, no diatom assemblage zone has been assigned. Diatoms recovered from the Zanclean sediments of Car Nicobar Island are more or less similar to that of low latitude equatorial assemblage at Site U1337, IODP Expedition 320/321 (Pälike et al. 2010). *Thalassiosira elliptipora* is a significant diatom taxon present in the assemblage that evolved at the upper part of early Pliocene and continued till middle Pleistocene (Gradstein et al. 2012). Another species of *Thalassiosira* i.e., *Thalassiosira leptopus* present in this assemblage is considered as a marker of upper Zanclean (Frydas, 1990; 1999). The present assemblage is devoid of *Thalassionema* group of diatoms that was characteristically present in the late Miocene sediments. Presence of *Azpeitia tabularis* in the Zanclean sediments is indicative of warm water condition (Barron, 1996).

Amongst the 23 diatom taxa, 12 are planktonic that provide good proxy for ecological interpretations. The planktonic diatoms are significant from ecological point of view because these can migrate from one region to another region. Out of the 12 planktonic taxa, eight are of warm water habitat, whereas, three of them are of cold water. During microscopic observation we have identified the marker radiolarians of RN10 (Sanfilippo and Nigrini, 1998). The base of the studied outcrop is demarcated by the base of *Didymocyrtis avita* belonging to the tropical low latitude radiolarian zone RN10 (Sanfilippo and Nigrini, 1998). Some of the commonly occurring radiolarian taxa are *Didymocyrtis avita*, *Rhopalastrum profunda*, *Siphocampe lineata*, *Artophormis* sp., *Semantis* sp. and *Stylochlamydium* sp. Radiolarian assemblage is also represented by warm water species with meagre representation of cold water forms. Silicoflagellates are represented by two species i.e., *Dictyocha fibula* and *D. fibula ausonia*, out of which *D. fibula* is present in most of the samples; however, *D. fibula ausonia* is dominant in upper part of the sequence. Eleven different forms of sponge spicules have been identified in the present siliceous microfossil assemblage. Diversified sponge spicules have also been observed in the studied samples represented commonly by Astrophorida, Proanatriaena, Pinulus and Sigma.

DISCUSSION AND CONCLUSION

The studies from the early to Middle Miocene sequence of Havelock Island yielded poorly preserved diatom valves which are predominantly broken. Jouse (1974) observed the same from the early Miocene diatom flora recorded from the Vityaz Station, central equatorial Pacific. Presence of *Coscinodiscus rhombicus*, *Rossiella paleacea* in the early to Middle Miocene sequence of Havelock Island are characteristics of the early Miocene. These species were also reported both by Jouse (1974) and Bukry (1973) from the lowermost Miocene of central equatorial Pacific. The assemblage from the Havelock Island in having these species also confirms that the age of the sediments as early Miocene. Evidences of poor and sporadic representation of diatoms from the Lower Miocene have also been provided by Schrader (1978) from the equatorial Atlantic Site 366 and subtropical Atlantic Site 369. However, calcareous nannofossils demarcates the age of the sediments of Havelock Island as early to middle Miocene owing to presence of marker calcareous nannofossils represented by NN4-NN5 zones of Martini, 1971. Middle Miocene is one of the warmest periods of the Miocene. This is also evident from the warm water dominant planktonic diatoms (Table 3) from the Havelock Island.

In the late Miocene dominance of the pennate diatom *Thalassionema nitzschioides* is indicative of intense ocean upwelling coupled with high primary productivity (Schuette and Schrader, 1981). Preponderance of *Thalassionema nitszchioides* in the late Miocene of Andaman Nicobar Basin also coincides with the increased upwelling and siliceous fertility in the equatorial Pacific (Barron, 1985; Barron and Baldauf, 1990). Moreover, *Thalassiothrix longissima* is also a very commonly occurring taxon in our study (Chakraborty and Ghosh, 2016). It is also comparable to that of Barron and Keller (1983) where *Thalassiothrix longissima* has been correlated to the high productivity upwelling regions. *Coscinodiscus marginatus* a cold water diatom is very rare in the present assemblage. Its presence is also indicative of intense ocean upwelling (Boersma and Mikkelsen, 1990). Preponderance of *Thalassionema* spp. and poor representation of *Actinoptychus senarius* and *Coscinodiscus marginatus* indicate the impact of polar cooling after the middle Miocene. This resulted the formation of strong thermal gradients and disjunction of high and low latitude surface water (Mikkelsen, 1990). During the late Miocene the appearance of a number of diatom taxa per unit time is generally high that possibly indicates increasing latitudinal thermal gradients coupled with enhanced provincialism Barron (1992). This also corroborates with our findings on the northern Indian Ocean specifically from the late Miocene sediments of Andaman and Nicobar Islands (Chakraborty and Ghosh, 2016). In the context of Asian monsoonal history, the late Miocene is considered as a crucial period. Monsoon intensification in the northern Indian Ocean is well documented during 10-8 Ma (Kroon et al., 1991; Gupta and Srinivasan, 1992; Gupta et al., 2004). On the other hand, in South Asia it intensified during 8-7 Ma (Tang et al., 2011). Diatom bloom as observed in the late Miocene of Neil Island signifies monsoon intensification. In view of this, it can be interpreted that the diatom bloom was possibly related to intensification of the monsoon circulation. The IODP Expedition 353 (Clemens et al., 2016) reveals that diatoms were abundant below the Miocene/Pliocene boundary. They came into conclusion that maximum diatom abundance was between 9.8 and 6.5 Ma at Site U1447. Our study (Chakraborty and Ghosh, 2016) carried out on the late Miocene (Tortonian) outcrop of Neil Island, Andaman-Nicobar Basin also corresponds to that of Clemens et al. (2016).

In the Zanclean sediments (early Pliocene) of Car Nicobar Island, a decline in siliceous microfossils was evident with a distinct pattern of variability amongst various groups of silicified organisms. The sediments of the studied outcrop in Car Nicobar Island yielded few diatom valves in comparison to the radiolarians. Quantitatively *Azpeitia nodulifera*, *Coscinodiscus radiatus*, *Stictodiscus nankoorensis* and *Triceratium favus* are present in most of the samples. The benthic diatom *Triceratium favus* is the most dominant species and almost consistently present in all the diatom yielding samples. The radiolarians *viz.*, *Didymocyrtis avita*, *D. tetrathalamus*, *Octopyle stenozona* and *Tetrapyle octacantha* present in this assemblage are indicative of subtropical region of the Pacific (Pisias and Moore, 1995) and are equivalent to the tropical assemblage of Moore (1978). It was observed that the radiolarian abundance is inversely proportional to the diatom abundance. Perhaps this situation might have occurred due to dissolution of more fragile nature of the diatom valves than the radiolarians or the radiolarians have acted as a predator and captured the diatoms (Swanberg and Anderson, 1985). In other way, this might have occurred due to the differential nutrient availability in the ocean. Amongst the siliceous microfossils, radiolarians were the main contributors to the biogenic opal fraction, followed by diatoms, sponge spicules and silicoflagellates. The assemblages of the radiolarians and diatoms together synergistically imply warm water condition in the northern Indian Ocean during the early Pliocene (Zanclean). The assemblage includes very few cold water diatoms which can be due to the exchange of cold water from the Indian and Pacific oceans.

The preservation of diatom valves is fair and the diversity is low. The decreasing trend of the siliceous microfossil abundance of this region in the early phase of Zanclean may be due to the closure of Pacific Pathway that caused nutrient depletion (Chakraborty and Ghosh, 2015). Opening and closing of the oceanic gateways have enormous influence on the faunal migratory pathways (Srinivasan and Sinha, 1998). Indonesian Gateway is one among them that altered the oceanographic setting of the Indo-Pacific region and its complete closure is intensively debated. Early evolution of this seaway started back at ~25 Ma with major tectonic events during the Pliocene including emergence of islands and mountain formation (Hill and Hall, 2003; Kuhnt et al., 2004). Later on, several studies on the plate tectonics in this region highlighted that a major re-organization took place during ~4-3 Ma. Prior to the closing of Indonesian Throughflow (ITF) this region of South-east Asia was very well connected by the deep water hydrographic communications from the Pacific Ocean, which is evidenced by high rate of BSi surface productivity (Cawthern et al., 2014). The primary productivity before the closure of ITF was pronounced owing to supply of nutrient rich cold water. This phenomenon is extensively visualised from our observation, which confirms that the late Miocene sediments contain well developed biogenic bloom of siliceous organisms (Chakraborty and Ghosh, 2016). Lazarus et al., (2009) suggested that the decrease of the robustness of the diatom valves in the Cenozoic directly reflects the reduction in the general level of silica in the oceans. Quantitative analysis of silicified microfossils indicates that at ~4.24 Ma there was a recovery phase of silica crash. It also corroborates with the findings of Cawthern et al., (2014) from Site 17 (Andaman Sea), NGHP Expedition 01. They commented that there was a silica crash at ~6 Ma and subsequently there was a recovery phase in the Andaman Sea. During ~4-2.5 Ma there was a short pulse in biogenic silica (BSi) productivity in the studied region due to riverine influx from nearby Irrawaddy and Ganges-Brahmaputra rivers (Cawthern et al., 2014). This impulse was responsible for regenerated nutrient supply that created a short recovery of biogenic silica as well as primary productivity. The decline of biogenic silica in the early Pliocene was caused by an early Pliocene shift of opal accumulation from equatorial region to the Antarctic Convergence.

Acknowledgements: We thank Prof. Sunil Bajpai, Director, Birbal Sahni Institute of Palaeosciences, Lucknow for his kind permission (BSIP/RDCC/ Pub. No. 79 (2016-17)) to communicate this manuscript and for providing the necessary laboratory facilities. A.C. thanks DST, New Delhi for DST-INSPIRE Fellowship (IF-120842).

References

BALL, V. (1870) Notes on the Geology of the vicinity of Port Blair Andaman Islands (also the Nicobar Archipelago). J. Asiatic Soc. Bengal, v.39(25), pp. 231-243.

BARRON, J.A. (1973) Late Miocene-Early Pliocene paleotemperatures for California from marine diatom evidence. Palaeogeogr. Palaeoclimatol. Palaeoecol., v.14, pp.277-291.

BARRON, J.A. (1976) Revised Miocene and Pliocene diatom biostratigraphy of upper Newport Bay, New Port Beach, California, Mar. Micropal., V.1, pp. 27-63.

BARRON, J.A. (1992) Neogene diatom datum levels in the equatorial and North Pacific. *In:* ISHIZAKI, K., SAITO, T. (Eds.), Neogene diatom datum levels in the equatorial and North Pacific. Centenary of Japanese Micropal., Tokyo, pp. 413-425.

BARRON, J.A. (1996) Diatom constraints on the position of the Antarctic Polar Front in the middle part of the Pliocene. Mar. Micropal., v.27, pp.195-213.

BARRON, J.A. and BALDAUF, J.G. (1990) Development of biosiliceous sedimentation in the North Pacific during the Miocene and early Pliocene. *In*: TSUCHI, R. (Ed.), Oji Inter. Seminar IGCP 246, 'Pacific Neogene Events'- Their Timing, Nature and Interrelationship. University of Tokyo Press, pp.43-63.

BARRON, J.A. and KELLER, G. (1983) Paleotemperature oscillations in the middle and late Miocene of the northeastern Pacific. Micropaleont., v.29, pp.150-181.

BOERSMA, A. and MIKKELSEN, N. (1990) Miocene age primary productivity episodes and oxygen minima in the central equatorial Indian Ocean. *In*: DUNCAN, R.A., BACKMAN, J. and PETERSON, L.C. (Eds.), Proceeding of the Ocean Drilling Program, Scientific Results 115. U.S. Government Printing Office, Washington, D.C., pp.589-609.

BOILEAU, V.H. (1950) Preliminary report on a geological reconnaissance in the Andaman Islands, Progress Rept. Geol. Suv. India (unpublished).

BRAUN, A., CHEN, J., WALOSZEK, D. and MAAS, A. (2007) First early cambrian radiolaria. Geol. Soc. London, Spec. Publ., v.286, pp.143-149.

BROECKER, W.S. and PENG, T.H. (1982) Tracers in the Sea, Lamont-Doherty Earth Observatory, Palisades, N.Y.

BUKRY, D. (1973) Phytoplankton stratigraphy, central Pacific Ocean, Deep Sea Drilling Project Leg 17. *In*: WINTERER, E.L. and EWING, J.L. (Eds.) Initial Reports, DSDP, v.17, pp.871-887.

CAMPBELL, A.S. (1954) Radiolaria. *In*: CAMPBELL, A.S. and MOORE, R.C. (Eds.), Protista 3: Lawrence, Kansas, Geol. Soc. America and Kansas Univ. Press, Treatise on Invertebrate Paleontology, pt. D, pp.11-163.

CAWTHERN, T., JOHNSON, J.E., GIOSAN, L., FLORES, J.A., ROSE, K. and SOLOMON, E. (2014) A late Miocene–Early Pliocene biogenic silica crash in the Andaman Sea and Bay of Bengal. Mar. and Petr. Geol., v.58, pp.490-501.

CHAKRABORTY, A. and GHOSH, A.K. (2015) *Acrobotrys disolenia* Haeckel from the late Miocene of Andaman and Nicobar Islands. Cur. Sci., v.108(11), pp.1990-1993.

CHAKRABORTY, A. and GHOSH, A.K. (2016) Ocean upwelling and intense monsoonal activity based on late Miocene diatom assemblages from Neil Island, Andaman and Nicobar Islands, India. Mar. Micropal., v.127, pp.26-41.

CHANDRA, A. and SAXENA, R.K. (1998) Lithostratigraphy of the Car Nicobar Island, Andaman and Nicobar Islands, India. Geophytology, v.26(2), pp.33-38.

CHANDRA, A., SAXENA, R.K. and GHOSH, A.K. (1999) Coralline algae from the Kakana Formation (Middle Pliocene) of Car Nicobar Island, India and their implication in biostratigraphy, palaeoenvironment and palaeobathymetry. Cur. Sci., v.76(11), pp.1498-1502.

CLEMENS, S.C., KUHNT, W., LEVAY, L.J. and the Expedition 353 Scientists (2016) Indian Monsoon Rainfall. Proc. Inter. Ocean Discovery Program, 353: College Station, TX (Intern. Ocean Discovery Program).

CURRAY, J.R, LAWYER, D.G., EMMET, F.J., RAITT, R.W. and HENRY, M., KIECKHEFER, R. (1979) Tectonics of the Andaman Sea and Burma, Geological and geophysical investigations of continental margins.

Mem. Am. Asso. Pet. Geol., v.29, pp.189-198.

CURRAY, J.R. (2005) Tectonics and history of the Andaman Sea region. J. Asian Earth Sci. v.25, 187-228.

DESHIKACHARY, T.V. (1989) Marine diatoms of the Indian Ocean region. *In:* DESHIKACHARY, T.V. (Ed.), Atlas of diatoms, Fascicle VI, Madras Science Foundation, Madras.

DESHIKACHARY, T.V. and MAHESWARI, C.I. (1958) Fossil diatoms from Colebrook Island. J. Indian Bot. Soc., v.37(1), pp.27-41.

DESHIKACHARY, T.V., GOWTHAMAN, S. and LATHA, Y. (1987) Diatom flora of some sediments from the Indian Ocean region. *In:* DESHIKACHARY, T.V. (Ed.), Atlas of diatoms, Fascicle VI, Madras Science Foundation, Madras.

DESHIKACHARY, T.V., LATHA, Y. and RANJITHA DEVI, K.A. (1984) *Rossiella* and *Bogorovia*, two fossil diatom genera. The Palaeobot., v.32(3), pp.337-340.

DOW, R.L. (1978) Radiolarian distribution and the Late Pleistocene history of the South-eastern Indian Ocean. Mar. Micropaleont., v.3, pp.203-227.

EHRENBERG, C.G. (1851) On the extensive rock formation of siliceous Polycystina from the Nicobar Island. Berlin Monatsbericht, pp.476-478.

EHRENBERG, C.G. (1854) Mikrogeologie, Leipzig: Voss, XXVIII+374 pp. Atlas 31 pp.41 pls. Fortsetzung, pp.1-88.

FAUQUETTE, S., SUC, J.P., JIMÉNEZMORENO, G., MICHEELS, A., JOST, A., FAVRE, E., BACHIRI TAOUFIQ, N., BERTINI, A., CLETPELLERIN, M., DINIZ, F., FARJANEL, G., FEDDI, N. and ZHENG, Z. (2007) Latitudinal climatic gradients in western European and Mediterranean regions from the Mid-Miocene (~15 Ma) to the Mid-Pliocene (~3.5 Ma) as quantified from pollen data. *In*: WILLIAMS, M., HAYWOOD, A., GREGORY, J. and SCHMIDT, D. (Eds.), Deep time perspectives on climate change: marrying the signal from computer models and biological proxies. Micropal. Soc., Geol. Soc. London, Sp. Pub., pp.481-502.

FENNER, J. (1985) Late Cretaceous to Oligocene planktic diatoms. *In*: BOLLI, H.M., SAUNDERS, J.B. and PERCH-NIELSEN, K. (Eds.), Plankton Stratigraphy, Cambridge (Cambridge Univ. Press), pp.713-762.

FINKEL, Z.V., KATZ, M. WRIGHT, J. (2005) Climatically driven macroevolutionary patterns in the size of marine diatoms over the cenozoic, Proc. Nat. Acad. Sci., USA, v. 102, pp. 8927-8932.

FRYDAS, D. (1999) Paleoecology, stratigraphy and taxonomy of the Pliocene marine diatoms from central Crete (Greece). Rev. de Micropaléont., v.42, pp.269-300.

GHOSH, A.K. and MAITRA, S.C. (1947) On the occurrence of diatomaceous earth in the Nicobar Islands. Trans. Natn. Inst. Sci. India, v.2, pp.441-450.

GHOSH, A.K., CHANDRA, A. and SAXENA, R.K. (2004) Middle Pliocene non-geniculate and geniculate coralline algae from the Car Nicobar Island, India. *In*: SRIVASTAVA, P.C. (Ed.), Vistas in Palaeobotany and Plant Morphology: Evolutionary & Environmental Perspectives. Prof. D.D. PANT Volume, Lucknow, India, pp.249-262.

GLADENKOV, A.Y. (2013) Neogene diatoms from the deep Hole Central-1, the Bering Sea Chukchi shelf. Micropaleont., v.59, pp.59-67.

GOWTHAMAN, S. and DESIKACHARY, T.V. (1989) Some new and interesting fossil diatoms. Phykos, v.28(1-2), pp.132-135.

GRADSTEIN, F.M, OGG, J.G. and SCHMITZ, M.D. (2012) The Geologic Time Scale 2012, Elsevier, Boston, USA. doi: 10.1016/B978-0-444-59425-9.00004-4.

GRUNOW, A. (1867) Algae Reise sciener Majestat Fregatte "Novara" um die Erde in den

Jahren, 1857, 1858, 1859. Bot. Teil., pp.1-104.

GUPTA, S.M. and FERNANDES, A.A. (1995) Supra-generic transfer function based late Miocene Paleomonsoonal changes in the northeastern Tropical Indian Ocean. Bull. Indian Geol. Assoc., v.28, pp.29-51.

GUPTA, S.M. and SRINIVASAN, M.S. (1992) Late Miocene radiolarian biostratigraphy and paleoceanography of Sawai bay Formation, Neill Island, Andamans, India. Micropal., v. 38(3), pp.209-235.

GUPTA, A.K., SINGH, R.K., JOSEPH, S. and THOMAS, E. (2004) Indian Ocean high-productivity event (10–8 Ma): linked to global cooling or to the initiation of the Indian monsoons? Geology, v.32(9), pp.753-756.

HILL, K. and HALL, R. (2003) Mesozoic-Tertiary Evolution of Australia's New Guinea Margin in a west Pacific context. *In:* R. HILLIS and R.D. MULLER (Eds.), The Evolution and Dynamics of the Australian Plate. Geological Society of Australia, Sp. Pub. 22 and Sp. Paper Geol. Soc. America, 372. Geol. Soc. Australia and Geol. Soc. America.

HEATH, G.R. (1974) Dissolved Silica and Deep-Sea Sediments. In: Studies in Paleo-Oceanography, EPM Special Publications, SEPM Society for Sedimentary Geology, v.20, pp.77-93.

HEDGES, S.B., CHEN, H., KUMAR, S., WANG, D.Y.C, THOMPSON, A.S. and WATANABE, H. (2001) A genomic time-scale for the origin of eukaryotes. BMC Evol. Biol., v.1(4).

JACOB, K. and SASTRI,V.V. (1951) Tertiary foraminifera from Sawai Bay, Car Nicobar Island. Sci. and Cult., v.17(4), pp.181-182.

JACOB, K. and SHRIVASTAVA, R.N. (1952) Fossil radiolaria and silicoflagellates from the Tertiary clays of Colebrook Island and Ritchie's Archipelago, Andaman Islands. Sci. and Cult., v.17(8), pp.346-348.

JOHNSON, D.A. and NIGRINI, C. (1980) Radiolarian biogeography in surface sediments of the western Indian Ocean. Mar. Micropaleontol., v.5, pp.111-152.

JOHNSON, D.A. and NIGRINI, C. (1982) Radiolarian biogeography in surface sediments of the eastern Indian Ocean. Mar. Micropaleontol., v.7, pp.237-281.

JOUSÉ, A.P. (1974) Diatoms in the Oligocene-Miocene biostratigraphic zones of the tropical areas of the Pacific Ocean. *In:* JOUSÉ (Ed.), Micropaleontology of the Ocean and Seas. Acad. Nauk. USSR, Moscow, pp.34- 48.

KANAYA, T. (1971) Some aspects of pre-Quaternary diatoms in the oceans. *In*: B.M. Funnell (Ed.), The Micropalaeontology of Oceans. Cambridge University Press, London, pp.545-565.

KARUNAKARAN, C., RAY, K.K., SEN, C.R., SAHA, S.S. and SARKAR, S.K. (1968) Tertiary sedimentation in the Andaman-Nicobar geosynclines. Jour. Geol. Soc. India, v.9, pp.32-39.

KLING, S.A. (1979) Vertical distribution of polycistine radiolarians in the central North Pacific. Mar. Micropal., v.4, pp.295-318.

KLOOTWIJK, C.T., GEE, J.S., PEIRCE, J.W., SMITH, G.M., and McFADDEN, P.L. (1992) An early India-Asia contact: paleomagnetic constraints from Ninetyeast Ridge, ODP Leg 121. Geology, v.20, pp.395-398.

KNOLL, A.H. (1992) The early evolution of eukaryotes: a geological perspective. *Sci.,* v.256(5057), pp.622-627.

KOLBE, R.W. (1957) Diatoms from equatorial Indian Ocean Cores. Rep. Swed. Deep-Sea Exped., v.9, pp.3-50.

KROON, D., STEENS, T. and TROELSTRA, S.R. (1991) Onset of monsoonal related upwelling in the western Arabian Sea as revealed by planktonic foraminifers. *In*: PRELL, W.L., and NIITSUMA, N. (Eds.), Proc. Ocean Drilling Program, Sci. Results, v.117. ODP, College Station, TX, pp.257-263.

KUHNT, W., HOLBOURN, A., HALL, R., ZUVELA, M. and KÄSE, R. (2004) Neogene history of the Indonesian Throughflow. *In*: CLIFT, P. et al. (Eds.), Continent-ocean interactions within east Asia marginal seas. Geophysical Monograph Series, v.149, pp.299-320, AGU, Washington, D. C.

LAZARUS, D., KOTRC, B., WULF, G. and SCHMIDT, D.N. (2009) Radiolarians decreased silicification as an evolutionary response to reduced Cenozoic ocean silica availability. Proc. Nat. Acad. Sci., USA, v.106, pp.9333-9338.

LAZARUS, D., BARRON, J., RENAUDIE, J., DIVER, P. and TÜRKE, A. (2014) Cenozoic Planktonic Marine Diatom Diversity and Correlation to Climate Change. PLoS ONE, v.9(1), pp. e84857.

LOZANO, J.A. and HAYS, J.D. (1976) Relationship of radiolarian assemblages to sediment types and physical oceanography in the Atlantic and Western Indian Ocean sectors of the Antarctic Ocean. *In*: CLINE, R.M. and HAYS, J.D. (Eds.), Investigation of Late Quaternary Paleoceanography and Paleoclimatology. Geol. Soc. of Amer. Mem., v.145, pp.303- 336.

MAHAPATRA, A.K. and SHARMA, V. (1994) Trissocyclid Radiolaria form late Early Miocene sequences of Colebrook, North Passage and Great Nicobar Islands, Northeast Indian Ocean. Micropal., v.40, pp.157-168.

MARTINI, E. (1971) Standard Tertiary and Quaternary calcareous nannoplankton zonation. Proc. II Plankt. Conf., Roma, v.2, pp.739-785.

MATHUR, K. (1973) Studies in the fossil micro flora of Andaman Islands-2: fossil diatoms from Havelock Island. Geophytology, v.3, pp.103-134.

MATHUR, K. (1981) Silicoflagellates, diatoms and other fossils from Outram Island, Andaman Sea. Andaman, India. Geosci. Jour., v.2(1), pp.31-42.

MATHUR, K. (1985) Middle Miocene diatoms from Nicholson Island, Andaman Sea, India. Geosci. Jour., v.6(I), pp.81-94.

MIKKELSEN, N. (1990) Cenozoic diatom biostratigraphy and paleoceanography of the western equatorial Indian Ocean. *In*: DUNCAN, R.A. and BACKMAN, J. (Eds.), Proc. Ocean Dril. Prog., Sci. Results, v.115, pp.411-432.

MOORE, T.C., Jr. (1978) The distribution of radiolarian assemblages in the modern and ice-age Pacific. Mar. Micropal., v.2, pp.229-266.

MORLEY, J.J. (1989) Radiolarian-based transfer functions for estimating paleoceanographic conditions in the South Indian Ocean. Mar. Micropal., v.13, pp.293-307.

NIGRINI, C. (1967) Rediolaria in pelagic sediments from the Indian and Atlantic Oceans. Bull. Scripps Inst. Oceanogr., v.11, pp.1-125.

PAL, T., CHAKRABORTY, P.P., DUTTA GUPTA, T. and SINGH, D.C. (2003) Geodynamic evolution of the outer-arc–forearc belt in the Andaman Islands, the central part of the Burma-Java subduction complex. Geol. Mag., v.140(3), pp.289-307.

PÄLIKE H., LYLE M., NISHI H., RAFFI I., GAMAGE K., KLAUS A. and the Expedition 320/321 Scientists (2010) Proceedings of the Ocean Drilling Program 320/321, Tokyo. doi:10.2204/Iodp.Proc. 320321. 109. 2010.

PETRUSHEVSKAYA, M.G. (1967) Radiolyarii otryadov Spumellaria i Nasselaria ntarkticheskoi oblasti (Antarctic spumelline and nasseline radiolarians), Issled. Fauny Morei 4 (12), Rel. biol. Issled. Sov. Antarkt. Eksped. 1955-1958, v.3, pp.5-186.

PISAS, N.G. and MOORE, T.C. (1995) Radiolarian response to oceanographic changes in the eastern equatorial Pacific at 2.3 and 4.8 Ma: Relationship between changing carbonate Deposition and surface oceanography. *In*: PISIAS, N.G., MAYER, L.A. and JANECEK, T.R. (Eds.) Proc. Ocean

Drilling Program, Sci. Results, v.138, pp.461-478, Texas A&M Univ., College Station TX, USA.

PLAUT G., GHIL M. and VAUTARD R. (1995) Interannual and interdecadal variability in 335 years of central England temperatures. Science, v.268, pp.710-713.

PREMA, P. and DESIKACHARY, T.V. (1989) New and interesting fossil diatoms from the Indian Ocean. Phykos, v.28(1-2), pp.127-131.

REIDEL, W.R. and SANFILIPPO, A. (1978) Stratigraphy and evolution of tropical Cenozoic radiolarians. Micropal., v.24(1), pp.61-96.

RENAUDIE, J. (2016) Quantifying the Cenozoic marine diatom deposition history: links to the C and Si cycles. Biogeosciences, v.13, pp.6003-6014.

RODOLFO, K.S. (1969a) Bathymetry and marine geology of the Andaman basin, and tectonic implications for southeast Asia. Geol. Soc. Am. Bull., v.80, pp.1203-1230.

RODOLFO, K.S. (1969b) Sediments of the Andaman basin northeastern Indian Ocean. Mar. Geol., v.7, pp.371-402.

ROY, T.K. (1983) Geology and hydrocarbon prospects of Andaman and Nicobar. *In*: BHANDARI, L. et al. (Eds.), Petroliferous Basins of India, Petroleum Asia Journal. KDMIPE, ONGC, Dehradun, pp.37-53.

SALZMANN, U., HAYWOOD, A.M. and LUNT, D.J. (2009) The past is a guide to the future? Comparing Middle Pliocene vegetation with predicted biome distributions for the twenty-first century. Philos. Trans. R. Soc. A, v.367, pp.189-204.

SANFILIPPO, A. and NIGRINI, C. (1998) Code numbers for Cenozoic low-latitude radiolarian biostratigraphic zones and GPTS conversion tables. Mar. Micropal., v.33(1-2), pp.109-117, 121-156.

SCHRADER, H.J. (1973) Cenozoic diatoms from the Northeast Pacific, Leg 18. Init. Rep. DSDP v.18, pp.673-797.

SCHRADER, H.J. (1974) Cenozoic Marine Planktonic Diatom Stratigraphy of the Tropical Indian Ocean. *In*: FISHER, R.L. et al. (Eds.), Initial Reports of Deep Sea Drilling Project, Leg 24. U.S. Govt. Printing office, Washington, D.C., pp.887-967.

SCHRADER, H.J. (1978) Diatoms in DSDP Leg 41 sites. *In*: LANCELOT, Y. and SEIBOLD, E. (Eds.) Initial Reports, Deep Sea Drilling Project, v.41, pp.791-812.

SCHUETTE, G. and SCHRADER, H. (1981) Diatoms in surface sediments: a reflection of coastal upwelling. *In*: RICHARDS, F.A. (Ed.), Coastal Upwelling. American Geophysical Union, Washington. D.C., pp.372-380.

SHACKLETON, N.J., BALDAUF, J.G., FLORES, J.A., IWAI, M., MOORE, T.C., Jr., RAFFI, I. and VINCENT, E. (1995) Biostratigraphic summary for Leg 138. Proc. Ocean Drilling Program, Sci. Results, v.138, pp.517-536.

SHARMA, V. and BAJPAI, S. (2011) Impact of the 2004-tsunami on the geology of Car Nicobar Island. Cur. Sci., v.100(8), pp.1145-1147.

SHARMA, V. and DANESHIAN, J. (1998) Miocene radiolaria from Nicholson and John Lawrence Islands, Andman Sea. Jour. Geol. Soc. India, v.52, pp.695-707.

SHARMA, V. and DANESHIAN, J. (2003) Early Neogene radiolarian changes in the Northeast Indian Ocean and paleoceanographic implications: Evidence from the Andaman-Nicobar Islands. *In*: RATANASATHIEN et al. (Eds.), Proc. 8th Inter. Congress Pacific Neogene Stratigraphy, Chiang Mai, Thailand, pp.126-158.

SHARMA, V., DANESHIAN, J. and DEVI, L. BHAGYAPATI (2011) Early Neogene radiolarian faunal turnover in the Northern Indian Ocean: Evidence from Andaman - Nicobar. J. Geol. Soc. India, v. 78, pp. 157-166.

SHARMA, V. and DEVI, L. BHAGYAPATI (2007b) Neogene oceanographic and climatic changes in the northern Indian Ocean: Evidence from radiolarian. Indian Journal of Marine Sciences, v.36(4), pp.361-368.

SHARMA, V. and RAM, M.P. (2003) Early to Middle Miocene radiolarian assemblages and biostratigraphy, Andaman Islands, Northeast Indian Ocean. J. Pal. Soc. India, v.48, pp.1-39.

SHARMA, V. and SHARMA, G.K. (1988) Radiolaria from Neill Island, Andaman Sea and their distributional characteristics. J. Pal. Soc. India, v.33, pp.7-19.

SHARMA, V. and SHARMA, G.K. (1989) Late Miocene to early Pliocene radiolarian biostratigraphy of Neil Island, Andaman Sea. J. Geol. Soc. India, v.34, pp.76-82.

SHARMA, V. and SINGH, S. (1993) Radiolarian biostratigraphy of early Pliocene sequences, Car Nicobar Island, Northeast Indian Ocean. J. Geol. Soc. India, v.41, pp.199-213.

SHARMA, V. and SINGH, S. (1997) Late Neogene radiolarian events in Andaman-Nicobar Islands, Northeast Indian Ocean. Micropal., v.43, pp.21-28.

SHARMA, V. and SRINIVASAN, M.S. (2007) Geology of Andaman- Nicobar: The Neogene, Capital Pub. Co., New Delhi.

SHARMA, V., SINGH, S. and RAWAL, N. (1999) Early Middle Miocene Radiolaria from Nicobar Islands, Northeast Indian Ocean. Micropaleont., v.45, pp.251-352.

SHARMA, V., SRINIVASAN, M.S. and MAHAPATRA, A.K. (1993) Early Miocene radiolarian and planktonic foraminiferal biostratigraphy, North Passage Island, Andaman Sea. J.Geol.Soc. India, v.41, pp.199-213.

SINGH, K. and SHARMA, J. (1996) Diatom and silicoflagellate assemblages from the Neogene sediments of Neill Island, Andaman Sea. *In*: PANDEY, J., AZMI, R.J., BHANDARI, A. and DAVE, A. (Eds.), Contrs. XV Indian Colloq. Micropal. Strat., Dehradun, pp.695-705.

SINGH, O.P., SRINIVASAN, M.S. and SHARMA, V. (2000) Early Neogene multiple microfossil biostratigraphy, John Lawrence Islands, Andman Sea. Micropal., v.46(4), pp.343-352.

SINGH, P. (1979) Early Pliocene calcareous nanoplankton from Neil Island, South Andaman, India. Bull. Ind. Geol. Assoc. v. 12(2), pp. 221-237.

SINGH, P. and VIMAL, K.P. (1973) A note on the geology and micropalaeontology of Neill Island, South Andaman. Curr. Sci., v.42(7), pp.239-241.

SINGH, P., VIMAL, K.P. and NAUTIYAL, D.D. (1978) Early Pliocene diatoms and silicoflagellates from Neill Island, South Andaman, India. J. Pal. Soc. India, v.21-22, pp.49-59.

SRINIVASAN, M.S. (1986) Geology of Andaman-Nicobar Islands. J. Andaman Sci. Assoc. Port Blair, v.2(1), pp.1-12.

SRINIVASAN, M.S. and AZMI, R.J. (1976a) Contribution to the Stratigraphy of Neil Island, Ritchie's Archipelago, Andaman Sea. Proc. VI Indian Colloq. Micropal. and Strat., pp.283-301.

SRINIVASAN, M.S. and AZMI, R.J. (1976b) New developments in the Late Cenozoic Lithostratigraphy of Andaman - Nicobar Islands, Bay of Bengal. Proc. VI Indian Colloq. Micropal. Stratgr., pp.302-327.

SRINIVASAN, M.S. and AZMI, R.J. (1979) Correlation of Late Cenozoic marine sections in Andaman-Nicobar, Northern Indian Ocean and Equatorial Pacific. J. Paleontol., v.53(6), pp.1401-1415.

SRINIVASAN, M.S. and CHATTERJEE, B.K. (1981) Stratigraphy and depositional environments of Neogene limestones of Andaman-Nicobar Islands, Northern Indian Ocean. J. Geol. Soc. India, v.22, pp.536-546.

SRINIVASAN, M.S. and SINHA, D.K. (1998) Early Pliocene closing of the Indonesian Seaway: Evidence from the northeast Indian Ocean and southwest Pacific deep sea cores. J. Southeast Asian Earth Sci., v.16, pp.29-44.

SRINIVASAN, M.S. and SRIVASTAVA, S.S. (1977) A new fossil diatom (*Annellus*) from Nicobar Islands, Bay of Bengal. *In*: VENKATACHALA, B.S. and SASTRI, W. (Eds.), Proc. 4th Colloq. Indian Micropalaeont. Stratigr., Dehradun (1974-75), pp.193-195.

SRINIVASAN, M.S., LOMBARI, G. and DAVE, A. (1983) Early Miocene planktonic foraminiferal and radiolarian zonation, Colebrook Island, Andaman Sea. J. Geol. Soc. Indian, v.24, pp.1-18.

SRINIVASAN. M.S. and SHARMA, V. (1973) Stratigraphy and Microfauna of Car Nicobar Island, Bay of Bengal. J. Geol. Soc. India, v.14, pp.1-11.

SWANBERG, N.R. and ANDERSON, O.R. (1985) The nutrition of radiolarians: Trophic activity of some solitary Spumellaria. Limnol. and Ocean., v.30(3), pp.646-652.

TANG, H., MICHEELS, A., ERONEN, J. and FORTELIUS, M. (2011) Regional climate model experiments to investigate the Asian monsoon in the Late Miocene. Clim. of the Past, v.7, pp.847-868.

TRÉGUER, P., NELSON, D.M., Van BENNEKOM, A.J., De MASTER, D.J., LEYNAERT, A. and QUÉGUINER, B. (1995) The silica balance in the world ocean: a reestimate. Science, v.268, pp.375-379.

TRÉGUER, P. and PONDAVEN, P. (2000) Silica control of carbon dioxide. Nature, v.406, pp.358-359.

VAN HEURCK, H. (1880-85) Synopsis des diatomees de Belgique. Atlas, pl. 1-30 (1880); pl. 31-77 (1881); pl. 78-103 (1882); pl. 104-132 (1883); pls. A, B, C (1884), Texte. Mtin., Browers Co. Anvers, 235 (1885), Decaju & Cie, Anvers. Table Alphabetique (Index) J.F. DIELTJENS ROP (1885).

Micropaleontology and its Applications
Pradeep K. Kathal, Rajiv Nigam & Abu Talib (Editors)
Scientific Publishers (India), 2018, 191-210 pp.

Foraminifera from Middle Jurassic Sediments of Jara Dome, Kachhch, Gujarat - Age and Palaeoecologic Implications

S.M. Wasim[2*], Y.P. Singh[2], Ainul H. Ansari[1] and Megha Prakasan[1]

[1]Department of Geology, Aligarh Muslim University, Aligarh- 202002, India
[2]Department of Geology, D. S. College, Aligarh- 202002, India
***Email:** w4wasimalig@gmail.com

Abstract

Benthic foraminiferal assemblages comprising twenty-six species are reported form Jara Dome, Kachhch. Of these, 12 are being recorded for the first time from the Indian region. The foraminiferal assemblages are dominated by the family Vaginulinidae comprising 22 species belonging to nine genera, covering 84.62% of the total species. The foraminiferal assemblages are employed to interpret age and depositional environment of these sediments. On the basis of the foraminiferal assemblages, Callovian age is assigned to the studied lithounit of the Jara Dome, Kachhch. Several contemporary techniques are employed to draw a fairly accurate picture of the palaeoenvironment which indicates that the overall deposition of the studied lithounit took place in an open ocean, shallow marine shelf environment ranging from mid to outer shelf with normal salinity and high level of dissolved oxygen.

Keywords: Foraminifera, first report, Callovian, shallow marine, Jara Dome, Kachhch, India.

INTRODUCTION

The geological investigations of the Jurassic rocks of Kachhch are important because many world oil producing regions like Middle-East and Russia have source rocks of Jurassic age (Nairn and Alsharhan 1997, Williams 2005 and Henderson

2013). In Peninsular India marine Jurassic rock are developed in Kachhch, Gujarat, but are well exposed only in the Kachhch Mainland, *i.e.*, Central Kachhch. The Kachhch region is widely acclaimed in the world as the storehouse of rich and varied megainvertebrates, including ammonites (Waagen 1873-1875, Spath 1927-1933, Prasad 1993, Bardhan et al. 1994 and Krishna et al. 2009). However, most of the previous studies on the Jurassic rock of Kachhch deal with stratigraphy and megafauna. Although considerable research has been carried out on the ammonites and other megafossils of these rocks, relatively little attention has been paid to their microfossils including foraminifera which only recently got attention following the encouraging results obtained in petroleum prospecting by the Oil and Nature Gas Corporation Limited (Pandey et al. 2009). However, most of the available studies deal only with the systematics and biostratigraphy with little emphasis on other aspects such as palaeoecology and palaeobiogeography (Bhalla and Abbas 1978; Bhalla and Talib 1991; Khosla *et al.* 1997, 2005; Pandey and Dave 1993; Gaur and Talib 2009; Talib *et al.* 2012 a, b and 2016). Compared to western countries very little work has been done on the Foraminifera of the Jurassic sediments of India. In view of Kachhch being placed in Category II basins with respect to its hydrocarbon prospects and lying in continuity to the South Indus Basin of Pakistan, where a number of oil and gas fields have already been discovered (Quadri 1986) as well as the discovery of oil and gas in the offshore part of the Kachhch Basin, foraminiferal study of the Jurassic rocks of Kachhch exposed in Jara Dome (Figure 1) is undertaken in the present study. Jara Dome is located between the latitudes 23°43'N to 23°45'N and longitudes 68°57'E to 69°00'E, south of village Jara about 196 km northwest of Bhuj, the district headquarter of the Kachhch region in Gujarat state of India.

GEOLOGICAL SETTING

The Jurassic sedimentary basin is a peri-continental basin in the extreme west of the Indian Peninsula. The basin covers an area of 26,400 sq km of which outcrop area includes 8000 sq km. Mesozoic sediments are over 2430m thick and fill the major part of the basin whereas 600m thick Tertiary sediments are exposed in the outer parts, bordering the Mesozoic uplifts (Mishra and Tiwari 2006). The Jurassic rocks of Kachhch are exposed as domed outcrops in three east-west trending ridges. The middle ridge is the most prominent where a number of large domed exposures are located. Traditionally the Jurassic rocks of Kachhch are divided into four formations, *viz.*, Patcham, Chari, Katrol and Umia in ascending order, ranging in age from the Bajocian to Tithonian (Pandey and Furscih 1998; Fursich et al. 2004a, 2004b; Bardhan et al. 2010; Alberti et al. 2015). Only Patcham and Chari formations from late Bathonian to Early Callovian rocks are exposed in the Jara Dome (Prasad 1998; Cariou and Krishna 1988; Rai 2006; Jain and Desai 2014). In preliminary study, we are collected 28 samples in Jara Dome, which is divisible in to six lithounits, viz., JR-1 to JR-7 (Figure 2), and the detail Lithologs study is in my future part of research. The lower unit of the present section lower unit of the present section Lithounit JR-1 is composed of 5m Shale intercalated with thin sandstone. Lithounit JR-2 is composed of 8m thick grey to yellowish fossiliferous shale, intercalated with limestone beds. Lithounit JR-3 is fine grained sandstone, reddish in colour and is 3.42 m thick. This is followed by thinly bedded orange colour sandstone with a thickness of 2.30m (Lithounit JR-4). The succeeding Lithounit JR-5 is composed of fine grained yellow colour sandstone and is 4.73m thick. The next unit is JR-6, which is a 0.43m thick fossiliferous limestone with pebbly conglomerate. The last Lithounit JR-7 of the Chari Formation exposed in the area is a 24.95m thick fine-grained sandstone bed, pink and reddish yellow in colour, with thick and thin beddings and having ripple marks as well as planer and trough cross-beds. However, only six samples of the Lithounit JR-2 yielded foraminifera and the rest of the sample are absent of foraminifera.

MATERIAL AND METHODS

The samples were coarsely crushed and boiled with sodium carbonate. The material was then washed through standard sieves of 60, and 120 mesh size. The disintegrated material was dried in an electric oven. Ten gram material from each mesh size was taken for picking the foraminiferal tests and all the tests were picked with the help of a fine sable hair brush under a stereozoom binocular microscope. Different foraminiferal species were arranged in assemblage microfaunal slides for identification and study. Some specimens were further cleaned in an ultrasonic cleaner. Superageneric and generic identifications were carried out following the classification of Loeblich and Tappan (1988) and the Ellis and Messina (1940-1990) catalogue of Foraminifera was consulted for species identification. All the photographs of the foraminiferal species for the plates were prepared with the help of Scanning Electron Microscope (SEM), Make JEOL, Model JSM-65.10 LV at University Sophisticated Instruments Facility (USIF), Aligarh Muslim University, Aligarh.

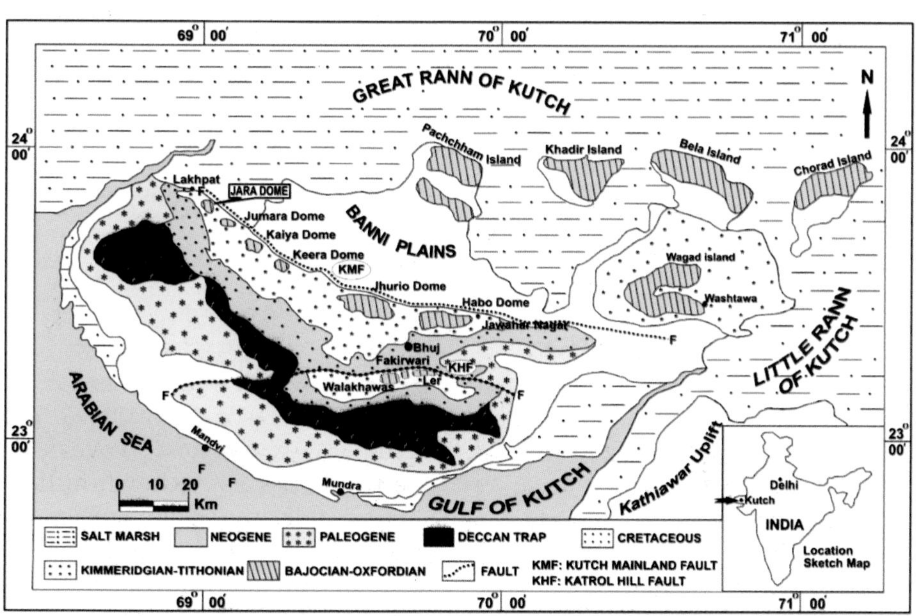

Fig.1. Geological map of Kachhch, Gujarat showing the study area (after Fürsich et al., 1991).

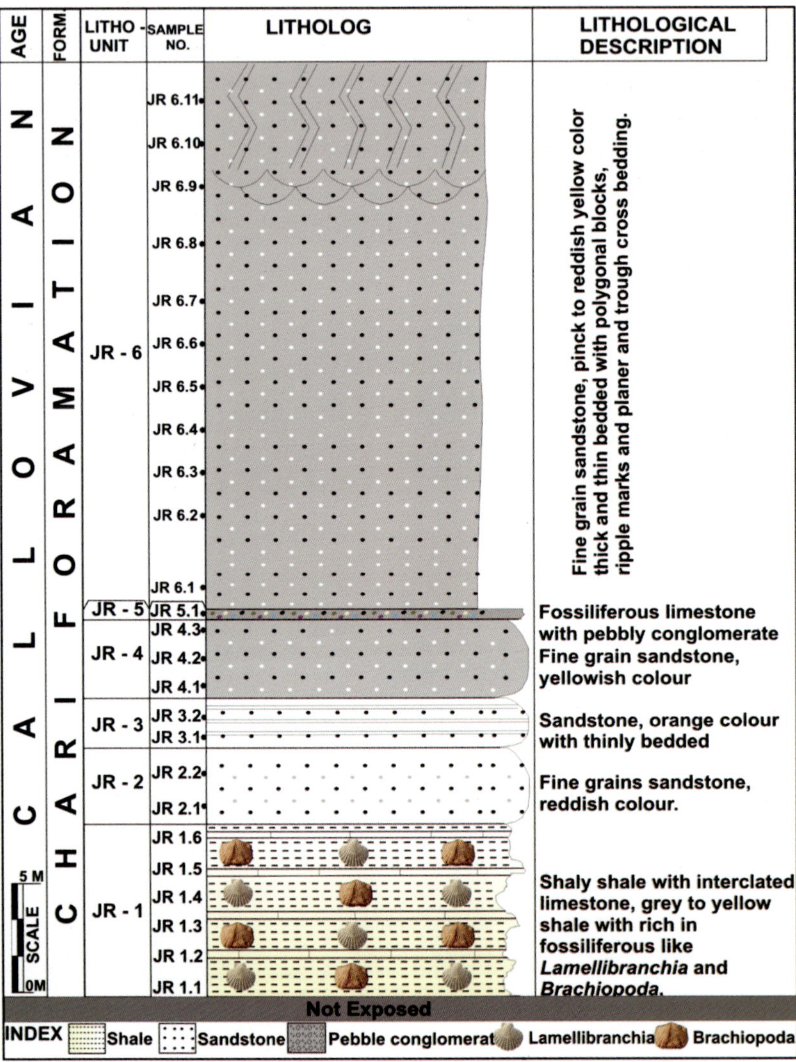

Fig.2. Measured sections of Middle Jurassic sequence exposed at Jara Village in area Dome, Kachhch showing sample locations.

FORAMINIFERAL COMPOSITION

A total of 1020 benthic foraminiferal specimens were obtained from the studied samples, most of them showing good preservation and belonging to twenty-six species (Figure 3). The foraminiferal assemblages include two suborders, *viz.*, Lagenina, having twenty-five species (96.15%) and Spirillinina, having one species (3.85%). Dominant family of the present assemblage is Vaginulinidae which comprises twenty-two species belonging to nine genera, covering 84.62% of the total species. Other families are Nodosariidae having three species belonging to three genera, covering 11.54% and Spirillinidae having one species belonging to one genus and constitutes 3.85% of the total foraminiferal species (Figure 4).

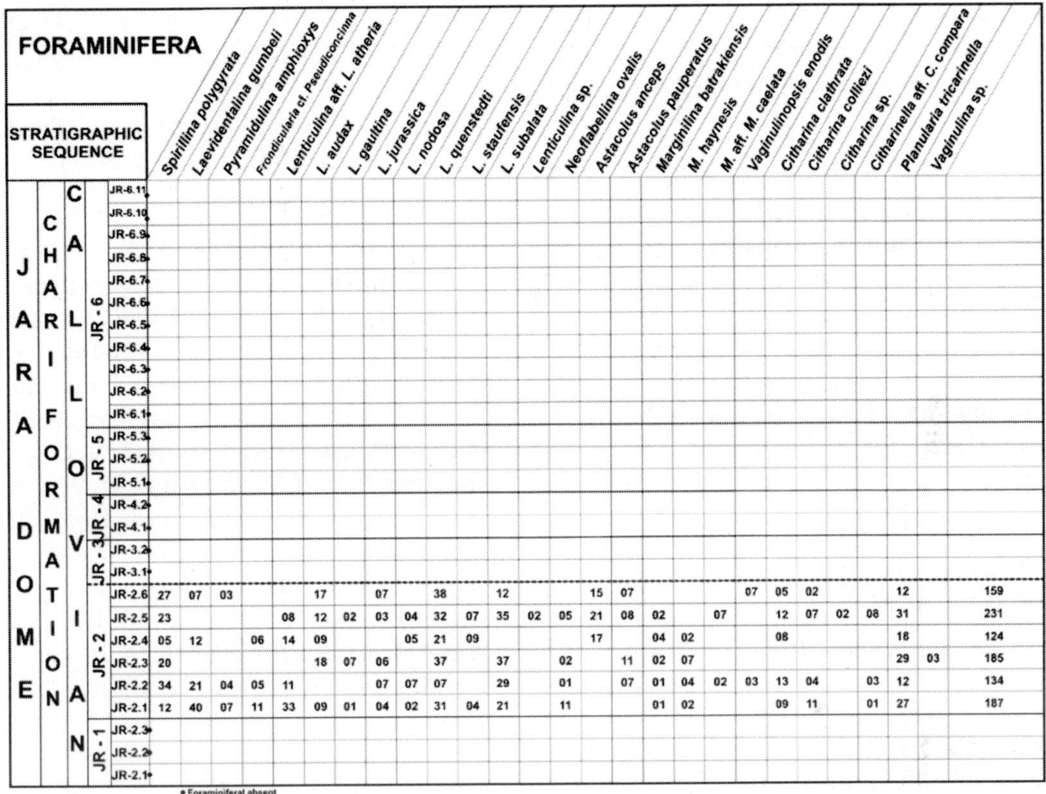

FORAMINIFERA

STRATIGRAPHIC SEQUENCE: JARA DOME — CHARI FORMATION — CALLOVIAN

Unit	Sample	Spirillina polygyrata	Laevidentalina gumbeli	Pyramidulina amphioxys	Frondicularia cf. Pseudoconcinna	Lenticulina aff. L. atheria	L. audax	L. gaultina	L. jurassica	L. nodosa	L. quenstedti	L. staufensis	L. subalata	Lenticulina sp.	Neoflabellina ovalis	Astacolus anceps	Astacolus pauperatus	Marginilina batrakiensis	M. haynesis	M. aff. M. caelata	Vaginulinopsis enodis	Citharina clathrata	Citharina colliezi	Citharina sp.	Citharinella aff. C. compara	Planularia tricarinella	Vaginulina sp.	
JR-6	JR-6.11•																											
	JR-6.10•																											
	JR-6.9•																											
	JR-6.8•																											
	JR-6.7•																											
	JR-6.6•																											
	JR-6.5•																											
	JR-6.4•																											
	JR-6.3•																											
	JR-6.2•																											
	JR-6.1•																											
JR-5	JR-5.3•																											
	JR-5.2•																											
	JR-5.1•																											
JR-4	JR-4.2•																											
	JR-4.1•																											
JR-3	JR-3.2•																											
	JR-3.1•																											
JR-2	JR-2.6	27	07	03			17		07		38		12			15	07				07	05	02			12		159
	JR-2.5	23				08	12	02	03	04	32	07	35	02	05	21	08	02		07		12	07	02	08	31		231
	JR-2.4	05	12		06	14	09			05	21	09				17		04	02			08				18		124
	JR-2.3	20					18	07	06		37		37		02		11	02	07							29	03	185
	JR-2.2	34	21	04	05	11			07	07	07		29		01		07	01	04	02	03	13	04		03	12		134
	JR-2.1	12	40	07	11	33	09	01	04	02	31	04	21		11			01	02			09	11		01	27		187
JR-1	JR-2.3•																											
	JR-2.2•																											
	JR-2.1•																											

• Foraminiferal absent

Fig.3. Frequency distribution of foraminifera in Jara Dome, Kachhch, Gujarat.

Following twelve species are being reported for the first time from the Indian region: *Pyramidulina amphioxys*, *Frandicularia* cf. *pseudoconcinna*, *Lenticulina atheria*, *Lenticulina audex*, *Lenticulina jurassica*, *Lenticulina nodosa*, *Neoflabellina ovalis*, *Vaginulinopsis enodis*, *Marginulina* aff. *M. caelata*, *Lenticulina* sp., *Citharina* sp., and *Vaginulina* sp.

SYSTEMATIC MICROPALAEONTOLOGY

A systematic account of the foraminiferal species recovered for the first time from the Indian subcontinent is presented here whereas others are only illustrated in the plates. We have followed the generic classification of foraminifera proposed by Loeblich and Tappan (1988). References concerning important shifts in generic names have only been cited. In order to avoid repetition, suffix *et syn.* has been added to the references which contain satisfactory synonymies. All the described and photographed specimens have been deposited in the micropalaeontological collection of the Department of Geology, Aligarh Muslim University, Aligarh.

Order: FORAMINIFERIDA Eichwald, 1830

Suborder: LAGENINA Delage and Hérouard, 1896

Superfamily: NODOSARIACEA Ehrenberg, 1838

Family: NODOSARIIDAE Ehrenberg, 1838

Subfamily: NODOSARIINAE Ehrenberg, 1838

Genus: PYRAMIDULINA Fornasini, 1894

Type species: *Pyramidulina eptagona* Fornasini, 1894

Pyramidulina amphioxys (Reuss, 1875)

Plate I, Figure 3

Nodosaria amphioxys Reuss 1874, p. 82, pl. (II)20, fig. 8.

Nodosaria amphioxys Reuss; Kozlo-wski 1957, p.65, pl.8, fig.4, *et syn*.

Dimensions (mm): Length 0.4-0.7, width 0.13 - 0.17.

Remarks: *Nodosaria* Lamarck, 1812 differs from the genus *Pyramidulina* Fornasini, 1894 in having ornamentation in the form of longitudinal costae. Some specimens belonging to *Pyramidulina amphioxys* originally described from the Cretaceous of Germany were recovered from the present material, which are comparatively small in size but identical in all other respect to the original form.

Material: Seven specimens.

Family ROBULOIDIDAE Reiss, 1963

Genus: ***Falsopalmula*** Bartenstein, 1948

Type species: *Flabellina tenuistriata* Franke, 1936

Falsopalmula deslongchampsi (Terquem, 1863)

Plate I, Figure 4

Flabellina deslongchampsi Terquem 1863, p.216, pl.10, fig.13.

Neoflabellina deslongchampsi (Terquem); Barnard 1952, p.340, fig.B-1.

Neoflabellina deslongchampsi (Terqu-em); Kalantari 1969, p.90, pl.3, figs.19, 20, *et syn.*

Falsopalmula deslongchampsi (Terqu-em); Coleman 198, p.118, pl.6.2.3, fig.1.

Dimensions (mm): Length 0.41-0.7, width 0.12-0.21, thickness 0.09-0.15.

Remarks. The present species was originally assigned to *Flabellina*, a generic name which Bartenstein (1948b) replaced with *Neoflabellina* on the basis that the former was already in use by a molluscan genus. Bartenstein (1948a) further stated that *Neoflabellina* was not known to occur before the Cretaceous and also differs in morphological features, possessing a coarse perforate surface. *Falsopalmula* ranged from Lower to Middle Jurassic whereas *Palmula* occurs in Palaeocene to Middle Miocene rocks (Loeblich & Tappan 1987). A few well-preserved specimens of *Falsopalmula deslongchampsi* Terquem (1863) are recorded in the present material which show close resemblance to the forms described by Barnard (1952) from the Oxford Clay (Middle Callovian to Lower Oxfordian) of England in overall shape of the test and nature of the sutures. This species has a parallel-sided test with a rounded proloculus, a lobate margin, and sutures ornate by low ribs and slightly rough surface.

Material: Nineteen specimens.

Subfamily FRONDICULARIINAE Reuss, 1860

Genus FRONDICULARIA Defrance, 1826

Type species: *Renulina complanata* Defrance, in de Blainvilla, 1824

Frondicularia aff. ***F. pseudoconcinna*** Bartenstein and Brand 1951

Plate I, Figure 5

Frondicularia pseudoconcinna Bartenstein and Brand 1951, p. 306, pl. 9, figs.213-215.

Frondicularia cf. *pseudoconcinna*, Bartenstein & Brand; 1969, Kalantari, p. 154, pl. 15, figs. 12a, b.

Dimensions (mm): Length 0.4-0.7, width 0.13-0.17, thickness 0.07-0.12

Remarks: *Frondicularia pseudoconcinna* was originally described from the Cretaceous of Germany. The present specimens are comparatively small in size with lesser number of chambers than the original form.

Material: Eleven specimens.

Family VAGINULINIDAE Reuss, 1860

Subfamily LENTICULININAE Chapman, Parr and Collins, 1934

Genus LENTICULINA Lamarck, 1804

Type species: *Lenticulina rotulatus* Lamarck, 1804

Lenticulina aff. ***L. atheria*** Dayn, 1958

Plate I, Figure 6

Lenticulina atheria Dayn 1958, p. 32, pl. 6, figs. 3a, b.

Dimensions (mm): Major diameter 0.23-0.35, minor diameter 0.17-0.28, thickness 0.13-0.21.

Remarks: This species closely resembles *Lenticulina subalatiformis* originally described by Reuss (1854). Our specimens are close to the original form described by Dayn (1958).

Material: 33 specimens.

Lenticulina audax Loeblich and Tappan, 1950

Plate I, Figure 7

Lenticulina audax Loeblich and Tappan 1950, p.43, Pl.11, figs. 18a-21b.

Lenticulina audax Loeblich and Tappan; Kalantari 1969, p. 34, pl. 11, figs. 30-33a-b.

Lenticulina audax Loeblich and Tappan; Kottachchi 2001, pl. 23, fig. 3.

Dimensions (mm): Length 0.5-0.7, width 0.38-0.56, thickness 0.21-0.28.

Remarks: *Lenticulina audax* is similar to *Cristellaria diademeta* Berthelin except in being approximately one - third larger and having twice the number of chambers in the final whorl. *Cristellaria gualtina* resembles with *L. audax* in size as well as the number of chambers in the final whorl but differs in the thickness of the umbo. *Cristellaria helios* is approximately one-half the size of *Lenticulina audax* and possesses limbate sutures while *Cristallaria mamillaris* differs in being nearly twice as large and in possessing fewer chambers in the final volution (Loeblich and Tappan 1950). This form frequently occurs in the upper part of the Callovian and ranges up to the Kimmeridgian.

Material: 65 specimens.

Lenticulina jurassica Kuznetsova, 1961

Plate 1, Figure 9

Lenticulina jurassica Kuznetsova 1961, p. 97, pl. 2, fig. 1a-c.

Dimensions (mm): Length 0.63-0.81, width 0.44-0.6, thickness 0.24-0.39.

Remarks: Present specimens belonging to *Lenticulina jurassica* from the Upper Jurassic (Callovian) of Russia are similar to the original form.

Material: 27 specimens

Lenticulina nodosa (Reuss, 1863)

Plate I, Figure 10

Cristellaria (Robulina) nodosa Reuss 1863, p. 78. pl. 9, fig. 6, 6a.

Cristellaria (Robulina) nodosa Reuss; Kalantari 1969, p. 147, figs. 3-7a, b, *et syn.*

Lenticulina nodosa (Reuss); Jendryka and Fuglewicz 1975, pl.12, figs.4-12.

Dimensions (mm): Major diameter 0.3-36, minor diameter 0.22-29, thickness 0.08-10.

Remarks: Considerable variation in this species and its tendency to uncoil has resulted in confusion regarding its taxonomy. Bartenstein (1974b) and Aubert and Bartenstein (1976) have identified within the *Lenticulina nodosa* plexus several subspecies and closely related species of stratigraphic and biogeographic relevance (Halbourn and Kaminski 1995).

Material: 18 specimens.

Lenticulina sp.

Plate I, Figure 13

Description: Test large, robust, planispiral, biconvex, rounded, biumbonate, periphery rounded; ten to fourteen well-marked chambers, triangular in shape, increasing gradually in size as added; sutures distinct, limbate, raised, gently curved, converge to form a non-elevated umbilical boss, sutures in initial chambers sharply elevated and later becoming less elevated but broader; aperture distinct, terminal radiate, at dorsal angle; wall calcareous perforate; surface smooth.

Dimensions (mm): Length 0.63 - 0.96, width 0.36 - 0.7, thickness 0.37 - 0.41.

Remarks: Two specimens of *Lenticulina* were found in the present material which could not be compared with any known species of the genus.

Material: Two specimens.

Subfamily MARGINULININAE Wedekind, 1937

Genus MARGINULINA d' Orbigny, 1826

Type species: *Marginulina raphanus* d'Orbigny, 1826

Marginulina* aff. *M. caelata Loeblich and Tappan, 1950

Plate II, Figure 5

Marginulina caelata Loeblich and Tappan 1950, p.46, pl.12, fig.10.

Dimensions (mm): Length 0.53-0.91, width 0.32-0.6, thickness 0.31-0.43.

Remarks: Present species somewhat resemble the original forms of *Marginulina caelata* described form the Oxfordian of USA. Our specimens have close pitting in the initial chambers unlike the original forms. *M. disparilis* also shows some resemblance to the present species but is much smaller in size, having greater width, and with a more prominent coil as compared to our specimens.

Material: Nine specimens.

Genus VAGINULINOPSIS Silvestri, 1904

Type species: *Vaginulina soluta* Silvestri var. *carinata* Silvestri, 1898

Vaginulinopsis enodis Loeblich and Tappan, 1950

Plate II, Figure 7

Vaginulinopsis enodis Loeblich and Tappan 1950, p. 46, pl.12, figs.11-16.

Dimensions (mm): Length 0.69 - 0.96, width 0.29-0.38, thickness 0.19-0.26.

Remarks: Our specimens are similar to the original forms described by Loeblich and Tappan (1950) from Jurassic of USA. *Vaginulinopsis enodis* is distinguished from *Cristellaria desorii* in being approximately twice as large and having depressed rather than limbate sutures.

Material: Ten specimens.

Subfamily VAGINULININAE REUSS, 1860

Genus CITHARINA d'Orbigny, 1839

Type species: *Vaginulina* (*Citharina*) *strigillata* Reuss, 1846

***Citharina* sp.**

Plate II, Figure 10

Description: Test medium to large, compressed; chambers six to nine following spherical proloculus; sutures distinct, thickened, depressed, gently curved, sub-parallel to one another; periphery entire, dorsal margin straight and thick; aperture terminal radiate, at dorsal angle; wall calcareous perforate; surface ornamented with seven to ten longitudinal costae, interrupted over sutures.

Dimensions (mm): Length 0.43 - 0.89, width 0.19 - 0.29, thickness 0.09-0.16.

Remarks: Present specimens of *Citharina* sp. do not resemble any described species of this genus. More well preserved specimens are required to assign a trivial name.

Material: Two specimens

Genus VAGINULINA d'Ornigny, 1826

Type species: *Nautilus leguman* Linne, 1758

***Vaginulina* sp.**

Plate I, Figure 12

Description: Test small, elongate, straight to arcuate, slightly ovate, compressed, most specimens with rounded margin, few with angular margin, at least initially; usually five to six chambers, may reach up to seven; initial end rounded or angular in profile, with a spherical or ellipsoidal proloculus; as many as three to four chambers, arranged in rectilinear or curvilinear fashion, chamber distinct, broader than high, increasing slightly in width and height as added; penultimate chamber slightly globular, smaller than final one; sutures oblique, flush, gently curved in coiled portion, oblique in later part; aperture marginal, terminal, radiate, placed on a small neck; wall calcareous perforate; surface smooth.

Dimensions (mm): Length 0.63-1.05, width 0.17-0.28, thickness 0.08-0.14.

Remarks: Some specimens of *Vaginulina* which could not be compared to any known species of the genus were recovered from the present material.

Material: Three specimens

AGE OF THE FORAMINIFERAL ASSEMBLAGES

The use of foraminifera for precise age determination of post-Mesozoic sediments is well established. Although foraminifera are excellent for biostratigraphy of Lower Jurassic sediments, the time resolution sometimes matching with that of ammonites (Herrero et al. 1996; Canales and Henriques 2008; Figueiredo et al. 2010; Canales et al. 2010; Figueiredo and Guterres 2012; Henriques and Canales 2013; Kiselev et al. 2013; Silva et al. 2014; Copestake and Johnson 2014; Levchuk et al. 2015) they are not reliable for precise dating of the Middle-Upper Jurassic rocks in the Indian region as species with restricted vertical range are rare in Middle to Upper Jurassic sediments of Kachhch (Bhalla and Abbas 1976; Kalia and Chowdhury 1983; Bhalla and Talib 1985, 1991; Talib and Faisal 2006, 2007; Gaur and Talib 2009; Talib *et al.* 2012a, b, 2016). The problem becomes further accentuated with dominance of Family Vaginulinidae and Nodosariidae during this period which include mostly fairly long-ranging species and exhibit a high degree of inter- as well as intra-specific variation. Such species are hardly of any stratigraphic value unless their range of morphologic variation with time is thoroughly investigated. Gradstein (1978), however, advocated the use of species of *Garantella*, *Reinholdella*, and reticulate *Epistomina* in Jurassic biostratigraphy. Unfortunately these forms are absent in the Present assemblages.

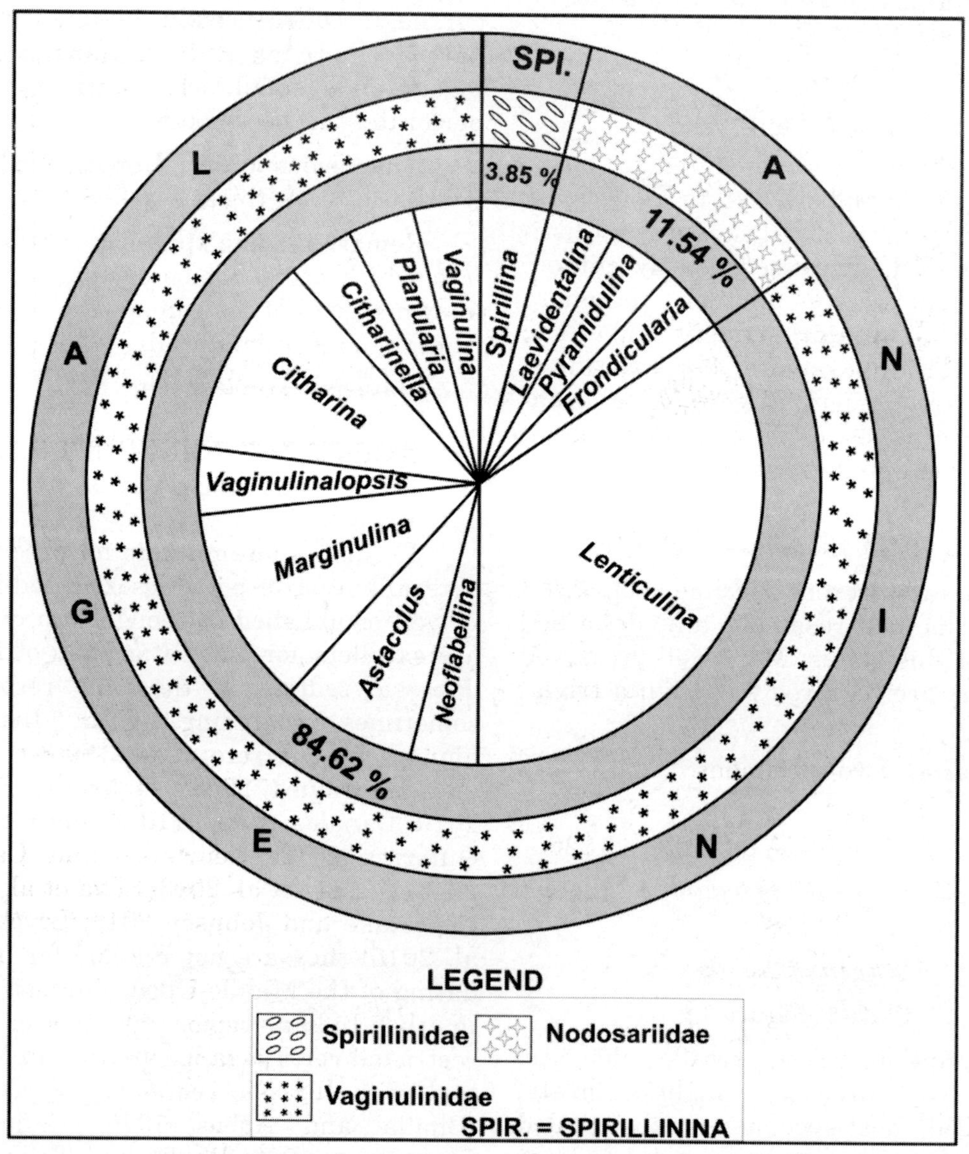

Fig.4. Composition of the foraminiferal assemblages, Chari foramation, Jara Dome, Kachhch.

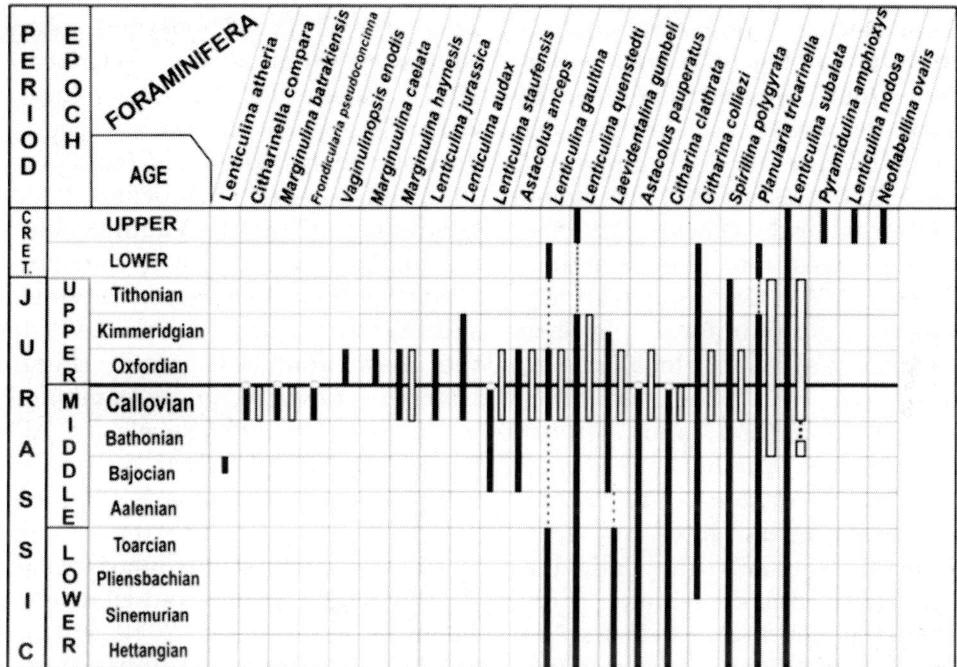

Fig.5. Known global and Indian ranges of foraminiferal species from Jara Dome, Kachhch.

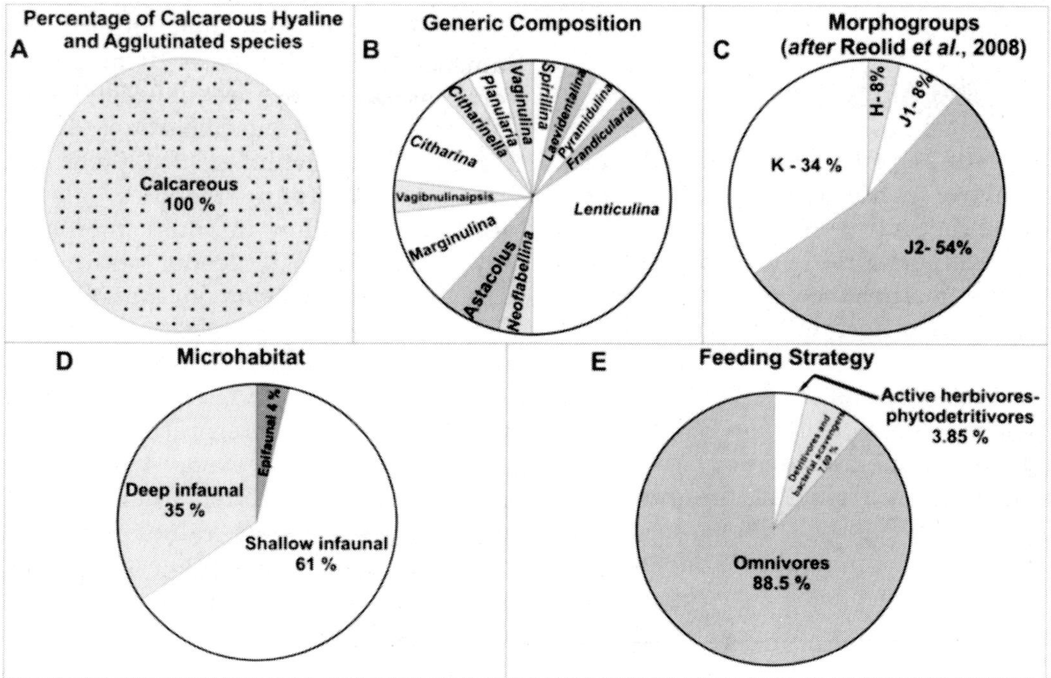

Fig.6. Pie diagrams displaying the characteristics of the foraminiferal assemblages, Jara Dome, Kachhch.

The Jara Dome foraminiferal assemblages contain a number of species which are reported from Jurassic sediment of various parts of world as well as the Indian region (Figure 5). It is apparent from Figure 5 that most of the Jara dome Jurassic foraminiferal species in the present assemblages are long-ranging. However, there are a few comparatively short ranging foraminiferal species restricted within Callovian globally, *viz.*, *Frondicularia* aff. *F. pseudoconcinna*, *Marginulina batrakiensis*, and *Citharinella* aff. *C. compara*. Furthermore, *Lenticulina staufensis*, *Astacolus pauperatus*, and *Citharina clatharata* have not been recorded in strata younger than Callovian globally. In the Indian region, three species *Citharina clathrata*, *Citharinella* cf. *C. compara*, and *Marginulina batrakiensis* are restricted in Callovian. In view of the above, Callovian age is assigned to the Shelly Shale (Lithounit JR 1) of the studied sequence.

PALAEOECOLOGY

Distribution of foraminifera is controlled by environmental factors and changes in environmental conditions such as salinity, temperature, and oxygenation lead to the disturbance of foraminiferal assemblage equilibrium (Murray 1991, Barrera *et al.* 1997, Mackensen and Bickert 1999, Thomas *et al.* 2000, Alegret and Thomas 2005, Ghoorchaei 2012). With gradual environment changes, structural reorganization occurring in foraminiferal assemblages is mainly evident in variations of numbers of individuals and species as well the test morphology.

The present foraminiferal assemblages are dominated only by calcareous hyaline (Figure 6A) forms which supporting shallow open marine environment above the CCD with normal salinity and well-oxygenated water (Barnard et al. 1981; Valchev 2003; Ghoorchaei et al. 2012; Smolen 2012) and absence of porcellaneous forms suggests middle to outer shelf environment (Valchev 2003). The dominance of family Vaginulinids indicates shallow water, open marine environment in the shelf region having normal salinity. At generic level, *Lenticulina* is the most abundant genus (Figure 6B). Diverse *Lenticulina* assemblages are indicative of high levels of dissolved oxygen and deeper shelf, open sea environment (Jones and Charnock 1985; Nagy 1992; Tyszka 1994; Nagy et al. 2009; Bernhard 1986; Koutsoukos et al. 1990; Szydlo 2004; Canales and Henriques 2008; Reolid et al. 2008a, b, 2010, 2012; Smolen 2012) and Middle to outer shelf environment (Reolid *et al.* 2008a, 2012). *Marginulina* along with *Citharina* is the second most abundant genus, indicating deeper part of the shelf (Meyer 2000; Samson 2001). The Jara dome foraminiferal assemblages are characterized by the dominance of morphogroups J2 (54%) (Figure 6C) of Reolid *et al.* (2008a, 2012). This morphogroup includes flattened forms with shallow infaunal microhabitat and active deposit feeder and grazing omnivore trophic behavior having genera such as *Frondicularia*, *Neoflabellina*, *Astacolus*, *Vaginulinopsis*, *Citharina*, *Planularia* and *Vaginulina*. Ornamentation is significant for interpreting dissolved oxygen levels and taxa with unornamented tests are related to environments with normal oxygen values while taxa with ornamented test suggest low oxygen conditions (Bernhard 1986). Of the eleven species belonging to Morphogroup J2 which are present in the Jara Dome assemblages, nine are unornamented suggesting normal oxygen values for the assem-

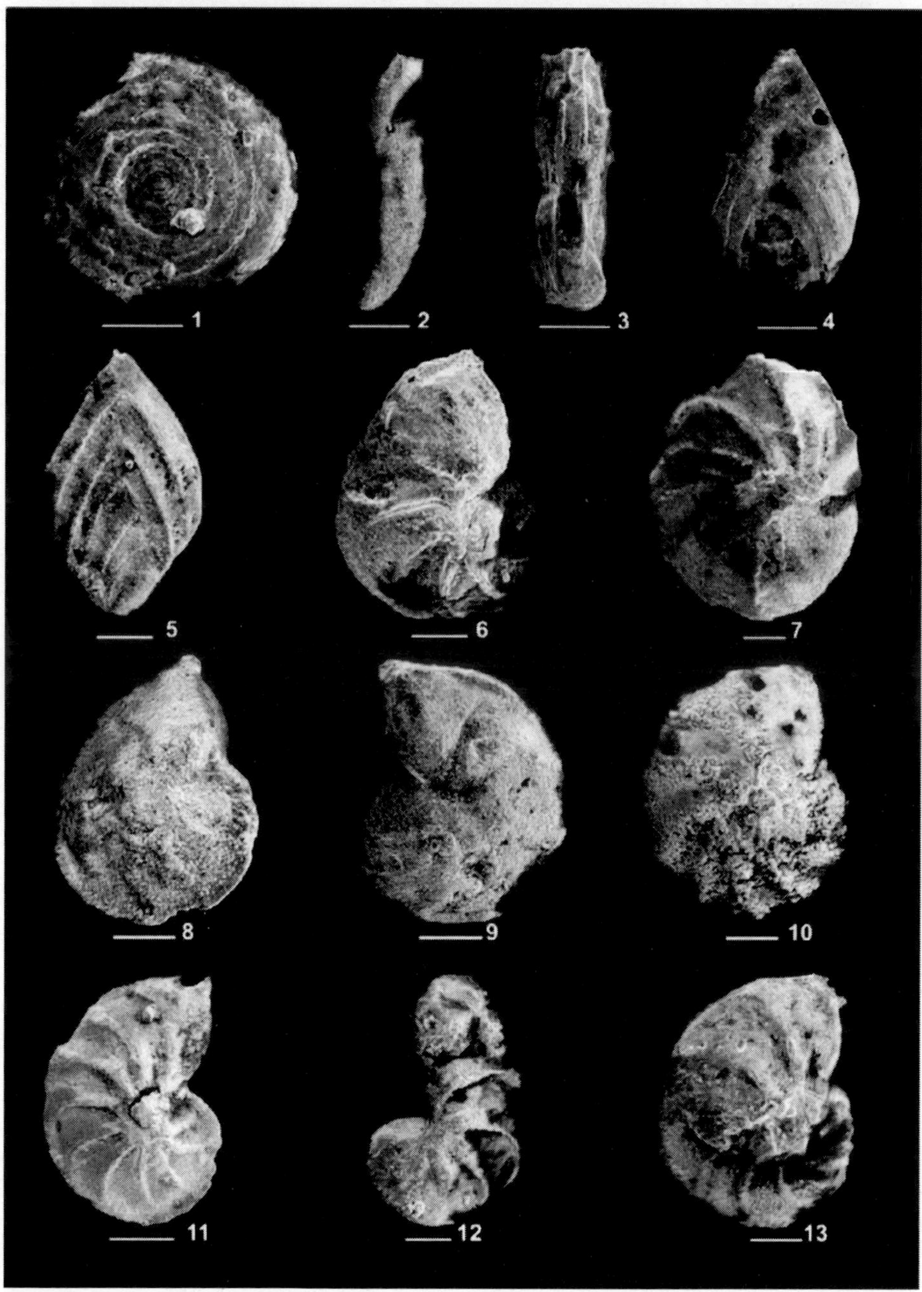

PlateI. Foraminiferal species recovered from Jara Dome, Kachhch, Gujarat (Bar Scale = 100 μm): **1.** *Spirillina polygyrata* Gümbel, 1862; **2.** *Laevidentalina gümbeli* (Schwager, 1865); **3.** *Pyramidulina amphioxys* (Reuss, 1874); **4.** *Falsopalmula deslongchampsi* (Terquem, 1863); **5.** *Frondicularia* cf. *pseudoconcinna* Bartenstein and Brand 1951; **6.** *Lenticulina* aff. *L. atheria* Dayn, 1958; **7.** *Lenticulina audax* Loeblich and Tappan, 1950; **8.** *Lenticulina gaultina* (Berthelin 1880); **9.** *Lenticulina jurassica* Kuznetsova, 1961; **10.** *Lenticulina nodosa* (Reuss, 1863); **11.** *Lenticulina quenstedti* (Gümbel, 1862); **12.** *Lenticulina staufensis* Paalzow, 1932; **13.** *Lenticulina* **sp.**

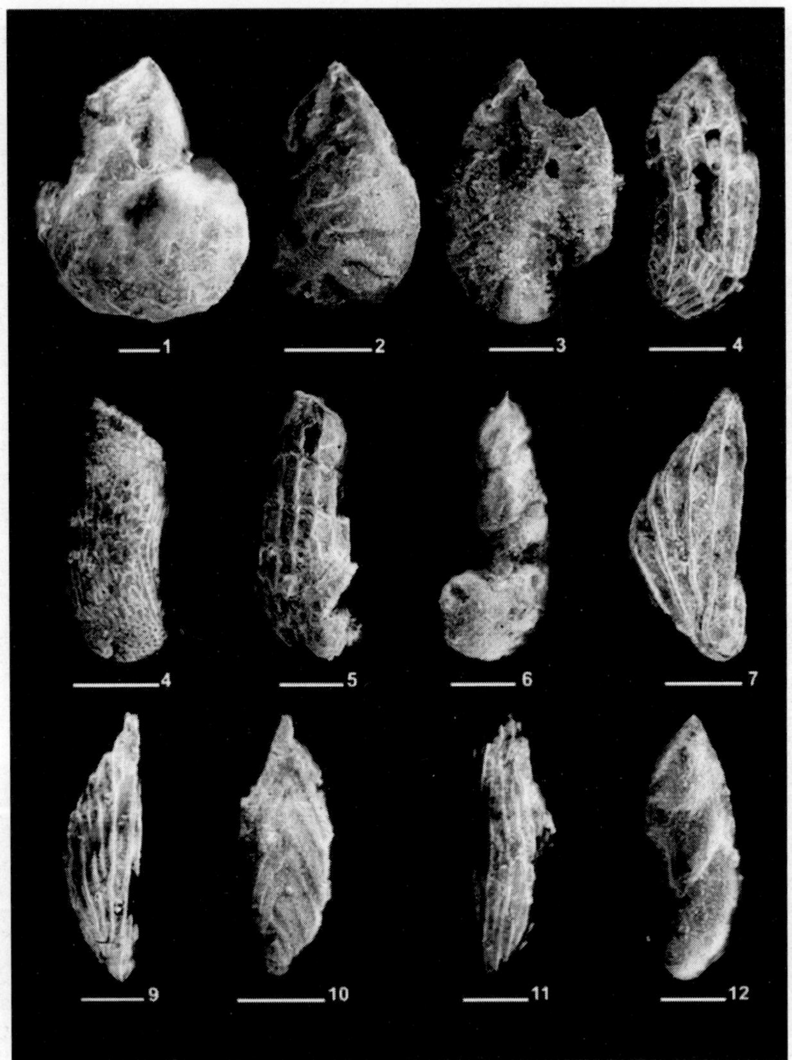

PlateII. Foraminiferal species from Chari Formation, Jara Dome, Kachhch (Bar Scale = 100 μm): **1.** *Lenticulina subalata* (Reuss, 1854); **2.** *Astacolus anceps* (Terquem, 1870); **3.** *Astacolus pauperatus* (Jones and Parker, 1860); **4.** *Marginulina batrakiensis* (Myatliuk, 1939); **5.** *Marginulina* aff. *M. caelata* Loeblich and Tappan; **6**. *Marginulina haynesis* Bhalla and Abbas 1978; **7.** Vaginulinopasis enodis Loeblich and Tappan, 1950; **8.** *Citharina clathrata* (Terquem), 1864; **9.** *Citharina colliezi* (Terquem, 1866); **10.** Citharina sp., **11.** *Citharinella* aff. *C. cobmpara* Loeblich and Tappan, 1950; **12.** Vaginulina sp.

blages. The life habit of foraminifera allow differentiation between epifaunal, shallow infaunal and potentially deep infaunal groups. In the studied samples, shallow infaunal group (61.54%) is dominant followed by potential deep infaunal group (34.62%) (Figure 6D), while epi-faunal group is insignificant (3.85%). Shallow infaunal group is dominant at all the specimen, species and generic levels. Dominancy of shallow infaunal and low concentration of epifaunal group suggest better oxygen and nutrient availability in the infaunal microhabitat. Feeding strategy allows differentiation into Omnivores (88.5%), detrivores and bacterial scavengers (7.69%), and grazing herbivores (3.85%) groups. In the studied samples (Figure 6E), Omnivores group is dominant at all the levels followed by detrivores and bacterial scavengers, whereas active herbivores phytodetritivores group is a minor constituent at all the levels. Dominancy of Omnivores group with higher proportions of *Lenticulina* suggests Middle to outer shelf environment for the present assemblages (Reolid *et al.* 2008a; 2012). The Fisher Index value for the entire assemblage is 5.75, which also indicate a deeper shelf sea with normal salinity (Wright 1972; Murray 1991; Boltvskoy and Wright, 1976 in Ujetz 1996; Valchev 2003).

CONCLUSION

The foraminiferal studies carried out on the Middle Jurassic rocks exposed at Jara Dome, Kachhch yielded assemblages comprising twenty-six species including twelve species reported for the first time from the Indian region. The assemblages are dominated by the family Vaginulinidae. In view of the presence of appreciably small number of short ranging species in the Jara Dome foraminiferal assemblages which are restricted to Callovian globally including the Indian region, some species which are rather long ranging but do not extend beyond Callovian globally, and some species considered as representative of Callovian by various authors, a Callovian age is assigned to the Lithounit JR 1 of the studied sequence of the Chari Formation exposed at Jara Dome, Kachhch. The Jara Dome foraminiferal assemblages are employed for palaeoenvironmental analysis which suggests that the overall deposition of the Chari sequence exposed in the area took place in a shallow, open marine environment in the Middle to outer shelf region, with normal salinity and well oxygenated waters.

Acknowledgements: We thank Prof. Abu Talib, Chairman, Department of Geology, A.M.U., Aligarh for providing various facilities. Thanks are also due to the Coordinator, USIF, A.M.U. for SEM photography. The district authorities of Kachhch provided logistic support during the fieldwork. The present study was funded by the Council of Scientific and Industrial Research, Government of India, New Delhi, in the form of a Senior Research Fellowship to the one of the author (SMW).

References

ALBERTI, M., PANDEY, D.K., HETHKE, M. and FÜRSICH, F.T. (2015) Ammonites of the subfamily Mayaitinae Spath, 1928 from the Oxfordian of Kachchh, Western India. Geobios, v.48 (2), pp. 85-130.

ALEGRE, T.L. and THOMAS. E. (2005) Cretaceous/Paleogene boundary bathyal paleo-environments in the central North Pacific (DSDP Site 465), the Northwestern Atlantic (ODP Site 1049), the Gulf of Mexico and the Tethys: The benthic foraminiferal record. Palaeogeogr., Palaeoclimatol., Palaeoecol., v. 224 (1), pp. 53-82.

AUBERT, J. and BARTENSTEIN, H. (1976) Lenticulina (L.) nodosa additional observations in the worldwide Lower Cretaceous. Bull. Cent. des Rech., Pau-SNPA, v. 10, pp. 1-33.

BARDHAN, S., DATTA, K., JANA, S.K. and PRAMANIK, D. (1994) Dimorphism in Kheraiceras Spath from the Callovian

Chari Formation, Kutch, India. J. Paleont., 287-293pp.

BARDHAN, S., JANA, S.K. and ROY, P. (2010) Sexual dimorphism and polymorphism in a Callovian Phlycticeras (Ammonoidea) assemblage of Kutch, India. Geobios, v.43(3), p.269-281.

BARNARD, T., CORDEY, W.G. and SHIPP, D.J. (1981) Foraminifera from the Oxford clay (Callovian - Oxfordian) of England; Rev. Espanol. Micropal., XIII. v. 3, p. 382-642.

BARRERA, E., SAVIN, S.M., THOMA, S.E. and JONES, C.E. (1997) Evidence for thermohaline circulation reversals controlled by sea-level change in the latest Cretaceous. Geol., v.25,p. 715-718.

BARTENSTEIN, H. (1974) Upper Jurassic-Lower Cretaceous primitive arenaceous foraminifera from DSDP Sites 259 and 261, Eastern Indian Ocean. In Veevers, J. J., Heirzler, J. R., et al. Init. Repts. DSDP, Washington (U.S. Government printing office), v. 27, p. 683-695.

BERNHARD, J.M. (1986) Characteristic assemblages and morphologies of benthic foraminifera from anoxic, organic-rich deposits - Jurassic through Holocene. J. Foram. Res., v. 16 (3), p. 207-215.

BHALLA, S.N., ABBAS, S.M. (1976) The age and paleogeographical significance of Jurassic Foraminifera from Kachhch, In 1st International Symposium on Benthonic Foraminifera of Continental Margins, Canada. Maritime Sediments, Special Publication, 1, pt. B, p. 537- 544.

BHALLA, S.N. and ABBAS, S.M. (1978) Jurassic foraminifera from Kachhch, India. Micropal., v. 24 (2), p. 160-209.

BHALLA, S.N. and TALIB, A. (1985) Lenticulina quenstedti (Gümbel) - A case study of variation in Foraminifera. Rev. Palèobiol., v.4, pp.9-13.

BHALLA, S.N. and TALIB, A. (1991) Callovian-Oxfordian foraminifera from Jhurio hill, Kachhch, Western India. Rev. Palèobiol., v.10, pp.85-114.

CANALES, M.L. and HENRIQUES, M.H. (2008) Foraminifera from the Aalenian and the Bajocian GSSP (Middle Jurassic) of Murtinheira section (Cabomondego, west Portugal): Biostratigraphy and paleoenvironmental implications. Mari. Micropal., v.67, pp.155-179.

CANALES, M.L., HENRIQUES, M.H., FIGUEIREDO, V.L. and GUTERRES, H.C. (2010) Foraminifera from the Lower-Middle Jurassic of the Lusitanian Basin (Portugal) - biostratigraphic and palaeoecological significance. In: Pena dos Reis, R. and Pimentel, N., Eds. Rediscovering the Atlantic: New Ideas for an Old Sea, v.1, pp.42-46. II Cen. & North Atlantic Conjugate Margins Conf., Lisbon.

CARIOU, E. and KRISHNA, J. (1988) The Tethyan Reineckiinae of Kachchh and Jaisalmer (West India): systematic, biostratigraphic, and biogeographic implications. Palaeontographica, A. v.203, pp.149-170.

COPESTAKE, P. and JOHNSON, B. (2014) Lower Jurassic Foraminifera from the Llanbedr (Mochras Farm) Borehole, North Wales, U.K. Palaeontogra. Soci., 403p.

DAYN, I.G. (1958) *In* BYKOYA, N.K. [New genera and species of foraminifera] [Russain] Leningrad, Vses. Neft. Nauchno- Isseled.

FIGUEIREDO, V.L. and GUTERRES, H.C. (2012) Análise quantitativa das associações de foraminiferos da passagem Jurássico Inferior-Médio do perfil de Maria Pares (Setor Norte da Bacia Lusitânica, Portugal) - implicações paleoecológicas. Memórias e Notícias de Geociências no Espaço Lusófono, Publicacão da Imprensa da Universidade de Coimbra, pp.1-8.

FIGUEIREDO, V.L.E., HENRIQUES, M.H.P. and CANALES, M.L.F. (2010) Foraminíferos bentonicos da passagem Jurássico Inferior-Médio do Sector Central da Bacia Lusitânica: o perfil de Zambujal de Alcaria. B. Geoci. Petrobras, v.19, pp.207-231.

FÜRSICH, F.T., PANDEY, D.K., CALLOMON, J.H., JAITLY, A.K. and SINGH, I.B. (2001) Marker beds in the Jurassic of the Kachhch Basin, Western India. J. Paleo. Soc. India, v.46, pp.173-198.

FÜRSICH, F.T., OSCHMANN, W., JAITLY, A.K. and SINGH, I.B. (1991) Faunal response to transgressive-regressive cycles: examples from the Jurassic of western India. Palaeogeogr., Palaeoclimatol., Palaeoecol., v.85, pp.149-159.

FÜRSICH, F.T., OSCHMANN, W., SINGH, I.B. and JAITLY, A.K. (1992) Hardgrounds, reworked concretion levels and condensed horizons in the Jurassic of western India: their significance for basin analysis. J. Geol. Soc. India, v.149, pp.313-331.

FÜRSICH, F.T., CALLOMON, J.H. PANDEY, D.K. and JAITLY, A.K. (2004a) Environments and faunal patterns in the Kachchh rift basin, western India, during the Jurassic. Rivista Italiana Paleontologia Stratigrafia, v.110, pp.181-190.

FÜRSICH, F.T., OSCHMANN, W., PANDEY, D.K., JAITLY, A.K., SINGH, I.B. and LIU, C. (2004b) Palaeoecology of middle to lower Upper Jurassic macrofaunas of the Kachchh Basin, western India: an overview. J. Paleo. Soc. India, v.49, pp.1-26.

GAUR, K.N., TALIB, A. (2009) Middle-Upper Jurassic Foraminifera from Jumara Hills, Kachhch, India. Rev. de Micropal., v.52, pp.227-248.

GHOORCHAEI, S., VAHIDINIA, M., ASHOORI, A. (2012) Late Maastrichtian benthic foraminiferal response to palaeo-environmental changes: a case study from the AbTalkh. Formation, west of Kopeh-Dagh Basin, Iran, J Geope, v.2(2), pp.25-42.

GRADSTEIN, F. (1978) Jurassic Grand Banks foraminifera. J. Foram. Res., v.8, pp.97-109.

HENDERSON, J. (2013) Tight Oil Developments in Russia, Oxford Inst. Energy Studies, v.52, pp.1-22.

HENRIQUES, M.H. and CANALES, M.L. 2013. Ammonite-benthic foraminifera turnovers across the Lower-Middle Jurassic transition in the Lusitanian Basin (Portugal). Geobios, v.46, pp.395-408.

HERRERO, C., COSPESTAKE, P. and JOHNSON, B. (1996) *Saracenella mochrasensis* sp. Nov (Foraminiferida) - A regional biostratigraphic marker species for upper Toarcian (Jurassic) of Europe. J. Foram. Res., v.26(3), pp.187-192.

HOLBOURN ANN, R.L. and KAMINSKI, M. A. (1995) Valanginian to Barremian benthic foraminifera from ODP Site 766 (Leg 123, Indian Ocean). Micropal., v.44(3), pp.197-250.

JAIN, S. and DESAI, B.G. (2014) Biostratigraphic implications of the discovery of late bathonian indonesian ammonite macrocephalites cf. Mantataranus boehm [m] from the core of Jara Dome, Kachchh, Western India. J. Pal. Soc. India, v.59(2), pp.169-180.

JONES T.R. and PARKER, W.K. (1860) On some fossil foraminifera from Chellaston near Derby. Quart. J. Geol. Soc. India, v.16 (1-2), pp.452-458.

KALANTARI, A. (1969) Foraminifera from the Middle Jurassic-Cretaceous successions of Koppet- Dagh Region (Iran). Nat. Iranian Oil Corp. Geol. Lab. Pub., v.3, pp.1-298.

KALIA, P. and CHOWDHURY, S. (1983) Foraminiferal biostratigraphy, biogeogr-aphy, and environment of Callovian sequence, Rajasthan, northwestern India. Micropal., v.29(3), pp.223-254.

KHOSLA, S.C., JAKHAR S.R. and MOHAMMED, M.H. (1997) Ostracodes from the Jurassic rocks of Habo hill, Kachchh, Gujarat. Micropal., v.43, pp.1-39.

KHOSLA, S.C., KUMARI, M., FLEX, A.D., JAKHAR, S.R. and NAGORI, M.L. (2005) Middle Jurassic ostracodes from the northern Island belt, Rann of Kachchh, Gujarat, India. J. Paleo. Soc. India, v.50, pp.p 17-64.

KISELEV, D., ROGOV, M., GLINSKIKH, L., GUZHIKOV, A., PIMENOV, M.,

MIKHAILOV, A. and TESAKOVA, E., (2013) Integrated stratigraphy of the reference sections for the Callovian /Oxfordian boundary in European Russia and Northern Caucasus. Volu. Jura., v.11, pp.59-96.

KOUTSOUKOS, E.A., LEARY, N. and HART, M.B. (1990) Latest Cenomonian-earliest Turonian low-oxygen tolerant benthonic foraminifera - A case study from the Sergipe basin (N. E. Brazil) and the western Anglo-Paris basin (southern England), Palaeogeogr., Palaeoclimatol., Palaeoecol., v.77, pp.145-177.

KRISHNA, J., PANDEY, B. and OJHA, J. R. (2009) Gregoryceras in the Oxfordian of Kachchh (India): Diverse eventful implications. Geobios, v.42, pp.197-208.

LEVCHUK, L.K. NIKITENKO, B.L. and MELEDINA, S.V. (2015) Callovian and Upper Jurassic foraminiferal and ammonite biostratigraphy of the Shaim petroleum region (West Siberia). Russ. Geol. Geophy., v.56, pp.932-946.

LOEBLICH, A.R. and TAPPAN, H. (1950) Foraminifera from the type Kiowa Shale, Lower Cretaceous of Kansas. University of Kansas Paleontological Contributions, Protozoa, v.3, pp.1-15.

LOEBLICH, A.R. and TAPPAN, H. (1988) Foraminiferal genera and their classification. Van Nostrand Reinhold Company, New York, I, 970 p.; II, 212p, 847pls.

MACKENSEN, A. and BICKERT, T. (1999) Stable carbon isotopes in benthic foraminifera: proxies for deep and bottom water circulation and new production. *In*: FISCHER, G., Wefer, G. (Eds.), Use of Proxies in Paleoceanography. Springer, Berlin, pp.229-254.

MEYER, M. (2000) Le complexe récifal Kimméridgeien-Tithonien du Jura meridional interne (France), evolution multifactorille, stratigraphie et tectonique. Terra and environment, v.24,179p.

MISHRA, D. and TIWARI, R.N. (2006) Lithofacies and depositional dynamics of golden Oolit. (Bathonian), Kachchh Mainland, Gujarat (India). J. Asi. Eart. Scie., v.26(5), pp.449-460.

MURRAY, J.W. (1991) Ecology and Palaeoecology of benthic foraminifera. Longman scientific and Technical, Harlow, 319 p.

NAGY, J. (1992) Environmental significance of foraminiferal morphogroups in Jurassic North Sea deltas. Paleo. Paleo., Paleo., v.95, pp.111-134.

NAGY, J. REOLID M. and RODRIGUEZ-TOVAR F.J. (2009) Foraminiferal morphogroups in dysoxic shelf deposits from the Jurassic of Spitsbergen. Pol. Resea., v.28, pp.214-221.

NAIRN A.E.M. and ALSHARHAN A.S. (1997) Sedimentary basins and petroleum geology of the Middle East. Elsevier, 1-811 pp.

PANDEY, J. and DAVE, A. (1993) Studies in Mesozoic foraminifera and chronostratigraphy of Western Kachhch, Gujarat. Palaeo. Indi., v.1, pp.1-221.

PANDEY, D.K. and Fürsich, F.T. (1998) Distribution and succession of Jurassic rocks in Gora Dongar, Pachchham “Island”, Kachchh, India. J. Geol. Soc. India, v.51, pp.331-344.

PANDEY, D.K. and FÜRSICH, F.T., 2001. Environmental distribution of scleractinian corals in the Jurassic of Kachchh, western India. J. Geol. Soc. India, v.57, pp.479-495.

PANDEY, D.K., FÜRSICH, F. T. and SHA, J. (2009) Inter-basinal marker intervals - A case study from the Jurassic basins of Kachchh and Jaisalmer, western India. Science China Series D-earth Sciences, v. 52 (12), p.1924-1931.

PANDEY, D., SINGH, S., SINH, A.M. and MACGREGOR, L. (2009) Structural imaging of Mesozoic sediments of Kachchh, India and their hydrocarbon prospects. Mari. Petrol. Geo., v.26, pp.1043-1050.

PATEL, S.J., JOSHI, P.N. and JOSEPH, J.K. (2012) Ammonite zonation of the Jurassic rocks of the Gangta Bet area, Wagad region, eastern Kachchh, India. J. Pal. Soc. India, v.57(2), pp.129-133.

PRASAD, S. (1993) A Note on the Middle Jurassic Stratigraphic Succession of Keera Dome, Kachchh District, Gujarat. J. Geol. Soc. India, v.41(2), pp.156-161.

PRASAD, S. (1998) Ammonite Biozonation of the Middle - Upper Jurassic Sediments with Special Reference to Keera and Jara Dome, Kachchh District, Gujarat. J. Geol. Soc. India, v.52, pp.25-40.

RAI, J. (2006) Reworked Pliensbachian-Aalenian nannofossils from Jara Dome, Kutch: Lower Jurassic Palaeobiogeography of western India at the XIth International Nannoplankton Association at Nebraska, U.S.A., pp.84-85.

REOLID, M., RODRÍGUEZ-TOVAR, F.J., NAGY, J. and OLÓRIZ, F. (2008a) Benthic foraminiferal morphogroups of mid to outer shelf environments of the Late Jurassic (Prebetic Zone, southern Spain) - Characterization of biofacies and environmental significance. Paleo. Paleo. Paleo., v.261, pp.280-299.

REOLID, M., RODRÍGUEZ-TOVAR, F.J., NAGY, J. and OLÓRIZ, F. (2008b) Benthic foraminiferal morphogroups of mid to outer shelf environments of the Late Jurassic (Prebetic Zone, southern Spain) - Characterization of biofacies & environmental significance. Paleo. Paleo. Paleo., v.261, pp.280-299.

REOLID, M., NAGY, J. and RODRÍGUEZ-TOVAR, F.J. (2010) Ecostratigraphic trends of Jurassic agglutinated foraminiferal assemblages as a response to sea-level changes in shelf deposits of Svalbard (Norway). Palaeogeogr., Palaeoclimatol., Palaeoecol., v.293, pp.184-196.

REOLID, M., SEBANE, A. RODRIGUEZ-TOVAR, F.J. and MAROK, A. (2012) Foraminiferal morphogroups as a tool to approach the Toarcian anoxic event in the Western Saharan Atlas (Algeria). Palaeogeogr., Palaeoclimatol., Palaeoecol., v.323, pp.87-99.

SAMSON, Y. (2001) Foraminiferes et reconstitution des variations bathymitriques, exemple du kimmiridgien de la rigiondu havre (Seine-Maritime, Normandie, France). Revue de Micropal., v.44 (1), pp.59-91.

SILVA, S.C., HERIQUES, M.H. and CANALES, M.L. 2014. High resolution ammonite-benthic foraminiferal biostratigraphy across the Aalenian-Bajocian boundary in the Lusitanian Basin (Portugal). Geolo. J., v.50, pp.477-496.

SMOLEN, J. 2012. Faunal dynamics of foraminiferal assemblages in the Bathonian (Middle Jurassic) ore-bearing clays at Gnaszyn, Kraków-Silesia Homocline, Poland. Acta Geol. Polon., v.62, pp.403-419.

SPATH, L.F. (1927-1933) Revision of the Jurassic cephalopod fauna of Kachh (Cutch). Parts I-VI. Memoirs of the Geological Survey of India, Palaeontologia Indica, new series 9 2, part I, 1-71, pls.1-7 [1927]; part II, 72-161, pls.8-19 [1928a]; part III, 162-278, pls. 20-47 [1928b]; part IV, 279-550, pls. 48-102 [1931a]; part v.551-658, pls. 103-124 [1931b]; part VI, 659-945, pls. 125- 130 [1933].

SZYDŁO, A. (2004) The distribution of agglutinated foraminifera in the Cieszyn Basin, Polish Outer Carpathians. Grzybowski Found. Sp. Pub., v.8, pp.461-470.

TALIB, A. and FAISAL, S.M.S. (2006) A preliminary note on the occurrence of foraminifera in the Middle to Upper Jurassic sediments of Fakirwari, Kachhch Mainland, Gujarat. J. Geol. Soc. India, v.68, pp.963-966.

TALIB, A. and FAISAL, S.M.S. (2007) On the occurrence of microfossils(Foraminiferida) in the Jurassic rocks of Ler dome,Kachhch Mainland, Gujarat. Curr. Sci., v.92(5), pp.595-596.

TALIB, A., GAUR, K. N., SISODIA, A. K., BILAL, A. B. and ROOHI, I. (2012a) Foraminifera from Jurassic Sediments of Keera Dome, Kachhch, J. Geol. Soc. India, v. 80, pp. 667-675.

TALIB, A., GAUR, K.N., SINGH, Y.P., WASIM, S.M. and ANJUM S. (2012b) Some Foraminiferal Taxa in the Middle to Late Jurassic Sediments of Kaiya Dome, Kachhch. J. Geol. Soc. India, v.79, pp.161-168.

THOMAS, E., ZACHOS, J.C. and BRALOWER, T.J. (2000) Deep-sea environments on a warm earth: latest Paleocene-early Eocene. In: Huber, B., MacLeod, K., Wing, S. (Eds.), Warm Climates in Earth History. Cambridge Univ. Press, Cambridge, pp.132-160.

TYSZKA, J. (1994) Response of Middle Jurassic foraminiferal morphogroups to dysoxic/anoxic conditions in the Pieniny Klippen Basin, Polish Carpathians. Paleo. Paleo. Paleo., v.110, pp.55-81.

UJETZ, B. (1996) Micropaleontology of Paleogene deep water sediments, Haut-Savoie, France. Published doctoral dissertation, Département de Géologie et de Paléontologie, Univ. of Geneva, v.22, pp.1-149.

VALCHEV, B. (2003) On the potential of small benthic foraminifera as paleoecological indicators, Recent advance. 50 years University of Mining and Geology "st. Ivan Rilski", Geol. Geoph., Sofia, v.46 (1), pp.189-194.

WAAGEN, W. (1873-1875) Jurassic fauna of Kutch. The Cephalopoda. Mem. Geol. Surv. India, Palaeo. Indica, series 91, part I, 1-22, pls.1-4 [1873]; part II, 23-76, pls.5-14 [1875a]; part III, 77-106, pls.15-24 [1875b]; part IV, pp.107-247; pls.25-60 [1875c].

WILLIAMS, E.C. (2005) Making sense of middle eastern petroleum. Geotimes, v.50(10), pp.22-25.

WRIGHT, C.A. (1972) Foraminidiferids from London Clay at Lower Swanwick and their paleoecological interpretation. Proc. Geologists' Assoc., v.83(3), pp.337-348.

Micropaleontology and its Applications
Pradeep K. Kathal, Rajiv Nigam & Abu Talib (Editors)
Scientific Publishers (India), 2018, 211-232 pp.

Preliminary Study of Microstructure and Taphonomy of Dinosaur Eggs and Nests of the Upper Cretaceous Lameta Formation near Padlya, Lower Narmada Valley, India

Harsha Dhiman[1*], Vishal Verma[2] and G.V.R. Prasad[1]

[1]Department of Geology, Center for Advanced Studies, University of Delhi, Delhi- 110007, India
[2]Higher Secondary School, Bakaner, Dhar District, Madhya Pradesh, India
***Email:** harshadhiman7@gmail.com

Abstract

Though a number of eggs and eggshell localities have been documented from India, taphonomic analysis is limited only to a few sites. A systematic study of dinosaur nests, eggs, and eggshell fragments from a nesting site in the Upper Cretaceous (Maastrichtian) Lameta Formation near Padlya village, Dhar District, Madhya Pradesh was carried out with the objectives of characterization of microstructural features in terms of parataxonomic classification and taphonomic analysis of the nesting sites. Microstructural study of the eggs and eggshells from this site revealed the presence of four titanosaurid sauropod dinosaur oospecies viz., *Megaloolithus jabalpurensis, M. cylindricus, Fusioolithus baghensis,* and *F. padiyalensis*, belonging to two oofamilies (Megaloolithidae and Fusioolithidae). Presence of as many as 30 nesting sites belonging to four oospecies on a single lithology and at the same stratigraphic level is suggestive of colonial nesting. The eggshells are found in a sandy carbonate horizon which was subsequently affected by pedogenic processes and their burial took place in a low energy environment. A multilayered egg-shell with a close similarity to a pathological condition called as 'ovum-inovo' has also been found in one of the studied eggs. Eggs in some nesting sites document collapse structures, resorption craters, while a few show signs of diagenetic alteration. The common occurrence of *M. jabalpurensis* and *F. baghensis* in India and South America and closely related species in southern

Europe points to a close palaeobiogeographic link between the southern continents and southern Europe through Africa.

Keywords: Dinosaurs, eggs, nests, Late Cretaceous, Lameta Group, Narmada valley.

INTRODUCTION

Eggshells are the biomineralized products i.e., minerals produced essentially by living organisms during their metabolic activities. As in modern birds, the eggshells of dinosaurs are composed of mineral calcite ($CaCO_3$) which gets fossilized efficiently. Damage to the microstructure of the eggshells can occur through the effects of diagenesis after burial (Grellet et al., 2010). Fossil eggshells have been documented from all continents except Antarctica (Hayward, 2005) and many of them show close similarity in microstructure especially those belonging to the former Gondwanaland (Vianey-Liaud et al., 2003, Chassagne-Manoukian et al., 2013, Fernandez and Khosla 2015). The Indian subcontinent is an important source of many oospecies of dinosaurs including those of sauropods and theropods and also those of lizards, chelonians, crocodilians and birds as well.

A reptilian eggshell is composed of a basic three-layered structure in which the inner membrane present in the lowermost layer contains aggregations of organic materials which are the nucleation sites for calcite crystals (Jackson et al., 2002). This forms the "mammillary layer". This is followed by columns of calcite crystals (aragonite in case of turtle eggshell) which comprise the "palisade layer". Outer surface is marked by rounded or ridge-like calcite crystals that form the "external layer". In between the junction of columnar calcite crystals pore canals are present which allow an efficient mechanism of gas exchange between the embryo and the atmosphere (Mikhailov, 1997) (Fig.1).

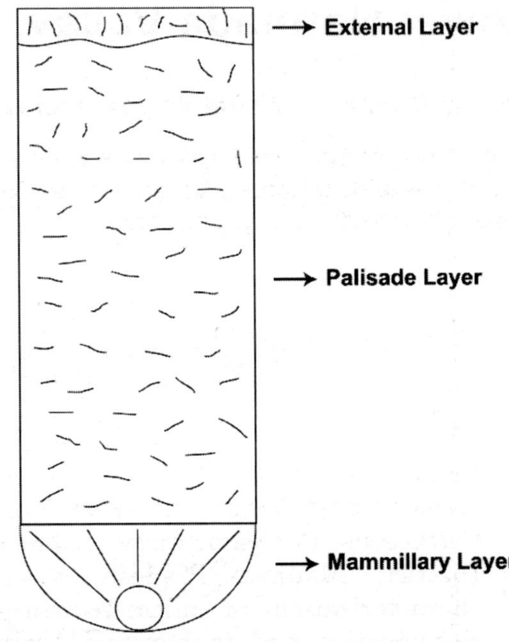

Fig.1. A general three-layered radial structure of an eggshell (after: Mikhailov, 1997).

The first fossil egg documented from India was that of a turtle from the Lower Cretaceous Uttatur Group of the Cauvery Basin (Sahni, 1957). The first dinosaur eggshell fragments were recorded from the Upper Cretaceous Lower Limestone horizon of the Lameta Formation, Jabalpur (Sahni and Gupta, 1982). Since then many sites within the Lameta Formation (e.g., Pisdura, Nand-Dongargaon, Jabalpur,

Balasinor, Bagh) (Jain and Sahni 1985, Sahni et al., 1994, Khosla and Sahni 1995, Mohabey 1996, Mohabey et al., 1998, Srivastava et al., 2015, Fernandez and Khosla 2015), the Deccan Intertrappean beds (Asifabad, Nagpur, Anjar, Mohgaon-kalan, Duddukuru, Gujri Gate, Kelapur, Piplanarayanwar) (Sahni et al., 1984; Vianey-Liaud et al., 1987, 2003; Srinivasan et al., 1996; Singh 2013; Sharma 2014) and the Upper Cretaceous Kallankuruchchhi Formation of Cauvery basin (Kohring et al., 1996) have yielded fossil eggs and eggshell fragments. Additionally, Sahni et al. (1994) made an attempt to understand the taphonomic processes involved in the preservation of dinosaur nests in the Lameta Formation exposed near Jabalpur. Till now no preserved embryo has been documented from any of these sites.

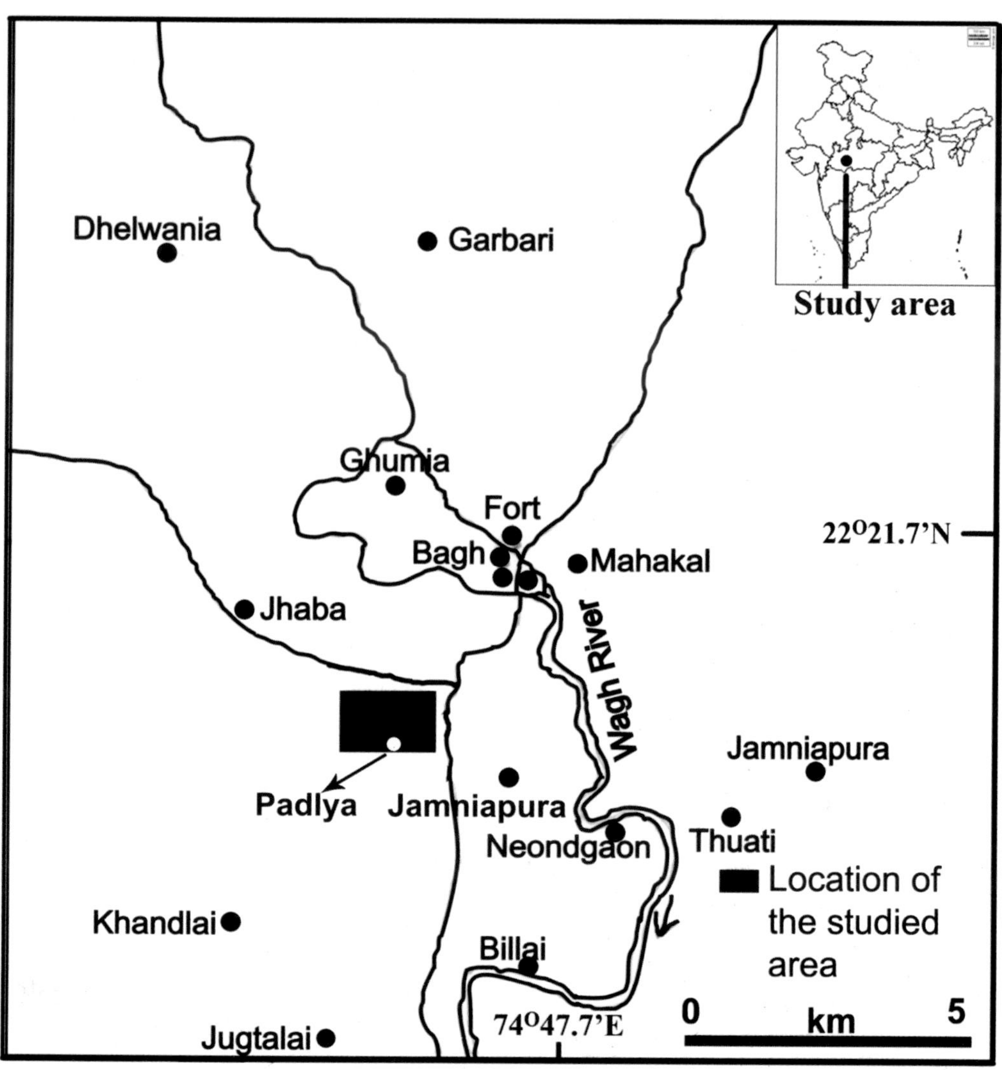

Fig.2. Map showing location of nesting sites in the Upper Cretaceous Lameta Formation near Padlya village, Dhar District, Madhya Pradesh.

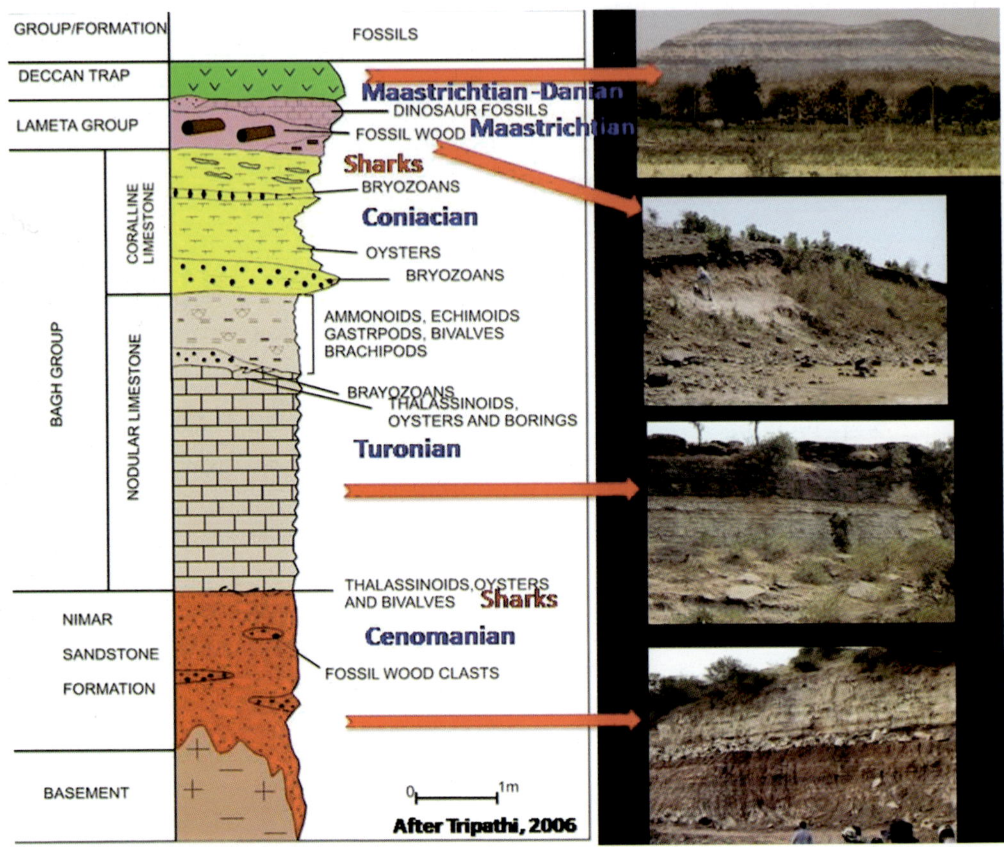

Fig.3. Composite lithocolumn of the geological formations exposed in lower Narmada valley (modified after: Tripathi, 2006).

Parataxonomy of the dinosaur eggshells from the Upper Cretaceous Lameta Formation has been provided by Khosla and Sahni (1995), Mohabey (1998), and Vianey-Liaud et al. (2003). Recently, Fernandez and Khosla (2015) revisited the ootaxonomy of dinosaur eggshells from the Late Cretaceous sites of peninsular India. They had synonymized some of the previously described species and described a new oogenus *Fusioolithus* under a new oofamily Fusioolithidae. *Fusioolithus* has a dinosaur-spherulitic basic type, tubospherulitic morphotype, tubocanaliculate pore system, and compactituberculate surface ornamentation (Fernandez and Khosla, 2015). Because of the presence of partially fused shell units, *Fusioolithus* has been placed in a separate oofamily. Hence at present there are nine valid Indian sauropod oospecies: *Megaloolithus cylindricus* (Khosla and Sahni, 1995), *M. jabalpurensis* (Khosla and Sahni, 1995), *M. megadermus* (Mohabey, 1998), *M. dhoridungriensis* (Mohabey, 1998), *M. khempurensis* (Mohabey, 1998), *Fusioolithus baghensis, F. padiyalensis, F. mohabeyi* and *F. dholiyaensis* (Khosla and Sahni, 1995; Fernàndez and Khosla 2015).

In the present study, the focus is on the dinosaur eggshell fragments documented from the Lameta Formation near

Padlya, Lower Narmada Valley (Madhya Pradesh), India (Fig.2). The objective is to identify the eggs and eggshells at oogenus and oospecies level within the framework of parataxonomic classification through the study of their microstructure under petrological microscope and Scanning Electron Microscope (SEM) and to understand the taphonomy of the nesting sites.

GEOLOGICAL SETTING

A narrow arm of the Tethys sea occupied the lower part of the Narmada valley (from Barwah near Indore in the east to Kawant (Gujarat) in the west during a marine transgression that took place between Cenomanian and Coniacian (Chiplonkar and Badve, 1973; Jafar, 1982). This area was first mapped by Blanford (1869) who divided the rock sequences informally into Nimar Sandstone, Nodular Limestone, Deola-Chirakhan Marl and Coralline Limestone. Following this work, Bose (1884) formally named these lithostratigraphic units. Since then different lithostratigraphic classification schemes have been suggested from time to time (Poddar, 1964; Raiverman, 1975; Guha, 1976; Singh and Srivastava, 1981). However, the formations proposed by Blanford (1869) and Bose (1884) are mostly followed except for the removal of the Deola-Chirakhan Marl as a separate lithostratigraphic unit. Some workers considered this unit to be a weat-hered part of the Nodular Limestone (Roy Chowdhury and Sastri, 1962; Sahni and Jain, 1966) or an integral part of the Coralline Limestone (Rode and Chiplonkar, 1935).

The Precambrian basement rocks of Mahakoshal Group are overlain with a nonconformity by the Nimar sandstone which is regarded as fluvial in origin in its lower part and brackish to marine in its upper part (Jafar, 1982; Badve and Nayak, 1984). The outcrops of Nimar Sandstone mostly consist of sandstone and conglomerate. The overlying Nodular Limestone is thin-bedded (~10 m thick) and yields a number of invertebrate fossils such as gastropods, bivalves, brachiopods, ammonoids, echinoids etc. The Coralline Limestone conformably overlies the Nodular Limestone and is rich in brachiopod shells and bryozoans. The bioturbated limestone, called as Deola-Chirakhan Marl, is highly fossiliferous and contains many invertebrate fossils. The Nimar Sandstone, the Nodular Limestone and the Coralline Limestone have been assigned Cenomanian, Turonian and Coniacian ages, respectively (Jaitley and Ajane, 2013). A 5 m thick green sandstone bed conformably overlying the Coralline Limestone and possibly representing the upper most part of the Bagh Group yielded oyster shells and shark teeth. Following the deposition of the Bagh Group, the marine sea way withdrew from the lower Narmada valley and continental sediments of Maastrichtian age representing the Lameta Formation were deposited with an erosional contact with the underlying Bagh Group. The Lameta Formation of the upper Narmada valley consisting of Green Sandstone, Lower Limestone, Mottled Nodular Beds and Upper Calcareous Sandstone, has been an important source of dinosaur skulls, postcranial bones, and nesting sites (Huene and Matley, 1933; Sahni et al., 1994). In the Bagh area, the Lameta Formation yields many silicified fossil wood logs, dinosaur nests and eggshell fragments. Following the deposition of the Lameta Formation, the flood basalts of Deccan Traps erupted and covered most of the pre-existing rocks and the marine formations of the Bagh Group are exposed only in areas where the overlying Deccan basalts have been removed by weathering and erosion. The Deccan basaltic flows are at many places intercalated with thin Intertrappean sedimentary beds.

METHODOLOGY

The fossilized dinosaur eggshell fragments were collected from Padlya village, Dhar District, Madhya Pradesh, India from the Upper Cretaceous Lameta Formation during a field visit in the month of March, 2014. The eggshell fragments were collected from both nests and scattered occurrences and were numbered accordingly. In the laboratory, the specimens first underwent cleaning in water following which they were dried for a couple of hours. Then they were subjected to cleaning in ultrasonic vibrator. The matrix left behind was cleaned under the microscope with fine needles and paint brush and the final cleaning was done through ultrasonic vibrator.

The large eggshell fragments or those embedded in the rock were cut to small size so that they are fit for SEM analysis. The specimens were selected for determination of the ultrastructure in the radial surface and morphology of the external and internal surfaces. The cleaned specimens were mounted on aluminium stubs and coated with gold-palladium metal. Scanning Electron Microscope (SEM) photomicrographs were taken and Energy Dispersion Spectra (EDS) analysis was carried out using Zeiss EVOMA 10 model at the Indian Agricultural Research Institute (IARI), Pusa, New Delhi.

Some eggshell fragments were further prepared for thin section study. The thin sections were made by applying standard histological techniques. The specimen was first grinded to a required flat surface using carborundum powder and was then fixed to glass slide with araldite. Following this, the slides were allowed to dry for about 2 days after which they were subjected to grinding on a lap using 400μm carborundum powder. This was followed by polishing on glass plate with 600μm, 800μm, and 1000μm carborundum powder until the thickness reached 0.003 mm. The thin sections made in this way were studied under Axio Imager A1m (Carl Zeiss) High Resolution Petrological Microscope in the Vertebrate Palaeontology Laboratory, Department of Geology, University of Delhi. Data on many of the taphonomic attributes of the eggs and nesting sites were collected in the field.

All the eggshell specimens and thin sections used in this study are deposited in Vertebrate Palaeontology Lab, Department of Geology, University of Delhi and bear DUGF/ catalogue numbers.

Systematic Description

Family MEGALOOLITHIDAE Zhao, 1979

Megaloolithus jabalpurensis Khosla and Sahni, 1995

Stratigraphic Horizon and Locality: Sandy carbonate horizon, Upper Cretaceous (Maastrichtian) Lameta Formation, Padlya, Dhar District, Madhya Pradesh.

Referred Specimens: Thin sections of eggshell fragments (DUGF 433, 435, 436, 438, 440, 445, 446) and a pathological eggshell (DUGF/443).

Description: Eggshell thickness varies between 1.22 mm and 1.68 mm. This morphotype is discretispherulitic and the surface ornamentation is compactituberculate. The average height of spherolith is 1.42 mm and width is 0.69 mm and height/width ratio is 2.1 : 1.0. Spheroliths are broad in the upper part, fan-shaped and show no fusion with adjacent spheroliths. Rather a distinct junction is present at the intersection of shell units. Smaller spheroliths are also present between the larger ones. The growth lines are clearly visible and moderately arched upwards. The pore canals are straight, vertical or subvertical (tubocanaliculate) (Fig.4).

DUGF/443 consists of spheroliths of variable length making it difficult to measure specific thickness range and height/ width ratio. The spheroliths are usually fan-shaped and are broader in upper part. Growth lines are not fully visible; however wherever found they appear to arch moderately upwards. Spheroliths are distinct and growth lines are not fused with those in the adjacent spherolith. This sample consists of two overlapping layers of spheroliths which has blocked the pore canals and shows condition of pathology.

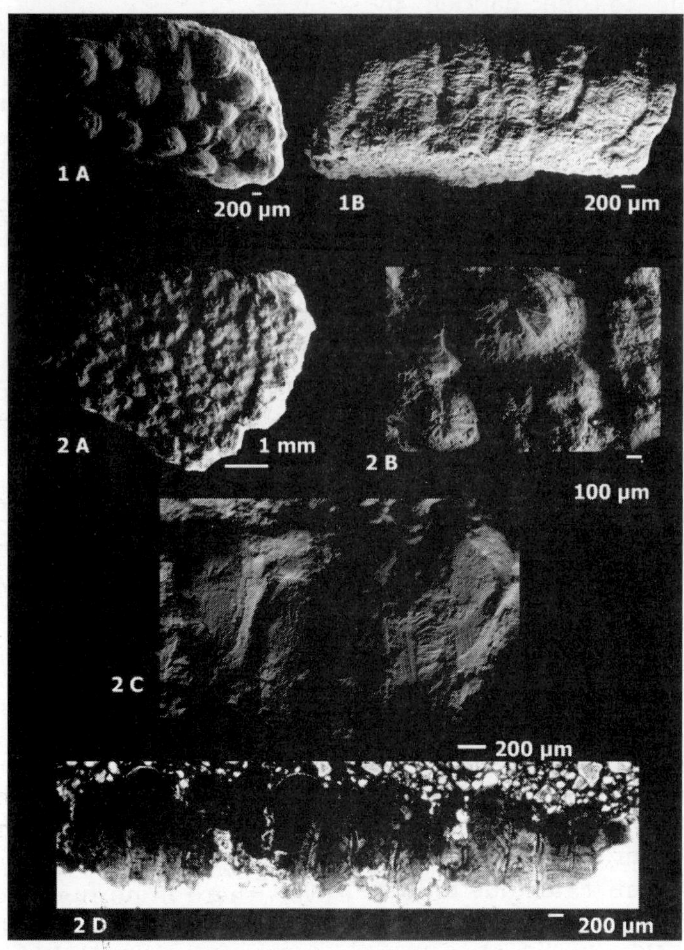

Fig.4. Eggshells of *Megaloolithus jabalpurensis* from the Upper Cretaceous Lameta Formation of Padlya, India. **1A**. SEM picture showing external surface (DUGF/438); **1B**. SEM picture of radial surface showing spheroliths and arching growth lines (DUGF/438); **2A**. SEM picture of internal surface showing mammillary knobs (DUGF/435); **2B**. SEM picture showing enlarged picture of the internal surface (DUGF/435); **2C**. SEM picture of radial surface showing spheroliths (DUGF/435); **2D**. Thin section of the radial surface showing fan-shaped spheroliths and arching growth lines (DUGF/435).

Remarks: Fernandez and Khosla (2015) have suggested that *M. jabalpurensis* is characterised by the absence of fusion among adjacent spheroliths and growth lines that are restricted to the individual shell unit. In the studied specimens, the growth lines are either visible or poorly visible. However, the spheroliths are distinctly separated from adjacent ones and do not show any fusion. The eggshell thickness and height/width ratio is comparable to one defined by Khosla and Sahni (1995). Pore canals appear to be subvertical but they cannot be properly observed due to diagenetic alteration. On the whole, morphostructural features and metrics of the studied specimens conform to that of *M. jabalpurensis*. Though DUGF/436, 438, 440, 445, and 446 have undergone significant diagenetic alteration it is still observable that the pattern of spheroliths and growth lines show no fusion. Therefore, these specimens possibly belong to *M. jabalpurensis*.

DUGF/443 has multiple layers of spheroliths that are tightly bound to each other and do not follow the pattern seen in the above described specimens. Mohabey (1984) documented similar morphostructure in some eggshells of the Lameta Formation, Kheda District, Gujarat and ascribed it to a pathological condition, such as ovum-in-ovo condition (Erben et al., 1979). This condition happens when there is an abnormal increase in thickness of the eggshell due to repetition of eggshell layers. This results in blocking of pore passage and embryo suffocates due to lack of oxygen. The specimen described in the present work also has layers of spheroliths superimposed over one another possibly indicating a pathologic condition. Otherwise the shape and arrangement of spheroliths in the shell, fan shape of spherolith and arching of growth lines favor its inclusion in *M. jabalpurensis*. However Grellet (2010) has attributed such morphology to diagenetic effects on the basis of cathodoluminescence analysis. As cathodoluminiscence could not be used for the present study, the multilayered shell of DUGF/ 443 is tentatively considered as a pathological eggshell.

Oofamily MEGALOOLITHIDAE Zhao, 1979

Megaloolithus cylindricus Khosla and Sahni, 1995

Stratigraphic Horizon and Locality: Sandy carbonate horizon, Upper Cretaceous (Maastrichtian) Lameta Formation, Padlya and Umrali, Dhar District, Madhya Pradesh.

Referred Specimens: Thin sections of eggshell fragments (DUGF/431, 434, 444).

Description: The eggshell thickness is 1.4-1.8 mm. The morphotype is discretispherulitic and surface ornamentation is compactituberculate. The height/width ratio is 3.07 : 1.0. The spheroliths are cylindrical, distinct from adjacent ones, subvertical and long. Growth lines are limited to individual shell units and show no fusion. Pore canals are straight and vertical, hence tubocanaliculate (Fig.5).

Remarks: These specimens are similar to those of *M. cylindricus* in the cylindrical outline of the spheroliths. The eggshell thickness and height/width ratio are also comparable to those of *F. dholiyaensis* Fernandez and Khosla, 2015. However, there is no fusion of the spheroliths and growth lines are limited to a specific shell unit as in *F. dholiyaensis*. The egg shells are strongly altered diagenetically thereby limiting our understanding of the nature of growth lines. However, the shape of spheroliths point to their close affinity to *M. cylindricus*.

Fig.5 Eggshells of *Megaloolithus cylindricus* from the Upper Cretaceous Lameta Formation of Padlya, India. **1A**. SEM picture showing external surface (DUGF/434); **1B**. SEM picture of radial surface showing long cylindrical spheroliths (DUGF/434); **2**. Thin section of radial surface showing long spheroliths. Because of extensive diagenesis it is difficult to mark growth lines (DUGF/431); **3**. Thin section of radial surface showing long and narrow spheroliths. Arching growth lines can be seen (DUGF/444).

Oofamily FUSIOOLITHIDAE Fernandez and Khosla, 2015

Fusioolithus baghensis (Khosla and Sahni, 1995)

Megaloolithus baghensis Khosla and Sahni, 1995: 91-92

Stratigrahic Horizon and Locality: Sandy carbonate horizon, Upper Cretaceous (Maastrichtian) Lameta Formation, Padlya, Dhar District, Madhya Pradesh

Referred Specimens: Thin sections of eggshell fragments (DUGF/430, 432).

Description: The eggshell thickness varies between 1.1 mm and 1.5 mm. The morphotype is discretispherulitic and surface ornamentation is compactituberculate. The height/width ratio of the spheroliths is 2.18 : 1.0. The spheroliths are fan-shaped, show partial fusion and growth lines tend to merge with adjacent spheroliths. Lateral margins of the spheroliths are sub-vertical. Each spherolith consists of a swollen basal cap unit. The pore canals with somewhat long and sub-vertical orientation are possibly tubocanaliculate (Fig.6).

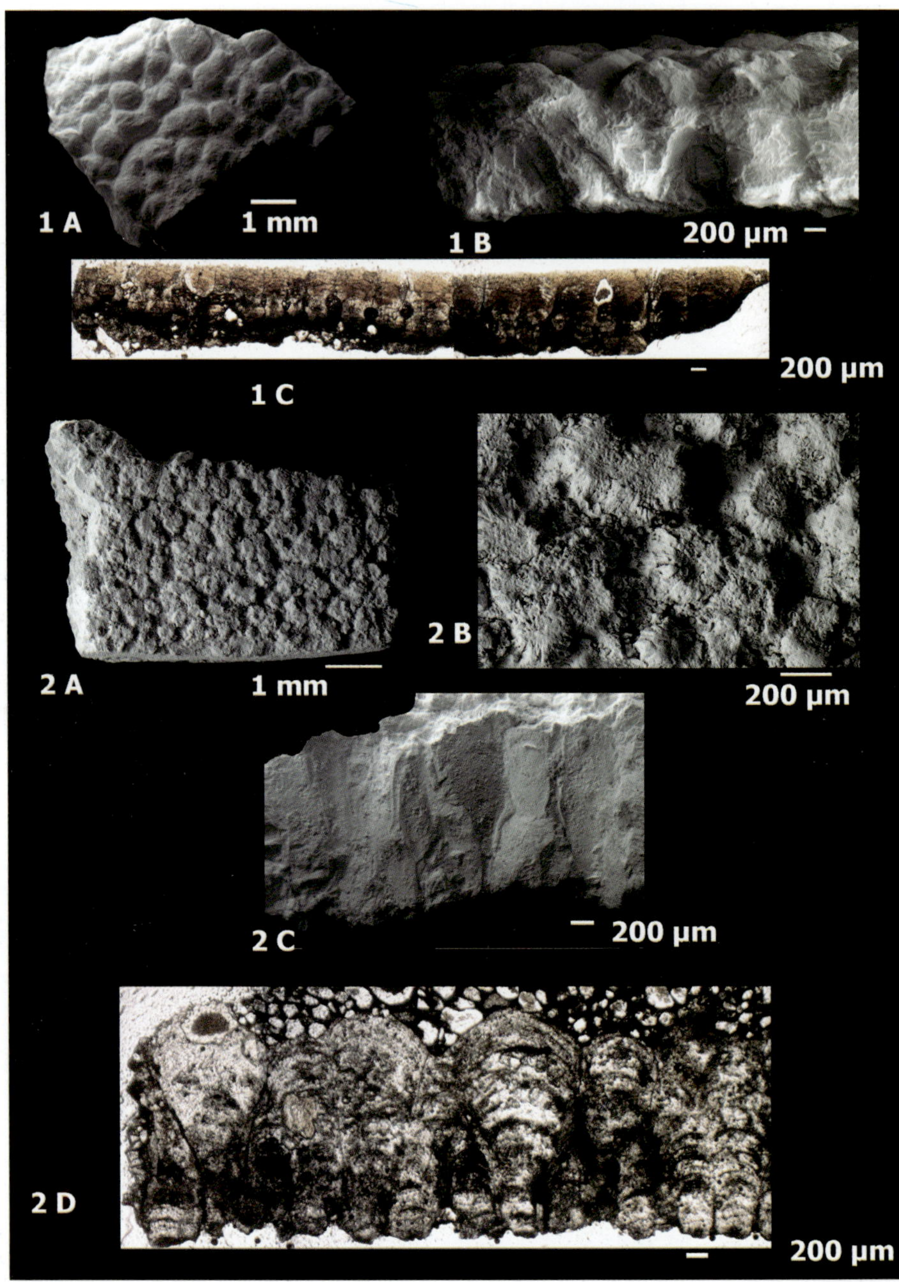

Fig.6. Eggshells of *Fusioolithus baghensis* from the Upper Cretaceous Lameta Formation of Padlya, India. **1A**. SEM picture showing external surface (DUGF/430); **1B**. SEM picture of radial surface showing fan-shaped spheroliths (DUGF/430); **1C**. Radial surface of thin section showing spheroliths and swollen basal cap units (DUGF/430); **2A**. SEM picture of internal surface showing mammillary knobs (DUGF/432); **2B**. SEM picture of enlarged internal surface showing resorption craters (DUGF/432); **2C**. SEM picture of radial surface (DUGF/432); **2D**. Thin section of radial surface showing fan-shaped spheroliths and swollen basal cap unit (DUGF/432).

Remarks: The thickness and height/ width ratio values are comparable to those reported by Khosla and Sahni (1995) for *M. baghensis*. More recently, *M. baghensis* is referred to the newly erected genus *Fusioolithus* (Fernandez and Khosla, 2015). The presence of swollen basal cap unit is the most characteristic feature and presence of fusion between adjacent spheroliths brings the specimen under oofamily Fusioolithidae. The specimens are thus referred to *F. baghensis*.

Oofamily FUSIOOLITHIDAE Fernandez and Khosla (2015)

Fusioolithus padiyalensis (Khosla and Sahni, 1995)

Megaloolithus padiyalensis Khosla and Sahni, 1995: 93-94

Stratigraphic Horizon and Locality: Sandy carbonate horizon, Upper Cretaceous (Maastrichtian) Lameta Formation, Padlya, Dhar District, Madhya Pradesh.

Referred Specimens: Thin sections of eggshell fragments (DUGF 437, 439, 441, 442)

Description: The eggshell thickness varies between 1.3 mm and 1.9 mm. The morphotype is discretispherulitic and surface ornamentation is compactituberculate. The height/width ratio is 3.8 : 1.0. The spheroliths are long, slender, fan-shaped, and elongated and of variable sizes. The fusion between spheroliths is prominent and growth lines merge with each other. The growth lines are also moderately arched upwards. Pore canals are long and subvertical, possibly tubocanaliculate (Fig.7).

Remarks: The prominent fusion of spheroliths brings these specimens under the oofamily Fusioolithidae. Though these eggshells are comparable to those of both *F. padiyalensis* and *M. megadermus*, the presence of fused shell units and growth lines confirm its assignment to *F. padiyalensis*. The thickness and height/ width ratio of the spheroliths also fall within the range of *F. padiyalensis* as described by Khosla and Sahni (1995). Table 1 shows the character states of the studied samples and their oospecies level affinities.

Taphonomy of the Nesting Sites

Taphonomy is the study of the processes that an organism undergoes following its death and until its discovery as a fossil. These processes include decomposition, post-mortem transport, burial, compaction, and other chemical, biological, and physical activities affecting the preservation of organisms as fossils. Through taphonomic analysis of nesting sites we can gain valuable insights into burial and preservational conditions, reproductive physiology of egg laying animals, palaeoenvironment, site selectivity and fidelity. The dinosaur eggs, nests, and eggshell fragments in the study area (Padlya) are documented from around 30 sites over an area of 960 m^2 in sandy carbonate horizon (Fig.8-10).

No skeletal remains, embryos, or juvenile dinosaur bones are found in association with the eggshells, eggs, and nests which precludes any effort to relate the eggs and eggshells to different sauropod dinosaur species. Hence it is unclear if the sauropod eggshell morphostructural diversity corresponds to taxonomic diversity or not. The eggshells have been characterized on the basis of morphostructural types which are based on megascopic and microstructural features. Sahni et al. (1994) attributed the absence of embryos or juveniles in Lameta dinosaur nesting sites of Jabalpur to early deep burial of eggs by sediments, rapid fluid loss from eggs due to high porosity of eggshells in a dry environment, or to pedogenic modification of sediments during which several episodes of brecciation and intrastratal reworking caused breakage of the eggs and embryonic bones. Similar explanations can be offered to the absence of embryos and hatchlings in Padlya nesting site. It is also possible that titanosaurid dinosaurs arrived in this area, which is possibly a high ground near a lake or pond, mainly for egg-laying and did not live in the area as such.

Brookfield and Sahni (1987), Tandon (1990), Tandon and Andrews (1995), and Mohabey et al. (1993) inferred that the Lameta Formation was deposited in a terrestrial palustrine or lacustrine setting. Brookfield and Sahni (1987) suggested semi-arid conditions leading to pedogenic modification of fluvial environments of the Lameta Formation. Tandon (1990) documented desiccation and shrinkage cracks, sheetwash events, distal and proximal alluvial fan deposits, honeycomb and pseudoanticlinal structures and other features related to pedogenesis in a semi-arid environmental setting.

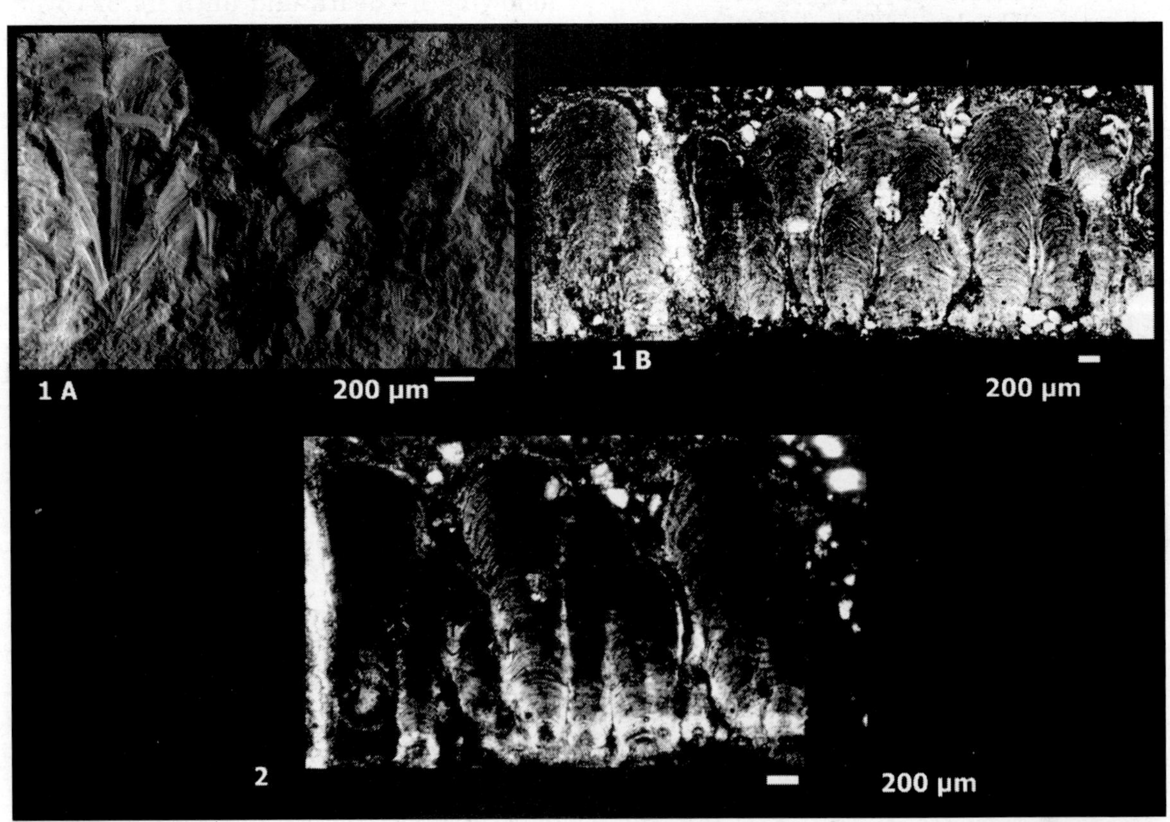

Fig.7. Eggshells of *Fusioolithus padiyalensis* from the Upper Cretaceous Lameta Formation of Padlya, India. **1A**. SEM picture showing radiating calcite crystals in radial view (DUGF/439); **1B** (DUGF/439) and **2** (DUGF/442). Thin section of radial section showing merged spheroliths and fused growth lines which are also arching upwards.

Fig.8. Field photographs of different nests: **1**: Nest 9; **2a-2b**: Nest 19; **3a-3b**: Nest 8.

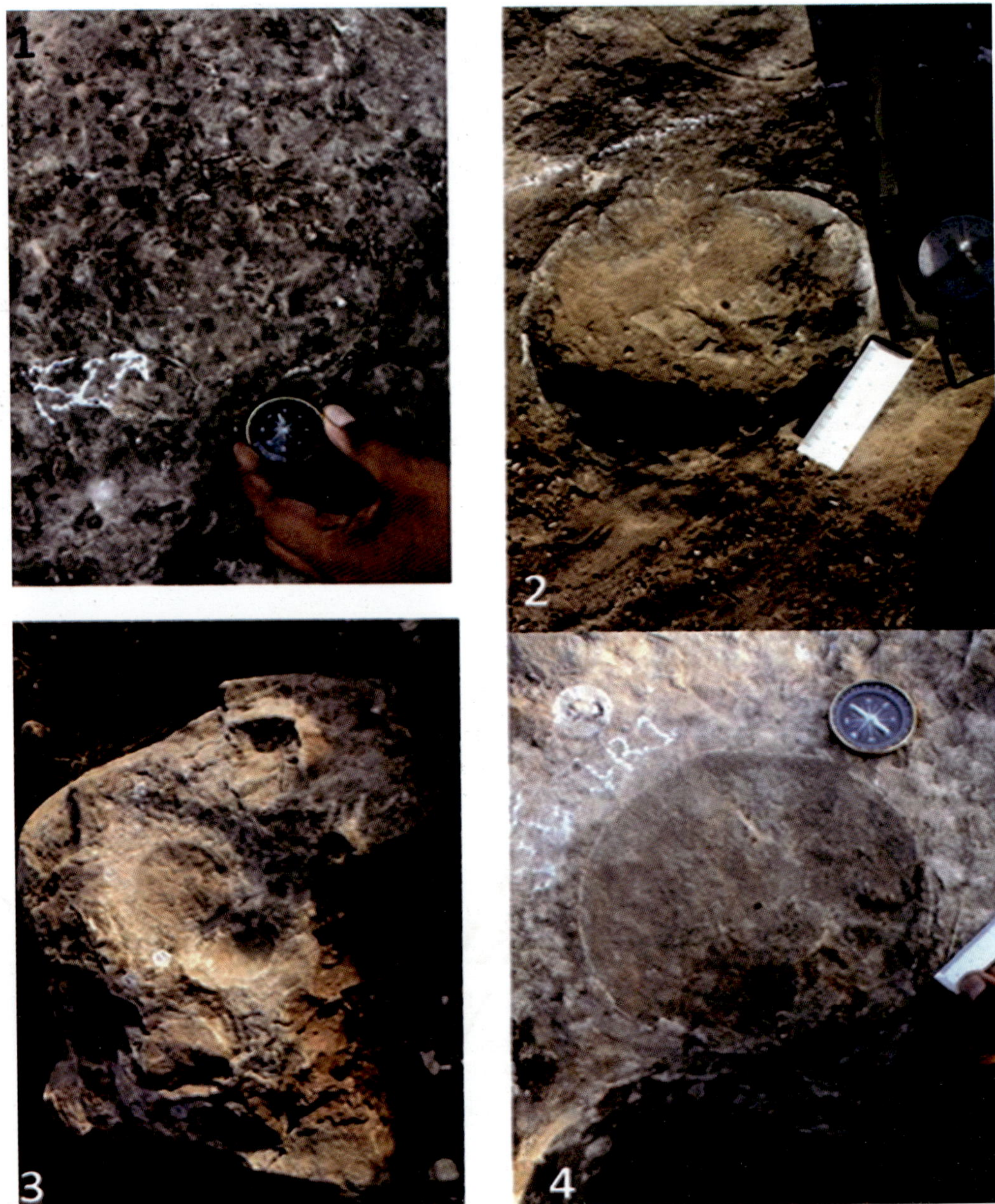

Fig.9. Field photographs of different nests: **1**: Nest 18; **2**: Nest 24; **3-4**: Nest 11.

Presence of nests and eggs in a sandy carbonate horizon at Padlya is suggestive of in situ fossilization and no or limited transport of the eggs. However, dense distribution of nests in some sites may indicate that the titanosaurid sauropod dinosaurs showed some tendency for site fidelity. The close placement of some nests may not have allowed the parents to care for the eggs and juveniles. The inter nest distance (IND) or the distance between adjacent nests of Padlya varies between

2.5 m to 72 m. This variable IND does not demonstrate any relationship between IND and size of egg laying animal. Presence of as many 30 nesting sites belonging to four oospecies on a single lithology and at same stratigraphic level is suggestive of colonial nesting.

Diagenetic effects are prevalent as demonstrated by majority of the thin sections which show some amount of silicification though it is not pervasive throughout the shell units. Rather it occurs randomly, particularly affecting internodal areas (pore canals).

Fragmentation of eggshells possibly occurred prior to burial. At one such site, concentrically distributed eggshells are found in a small area and the external surfaces of the eggshells are facing in opposite directions. This kind of preservation may indicate 'collapsing of eggs' resulting in concentric layering at the base of the egg (Sahni and Khosla, 1994) (Fig.11).

Fig.10. The largest nest with 15 eggs in one clutch arranged in a circle.

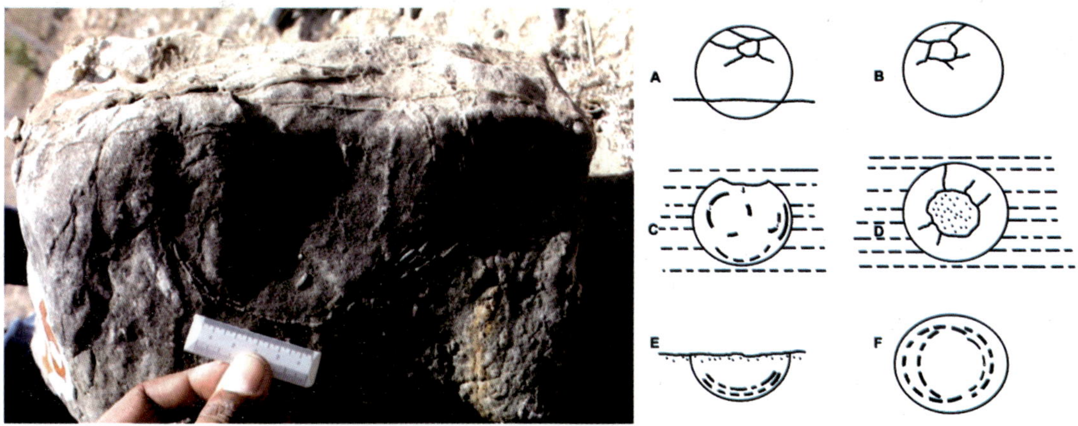

Fig.11. On left are the concentrically distributed collapsed eggshells at the bottom of the egg. **A-F**. Schematic diagram showing the stages during 'collapsing of eggs'. **A**. profile view, **B**. planar view, **C**. shows the eggshell fragments settling down due to gravity, **D**. same in plan view, **E**. preser-vation of fragments in concentric manner, **F**. same in plan view (after: Sahni and Khosla, 1994).

Fig.12. SEM picture of resorption craters on the inner surface of the eggshell in *F. baghensis* (Scale: 200µm).

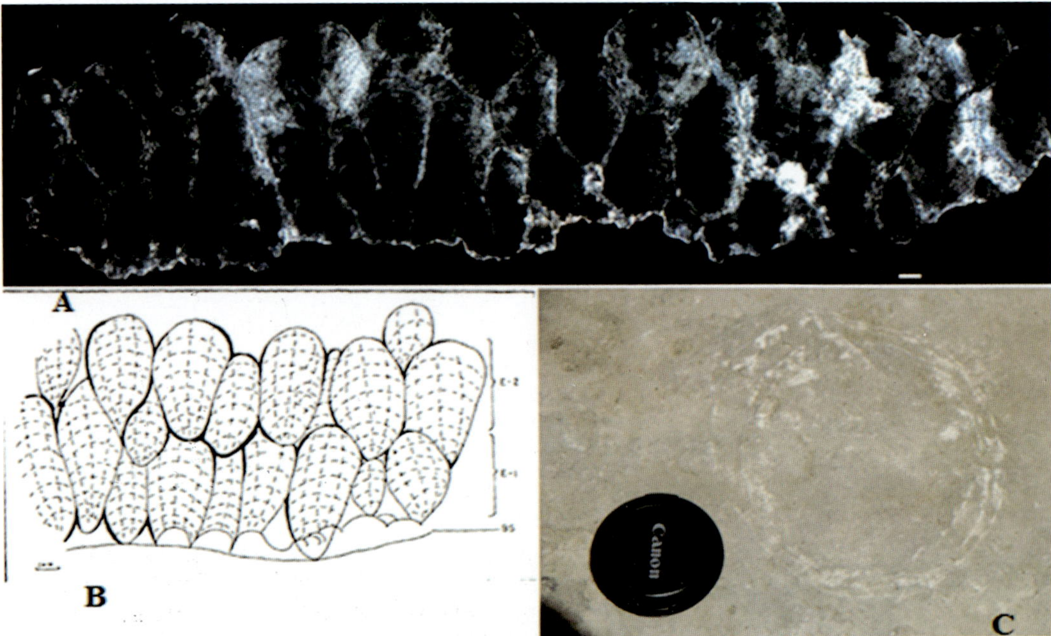

Fig.13. A. Shows the multilayered eggshell in thin section with a microstructure similar to that of *M. jabalpurensis* (Scale: 200µm). **B.** Shows the supposed pathologic 'ovum-inovo' condition in dinosaur eggshell from Rahioli (after Mohabey, 1984). **C**. A pathological 'ovum-inovo' egg documented from the field.

Table 1. Egg shell characteristics of sauropod dinosaur oospecies, Padiya District, Dhar.

Specimen No.	Shell thickness	Height/ Width	Shape of shell unit	Nature of growth lines	Determined oospecies
DUGF 430	1.1-1.5mm	2.18:1	Spheroliths are broader on upper part and are partially fused. There are characteristic swollen shaped basal caps.	The growth lines enter into adjacent spheroliths.	*Fusioolithus baghensis*
DUGF 431	1.4-1.8mm	3.07:1	Spheroliths are long and straight. Samples are highly altered. Spheroliths appear to fuse.	High alteration makes it difficult to view growth lines.	*Megaloolithus cylindricus*
DUGF 432	1.5-1.9mm	1.84:1	Spheroliths are fan shaped and partially fused. Swollen shaped basal units present.	Moderately arched upwards and show fusion with adjacent spheroliths.	*Fusioolithus baghensis*
DUGF 433	1.22-1.68mm	2.1:1	Spheroliths are fan shaped and don't show fusion. Smaller spheroliths are present in between.	Moderately arched upwards.	*Megaloolithus jabalpurensis*
DUGF 434	1.68-1.86mm	3.62:1	Spheroliths are long and subvertical. The sample is highly altered.	Growth lines are difficult to identify due to high alteration.	*Megaloolithus cylindricus*
DUGF 435	1-1.6mm	2.42:1	Fan-shaped spheroliths not fusing with each other.	Moderately arched upwards and distinct within a shell unit.	*Megaloolithus jabalpurensis*
DUGF 436	1-1.24mm	1.77:1	Spheroliths are broader upwards and distinct from each other.	No fusion between adjacent spheroliths.	*Megaloolithus jabalpurensis*
DUGF 437	1.3-1.9mm	3.8:1	Spheroliths are long and slender, of variable size, showing fusion with each other.	Moderately arched showing fusion with each other.	*Fusioolithus padiyalensis*
DUGF 438	1-1.7mm	2.0:1	Spheroliths are fan sphaped and distinct.	Moderately arched upwards and distinct within shell units.	*Megaloolithus jabalpurensis*

Specimen No.	Shell thickness	Height/ Width	Shape of shell unit	Nature of growth lines	Determined oospecies
DUGF 439	1.1-1.84mm	2.3:1	Spheroltihs are fan shaped and show fusion.	Moderately arched upwards and merges into adjacent spherolith.	*Fusioolithus padiyalensis*
DUGF 440	1.14-1.66mm	1.97:1	Spheroliths are broader on upper part and are distinct.	Alteration is high but growth lines seem to arch upwards.	*Megaloolithus jabalpurensis*
DUGF 441	0.8-2mm	2.4:1	Spheroliths are fan shaped and show fusion.	Moderately arched upwards and merges into adjacent spherolith.	*Fusioolithus padiyalensis*
DUGF 442	1.06-1.8mm	2.5:1	Spheroliths are fan shaped and show fusion.	Moderately arched upwards and merges into adjacent spherolith.	*Fusioolithus padiyalensis*
DUGF 443	Pathological sample. Difficult to find a range.	Pathological sample. Difficult to find a range.	Speroliths are tightly packed with each other. No fusion.	Moderately arched upwards and distinct within shell units.	*Megaloolithus jabalpurensis*
DUGF 444	1.4-2mm	3.65:1	Spheroliths are long and subvertical. High alteration. No fusion.	Growth lines arch upwards.	*Megaloolithus cylindricus*
DUGF 445	1.1-1.8mm	2.74:1	Spheroliths are some- what fan shaped and show no fusion.	Not readily visible due to alteration.	*Megaloolithus jabalpurensis*
DUGF 446	0.9-1.02mm	1.5:1	Spheroliths appear broader and have fan shape. Their size is small.	Moderately arching upwards.	*Megaloolithus jabalpurensis*

A pathological condition called as 'Ovum-in-ovo' has also been found in one of the thin sections and a similar feature is also noticed in one of the eggs in the field (Fig.13). Such cases have been reported in the past (Erben et al., 1979; Kerourio, 1981; Hirsch et al., 1989; Zelenitskyand Hills, 1997; Vianey-Liaud et al., 1994; Jackson et al., 2003). Such abnormal multilayered eggs in hard-shelled eggs of turtles are attributed to stress (Erben et al., 1979) resulting from physiological to environmental stresses, sometimes induced by lack of suitable nesting site or materials, drought, flooding, high population density and disease. However,

Sahni et al. (1994) are skeptical about the reported 'ovum-in-ovo' condition from Rahioli documented by Mohabey (1984) and felt that the multilayered appearance of shell layers could be an artifact of inappropriate sectioning of the eggshells. One nest from Padlya contains a single egg which contains two shell layers one inside the other. This appears to be the best case for 'ovum-in-ovo' condition as it is noticed both in thin section and also in the field.

Apart from this resorption craters have been found from SEM figures on the basal mammillary caps of the inner surface that indicate actively growing embryos as they utilize calcium carbonate from the shell for the building of skeleton (Fig.12).

PALEOBIOGEOGRAPHIC SIGNIFICANCE OF INDIAN SAUROPOD EGGSHELLS

At present nine valid Indian sauropod oospecies, viz; *Megaloolithus cylindricus* (Khosla and Sahni, 1995), *M. jabalpurensis* (Khosla and Sahni, 1995), *M. dhoridungriensis* (Mohabey, 1998), *M. khempurensis* (Mohabey, 1998), *M. megadermus* (Mohabey, 1998), *Fusioolithus mohabeyi* (Khosla and Sahni, 1995), *F. padiyalensis* (Khosla and Sahni, 1995), *F. baghensis* (Fernàndez and Khosla 2015), and *F. dholiyaensis* (Khosla and Sahni, 1995), are known from India. Vianey-Liaud et al. (2003) while comparing the Megaloolithus species from Europe and India observed that there are close similarities between the oospecies of India and France that reveal four groupings: 1) *Megaloolithus microtuberculata* and *M. cylindricus* and *F. padiyalensis*, 2) *M. mammillare* and *M. jabalpurensis*, 3) *M. siruguei* and *M. khempurensis*, and 4) *M. pseudomammillare* and *F. baghensis*. Fernàndez and Khosla (2015) compared Indian megaloolithid species from the Late Cretaceous of India and Argentina and synonymized *M. patagonicus* Calvo et al., 1997 from Argentina with *M. jabalpurensis*, Tipo 1d of Fernàndez (2013) from Argentina with *M. cylindricus* and Tipo 1e of Fernandez (2013) with *M. megadermus*, and *Patagoolithus salitralensis* (Simon, 2006) with *Fusioolithus baghensis* (Khosla and Sahni, 1995, Fernàndez and Khosla, 2015). Recently a new oospecies *Pseudomegaloolithus atlasi* (Vianey-Liaud et al. 2003) was reported from Late Cretaceous 'red beds' of Tendrara (high Plateaus of Morocco) (Chassagne-Manoullian et al., 2013) and it was considered closely related to South American and Indian megaloolithid eggshells. These close similarities in the megaloolithid oospecies between India, South America, Southern Europe (Spain to Romania through Southern France) and Northern Africa point to ancient Gondwanan lineage that was widely distributed in the Gondwanan continents (S. America, Africa and India) and insular areas of Southern Europe.

CONCLUSION

Morphostructural analysis indicated that amongst 30 nesting sites documented from Padlya village, Dhar district, Madhya Pradesh, India, seven belong to *M. jabalpurensis*, three are of *M. cylindricus*, two are related to *F. baghensis*, four represent *F. padiyalensis*, and one that shows pathological condition is possibly of *M. jabalpurensis*.

Therefore, two families of titanosaurid sauropod dinosaurs are nested at the investigated site. The presence of four oospecies in a small area on a single lithological unit may indicate adoption to colonial nesting behavior by titanosaurid sauropod dinosaurs. Relatively large number of nests in the study area implies that a suitable paleoenvironment for laying of eggs by dinosaurs was present and also the preservation potential was good resulting in the occurrence of a number of eggshell fragments in this locality. The oospecies *M. jabalpurensis* and *F. baghensis* recorded from Padlya site are also known from South America (Fernandez and Khosla, 2015) and similar oospecies were described from southern Europe (Vianey Liaud et al. 2003), which suggests that there was a close palaeobiogeographic link between the southern continents and southern Europe through Africa in the Late Cretaceous.

Acknowledgements: HD thanks University of Delhi for supporting this work through UGC non-NET Fellowship and GVRP acknowledges financial support to this work from J.C. Bose National Fellowship.

References

BADVE, R.M. and NAYAK, K.K. (1984) Occurrence and significance of the algae genus Hallmeda from Nimar Sandstone, Bagh Beds, Jhabua, M.P. Biovigyanam, 9, pp. 137-148.

BLANFORD, W.T. (1869) On the geology of Tapti and lower Narbada valley and some adjoining districts. Mem. Geol. Sur. India 6(3), 163-184.

BISWAS, S.K. (1982) Rift basins in western margin of India and their hydrocarbon prospects with special reference to Kutch basin: Am. Assoc. Petrol. Geologists (AAPG) Bull., v. 66, p. 1497-1513.

BOSE, P.N. (1884) Geology of lower Narbada valley between Nimavar and Kawant. Mem. Geol. Sur. India, v.2, 1-72.

BROOKFIELD, M.E. and SAHNI, A. (1987) Palaeoenvironments of the Lameta Beds (Late Cretaceous) at Jabalpur, Madhya Pradesh, India: Soils and Biotas of a Semi-arid Alluvial Plain. Cretaceous Res., v.8, pp.1-14.

CALVO J.O., ENGELLAND S., HEREDIA S.E., and SALGADO, L. (1997) First record of dinosaur eggshells (?Sauropoda–Megalolithidae) from Neuque´n, Patagonia, Argentina, Gaia, v.14, pp. 23–32.

CHASSAGNE, M., HADDOUMI, H., CAPPETTA, H., CHARRIÈRE, A., FEIST, M., TABUCE, R. and VIANEY-LIAUD, M. (2013) Dating the "red beds" of the Eastern Moroccan High Plateaus: Evidence from late Late Cretaceous charophytes and dinosaur eggshells, Geobios, v.46(5), pp.371-379.

CHIPLONKAR G.W. and BADVE, R.M. (1973) Age and affinities of the Bagh fauna - a reassessment. *In*, Proc. Indian Nat. Acad. Sci., v.45, pp.19-29.

ERBEN, H. K., HOEFS, J. and WEDEPOHL, K.H. (1979). Paleobiological and isotopic studies of eggshells from a declining dinosaur species. Paleobiology, v.5(04), pp.380-414.

FERNA´NDEZ, M.S. (2013) Ana´lisis de ca´scaras de huevos de dinosaurios de la Formacio´n Allen, Creta´cico Superior de R□o Negro (Campaniano-Maastrichtiano): utilidades de los macrocaracteres de intere´s parataxono´mico. Ameghiniana, v.50(1), pp.79–97.

FERNANDEZ, M.S. and KHOSLA, A. (2015) Parataxonomic review of the Upper Cretaceous dinosaur eggshells belonging to the oofamily Megaloolithidae from India and Argentina, Historical Bio., v.27(2), pp.158-180.

GRELLET TINNER, G., CORSETTI F. and BUSCALIONI, A.D. (2010) The importance of microscopic examinations of eggshells: Discrimination of bioalteration and diagenetic overprints from biological features. J. Iberian Geol., v.36 (2), pp.181-192.

GUHA, A. K. (1976) Lithostratigraphic classification of Bagh Group (beds), Madhya

Pradesh. *In* Proc. IVth Colloq. Micropal. Stratigr., Dehra Dun, pp.209-216.

HAYWARD, J.L. (2005) A closer look at dinosaurs. J. Adventist Education, pp.29-36.

HIRSCH, K.F., STADTMAN, K.L., MILLER, W.E. and MADESEN Jr., J.H. (1989) Upper Jurassic dinosaur egg from Utah. Science, v.243, pp.1711-1713.

HUENE, F.R. and MATLEY, C.A. (1933) The Cretaceous Saurischia and Ornithischia of the Central provinces (No. 1-4). Manager of Publications.

JACKSON, F.D., SCHWEITZER, M.H., and SCHMITT, J.G. (2002) Dinosaur eggshell study using scanning electron microscopy, *SCANNING-New York and Baden Then Mahwah-*, v.24(5), pp.217-223.

JACKSON, F.D. and VARRICCHIO, D.J. (2003) Abnormal, multilayered eggshell in birds: implications for dinosaur reproductive anatomy. J. Vertebr. Paleontol., v.23(3), pp.699-702.

JAFAR, S.A. (1982) Nannoplankton evidence of Turonian transgression along Narmada valley, India and Turonian-Coniacian boundary problem. J. Pal. Soc. India, v.27, pp.17-30.

JAIN, S.L. and SAHNI, A. (1985) Dinosaurian eggshell fragments from the Lameta Formation at Pisdura, Chandrapur district, Maharashtra. Geosci. J. (India), v.2, pp.211-220.

KEROURIO, P. (1981) La distribution des "Coquilles d' oeufs de Dinosauriensmultistratiflees" dans le Maestrichtien continental du Sud de la France. Geobios., v.14,pp.533-536.

KHOSLA, A. and SAHNI, A. (1995) Parataxonomic classification of Late Cretaceous dinosaur eggshells from India. J. Pal. Soc. India, v.40, pp. 87-102.

KOHRING, R., BANDEL, K., KORTUM, D., and PARTHASARARTHY, S. (1996) Shell structure of a dinosaur egg from the Maastrichtian of Ariyalur (Southern India). N. Jb. Geol. Palaont. Mh., pp.48-64.

MEDLICOTT, J.G. (1860) On the geological structure of the central portion of the Nerbudda district. Mem. Geol. Sur. India, v.2, pp.97–278.

MIKHAILOV, K. (1997) Fossil and recent eggshells in amniotic vertebrates: fine structure, comparative morphology and classification, The Pal. Assoc., Sp. Papers Pal., pp.56:1–80.

MOHABEY, D.M. (1984) Pathologic dinosaurian eggshells from Kheda District, Gujarat. Curr. Sci., v.53(13)

MOHABEY, D.M., and UDHOJI, S.G. (1993) Palaeoenvironmental interpretation of Lameta Formation (Late Cretaceous) of Nand area, Nagpur district, Nagpur. Gondwa Geol. Mag., v.4-5, pp.13-22.

MOHABEY, D.M. (1996) A new oospecies, *Megaloolithus matleyi*, from the Lameta Formation (Upper Cretaceous) of Chandrapur district, Maharashtra, India, and general remarks on the paleoenvironment and nesting behavior of dinosaurs. Cretaceous Res., v.17, pp.183–196.

MOHABEY, D.M. (1998) Systematics of Indian Upper Cretaceous dinosaur and chelonian eggshells. J. Vertebrate Pal., v.18, pp.348-362.

PODDAR, M.C. (1964) Mesozoic of Western India-their geology and oil prospects. Rept 22nd Int. Geolo. Cong.,127.

RAIVERMAN, V. (1975) Facies transition among Nimar, Bagh and Lameta beds. Rec. Res. Geol., v.2, pp.123-139.

RODE, K.P. and CHIPLONKAR, G.W. (1935) A contribution to the stratigraphy of Bagh Beds. Curr. Sci., v.2, pp.322-323.

ROY CHOWDHURY, M.K. and SASTRI, V.V. (1962) On the revised classification of the Cretaceous and associated rocks of the Man River section of lower Narbada valley. Rec. Geol. Sur. India, v.91, pp.283-301.

SAHNI, M. (1957) A fossil reptilian egg from the Uttatur (Cenomanian) of south India, being the first record of a vertebrate fossil egg in India. Rec. Geol. Sur. India, v.87, pp. 671-674.

SAHNI, M.R. and JAIN, S.P. (1966) Note on a revised classification of the Bagh Beds,

Madhya Pradesh. J. Pal. Soc. India, v.11, pp.24-25.

SAHNI, A. and GUPTA, V.J. (1982) Cretaceous eggshell fragments from the Lameta Formation, Jabalpur, India. Bull. Indian Geol. Assoc., v.15(1), pp.85-88.

SAHNI, A., RANA, R. S., and PRASAD, G.V.R. (1984) SEM studies of thin eggshell fragments from the Intertrappeans (Cretaceous-Tertiary transition) of Nagpur and Asifabad, peninsular India. J. Palaeont. Soc. India, v.29, pp.26-33.

SAHNI, A. and KHOSLA, A. (1994) Paleobiological, taphonomical and paleoenvironmental aspects of Indian Cretaceous sauropod nesting sites. Gaia, v.10, pp.215–223.

SAHNI, A., TANDON, S., JOLLY, A., BAJPAI, S., SOOD, A. and SRINIVASAN, S. (1994) Upper Cretaceous dinosaur eggs and nesting sites from the Deccan volcano-sedimentary province of peninsular India. *In* CARPENTER, K., HIRSCH, K.F., and HORNER, J.R. (eds.), Dinosaur eggs and babies, pp. 204-226. New York: Cambridge University Press.

SHARMA, A. (2014) Microstructural and ultrastructural study of Cretaceous eggshells from peninsular India, M.Sc. Dissertatio, Department of Geology, University of Delhi, India, pp.1-209.

SIMO´N ME. (2006) Ca´scaras de huevos de dinosaurios de la Formacio´n Allen (Campaniano-Maastrichtiano), en Salitral Moreno, provincial de R□o Negro, Argentina. Ameghiniana. v.43(3), pp.513–518.

SINGH, S.K. and SRIVASTAVA, H.K. (1981) Lithostratigraphy of the Bagh Beds and its correlation with Lameta Beds. J. Pal. Soc. India, v.26, pp.77-85.

SINGH, D. (2013) Palaeontological investigations of the Intertrappean beds of Narmada Valley, M. Sc. Dissertation, Department of Geology, University of Delhi, pp.1-123.

SRIVASTAVA, R., PATNAIK, R., SHUKLA, U.K. and SAHNI, A. (2015) Crocodilian nest in a Late Cretaceous sauropod hatchery from the type Lameta Ghat locality, Jabalpur, India, PloS one, v.10(12).

TRIPATHI, S.C. (2006) Geology and Evolution of the Cretaceous Infratrappean Basins of Lower Narmada Valley, Western India. J. Pal. Soc. India, v.67, pp.459-468.

TANDON, S.K., VERMA, V.K., JHINGRAN, V., SOOD, A., KUMAR, S., KOHLI, R.P. and MITTAL, S. (1990) The Lameta Beds of Jabalpur, Central India: deposits of fluvial and pedogenically modified semi-arid fan-palustrine flat systems. *In:* Cretaceous event stratigraphy and the correlation of the Indian nonmarine strata, Contributions from the Seminar Cum Workshop IGCP, v.216, pp.27-30.

TANDON, S.K., SOOD, A., ANDREWS, J.E. and DENNIS, P.F. (1995) Palaeoenvironments of the dinosaur-bearing Lameta Beds (Maastrichtian), Narmada Valley, Central India; Elsevier; Palaeogeo. Palaeoclimat. Palaeoeco., v.117, pp.153 184.

VIANEY-LIAUD, M., MALLIAN, P., BUSCAIL, O., and MONTGELARD, C. (1994) Review of French dinosaur eggshells, morphology, structure, mineral, and organic composition; pp. 151-183 in K. *In* CARPENTER, K., HIRSCH, K.F., and HORNER, J.R. (eds.). Dinosaur Eggs and Babies. Cambridge Univ. Press, New York.

VIANEY-LIAUD, M., KHOSLA, A. and GARCIA, G. (2003) Relationships between European and Indian dinosaur eggs and eggshells of the oofamily Megaloolithidae. J. Vert. Pal., v.23(3), pp.575-585.

ZELENITSKY, D.K. and HILLS, L.V. (1997) Normal and pathological eggshells of Spheroolithus albertensisoosp. Nov., from the Oldman FORMATION (Judith River Group, Late Campanian), southern Alberta, J. Vert. Pal., v.17, pp.167-171.

ZHAO, Z.K. (1979) Progress in the research of dinosaur eggs. *In*: Mesozoic and Cenozoic red beds of South China. Beijing: Sci. Press., pp. 330–340.

Micropaleontology and its Applications
Pradeep K. Kathal, Rajiv Nigam & Abu Talib (Editors)
Scientific Publishers (India), 2018, 233-248 pp.

Study of Pliocene Radiolaria from Kerguelen Plateau Site 738B (Sections 2H4 to 2H5): Distribution and Systematics

G. K. Sharma* and Deepti

Department of Geology, Kumaun University, Nainital, India
***Email**: gksharma61@yahoo.com

Abstract

Thirty radiolarian were identified and described from twenty eight samples of two sections viz. (2H4 and 2H5) from Kerguelen Plateau Site 738B. The sections show wide diversity and sparsely distribution of radiolarian. From the total identified taxa, 11 are Spumalleria and 19 Nassellarians. The systematics along with their distribution in the region of important taxa is given. The assemblage is of typical Antarctic, sub-antarctic, cosmopolitan and bipolar. On the basis of total assemblage, the sections are of Pliocene age

Keywords: Pliocene Radiolaria, systematic, distribution and Southern Ocean.

INTRODUCTION

This study first described about systematics and distribution of late Pliocene radiolarian in two sections viz., 2H4 and 2H5 of Leg 119 Site 738B. The area lies on a large sediment drift at the base of Southern slope of the Kerguelen Plateau between latitude 62°42.54'S and longitude 82°47.25'E (Figure 1). The sediments consist of moderately, well preserved micro faunas of Radiolaria, Foraminifera, Diatom and Calcareous Nannoplanktons. Twenty-eight samples of length 2.895 m was used and collected at a water depth of 2252.5 m. for this study (Fig. 2).

Ehrenberg (1844a) started the study on the Southern ocean radiolarian and described 21 species. Haeckel (1887)

noticed a few radiolarian species from the Indian Ocean sector of Antarctica. Popofsky (1908) described radiolarian from the Wilhelm II Coast of Antarctica and Kerguelen Island. Riedel (1958) carried out the study on the B.A.N.Z. Antarctic Research Expedition sediments and studied the radiolarians in detail.

Hays (1965) established radiolarian biostratigraphy of late the Tertiary Antarctic Seas of the Southern Ocean Region. The detail investigation on the Neogene radiolarian stratigraphy were succeeded by many workers (Hays, 1967, 1970; Chen, 1974, 1975a, 1975b; Hays and Opdyke, 1967; Keany and Kennett, 1972,1975; Keany, 1976, 1979; Lombari and Lazarus, 1988; Petrushevskaya, 1967, 1975; Weaver, 1976a, 1976b; Riedel and Sanfilippo, 1970, 1971; Kling, 1971; Theyer et al., 1978; Riedel and Westberg, 1982; Bandy and Casey, 1969; Hays and Berggren, 1971; and Weaver 1973, 1983). Further, Nakaseko and Nishimura (1982) examined the bottom sediments of the Bellingshausen Basin in the Antarctic Sea. Abelmann and Gersonde (1988) and Lazarus (1990) described Neogene biostratigraphy on cored sections of Falkland Plateau and Weddell Sea. Lazarus (1992) presented radiolarian stratigraphy of middle Miocene to Holocene sediments of different sites from ODP Legs 119 and 120. McIntyre and Kaczmarska (1996) worked on sections of ODP, Site 745 in the Kerguelen Plateau and placed *Stylatractus universus* (Hays) at 24.8- 24.4 mbsf. Lazarus (2002) described environmental control in the course of radiolarian diversity, evolutionary rates and taxa longevities of Antarctic Neogene radiolaria from ODP Leg 119. Sharma et al. (2004, 2006) carried out the detail study on the Pleistocene sediments and reported 83 radiolarian taxa and established two radiolarian zones of Tasman region. Stepanjants et al. (2006) discussed bipolar distribution of Phaeodaria, Radiolaria and Medusozoa and identified 46 bipolar radiolarian species. Sharma and Takahashi (2007, 2008) reported 75 radiolarian species from the Pleistocene sediments of the SE Indian area of the Antarctic Continental margin and established one radiolarian zone i.e. lower and upper Chi for the Antarctic Continental margin. Renaudie and Lazarus (2012) described a few radiolaria species from Southern Ocean region. Renaudie and Lazarus (2013) studied new species of Neogene radiolaria from Southern Ocean region. Sharma (2013) gave detailed biostratigraphic zonation of 21 radiolarian species and established Pliocene- Pleistocene boundary from ODP leg 119, Site 745. Sharma and Kanojia (2013) worked on Pleistocene radiolarian from leg 119 site 738B of the Southern Ocean region and on the basis of appearance and disappearance of taxa two zones Omega and Psi were established.

Sharma and Dayal (2015a) studied Pliocene radiolarian taxa from the Antarctic region 119 Site 745 and identified and described 26 species from two sections while recognizing one radiolarian zone. Sharma and Dayal (2015b) described Pliocene radiolarias from sections (14H-15H) from Kerguelen Plateau and proposed one Omega Zone.

METHODOLOGY

Samples of about 2.5-4 g were disaggregated in 35 vol% Hydrogen peroxide (H_2O_2) followed by boiling for about 4-8 hours and washing through a sieve of 63 milimicron (µm) size with water. The residue is then dried in an oven and is used for the preparation of strewn slides

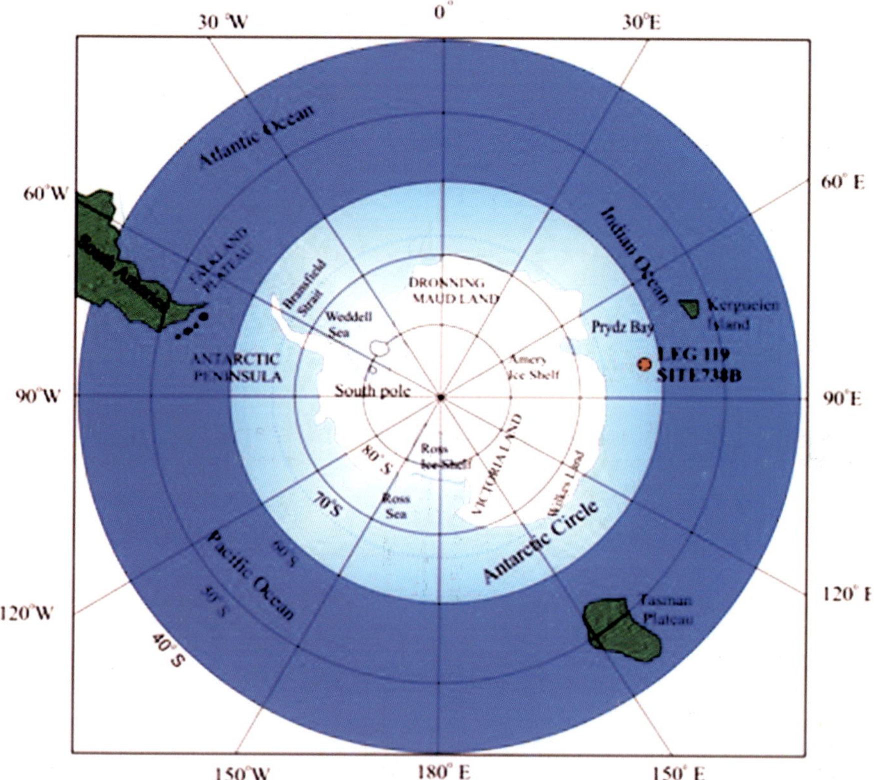

Fig.1. Location map showing the study area of Leg119 Site 738B.

by using Canada balsam as a mounting medium. Generally, at least 2 - 3 slides (of cover slip size 22:22 mm) were prepared for taxonomic and stratigraphic work.

SYSTEMATIC PALAEONTOLOGY

The classification of the Subclass Radiolaria followed is of De Weaver et al. (2001); Nigrini and Lombari (1984) and Lazarus (1990). Remarks on observed morphological features and their modifications have been added for many taxa. Species within a genus and genera within a family are arranged alphabetically. The characteristic morphological features for each new species and those given in open nomenclature are described. The synonymy for each taxon is incomplete and consists of references of interest to the present study. The distribution of each species is estimated as percentage of the total number of radiolaria present in the slides of each sample is shown in Table 1. The microphotographs of all the identified species are illustrated in Plate I.

Phylum SARCODINA Hertwig and Lesser, 1874

Class ACTINOPODA Calkins, 1909

Subclass RADIOLARIA Müller, 1858

Order POLYCYSTINA Ehrenberg, 1838, *emend.* Riedel, 1967b

Suborder SPUMELLARIA Ehrenberg, 1875

Family SATURNALIDAE Deflandre, 1953

Genus SATURNALIS Haeckel, 1881

Saturnalis circularis Haeckel
(Pl.1, Fig. 9)

1887 *Saturnalis circularis*- Haeckel, p.131.

1975a *Saturnalis circularis* Haeckel-Chen, p. 454, pl.24, fig.2.

2008 *Saturnalis circularis* - Haeckel, Sharma and Takahashi, p.71, pl.4, fig.D.

Distribution: Sharma and Takahashi (2008) reported its presence from very rare to rare in the Antarctic region.

Abundance: Very rare

Range: Chen (1975a) and Keany (1979) reported its occurrence from Oligocene to lower Pleistocene. Lazarus (2002) showed its occurrence from 0-17.5 Ma.

Table 1. Distribution of each species estimated as percentage (%) of the total number of radiolaria from slides of each sample: **VA**=Very Abundant (>50%); **A**=Abundant in % (20-50); **C**=Common (5-20), **F**=Few (0.5- 5), **R**=Rare (<0.5), but more than single specimen; **+** =Very Rare (single specimen), **−** =absent) and preservation **G**=Good, **M**=Moderate and **P**=Poor.

a	**Sample Name (Hole, Core, Section, Interval (cm) 738B-**	**2H-4,7-9**	**2H-4,17-19**	**2H-4,27-29**	**2H-4,37-39**	**2H-4,47-49**	**2H-4,57-59**	**2H-4,67-69**	**2H-4,77-79**	**2H-4,87-89**	**2H-4,97-99**	**2H-4,107-109**	**2H-4,117-119**	**2H-4,137-139**	**2H-5,7-9**	**2H-5,17-19**	**2H-5,27-29**	**2H-5,37-39**	**2H-5,49-51**	**2H-5,57-59**	**2H-5,67-69**	**2H-5,77-79**	**2H-5,87-89**	**2H-5,96-98**	**2H-5,107-109**	**2H-5,117-119**	**2H-5,129.5-131.5**	**2H-5,137-139**	**2H-5,146.5-148.5**
b	Depth (mbsf)	8.57	8.67	8.77	8.87	8.97	9.07	9.17	9.27	9.37	9.47	9.57	9.67	9.87	10.07	10.17	10.27	10.37	10.49	10.57	10.67	10.77	10.87	10.96	11.07	11.17	11.295	11.37	11.465
c	Total counts	1005	1030	1010	1005	1005	1020	1005	1050	1040	1020	1020	1050	1010	1005	1030	1010	1005	1005	1020	1005	1050	1040	1020	1020	1050	1010	1010	1010
d	Age		P				L			I				O			C				E				N			E	
e	Abundance	C	C	C	C	C	C	C	C	C	C	C	C	C	C	C	C	C	C	C	C	C	C	C	C	C	C	C	C
f	Preservation	G	G	G	G	G	G	G	G	G	G	G	G	G	G	G	G	G	G	G	G	G	G	G	G	G	G	G	G
h	Name of Species																												
1	*Antarctissa cylindrica*	F	F	F	F	F	F	F	F	F	F	R	F	R	F	R	F	R	R	R	R	R	R	F	F	F	R	R	R
2	*Antarctissa denticulata*	C	C	C	F	F	F	F	F	F	F	F	F	F	C	C	C	C	C	C	C	C	C	C	C	F	F	F	F
3	*Antarctissa strelkovi*	-	-	-	-	-	-	-	-	-	+	-	-	R	+	R	+	F	F	F	F	F	F	F	R	F	F	F	F
4	*Bathropyramis sp.*	-	+	-	+	-	-	-	+	-	-	-	-	-	R	+	+	-	+	-	-	-	-	-	-	+	-	+	+
5	*Cenosphaera sp.2*	+	-	R	R	+	-	+	+	R	+	+	+	+	R	+	R	R	R	-	R	R	R	R	R	R	+	R	R
6	*Cenosphaera sp.3*	-	-	-	-	R	-	R	-	-	R	+	-	R	F	R	R	R	R	R	R	R	R	R	R	R	R	+	R
7	*Cornutella profunda*	F	F	+	R	R	+	R	R	+	+	R	R	R	R	R	R	+	R	R	+	-	+	+	R	+	R	R	R
8	*Cycladophora pliocenica*	+	+	+	+	-	+	-	+	-	-	-	-	-	R	-	R	+	+	-	R	R	R	+	+	R	+	-	+

9	*Cycladophora robusta*	-	+	-	-	-	+	-	+	+	R	-	-	-	R	+	+	-	-	-	-	-	-	-	-	-	-	-	-
10	*Desmospyris spongiosa*	-	-	+	-	+	+	+	+	-	+	F	-	F	-	-	-	F	F	F	F	F	F	F	F	F	F	F	F
11	*Dictyocoryne sp.*	-	-	-	-	-	-	-	-	+	-	-	-	-	+	-	-	-	-	-	-	-	-	-	-	-	-	-	-
12	*Eucyrtidium calvertense*	-	-	-	-	-	-	-	-	+	+	R	-	+	-	+	-	+	+	R	R	R	+	R	R	R	+	+	+
13	*Helotholus vema*	-	-	-	-	-	-	-	-	-	-	+	-	+	-	-	-	+	R	R	R	R	R	R	R	R	+	+	R
14	*Lithelius minor*	+	R	-	-	-	-	+	R	+	-	-	-	-	R	R	R	R	R	+	+	-	R	R	+	+	R	-	-
15	*Lithelius nautiloides*	R	R	-	R	+	+	R	R	R	+	+	-	-	+	R	R	+	R	R	+	R	+	+	R	R	+	-	-
16	*Lithomitra lineata*	-	+	-	-	+	-	-	+	+	-	-	-	-	+	+	+	-	-	-	-	-	-	-	+	-	R	R	+
17	*Peripyramis circumtexta*	R	R	+	+	+	+	-	+	-	+	-	+	-	+	R	R	R	R	+	+	+	-	-	+	-	-	+	R
18	*Phorticium pylonium*	-	-	-	-	-	-	-	-	-	-	-	-	-	-	-	R	+	+	+	+	R	-	+	R	+	-	-	+
19	*?Prunopyle antarctica*	-	-	R	-	-	-	R	-	-	-	+	-	-	R	+	R	R	+	+	R	R	+	-	-	+	-	R	-
20	*Saccospyris antarctica*	R	R	R	+	R	R	+	R	R	+	+	R	-	R	F	R	F	R	R	R	R	R	+	R	R	R	-	-
21	*Saccospyris preantarctica*	+	R	+	R	R	R	R	R	R	R	F	F	+	F	F	F	R	F	F	F	F	R	R	R	F	F	R	R
22	*Saturnalis circularis*	-	-	-	-	-	-	-	-	-	-	-	-	+	-	-	+	-	-	-	+	-	+	-	-	-	-	-	-
23	*Spongopyle osculosa*	R	R	R	R	+	R	+	+	+	R	+	R	R	R	R	R	R	R	R	R	R	+	R	R	R	-	R	R
24	*Spongotrochus glacialis*	F	F	F	F	F	F	F	F	F	F	F	F	F	F	F	F	F	F	F	F	F	F	F	F	F	F	F	F
25	*Spongurus sp.*	R	R	+	-	+	-	+	+	-	-	-	-	-	-	+	-	-	-	-	-	-	-	-	-	-	-	-	-
26	*Spongurus pylomaticus*	R	R	R	-	+	-	-	-	-	-	-	+	-	R	R	R	+	+	+	-	-	-	-	+	+	+	-	+
27	*Stylochlamydium venustum*	+	R	R	R	R	R	+	R	R	+	+	R	+	R	R	+	R	R	+	-	+	+	R	+	+	+	+	+
28	*Stylodictya aculeata*	R	R	-	R	-	-	-	-	+	+	+	+	-	R	R	R	R	R	R	-	R	-	R	R	R	R	R	R
29	*Stylodictya validispina*	R	+	-	-	+	-	+	-	+	-	-	-	-	-	+	-	-	+	+	+	-	-	+	R	-	+	-	+
30	*Zygocircus piscicaudatus*	-	-	+	-	-	-	-	-	-	-	+	-	-	-	+	-	R	-	-	-	-	-	-	-	-	-	-	-

Family **SPONGODISCIDAE** Haeckel, 1862, *emend.* Riedel, 1967b

Genus SPONGOPYLE Dreyer, 1889

Spongopyle osculosa Dreyer
(Pl.1, Fig.20)

1889 *Spongopyle osculosa* - Dreyer, p.42, pl.11, figs.99,100.

1967 *Spongodiscus* (?) *osculosus* (Dreyer) - Petrushevskaya, p.42, figs.20-22.

1975a *Spongodiscus osculosus* (Dreyer) - Chen, p.455, pl.24, fig.4.

Distribution: Sharma and Takahashi (2008) reported its presence from very rare to few in the Antarctic region.

Abundance: Very rare to rare.

Genus SPONGOTROCHUS Haeckel, 1860

Spongotrochus glacialis Popofsky group (Pl.1, Fig. 2)

1908 *Spongotrochus glacialis* - Popofsky, p.228, pl.26, fig.8, pl.27, fig.1, pl.28, fig.2.

1975 *Spongotrochus glacialis* Popofsky group - Petrushevskaya, p.575, pl.5, fig.8, pl.35, figs.1-6, (with synonymy).

Distribution: Riedel (1958) reported *S. glacialis* in the Antarctic region, while Lozano (1974) showed its presence in the Sub Antarctic region. Benson (1966) considered it as a cosmopolitan species.

Abundance: Few

Range: Miocene to Recent (Petrushevskaya, 1975); Neogene and Oligocene? (Chen, 1975a); Neogene (Keany, 1979).

Family PYLONIIDAE Haeckel, 1881

Genus PHORTICIUM Haeckel, 1881, p.964; 1887, p.709

Phorticium pylonium (Clevei) Haeckel (Pl.1, Fig.30)

1887 *Phorticium pylonium* - Haeckel, p. 709, pl.49, fig.10.

1958 *Phorticium pylonium* - Haeckel, Riedel, p.229, pl.2, fig.5.

1966 non *Phorticium pylonium* - Haeckel, Cleve in Benson, p.252, pl.16, figs.5-9, pl.17, figs.1-3.

2003 *Phorticium pylonium* (Clevei) - Haeckel, Cortese et al., pl.2, figs.12-13; pl.3, fig.5; pl.4, figs.18-19.

Distribution: Riedel (1958) showed presence of *P. pylonium* (Clevei) Haeckel, in the Antarctic region.

Abundance: Very rare to rare.

Range: Miocene to Recent (Petrushevskaya, 1975).

Family PYLONIIDAE Haeckel, 1881

Genus PRUNOPYLE Dreyer, 1889

?Prunopyle antarctica Dreyer (Pl.1, Fig. 10)

1889 *?Prunopyle Antarctica* - Dreyer, p. 24-25, pl.5, fig.75.

1975a ?*Prunopyle antarctica* (Dreyer) - Chen, p.454, pl.23, fig.5, 6.

Distribution: Riedel (1958) and Petrushevskaya (1967) reported this species from Antarctic and sub polar region.

Abundance: Very rare to rare.

Range: Chen (1975a) mentioned presence of *P. antarctica* from Pleistocene to Recent.

Family LITHELIIDAE Haeckel, 1862

Genus LITHELIUS Haeckel, 1862

Lithelius minor Jörgensen (Pl.1, Fig.15)

1900 *Lithelius minor* - Jörgensen p.65, pl.5, fig.24.

1979 *Lithelius minor* Jörgensen - Nigrini and Moore, p.S135, pl.17, figs.3,4a,b.

Distribution: Lombari and Boden (1985) showed the presence of *L. minor* near the Antarctic region.

Abundance: Very rare to rare.

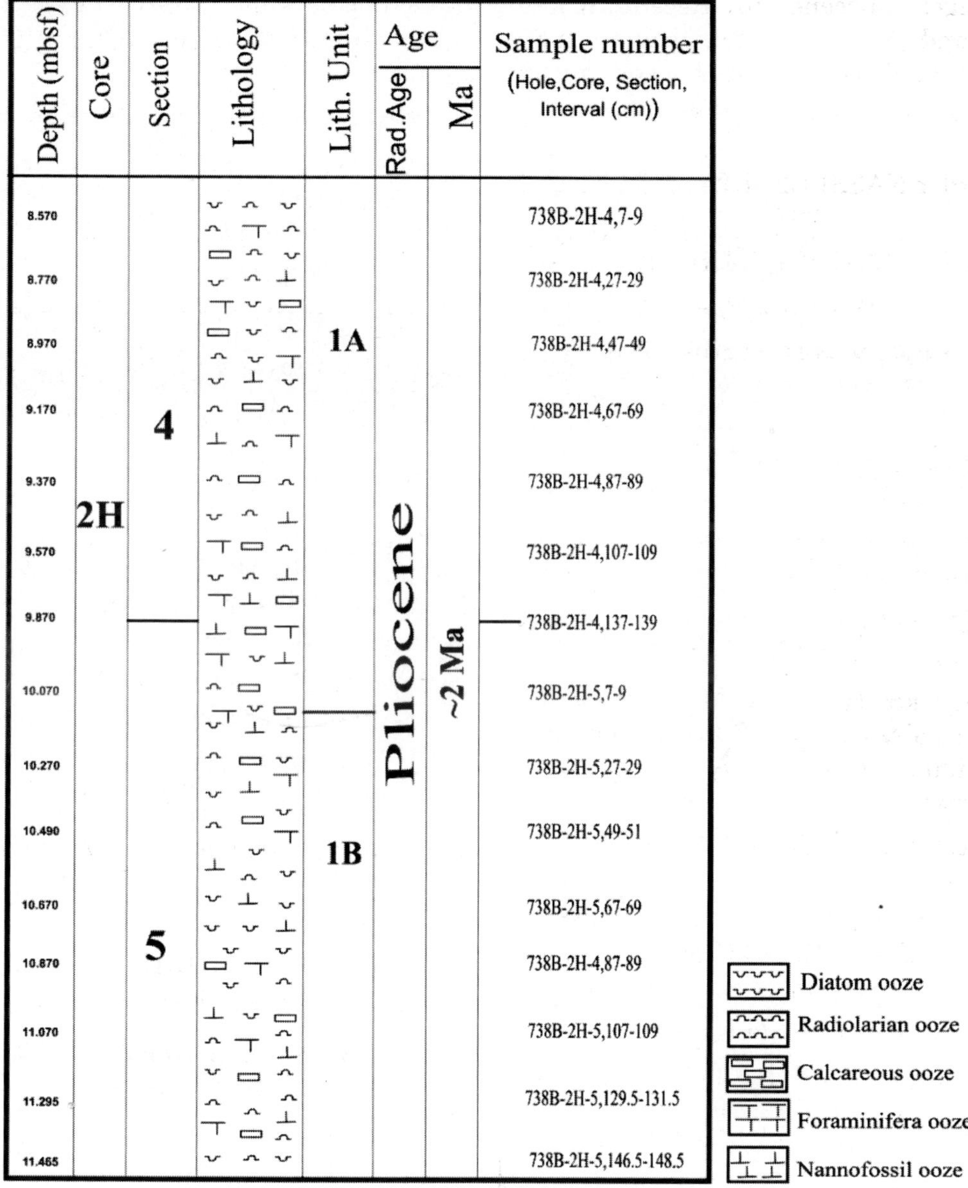

Fig.2. Age, position, lithostratigraphic succession and depth of section (2H4 to 2H5 of Leg 119 Site 738B).

Lithelius nautiloides Popofsky
(Pl.1, Fig.16)

1908 *Lithelius nautiloides* – Popofsky, p.230-231, pl.27, fig.4.

1975 *Lithelius(?) nautiloides* Popfsky - Petrushevskaya, p.572, pl.3, figs.1,3,5, pl.33, figs.3,4.

Abundance: Very rare to rare.

Range: Pliocene to Recent (Chen, 1975a and Keany, 1979). Hays (1965) reported *L. nautiloides* to be endemic to the present day Antarctic fauna.

Suborder NASSELLARIA Ehrenberg, 1875

Family ACANTHODESMIIDAE Haeckel

Genus *Desmospyris* Haeckel,1881

Desmospyris spongiosa Hays
(Pl.1, Fig.26)

1965 *Desmospyris spongiosa* – Hays, p.173-175, pl.ll,fig.1.

1975 *Desmospyris spongiosa* – Hays, Chen, pl.15, figs.1-2.

1990 *Desmospyris spongiosa* – Hays, Lazarus, p.716, pl.4. figs.9-11.

Remarks: Hays (1965) reported this species as abundant in Phi zone and has a range similar to the range of *Eucyrtidium calvertense.*

Abundance: Very rare to few.

Family PLAGONIIDAE Haeckel, 1881, *emend.* Riedel, 1967b

Genus ANTARCTISSA Petrushevskaya, 1967, p.85, 1971b, p.120.

Antarctissa cylindrica Petrushevskaya
(Pl.1, Fig.19)

1967 *Antarctissa denticulata* var. *cylindrica* - Petrushevskaya, p.89, pl.49, fig.6, pl.50, fig.1.

1972a *Antarctissa cylindrica* - Petrushevskaya, pl.1, fig. 8, pl.2, fig. 6.

Abundance: Rare to few.

Range: Lazarus (1990) reported *A. cylindrica* as its last appearance in the Psi zone. Lazarus (2002) showed its occurrence from 0.6-6.4 Ma in the Antarctic region.

Antarctissa denticulata (Ehrenberg)
(Pl.1, Fig.6)

1844b *Lithobotrys denticulata* – Ehrenberg, p. 203.

1873 *Lithopera denticulata* (Ehrenberg) – Ehrenberg, pl.12, fig.7.

1968 *Antarctissa denticulata* (Ehrenberg)- Petrushevskaya, p.84-86, fig.49, 1-1V.

Abundance: Few to common.

Range: Petrushevskaya (1975) reported its range from the Pleistocene to the Recent whereas Chen (1975a) assigned it from the Pliocene to the Pleistocene. Lazarus (2002) showed its presence from 0 -2.6 Ma in the Antarctic region.

Antarctissa strelkovi Petrushevskaya
(Pl.1, Fig. 29)

1908 *Helotholus histricosa* Jörgensen – Popofsky, p.279-281, pl.32, fig.1-5, pl.36, fig.2.

1968 *Antarctissa strelkovi* – Petrushevskaya, p.88-90, fig.51, 111-V1.

Range: Petrushevskaya (1975) reported its occurrence from the Miocene-Recent, Lazarus (2002) showed its occurrence from 0-2.5 Ma in the Antarctic region.

Genus HELOTHOLUS Jörgensen, 1905

Helotholus vema Hays
(Pl.1, Fig. 27)

1965 *Helotholus vema* – Hays, p.176, pl.II, fig.3.

1971 *Pseudocubus vema* - Petrushevskaya, p.46.

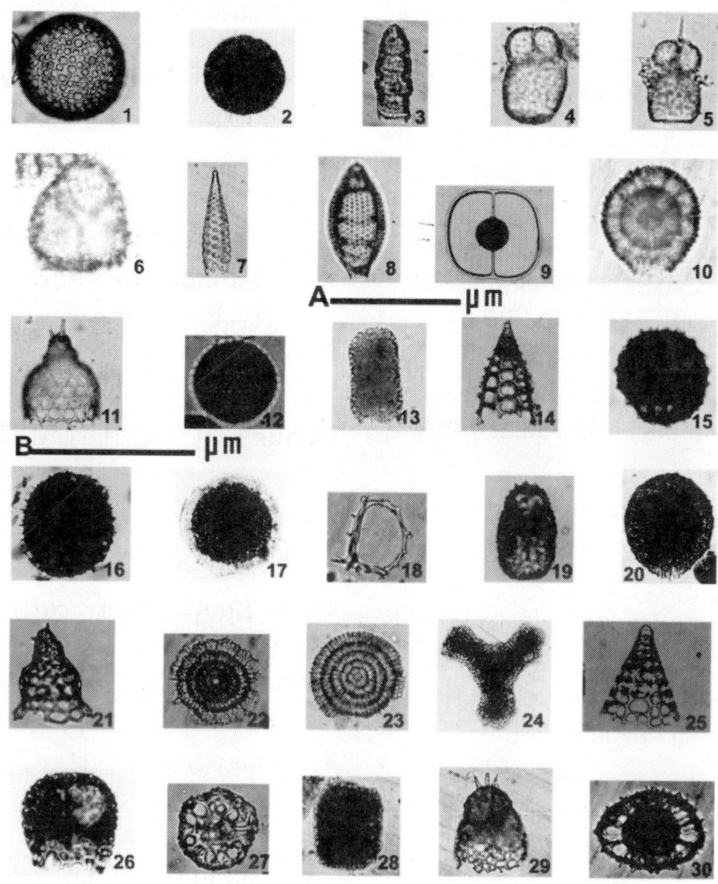

Plate1.1. *Cenosphaera* sp. 2, 745B-2H-5, 77-79; focused on OS, scale bar B; **2**. *Spongotrochus glacialis*. 738B-2H-4, 27-29, focused on OS, scale bar A; **3**. *Lithomitra lineata*. 738B-2H-5, 129-131; focused on OS, scale bar B; **4**. *Saccospyris Antarctica*. 738B-2H-5, 37-39; focused on IS, scale bar A; **5**. *Saccospyris preantarctica*. 738B-2H-5, 17-19; focused on IS, scale bar A; **6**. *Antarctissa denticulata*. 738B-2H-4, 7-9; focused on OS, scale bar B; **7**. *Cornutella profunda*. 738B-2H-4, 17-19; focused on OS, scale bar B; **8**. *Eucyrtidium calvertense* 738B-2H-5, 77-79, focused on OS, scale bar A; **9**. *Saturnalis circularis*. 738B-2H-5, 27-29; focused on OS, scale bar A; **10**. ?*Prunopyle antarctica*. 738B-2H-5, 67-69; focused on IS, scale bar A; **11**. *Cycladophora pliocenica*. 738B-2H-5, 7-9; focused on IS, scale bar B; **12**. *Cenosphaera* sp.3 745B-2H-5, 7-9; focused on IS, scale bar B; **13**. *Spongurus pylomaticus* 738B-2H-5, 7-9; focused on OS, scale bar B; **14**. *Peripyramis circumtexta*.738B-2H-5, 146-148; focused on IS, scale bar A; **15**. *Lithelius minor*. 738B-2H-4, 77-79; focused on OS, scale bar A; **16**. *Lithelius nautiloides*. 738B-2H-4, 87-89; focused on IS, scale bar A; **17**. *Stylochlamydium venustum*.738B-2H-4, 57-59; focused on OS, scale bar A; **18**. *Zygocircus piscicaudatus* 745B-2H-5,37-39; focused on OS, scale bar A; **19**. *Antarctissa cylindrical*. 738B-2H-4, 37-39; focused on OS, scale bar B; **20**. *Spongopyle osculosa*. 738B-2H-5, 37-39, focused on OS, scale bar A; **21**. *Cycladophora robusta*. 738B-2H-5, 27-29; focused on IS, scale bar B; **22**. *Stylodictya aculeata*. 738B-2H-5, 49-51; focused on OS, scale bar A; **23**. *Stylodictya validispina*. 738B-2H-4, 7-9; focused on IS, scale bar A; **24**. *Dictyocoryne sp*. 738B-2H-4, 87-89; focused on OS, scale bar B; **25**. *Bathropyramis sp*. 738B-2H-5, 7-9; focused on OS, scale bar B; **26**. *Desmospyris spongiosa*. 738B-2H-5, 49-51; focused on OS, scale bar B; **27**. *Helotholus vema*. 738B-2H-4, 107-109; focused on IS, scale bar B; **28**. *Spongurus* sp. 738B- 2H-4, 17-19; focused on OS, scale bar A; **29**. *Antarctissa strelkovi*. 738B- 2H-5, 67-69; focused on OS, scale bar B; **30**. *Phorticium pylonium*. 745B-2H-5, 27-29; focused on OS, scale bar A. (OS= outer surface; IS=inner surface).

1990 *Helotholus vema* – Hays, Lazarus, pl.7, figs.1-5.

Abundance: Very rare to rare.

Remarks: Lazarus (2002) reported its first occurrence at 4.9 Ma and last occurrence at 2.2 Ma.

Family THEOPERIDAE Haeckel, 1881, *emend.* Riedel, 1967b, p.296.

Genus CYCLADOPHORA Ehrenberg, 1872b, *emend.* Lombari and Lazarus,1988

Cycladophora pliocenica (Hays)
(Pl.1, Fig. 11)

1965 *Clathrocyclas bicornis* - Hays, p.179, pl.3, fig.3.

1988 *Cycladophora pliocenica* - (Hays) Lombari and Lazarus, p.104.

2008 *Cycladophora pliocenica* - (Hays) Sharma and Takahashi, p.76, pl.5, figs.N,T.

2013 *Cycladophora pliocenica* - (Hays) Sharma, p.180, fig.5-O,P.

Distribution: Keany and Kennett (1975) reported *C. bicornis* occurrences from Pliocene to Pleistocene. Sharma et al. (2004) and Sharma and Takahashi (2008) showed its presence in the Pleistocene sediments of Tasman and Antarctic regions.

Abundance: Very rare to rare

Range: The last appearance of *Clathrocyclas bicornis* (Hays) = *Cycladophora pliocenica*, Lombari and Lazarus, 1988) was noted by Hays and Opdyke (1967) within the Chi zone.

Lazarus (1990) also reported its presence in the Chi zone.

Cycladophora robusta Lombari and Lazarus
(Pl.1, Fig. 21)

1988 *Cycladophora robusta* - Lombari and Lazarus, p.105, pl.2, fig.1-14.

2013 *Cycladophora robusta* – Lombari and Lazarus, Sharma, p.180,fig.4-G,P.

Abundance: Very rare to rare.

Range: Miocene (Lombari and Lazarus, 1988).

Genus EUCYRTIDIUM Ehrenberg, 1847b, *emend.* Nigrini, 1967

Eucyrtidium calvertense Martin
(Pl.1, Fig.8)

1904 *Eucyrtidium calvertense* - Martin, p.450, pl.80, fig.5.

1990 *Eucyrtidium calvertense* Martin - Lazarus, p.716, pl.6, figs.4-6.

Distribution: Hays (1965) and Lazarus (1990) reported its presence in Antarctic region.

Abundance: Very rare to rare.

Range: Pliocene (Chen, 1975a; Keany, 1979)

Family CANNOBOTRYIDAE Haeckel, 1881

Genus SACCOSPYRIS Haecker, 1908

Saccospyris antarctica Haecker
(Pl.1, Fig.4)

1908 *Saccospyris antarctica* - Haecker, p.447, pl.84, figs.584, 589,590.

1975 *Saccospyris antarctica* Haecker - Petrushevskaya, p.589, pl.13, figs. 21, 28.

Abundance: Very rare to few.

Range: Pliocene-Pleistocene (Petrushevskaya, 1975).

SACCOSPYRIS PREANTARCTICA
Petrushevskaya
(Pl.1, Fig.5)

1972a *Saccospyris* sp. – Petrushevskaya, pl.2, fig.7.

1975 *Saccospyris preantarctica* - Petrushevskaya, p.589, pl.13, figs.19,20.

2008 *Saccospyris preantarctica* - Petrushevskaya, Sharma and Takahashi, p.79, fig.7E.

Distribution: Sharma and Takahashi (2008) reported its presence from very rare to few in the Antarctic region.

Abundance: Very rare to few.

Range: Miocene to Recent (Petrushevskaya, 1975).

Remarks: Petrushevskaya (1975) showed the presence of *S. preantarctica* in Miocene of Antarctic region. While, Weaver (1983) reported this species from the Pliocene sediments of Falkland Plateau (DSDP Leg 71).

Family STEPHANIIDEA Haeckel, 1887

Genus *Zygocircus* Bütschli, 1882

Zygocircus piscicaudatus Popofsky
(Pl.1, Fig.18)

1913 *Zygocircus piscicaudatus* - Popofsky, pp.287-288, pl.28, fig.3.

1971b *Zygocircus piscicaudatus* - Petrushevskaya, pl.145, fig.4 (not 5-9).

Abundance: Very rare to rare.

CONCLUSION

Well preserved, abundant and highly diverse radiolarian assemblages were found in the sections 2H-4 and 2H-5 of Leg 119 site 738B. Thirty radiolarian taxa were identified and taxonomy notes of stratigraphically important taxa were given. The assemblage belongs to a typical Antarctic. The dominant species present in the sections are, namely- *Antarctissa denticulata, Antarctissa cylindrica, Antarctissa strelkovi, Lithelius minor, Saccospyris antarctica, Saccospyris preantarctica, Spongopyle osculosa* and *Spongotrochus glacialis.*

Acknowledgements: GKS thankfully acknowledge receiving samples from the Lamont-Doherty Geological Observatory, USA (Request Number-18214 A). Authors are grateful to UCOST, Dehradun, India for providing the financial assistance in the form of major project (Project no. UCS &T/R&D/PHY SC.-03/11-12/3993/1\2011) GKS. We are also thankful to Head, CAS, Department of Geology for providing the facilities.

References

ABELMANN, A. and GERSONDE, R. (1988) *Cycladophora davisiana* stratigraphy in Pliocene- Pleistocene cores from the Antarctic Ocean (Atlantic Sector). Micropal., v.34, pp.268 -276.

BANDY, O.L. and CASEY, R.E. (1969) Major Late Cenozoic planktonic datum planes, Antarctic to the Tropics. Antarctica, v.4(5), pp.170-171.

BENSON, R.N. (1966) Recent Radiolaria from the Gulf of California *(Ph.D. dissertation).* Minnesota University, 577.

BÜTSCHLI, O. (1882). Beitrage zur Kenntnis der Radiolarienskelette, insbesondere der Cyrtida. Z. Wissenschaft. Zoology, v.36, pp. 485-540.

CALKINS, G.N. (1909) Protozoology. Lea and Febiger Co., p.349.

CHEN, P.H. (1974) Some new Tertiary Radiolaria from Antarctic deep sea sediments. Micropal., v.204, pp.480-492.

CHEN, P.H. (1975a) Antarctic radiolaria. *In*: Initial Reports of Deep Sea Drilling Project Eds. Hayes, D.E., Frakes, L.A., et al., U.S. Government Printing office, Washington, v.28, pp.437-513.

CHEN, P.H. (1975b) Post Paleocene radiolaria: their taxonomy, boistratigraphy and phylogeny and the development of late Neogene cold water faunas. (Ph. D. dissertation), Columbia University, New York.

CORTESE, G., BJORKLUND, K.R. and DOLVEN, J.K. (2003) Polycystine radiolarians in the Greenland-Iceland-Norwegian (GIN) Seas: species and assemblage distribution. Sarsia, v. 8, pp. 65-88.

DEFLANDRE, G. (1953) Radiolaries fossiles. In: Traite de Zoologie, (Ed. Grasse, P.P.), fasc. 2, Masson, Paris, v.1, pp.389-433.

De WEAVER, P., DUMITRICA, P., CAULET, J. P., NIGRINI, C., CARIDROIT, M. (2001) Radiolarians in the sedimentary record. The Netherlands, Gordon and Breach Science Publishers. 533p.

DREYER, F. (1889) Morphologische Rardiolarien studien. I.Die Pylombildungen in vergleichend-anatomischer undentwicklungs geschichtlicher Beziehung bei Radi-olarien und bei Protisten uberhaupt, nebst System und Beschreibung neuer und des bis jetzt bekannten pylomatischen Spumellarien. Jeinaische Zeitschrift Naturwissenschaft, v.23, pp.1-138.

EHRENBERG, C.G. (1838) Uber die Bildung der Kreidefelsen und des Kreidemergels durch unsichtbare Organismen. Konigliche Preussischen Akademie der Wissenschaften zu Berlin, Abhandlungen, Jahrgang, v.1838, pp.59-147.

EHRENBERG, C.G. (1844a) Uber 2 neue Larger von Gebirgsmassen aus Infusorien als Meeres- Absatz in Nord-Amerika und eine Vergleichung derselben mit den organischen Kreide- Gebilden in Europe und Afrika. Monatsber. KOnigliche Preussischen Akademie der Wissenschaften zu Berlin, pp.57-95.

EHRENBERG, C.G. (1844b) Einige vorlaufige Resultate seiner Untersuchungen der ihm von der Sudpolreise des Capitain Ross, so wie von den Herren Schayer und Darwin zugekommenen Materialien uber das Verhalten des kleinsten Lebens in den Oceanen und den grossten bisher zuganglichen Tiefen Weltmeeres: Monatsber. Konigliche Preussischen Akademie Wissenschaften Berlin Jahre, v.1844, pp.: 182-207.

EHRENBERG, C.G. (1847b) Uber die mikroskopischen kieselschaligen Polycystinen als machtige Gebirgmasse von Barbados: Monatsber. Konigliche Preussischen Akademie der Wissenschaften zu Berlin, Jahrgang, v.1847, pp.40-60.

EHRENBERG, C.G. (1872b) Mikrogeologische Studien uber das Kleinste Leben der Meeres-Tiefgrunde aller Zonen und dessen geologischen Einfluss. Konigliche Preussischen Akademie Wissenschaften Berlin, Abh. Jahre, v.1872, pp. 331-399.

EHRENBERG, C.G. (1873) Grossere Felsproben des Polycystinen-Mergels von Barbados mit weiteren Erlauterungen. Konigliche Preussischen Akad. Wissenschaften Berlin, Monat., v.1873, pp.263-312.

EHRENBERG, C.G. (1875) Fortsetzung der mikrogeolgischen Studien als Gesammt-Uebersicht der mikroscopischen Palaontologie gleichartig analysierter Gebirgsartender Erde, mit specieller Rucksicht auf den Polycystinen Mergel von Barbados. Konigliche Preussischen Akademie Wissenschaften zu Berlin, Abh., Jahre, v.1875, pp.1-225.

GOLL, R.M. (1968) Classification and phylogeny of Cenozoic Trissocyclidae (Radiolaria) in the Pacific and Caribbean Basins. Part 1. Palaeon., v.42(2), pp.1409-1432.

HAECKEL, E. (1860a) Uber neue, lebende Radiolarien des Mittelmeeres: Monat. Kongliche Preussischen Akademie Wissenschaften Berlin, Jahrg., pp.794-817.

HAECKEL, E. (1860b) Fernere Abbildungen und Diagnosen neuer Gattungen und Arten von lebenden Radiolarien des Mittelmeeres. Kongliche Preussischen Akademie, Jahre, v.1860, pp.835-845.

HAECKEL, E. (1862) Die Radiolarien (Rhizopoda Radiolaria), eine Monographie, Reimer, Berlin, 572 p.

HAECKEL, E. (1881) Entwurf eines Radiolarien-System auf Grund von Studien der Challenger Radiolarien. Jenaische Zeitschrift Naturwissenschaft, v.15(8), pp. 418-472.

HAECKEL, E. (1887) Report on the Radiolaria collected by H. M. S. Challenger during the years 1873-1876. Reports of Scientific Results, Voyage H.M.S. Challenger, Zoology, v.18, pp.1-1803.

HAECKER, V. (1908) Tiefsee Radiolarien. Spezieller Teil: Wissensch Erqeb. Deutsche Teifsee Expedition, Jena, v.14, pp. x-476.

HAYS, J.D. (1965) Radiolaria and late Tertiary and Quaternary history of Antarctic seas. *In*: Bio. Antarctic Seas II. American Geophy. Union, Antarc. Res., Ser. 5, pp.125-184.

HAYS, J.D. (1967) Quaternary sediments of Antarctic Ocean. *In*: Progress in Oceanography, v.4, pp.117-131.

HAYS, J.D. (1970) Stratigraphy and evolutionary trends of Radiolaria in North Pacific deep-sea sediments. *In*: Geol. Investigations North Pacific, HAYS, J.D., (Eds.), Geol. Soc. America, Mem. 126, pp. 185-218.

HAYS, J.D. and BERGGREN, W.A. (1971) Quaternary boundaries and correlations. *In*: Micropaleontology of the oceans, FUNNEL, B. M. and RIEDEL, W.R., (Eds.), Cambridge Univ. Press, p.669-692.

HAYS, J.D. and Opdyke, N.D. (1967) Antarctic Radiolaria, magnetic reversals and climatic change. Science, v.158, pp.1001-1011.

HERTWIG, R. LeESSER, E. (1874) Ueber Rhizopoden und denselben nahestenden Organismen. Arch. Mikr. Anat., 10 (Suppl.), 35 p.

JORGENSEN, E. (1900) Protophyten und Protozoen im Plankton aur der Norweigischen Westkuste, Bergen Museums Aarbog, v.1899, pp.51-95.

JORGENSEN, E. (1905) The protist plankton and the diatoms in the bottom samples. Bergens Museum Skr., pp.49-225.

KEANY, J. (1976) Early Pliocene paleoclimatology and radiolarian biostratigraphy of the Southern Ocean. Antarctic Research, v.11, pp.171-173.

KEANY, J. (1979) Early Pliocene radiolarian taxonomy and biostratigraphy in the Antarctic region. Micropal., v.25(1), pp.50-74, pls.1-5.

KEANY, J. and KENNETT, J.P. (1972) Pliocene-early Pleistocene paleoclimatic history recorded in Antarctic-Subantarctic deep-sea cores. Deep Sea Research, v.19, pp.529-548.

KEANY, J. and KENNETT, J.P. (1975) Pliocene-Pleistocene radiolarian biostratigraphy and paleoclimatology at DSDP site 278 on the Antarctic convergence. In: Initial Reports of Deep Sea Drilling Project, U.S. Government Printing Office, Washington, v.29, pp.757-767.

KLING, S.A. (1971) "Radiolaria: Leg 6 of Deep Sea Drilling Project. In: Initial Reports of Deep Sea Drilling Project," U. S. Govt Printing Office, Washington, v.6, pp.1069-1117.

LAZARUS, D. (1990) Middle Miocene to Recent radiolarians from the Weddell Sea, Antarctica, ODP Leg 113, *In*: Proc. Ocean Drilling Prog. Sci. Results, BARKAR, P.F. and KENNETT, J.P., (Eds.), p. 709-728.

LAZARUS, D. (1992) Antarctic Neogene radiolarians from the Kerguelen Plateau, ODP Legs 119 and 120. *In*: Proc. Ocean Drilling Prog. Sci. Results, WISE, S.W. and SCHLICH, R., (Eds.), 120: 785-810.

LAZARUS, D. (2002) Environmental control of diversity, evolutionary rates and taxa longevities in Antarctic Neogene radiolaria. Paleon. Electronica, v.5, 32 p.

LOMBARI, G. and BODEN, G. (1985) Modern Radiolarian global distribution. Cushman Found. Foram. Res. sp. Pub. 16A, pp.1-125.

LOMBARI, G. and Lazarus, D.B. (1988) Neogene Cycladophorid Radiolaria from North Atlantic, Antarctic and North Pacific Deep Sea sediments. Micropal., v.34, pp.97-135.

LOZANO, J.A. (1974) Antarctic sedimentary, faunal and sea surface temperature responses during the last 230,000 years with emphasis in comparison between 8,000 years ago and today. (Ph. D. dissertation), Columbia University, New York, 400 p.

MARTIN, G.C. (1904) Radiolaria. *In*: Systematic Paleontology of the Miocene Deposits of Maryland, CLARK, W.B., EASTMAN, C.R., GLENN, L.C., BAGG, R.M., BASSLER, R.S., BOYER, C.S., CASE, E.C., HOLLICK, C.A. and BALTIMORE (Eds.), Maryland Geol. Sur. Johns Hopkins Press, pp. 477-459.

McINTYRE, L. and KACZMARSKA, I. (1996) Improved resolution of the Pleistocene extinction level of *Stylatractus universus* (Hays) in ODP Hole 75B, Kerguelen Plateau. Micropal., v.42, pp.375-379.

MÜLLER, J. (1858) Uber die Thalassiocollen, Polycystien Acanthometren Mittelmeers. Wissenschaften Berlin, Abhandlungen, Jahre Fur., v.1858, pp.1-62.

NAKASEKO, K. and NISHIMURA, A. (1982) Radiolaria from the bottom sediments of the Bellingshausen Basin in the Antarctic Sea. Rep. Tech. Res. Cen., Japan Nat. Oil Corp., v.16, pp. 91-244.

NIGRINI, C.A. (1967) Radiolaria in Pelagic Sediments from the Indian and Atlantic Oceans: Scripps Institution of Oceanography, California University, California, Bull. v.11, pp.1-25.

NIGRINI, C.A. and LOMBARI, C. (1984) Miocene Radiolaria. Cushman Found. Foram. Res. sp. pub. 22, S1-102, N1-206.

NIGRINI, C.A. and MOORE, T.C. Jr. (1979) A Guide to Modern Radiolaria. Cushman Found. Foram. Res. sp. pub. p. S1- 142, N1-106.

PETRUSHEVSKAYA, M.G. (1968) Radiolarians of orders of Spumellaria and Nassellaria of the Antarctic regions. *In*: Biological reports of the Soviet Antarctic Expedition (1955-58), (Translated from Russian by the Israel Program for Scientific Translations), ANDRIYASHEV, A.P. (Ed.), Published for the National Science Foundation, Washington, D.C.), v.3(1).

PETRUSHEVSKAYA, M.G. (1971) Radiolaria in the plankton and Recent sediments from the Indian Ocean and Antarctic. *In*: The Micropaleontology of Oceans. FUNNELL, B.M. and RIEDEL, W.R. (Eds.) Cambridge University Press, pp. 319-329.

PETRUSHEVSKAYA, M.G. (1971b) Radiolyarii Mirovogo Okeana po materialam Sovetskikh ekspeditisii. In: Issledovanii Fauny Morey Leningrad BYKHOVSHII, B.E. (Eds.), Nauka, v.9, pp.5-294.

PETRUSHEVSKAYA, M.G. (1972a) Biostratigraphy of deep-sea Quaternary sediments on the radiolarian data. Okeanologia, v.12, pp.71-86.

PETRUSHEVSKAYA, M.G. (1975) Cenozoic radiolarians of the Antarctic, Leg 29, DSDP. *In*: Initial Reports of Deep Sea Drilling Project, KENNETT, J.P. and HOUTZ, R.E. (Eds.) U.S. Government Printing Office, Washington, v.29, pp.541-675.

POPOFSKY, A. (1908) Die Radiolarien des Antarktis (mit Ausnahme der Tripyleen). Deutsche Südpolar-Expedition 1901-1903, Part 10, Zoology, v.2(3), pp.183-305.

POPOFSKY, A. (1913) Die Nassellarien des Warmwassergebietes. Deutsche Südpolar-Expedition, Zoology, v.14(6), pp.217-416, pls.28-38.

RENAUDIE, J. and LAZARUS, D. (2012) New species of Neogene radiolarians from the Southern Ocean. Micropal., v.31, pp.29-52.

RENAUDIE, J. and LAZARUS, D. (2013) On the accuracy of paleodiversity reconstructions: a case study in Antarctic Neogene radiolarians. Paleobio, v.39(3), pp.491–509.

RIEDEL, W. R. (1958) Radiolaria in Antarctic Sediments. Reports B.A.N.Z. Antarctic Research Expedition, v.6, pp.217-255.

RIEDEL, W.R. (1967b) Subclass Radiolaria. *In*: The Fossil Record, HARLAND, W.B. (Eds.), Geological Society of London, pp. 291-298, text figs.8.1 A, B.

RIEDEL, W. R. and SANFILIPPO, A. (1970) Radiolaria Leg 4, DSDP, In: Initial report of Deep Sea Drilling Project, BADER, R.G. (Ed.) U. S. Government Printing Office, Washington, v.4, pp.503-573.

RIEDEL, W.R. and SANFILIPPO, A. (1971) Cenozoic Radiolaria from the western tropical Pacific, Leg 7. *In*: Initial Reports of Deep Sea Drilling Project, WINTERER, (Ed.), U.S. Govt. Printing office, Washington, v.7, pp.1529-1672.

RIEDEL, W.R. and WESTBERG, M. J. (1982). Neogene radiolarians from the eastern tropical Pacific and Caribbean, DSDP Leg 68, p.289-300, *In*: Initial Reports of Deep Sea Drilling Project, Washington, D.C., U.S. Govt. printing office, v. 68, pp.289-300.

SHARMA, G.K. (2013) Neogene Radiolarian Biostratigraphy from ODP Leg 119, Site745. Ofioliti, v.38(2), pp.175-203.

SHARMA, G.K. and DAYAL, D. (2015a) Distribution of Pliocene Radiolaria at Site 745 (Section 14-15 H, Leg 119, from the Kergulean Plateau and Biostratigraphy, v.23(7), pp.56-66.

SHARMA, G.K. and DAYAL. D. (2015b) Pliocene Radiolaria from Leg 119 Site 745, Sec. (12H-13H), *In*: News of Osaka Micropal., sp. v.15, pp.125-138.

SHARMA, G.K. and KANOJIA. S., (2013) Pleistocene radiolaria from Leg 119 Site 738B Section (1H1-1H2): Systematics and Biostratigraphy. OJG, USA, v.3, pp.427-438.

SHARMA, G.K. and TAKAHASHI, K. (2007) Pleistocene radiolarian from the Antarctic Continental margin: distribution and biostratigraphy. J. Geol. Soc. India, v.69, pp.813-826.

SHARMA, G.K. and TAKAHASHI, K. (2008) Distribution and Taxonomy of Pleistocene Radiolarians from KH94-4-AMR-2PC core of the Southern Ocean region. Palaeoworld, v.17, pp.57-82.

SHARMA, G.K. TAKAHASHI, K. and DALAKOTI, V.S. (2004) Taxonomy and distribution of Pleistocene Radiolarians from the Tasman region. *N. Jah. Geol. Palaon.*, v.231, pp.297-347.

SHARMA, G.K, TAKAHASHI, K. and DALAKOTI, V.S. (2006) Biostratigraphy and distribution of Pleistocene Radiolaria from the Tasman region. Acta Micropal. Sinica, v.23(1), pp.31-50.

STEPANJANTS, S.D., CORTESE, G., KRUGLIKOVA, S.B. and BJORKLUND, K.R. (2006) A review of bipolarity concept: history and examples from Radiolaria and Medusozoa (Cnidaria). Mar. Bio. Res., v.2, pp.200-241.

THEYER, F., MATO, C.Y. and HAMMOND, S.R. (1978) Paleomagnetic and geochronologic calibration of latest Oligocene to Pliocene radiolarian events, Equatorial Pacific. Mar. Micropal., v.3, pp.377-95.

WEAVER, F.M. (1973) Pliocene paleoclimatic and paleoglacial history of East Antarctica recorded in deep-sea piston cores. Florida State Univ., Sediment Res. Lab. Contri., v.36, p.142 p., pl. 1, text-fig. 1-21.

WEAVER, F.M. (1976a) Antarctic radiolaria from the southern Pacific Basin, DSDP, Leg 35. *In*: Initial Reports of Deep Sea Drilling Project, (HOLLISTER, C.D. and CRADDOCK, C. (Eds.) U.S. Govt. Printing Office, Washington, v.35, pp.569-603.

WEAVER, F.M. (1976b) Late Miocene to Pliocene radiolarian paleobiogeography and biostratigraphy of the Southern Ocean (Ph.D. dissertation) Florida State University.

WEAVER, F.M. (1983) Cenozoic radiolarians from the Southwest Atlantic, Falkland Plateau region, DSDP leg 71. *In*: Initial Reports of Deep Sea Drilling Project, LUDWIG, W.J. and KRASHENNINIKOV, V.A. (Eds.), U.S. Govt. Printing Office, Washington, v.71, pp.667-686.

Micropaleontology and its Applications
Pradeep K. Kathal, Rajiv Nigam & Abu Talib (Editors)
Scientific Publishers (India), 2018, 249-264 pp.

A Comparative Study of Foraminifera and its Ecological Significance Between Arasalar and Thirumalairajanar Estuaries in Karaikal, East Coast of India

M. Suresh Gandhi[1*], M. Manjima[1], P. Jeshma[1], R. Tharun[1] and N. Rajeshwara Rao[2]

Department of Geology, University of Madras, Guindy Campus, Chennai - 600 025, India
Department of Applied Geology, University of Madras, Guindy Campus, Chennai- 600 025, India
***Email:** msureshgandhi@gmail.com

Abstract

The present study, carried out in Arasalar and Thirumalairajanar estuaries, Karaikal region, Nagapattinam and Thiruvarurdistricts, Tamil Nadu, India, has revealed rich living populations of Foraminifera dominated by calcareous species from Arasalarand Thirumalairajanar estuaries.

Relation of the organic matter contents, calcium content and the rate of sedimentation with populations of the foraminiferainin the estuarine sediments has been worked out. The source and maturity status of the organic matter in the sediments have been worked out.The diversity of the species has been correlated to the presence of high amount of fine sand and clay in the sediment composition.

The relation of trace element pollution has been compared with the morphological deformities/broken tests of Foraminifera in Thirumalairajanar estuary.

Keywords: Benthic Foraminifera, organic matter, trace element geochemistry, pollution, estuarine environment, southeast coast, India.

INTRODUCTION

Microorganisms, like Foraminifera are ideally suited to environmental studies because their sensitivity to environmental change and the short duration of life allow them to respond rapidly to environmental changes. Their fossil-counterparts also provide important clues in global environmental changes on long-time scales, particularly from deep-sea sediments. Across vast area of the ocean floor, the shells of planktonic micro-organisms sinking from surface waters provide the dominant source of sediment, and they continuously accumulate. Study of changes in assemblages of microfossils and of change in their shell chemistry (e.g., oxygen isotope composition) is fundamental criteria in the identification climate changes in the geological past. Being the good ecological indicators, the Foraminifera are most widely used to understanding ecology, palaeoclimate and oceanography, while as fossils they are used in the study of biostratigraphy, age dating and correlation of sediments and palaeoenvironmental interpretation, palaeotemperature, glaciation and other palaeogeographic features (Murray, 2006). Their good preservation and fossilization potential is extremely useful in pollution studies all over the world (Saraswat et al. 2004). Foraminiferal ecosystem in relation to coastal and estuarine Sediments of Kovalam-Tuticorin, South India was studied by Jayaraju and Reddy (1995). Rao and Balasubramanian (1996) studied the distribution of foraminifera in the Cochin estuary. A study by Nigam et al. (2005) revealed drastic fall in total foraminiferal number in the lower reaches of the Mandovi River estuary probably due to mining activities in the catchment area of the river, was suggested as the plausible reason for the decline of fauna.Ghosh et al. (2009) have studied the distribution of foraminifera in the Narmada and Tapti estuaries of the Gulf in order to use these as analogues for the study of palaeo-macro-tidal estuarine environments and as a means of recording the extent of sea-level change in estuarine settings. Ghosh (2012) has discussed about the systematics and distribution of the foraminiferal species in the macrotidal estuaries of the Gulf of Cambay. The aim of the present work is to gain a better comparative study of two estuarine sediment samples in Thirumalairajanar and Arasalar estuaries, south coast of Tamil Nadu to delineate the environmental conditions.

STUDY AREA

The Thirumalairajanar and Arasalar estuaries covered in the present study fall in the Karaikal region in Nagapattinam and Thiruvarur Districts, Tamil Nadu, India (Fig.1), and located about 140 kms south of Pondicherry and 300 kms south of Chennai along the Coromandel Coast of the Bay of Bengal. The region is drained by seven distributaries of Cauvery viz. Nandalar, Vettar, Vanjiyar, Nular, Arasalar, Tirumalairajanar and Puravadaiyanar and all these rivers confluence into the Bay of Bengal.

The Arasalar, a tributary of the river Cauvery, having a total run of 24 km, enters Karaikal region, a little east of Akalanganni. Thirumalairajanar flows in the Thanjavur, Thiruvarur and Nagapattinam districts of Tamil Nadu and splits from Kudamuruthi, a tributary of Cauveri River at Rajagiri near Papanasam, Thanjavur district and enters the Bay of Bengal at Tirumalairayam-pathinam near Karaikal. The estuary is located at latitude between the latitude N 10°52′ to

10°53′ and longitude E 79°48′ to 79°51′. It is limited on the north by the Nandalar and on the south-east by the Vettar. The separation between these two estuaries is less than 4 km (Fig. 1).

METHODOLOGY

The samples, both surface sediment and water samples were collected from Arasalar and Thirumalairajanar estuaries at 24 sites (12 surface water, 12 sediment samples in each estuary) during the month of September 2014 using the Global Positioning System (GPS). Surface water samples were collected using Aqua trap water sampler in plastic bottles, which were pre-cleaned with 1 N HCl and the samples were kept in icebox onboard. 2 ml of concentrated HNO_3 was added to each sample (1 litre), which was used for trace element studies. The values of pH were determined onboard immediately after collection using calibrated thermometer with a resolution of 0.1 and Elico portable water quality analyzer, respectively.

Each sediment samples was washes (2 minutes) and sieved (ASTM 230 mesh) in order to remove the suspended impurities as well as the fine particles and kept immersed in a solution of rose Bengal for at least 6 hours (Walton, 1952). The foraminiferal tests were separated using standard methods of separation. Loeblich and Tappan (1987) are widely utilized for generic classification of foraminifera.

RESULTS AND DISCUSSION

In Arasalar estuary, a total of 38 benthic foraminiferal species belonging to 21 genera, 10 super-families, 12 families, 16 sub-families and 5 sub-orders are identified. In Thirumalairajanar estuary, a total of 22 benthic foraminiferal species belonging to 14 genera 7 super-families, 8 families, 9 sub-families and 3 sub-orders are identified (Table 1).

Total foraminifera in each estuary

In Arasalar estuary, 38 foraminiferal species belonging to 21 genera have been identified. Samples from Location 10 and 11 yielded more while sample number 3 and 5 yielded lesser number of species. The species are almost equally distributed in all other locations. *Ammonia beccarii, Ammonia dentata, Ammonia tepida, Elphidium crispum, Quinqueloculina seminulum,* and *Nonionides elongatum* are noticed dominantly in the Arasalar estuary. Foraminifera are abundant at location 11 near the seaward side and low at location 5. The Miliolina and Rotaliina are the dominant suborders in the study area. The majority of the fauna is found to be dead and living specimens are rare (living species recorded in each estuary is 96 and 65 only). The total foraminiferal specimens (living and dead, Walton, 1952) from the samples collected from the study area ranges from 210 to 1008.

In Thirumalairajanar estuary, 22 foraminiferal species belonging to 14 genera have been identified, which are equally distributed throughout the estuary. The benthic foraminiferal species *Ammonia beccarii, A. dentata, A. tepida, Nonionides elongatum,* and *Cribromilioline lamilletti* had been observed to be widespread and abundant during the collection. Foraminifera are abundant at location 2 and shows a low concentration at location 6. The suborders Miliolina and Rotaliina are dominant in the study area. The total foraminiferal specimens (living + dead) from the samples collected from the study area ranges from 214 to 2592.

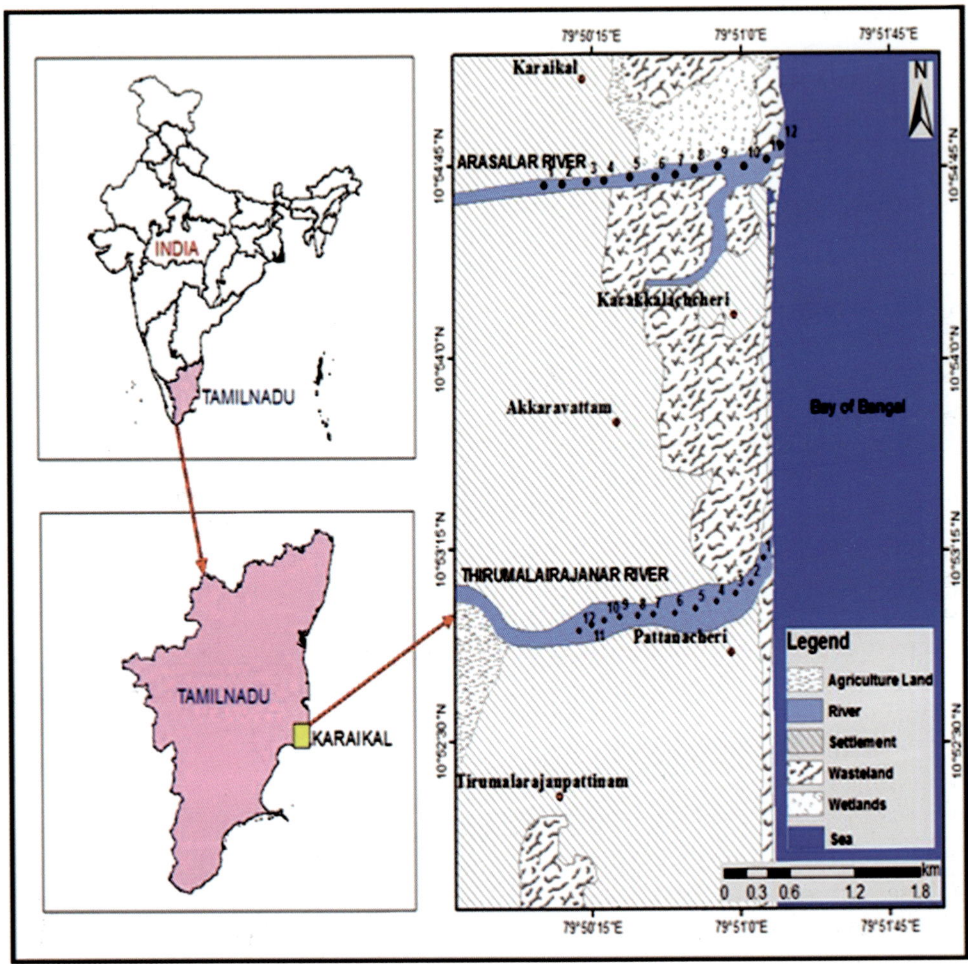

Fig.1. Map of the study area.

Organic Matter

Generally, organic carbon is found to be enriched in fine sediment (Paropkari, 1979; Demaison and Moore, 1980). In Arasalar estuary, the organic matter content ranges from 0.2% to 3.3% (Table 2). The lowest value was recorded at location 12 near the river mouth and highest value was recorded in the middle part of the study area (atlocation 5).

In Thirumalairajanar estuary, the organic matter content ranges from 0.2% to 2.4% (Table 2). The highest value was recorded atlocation 3, and lowest value was recorded atlocation 7. The abundance of organic matter depicts that the sediments are immature and appears to have been derived from marine sources. In addition to the above, other controlling factor may be rate of sedimentation, which is variable in different areas of the river. The high inorganic sedimentation will dilute the environment of organic matter in sediments and increase of organic matter with decreasing the grain size has been reported from many areas.

Table 1. Checklist of the foraminifera recorded in the Arasalar and Thirumalairajanar estuaries.

S.No.	Arasalar estuary	Thirumalairajanar estuary
1	*Textularia inflate*	*Ammobaculites agglutinans*
2	*Ammobaculites laevigata*	*A. globigeriniforms*
3	*A. intricate*	*A. elegans*
4	*Spiroloculina orbis*	*Spiroloculina orbis*
5	*S. communis*	*S. henbesti*
6	*S. nitida*	*Quinqueloculina lamarckiana*
7	*Quinqueloculina lamarckiana*	*Q. seminula*
8	*Q. seminulum*	*Q. elongate*
9	*Q.* sp.1	*Q. oblonga*
10	*Q.* sp.2	*Q. patagonica*
11	*Triloculina schreiberiana*	*Cymbaloporeta milletti*
12	*Legina perlucida*	*Triloculina tricarinata*
13	*Fissurina cucullata*	*H. percucida*
14	*F. laevigata*	*Cibicides refulgens*
15	*Globigerina bulloides*	*Hynessina depressula*
16	*G. sacculifer*	*Nonion commune*
17	*B.* sp.1	*N. elongatum*
18	*Bolivina marginata*	*Ammonia beccarii*
19	*S. proboscidea*	*A. dentate*
20	*R. globularis*	*A. tepida*
21	*Amphistegina radiata*	*A. convexa*
22	*Nonion elongatum*	*Pararotaliua schroeteriana*
23	*N. grateloupi*	
24	*N. scaphum*	
25	*N.* sp. 1	
26	*N.* sp.1	
27	*Pararotalia calcar*	
28	*P. nipponica*	
29	*Ammonia beccarii*	
30	*A. dentata*	
31	*A. tepida*	
32	*A. inflate*	
33	*A. trispinosa*	
34	*Cibicides bradyi*	
35	*Elphidium crispum*	
36	*E. excavatum*	
37	*E. hispidulum*	
38	*E. norvangi*	

Hydrogen Ion Concentration (pH)

It influences the solubility of mineral matter and other anthropogenic pollutants. Arasalar and Thirumalairajanar estuarine waters are characterized by the pH ranges from 8.5 to 8.8 with an average of value 8.7 (Table 2). The pH of these waters indicates that the waters are alkaline reflecting the dominance of marine environment.

Salinity

The salinity of the Arasalar estuary varies from 22.12 to 32.95 ppt with an average of 30.58 ppt. The highest value of salinity is observed in location 9 and lowest value in location 4 (Table 2). In Thirumalairajanar estuary, the value ranges from 32.30 to 34.31 ppt with an average of 33.34 ppt (Table 2). In both these estuaries, the salinity is found to be increasing towards the Bay of Bengal which is a clear indication of dominance of saline water

Calcium Carbonate

The calcium carbonate content of Arasalar estuary ranges from 0.8% to 3.3% with an average value of 2.3% (Table 2). The low values of carbonate in this area attribute to the nature of the substratum. The fine particles, which dominate the river and do not support biological activity, resulting in low carbonate values. Riverine mixing processes such as sudden fluctuations in salinity, oxygen content and temperature are possible due to horizontal and vertical mixing of water. These processes are also capable of lowering biogenic activity and hence result in low carbonate values. The eastern and western parts of the study area have high $CaCO_3$ and middle part of the study area has low $CaCO_3$.

In Thirumairajanar estuary, the $CaCO_3$, ranges from 0.6% to 3.0% and with an average value of 2.4% (Table 2). The higher content of $CaCO_3$ is found in the middle part of river which may be due to turbidity current/gravity flow or from the influx of calcareous biogenic materials. Whereas, the lower content in the riverine side and sea side should have been due to the emplacement of siliceous sand under gravity flow activity.

Substrate

Estimated sand content in Arasalar estuarine sediment sample ranges from 6.8 to 88.2% with an average of 52.7%; the silt content ranges from 11.4 to 92.8% with an average of 46.8% and the clay percentage ranges from 0.2 to 1.0% with an average of 0.5% (Table 2). In Thirumalairajanar estuary, percentage of sand content ranges from 6.4% to 89.4% with an average of 74.9%; the silt content ranges from 10.4% to 93% with an average of 24.6%; and clay percentage ranges from 0.2% to 0.8% with an average of 0.4% (Table 2). It indicates that the most of the sediments are sand and silty sand in nature.

Total Foraminiferal Number (TFN)

The TFN denotes the total number of Recent benthic foraminiferal specimens recovered from 1gm sediment. Table 4 illustrates the spatial variation of TFN in the study area. In the case of Arasalar estuary, the TFN increases towards both eastern and western side. Location 11 and 2 noticed the higher densities of TFN. The values vary greatly from a minimum of 210 foraminiferal species at location 5 to a maximum of 1008 at location 11. However, only sample Location 11 has got a TFN more than 1,000.In the Thirumalirajanar estuary, the TFN increases towards the seaward side of the study area. The highest density is found at location 1. The values vary greatly from a minimum of 214 foraminiferal species at location 6 to a maximum of 2592 at location 2. Five locations (Location 1,3,4,5 and 11) had TFN more than 1,000 of which Location 2 had more than 2,000 TFN. At the station no 12, the total distribution is lesser in amount (177 nos.) may be due to the development of sand spits across the river mouths from south to north direction indicated that the littoral drift predominantly moves from south to north direction in the study area.

Living/Dead Ratio

In Arasalar estuary, the maximum ratio observed is 15.82% in location 5 where in Thirumalairajanar estuary, the maximum ratio is 16.50 observed in location 9 (Table 3). The higher values of living/dead ratio may be attributed to the higher rate of sedimentation.

COMPARISON BETWEEN ARASALAR AND THIRUMALAIRAJANAR ESTURINE FORAMINIFERAL DIVERSITY

Considerable differences in foraminiferal assemblages were observed in Arasalar and Thirumalairajanar estuaries. High densities and diversities were observed in all the sites of the Arasalar estuary when compared to the Thirumalairajanar estuary. In the present study, a total of 60 species of foraminifera were recorded from both estuaries. Out of which 38 species of foraminifera, belonging to 21 genera, 12 families and 16 sub-orders were recorded from Arasalar estuary, and 22 species of foraminifera were observed from the sediment samples of Thirumalairajanar estuary. They belonged to 14 genera, 8 families and 3 sub-orders. Seven species were common in both estuaries. Thirty one and fifteen species were found to be unique in Arasalar and Thirumalairajanar estuaries respectively. Foraminiferal abundance were also less in Arasalar estuary compared to Thirumalairajanar estuary and but species diversity is high in Arasalar estuary. The species belonging to sub-orders Lagena and Globigerina are only seen in Arasalar estuary. The suborders Textulariina, Miliolona and Rotaliina show higher abundance in Thirumalairajanar estuary as compared to Arasalar estuary. Among the living foraminiferal population in both the estuaries, calcareous forms are found to predominant. In both the area living foraminiferal population belonging to the genus Ammonia (especially *A. beccarii, A. tepida* and *A. dentata*) are found to be predominant. In Arasalar estuary living/dead ratio is higher in middle part of the estuary and in Thirumalairajanar estuary, living/dead ratio is higher in eastern side.

DISTRIBUTION OF TRACE ELEMENTS IN SEDIMENT SAMPLES

Trace elements in sediments are normally derived from the weathering of source rocks and are introduced into the marine environment by various geological processes. The geochemical analysis of the trace elements such as Fe, Mn, Cr, Cu, Ni, Co, Pb and Zn, has been carried with the help of Atomic Absorption Spectroscopy (AAS) (Perkin Elmer AA700) facility available in the Department of Applied Geology, University of Madras.

Table 2. Estimated values of SSC, $CaCo_3$, OM, pH and salinity, Arasalar and Thirumalairajanar estuaries.

Sl. No.	Sand	Silt	Cl-ay	Mud	Ca Co_3	OM	pH	Sali-nity	Sediment Type	Sand	Silt	Mud	Clay	Ca Co_3	OM	Sediment Type
				Arasalar Estuary							Thirumalairajanar Estuary					
1	79.8	19.8	0.4	20.2	3.2	0.7	8.7	30.25	Silty Sand	75.8	23.6	24.2	0.6	2.1	1.5	Silty Sand
2	45.8	53.2	1.0	54.2	3.3	0.5	8.8	31.51	Sandy Silt	89.4	10.4	10.6	0.2	2.9	0.8	Sand
3	87.2	12.4	0.4	12.8	3.0	0.3	8.7	30.25	Sand	88.0	11.8	12.0	0.2	2.8	0.2	Sand
4	12.4	87.0	0.6	87.6	1.2	2.5	8.5	22.12	Silt	78.4	21.2	21.6	0.4	2.8	1.0	Silty Sand
5	6.8	92.8	0.4	93.2	0.8	3.3	8.7	29.34	Silt	89.0	10.4	11.0	0.6	2.9	1.1	Sand
6	11.4	88.2	0.4	88.6	1.1	2.7	8.8	32.05	Silt	86.8	13.0	13.2	0.2	3.0	0.7	Sand
7	81.8	18.0	0.2	18.2	3.2	0.8	8.7	32.05	Sand	6.4	93.0	93.6	0.6	0.6	2.4	Silt
8	66.0	33.4	0.6	34.0	1.9	1.3	8.8	30.25	Silty Sand	77.4	22.0	22.6	0.6	2.4	0.6	Silty Sand
9	49.2	50.4	0.4	50.8	2.6	1.2	8.8	32.95	Sandy Silt	76.4	23.0	23.6	0.6	2.5	0.5	Silty Sand
10	66.8	32.8	0.4	33.2	3.0	1.3	8.7	32.05	Silty Sand	78.6	21.0	21.4	0.4	2.6	0.8	Silty Sand
11	37.2	62.4	0.4	62.8	1.2	1.6	8.8	32.05	Sandy Silt	81.4	17.8	18.6	0.8	2.4	1.2	Sand
12	88.2	11.4	0.4	11.8	2.8	0.2	8.8	32.05	Sand	71.6	28.0	28.4	0.4	1.8	1.3	Silty Sand
Min.	**52.7**	**46.8**	**0.5**	**47.3**	**2.3**	**1.4**	**8.5**	**22.12**		**74.9**	**24.6**	**25.1**	**0.5**	**2.4**	**1.0**	
Max.	**6.8**	**11.4**	**0.2**	**11.8**	**0.8**	**0.2**	**8.8**	**32.95**		**6.4**	**10.4**	**10.6**	**0.2**	**0.6**	**0.2**	
Avg.	**88.2**	**92.8**	**1.0**	**93.2**	**3.3**	**3.3**	**8.7**	**30.58**		**89.4**	**93.0**	**93.6**	**0.8**	**3.0**	**2.4**	

Table 3. Living/dead foraminiferal distribution in Arasalar and Thirumalairajanar estuaries.

Location	Living	Dead	Living sp. in %	Dead sp. in %	Living/ Dead Ratio	Living	Dead	Living sp. in %	Dead sp. in %	Living/ dead ratio
		Arasalar Estuary					Thirumalairajanar Estuary			
1	18	380	4.5	7.27	4.7	54	1376	3.8	14.83	3.9
2	40	565	6.6	10.80	7.1	65	1937	3.2	20.87	3.4
3	12	359	3.2	6.86	3.3	25	1003	2.4	10.81	2.5
4	19	527	3.2	10.08	3.6	31	728	4.1	7.84	4.3
5	96	114	3.2	2.18	14	50	913	5.2	9.84	5.5
6	18	474	3.2	9.06	3.8	24	130	15.6	1.40	18.5
7	19	237	7.4	4.53	8	18	655	2.7	7.06	2.7
8	11	548	1.9	10.48	2	15	430	3.4	4.62	3.5
9	11	666	1.6	12.73	1.7	29	315	8.4	3.39	9.2
10	14	544	2.5	10.40	2.6	30	651	4.4	7.02	4.6
11	17	837	1.9	16.00	2	43	639	6.3	6.89	6.7
12	58	119	13	2.28	15	18	523	3.3	5.62	3.4
Total	**213**	**5230**	**52.2**	**100**	**67.8**	**402**	**9300**	**62.8**	**100**	**68.2**
Max.	**96**	**837**	**7.4**	**16.00**	**14**	**65**	**1937**	**15.6**	**20.87**	**18.5**
Min.	**11**	**114**	**1.6**	**2.18**	**1.7**	**15**	**130**	**2.4**	**1.40**	**2.5**
Avg.	**17.8**	**447.5**	**4.35**	**8.56**	**5.7**	**33.5**	**775**	**5.2**	**8.35**	**5.7**

Table 4. Trace elements concentration in Arasalar and Thirumalairajanar estuarine water samples.

S. No.	Na	K	Ca	Mg	Cl	Fe	Mn	Cu	Ni	Co	Pb	Zn	Na	K	Ca	Mg	Cl	Fe	Mn	Cu	Ni	Co	Pb	Zn
	Arasalar Estuary												Thirumalairajanar Estuary											
	mg/L					µg/L							mg/L					µg/L						
1	9640	3200	144	652	12410	3.06	0.92	0.22	1.45	1.99	1.69	0.77	10290	468	88	892	18500	4.53	1.27	0.10	0.44	1.04	3.76	0.73
2	10290	3640	189	720	18720	2.74	0.62	0.10	0.89	1.02	3.53	2.37	10220	466	94	886	18000	4.36	1.63	0.12	0.51	1.29	3.52	0.63
3	10090	3400	210	743	15800	2.41	0.33	0.08	0.80	1.92	4.14	2.78	10666	492	120	881	18700	6.41	1.77	0.09	0.60	0.64	3.98	2.64
4	10180	3650	88	606	18110	5.43	1.33	0.13	1.21	1.57	4.05	2.52	10374	469	42	869	18500	6.12	1.44	0.10	0.44	1.26	3.88	1.08
5	10580	4053	148	646	20500	8.56	2.16	0.09	1.15	1.10	3.79	2.21	10480	467	134	879	18100	6.50	1.72	0.20	0.87	2.20	4.53	2.82
6	10480	3767	106	623	19020	5.56	1.67	0.09	1.01	1.33	4.35	2.80	10470	465	182	903	18500	4.16	1.35	0.12	0.33	0.75	3.80	0.40
7	10470	3723	174	685	19020	6.23	1.41	0.09	1.13	1.99	4.06	2.15	10325	450	110	899	18000	2.86	0.64	0.11	0.55	0.32	3.99	2.01
8	10480	3873	152	653	20740	5.74	1.17	0.06	1.02	1.99	3.86	2.73	10580	480	74	870	18500	3.60	1.28	0.10	0.67	0.66	3.72	0.56
9	10170	3705	204	725	19520	5.58	1.49	0.02	0.81	1.52	4.20	2.97	9840	440	90	881	17600	2.63	0.75	0.10	0.55	0.77	3.48	0.62
10	10180	3628	176	694	19020	5.30	0.92	0.06	0.94	1.64	3.91	1.95	10459	460	224	913	18400	2.63	0.89	0.09	0.29	0.85	3.91	0.68
11	10220	3705	174	698	19020	5.29	0.75	0.03	0.58	1.65	3.67	1.66	10051	450	114	875	17600	4.33	0.92	0.16	1.01	1.56	4.34	3.30
12	10290	3800	187	711	19020	3.48	1.19	0.02	0.68	1.04	3.70	1.68	9640	442	144	882	17600	9.72	2.23	0.22	1.00	1.98	4.09	1.63
Min	**9640**	**3200**	**88**	**606**	**12410**	**2.41**	**0.33**	**0.02**	**0.58**	**1.02**	**1.69**	**0.77**	**9640**	**440**	**42**	**869**	**17600**	**2.63**	**0.64**	**0.09**	**0.29**	**0.32**	**3.48**	**0.40**
Max	**10580**	**4053**	**210**	**743**	**20740**	**8.56**	**2.16**	**0.22**	**1.45**	**1.99**	**4.35**	**2.97**	**10666**	**492**	**224**	**913**	**18700**	**9.72**	**2.23**	**0.22**	**1.01**	**2.20**	**4.53**	**3.30**
Avg	**10256**	**3679**	**163**	**680**	**18408**	**4.95**	**1.16**	**0.08**	**0.97**	**1.56**	**3.75**	**2.22**	**10266**	**462**	**125**	**886**	**18150**	**5.11**	**1.32**	**0.13**	**0.63**	**1.13**	**3.98**	**1.62**

Table 5. Trace element concentrations in Arasalar and Thirumalirajanar estuarine sediment samples.

S. No.	Fe	Mn	Cr	Cu	Pb	Zn	Co	Ni	Fe	Mn	Cr	Cu	Pb	Zn	Co	Ni
	Arasalar Estuary								Thirumalairajanar Estuary							
	µg/g								µg/g							
1	31000	394	32.0	20.9	45.8	56.9	48.2	73.5	41623	563	53.2	67.1	53.2	60.1	92.8	62.5
2	38920	351	53.2	29.7	61.2	67.3	65.4	86.8	35300	552	41.8	49.3	46.1	43.5	67.3	54.9
3	28900	275	31.7	18.3	53.6	63.0	65.3	84.0	26000	391	30.0	28.6	20.9	19.5	75.6	33.0
4	66800	678	62.0	32.6	70.2	77.1	48.3	16.7	44863	476	57.1	58.2	64.5	35.1	69.1	48.5
5	70000	694	54.7	37.0	55.9	55.2	39.6	66.0	51729	568	51.8	65.3	58.3	48.5	59.4	66.4
6	60600	624	47.5	35.1	64.1	64.7	58.5	81.2	38300	490	38.3	41.1	45.9	47.3	68.5	44.2
7	28000	355	29.8	18.6	69.7	83.9	45.0	51.8	68000	776	83.0	79.3	92.3	66.4	83.4	74.8
8	44900	490	51.5	33.7	58.4	61.2	30.5	30.3	36800	440	26.2	41.5	17.8	31.0	88.7	46.6
9	62453	511	62.6	31.1	49.3	44.6	87.6	95.7	27100	483	31.8	31.7	11.5	17.0	69.5	38.3
10	34400	478	39.8	30.5	60.4	52.1	77.2	84.4	41400	542	20.1	40.7	30.9	40.0	76.5	52.5
11	48576	564	43.9	36.4	68.5	76.3	78.1	94.1	49376	513	35.7	46.2	36.1	40.2	65.2	58.3
12	25000	317	22.0	18.4	59.9	65.4	45.7	27.9	42756	607	44.2	52.8	77.4	55.7	70.4	64.5
Min.	**25000**	**275**	**22.0**	**18.3**	**45.8**	**44.6**	**30.5**	**16.7**	**26000**	**391**	**20.1**	**28.6**	**11.5**	**17.0**	**59.4**	**33.0**
Max.	**70000**	**694**	**62.6**	**37.0**	**70.2**	**83.9**	**87.6**	**95.7**	**68000**	**776**	**83.0**	**79.3**	**92.3**	**66.4**	**92.8**	**74.8**
Avg.	**44962**	**478**	**44.2**	**28.5**	**59.8**	**64.0**	**57.5**	**66.0**	**41937**	**533**	**42.8**	**50.2**	**46.2**	**42.0**	**73.9**	**53.7**

Enrichment Factor analysis of Arasalar and Thirumalairajanar estuaries shows that Co and Pb contamination is found to be extremely severe in most of the locations (Tables 4 & 5). The sediments of these locations derive these contaminants probably from terrestrial surface flows. In Arasalar estuary, enrichment factor with regards to Mn, Cr, Cu, and Zn shows no enrichment, Ni shows minor enrichment, Pb and Co has minor and minor to moderate enrichment respectively. In Thirumalairajanar estuary, enrichment factor with regards to Mn, Cr and Zn shows no enrichment, Ni and Cu shows minor enrichment, Pb and Co has minor to moderate and moderate enrichment respectively (Table 6). Elements Cu displays quite similar pattern of distribution, these elements are used as markers of metal industries (Kumar et al. 2001; Loska et al. 2004). These high levels of the enrichment factors might be related to the anthropogenic effect from several sources (Nasr et al. 2006). The Cr and Zn show low enrichment in Arasalar and Thirumalairajanar estuary.

METHODS FOR ESTIMATING POLLUTION IMPACT

A number of calculation methods (Ridgway and Shimmield, 2002) have been put forward for quantifying the degree of metal enrichment in sediments. Various authors (Salomons and Förstner 1984; Müller 1969; Hakanson 1980) have proposed pollution impact scales (or ranges) to convert the calculated numerical results into broad descriptive bands of pollution ranging from low to high intensity.

Enrichment Factor (EF)

A common approach to estimating the anthropogenic impact on sediments is to calculate a normalized enrichment factor (EF) for metal concentrations above uncontaminated background levels. The enrichment factor (EF) (Szefer et al. 1998) permits to calculate the heavy metal contamination and it was calculated according to the following formula:

$$EF = \frac{C_{sample} / Al_{sample}}{C_{crust}/Al_{crust}}$$

C_{sample} - trace element concentration in the sample;

C_{crust} - trace element concentration in the continental crust (Taylor and McLeanan, 1995);

Al_{sample} - Al content in the sample;

Al_{crust} - Al content in the continental crust (Taylor and McLeanaen, 1995).

Al has been chosen as normalization element because of its origin being exclusively lithospheric. EF values were interpreted according to Acevedo-Figueroa et al. (2006) where EF <1 indicates no enrichment; 1~3 is minor; 3~5 is moderate; 5~10 is moderately severe; 10~25 is severe; 25~50 is very severe; and >50 is extremely severe.

Contamination Factor (CF)

Sediments are used as environmental markers and this ability to trace the contamination sources and to monitor the contaminants is well recognized. Thus, the accumulation of metals in the sediments is strongly controlled by the nature of the substrate as well as the physico-chemical conditions controlling dissolution and precipitation (Muthuraj and Jayaprakash, 2007). The trace metals themselves may have a synergistic or antagonistic effect on the environment and the level of contamination is expressed by CF (Pekey et al. 2004; Savvides et al. 1995).

Table 6. Enrichment factor of surface sediments of Arasalar and Thirumalairajanar estuary.

S. No.	Mn	Cr	Cu	Pb	Zn	Co	Ni	Mn	Cr	Cu	Pb	Zn	Co	Ni
	Arasalar Estuary							**Thirumalairajanar Estuary**						
1	0.46	0.36	0.46	2.29	0.60	2.54	1.08	0.66	0.59	1.49	2.66	0.63	4.88	0.92
2	0.41	0.59	0.66	3.06	0.71	3.44	1.28	0.65	0.46	1.10	2.31	0.46	3.54	0.81
3	0.32	0.35	0.41	2.68	0.66	3.44	1.24	0.46	0.33	0.64	1.05	0.21	3.98	0.49
4	0.80	0.69	0.72	3.51	0.81	2.54	0.25	0.56	0.63	1.29	3.23	0.37	3.64	0.71
5	0.82	0.61	0.82	2.80	0.58	2.08	0.97	0.67	0.58	1.45	2.92	0.51	3.13	0.98
6	0.73	0.53	0.78	3.21	0.68	3.08	1.19	0.58	0.43	0.91	2.30	0.50	3.61	0.65
7	0.42	0.33	0.41	3.49	0.88	2.37	0.76	0.91	0.92	1.76	4.62	0.70	4.39	1.10
8	0.58	0.57	0.75	2.92	0.64	1.61	0.45	0.52	0.29	0.92	0.89	0.33	4.67	0.69
9	0.60	0.70	0.69	2.47	0.47	4.61	1.41	0.57	0.35	0.70	0.58	0.18	3.66	0.56
10	0.56	0.44	0.68	3.02	0.55	4.06	1.24	0.64	0.22	0.90	1.55	0.42	4.03	0.77
11	0.66	0.49	0.81	3.43	0.80	4.11	1.38	0.60	0.40	1.03	1.81	0.42	3.43	0.86
12	0.37	0.24	0.41	3.00	0.69	2.41	0.41	0.71	0.49	1.17	3.87	0.59	3.71	0.95
Min.	**0.32**	**0.24**	**0.41**	**2.29**	**0.47**	**1.61**	**0.25**	**0.46**	**0.22**	**0.64**	**0.58**	**0.18**	**3.13**	**0.49**
Max.	**0.82**	**0.70**	**0.82**	**3.51**	**0.88**	**4.61**	**1.41**	**0.91**	**0.92**	**1.76**	**4.62**	**0.70**	**4.88**	**1.10**
Avg.	**0.56**	**0.49**	**0.63**	**2.98**	**0.67**	**3.04**	**0.95**	**0.64**	**0.49**	**1.13**	**2.35**	**0.44**	**3.90**	**0.79**

Table 7. Contamination factor of surface sediments of Arasalar and Thirumalairajanar estuary.

S. No	Fe	Mn	Cr	Cu	Pb	Zn	Co	Ni	Fe	Mn	Cr	Cu	Pb	Zn	Co	Ni
	Arasalar Estuary								**Thirumalairajanar Estuary**							
1	0.66	0.46	0.36	0.46	2.29	0.60	2.54	1.08	0.88	0.66	0.59	1.49	2.66	0.63	4.88	0.92
2	0.82	0.41	0.59	0.66	3.06	0.71	3.44	1.28	0.75	0.65	0.46	1.10	2.31	0.46	3.54	0.81
3	0.61	0.32	0.35	0.41	2.68	0.66	3.44	1.24	0.55	0.46	0.33	0.64	1.05	0.21	3.98	0.49
4	1.42	0.80	0.69	0.72	3.51	0.81	2.54	0.25	0.95	0.56	0.63	1.29	3.23	0.37	3.64	0.71
5	1.48	0.82	0.61	0.82	2.80	0.58	2.08	0.97	1.10	0.67	0.58	1.45	2.92	0.51	3.13	0.98
6	1.28	0.73	0.53	0.78	3.21	0.68	3.08	1.19	0.81	0.58	0.43	0.91	2.30	0.50	3.61	0.65
7	0.59	0.42	0.33	0.41	3.49	0.88	2.37	0.76	1.44	0.91	0.92	1.76	4.62	0.70	4.39	1.10
8	0.95	0.58	0.57	0.75	2.92	0.64	1.61	0.45	0.78	0.52	0.29	0.92	0.89	0.33	4.67	0.69
9	1.32	0.60	0.70	0.69	2.47	0.47	4.61	1.41	0.57	0.57	0.35	0.70	0.58	0.18	3.66	0.56
10	0.73	0.56	0.44	0.68	3.02	0.55	4.06	1.24	0.88	0.64	0.22	0.90	1.55	0.42	4.03	0.77
11	1.03	0.66	0.49	0.81	3.43	0.80	4.11	1.38	1.05	0.60	0.40	1.03	1.81	0.42	3.43	0.86
12	0.53	0.37	0.24	0.41	3.00	0.69	2.41	0.41	0.91	0.71	0.49	1.17	3.87	0.59	3.71	0.95
Min.	**0.53**	**0.32**	**0.24**	**0.41**	**2.29**	**0.47**	**1.61**	**0.25**	**0.55**	**0.46**	**0.22**	**0.64**	**0.58**	**0.18**	**3.13**	**0.49**
Max.	**1.48**	**0.82**	**0.70**	**0.82**	**3.51**	**0.88**	**4.61**	**1.41**	**1.44**	**0.91**	**0.92**	**1.76**	**4.62**	**0.70**	**4.88**	**1.10**
Avg.	**0.96**	**0.56**	**0.49**	**0.63**	**2.98**	**0.67**	**3.04**	**0.95**	**0.90**	**0.64**	**0.49**	**1.13**	**2.35**	**0.44**	**3.90**	**0.79**

CF is calculated from the ratio of the sediment metal content at a given station to the normal concentration levels (NCLs); CF values indicate the enrichment of metals in the sediment.

$$CF = \frac{\text{Metal value in polluted sediment}}{\text{Background value of the metal}}$$

CF >1 illustrates that the sediment is contaminated with respect to that metal; if CF <1, there is no metal enrichment by natural or external inputs. CF of the sediments in the study area was derived using the world crustal average reported by Wedephol (1995) as background values. According to Pekay et al. (2004), the CF is classified into four groups:

1 ≤CF low contamination factor

1 ≤CF < 3 moderate contamination factor

3 ≤ CF< 6 considerable contamination factor

6 ≥ CF very high contamination factor

The values of CF in Arasalar and Thirumalairajanar estuaries are shown in Table 7. The contamination factor of the metals in the Arasalar estuary is found to be of the order of Co> Pb> Fe> Ni> Zn> Cu> Mn> Cr and in Thirumalairajanar estuary Co> Pb> Cu> Fe> Ni> Cr> Mn> Zn is found. In these two estuaries, Co and Pb show considerable contamination factor, suggesting that the contamination zone with regard to these elements is very high in the study area. This may be due the presence of suspended particles in aquatic environment bring more fossil fuel and garbage dumping. While, the contamination factor of (1< CF) is shown by Fe, Mn, Cr, Cu, Zn, and Ni and these elements are not contaminating the marshy sediments.

FACTOR ANALYSIS

Factor Analysis attempts to explain the correlations between the observations in terms of the underlying factors, which are not directly observable. Factor analysis is used to reduce the number of variables and to detect structure in the relationship between the variables. The complex hydrogeochemical data can be reduced to an easily interpretable form using factor analysis. In this work, R-mode factor analysis was chosen as it provides several positive features that allow interpretation of the data set more scientifically.

In order to evaluate the variables belonging to a specific chemical process and also to find out the dominance and contribution of the major elements in terms of the total data set, factor loadings, communalities, and Eigen values have been exanimate (Table 8 to 9).

TRACE ELEMENTS AND FORAMINIFERA CORRELATIONS

The distribution of foraminifera is also compared and correlated with trace elements based on this trace elements are unlikely to favour any particular species of benthic foraminifera. In practice, however, it is often difficult to separate effects caused by trace elements from those caused by organic material because most polluted areas are subjected to some kind of organic enrichment (Alve, 1995). Brocken species are noticed in some locations in Thirumalairajanar estuary, where the trace elements concentration is slightly higher in these regions this is also one of the reasons for the morphologic deformities abnormal study in this regions not only churning action and rate of sedimentation. Therefore, they can be used as environmental indicators in highly impacted areas, where other potential bio-indicators disappeared due to contamination. The high diversity of foraminiferal assemblages allowed a diversified response of species to different pollutants. The results from scientific literature indicate that benthic foraminifera are excellent indicators to monitoring the coastal environments.

Table 8. Factor analysis, trace elements, ecological parameters, foraminifera assemblages of Arasalar and Thirumalairajanar estuary.

Parameters	**Arasalar**			**Thirumalairajanar**		
	Component			**Component**		
	Factor 1	**Factor 2**	**Factor 3**	**Factor 1**	**Factor 2**	**Factor 3**
pH	0.37	0.62	-0.24	-0.28	-0.43	0.48
Salinity	0.25	0.65	-0.38	-0.75	0.54	-0.25
Na	**0.89**	-0.25	0.03	-0.67	0.49	-0.20
K	**0.92**	-0.28	0.01	-0.49	**0.81**	-0.17
Ca	0.11	**0.80**	-0.44	-0.04	-0.24	-0.40
Mg	0.07	**0.86**	-0.36	-0.43	-0.38	-0.03
Cl	**0.95**	-0.04	0.20	-0.75	0.55	-0.25
Fe	0.61	-0.56	0.14	0.67	0.51	-0.31
Mn	0.52	-0.71	-0.10	0.45	0.70	-0.24
Cu	-0.81	-0.49	0.04	**0.92**	0.03	-0.15
Ni	-0.51	-0.75	0.02	**0.91**	0.07	-0.12
Co	-0.45	.042	0.14	**0.83**	0.28	0.10
Pb	**0.80**	0.04	0.23	0.59	0.16	-0.30
Zn	0.60	-0.001	0.31	0.59	0.21	-0.13
Living	-0.31	0.17	0.47	0.19	0.36	**0.85**
Dead	-0.05	0.50	**0.82**	-0.00	0.58	**0.75**
Total pop.	-0.14	0.47	**0.84**	0.05	0.54	**0.80**
Total individuals	0.07	0.49	0.45	-0.08	0.59	-0.30

Table 9. Factor analysis for trace elements, textural and foraminiferal assemblages, Arasalar and Thirumalairajanar estuaries.

Parameters	**Arasalar**			**Thirumalairajanar**		
	Component			**Component**		
	Factor 1	**Factor 2**	**Factor 3**	**Factor 1**	**Factor 2**	**Factor 3**
Sand	-0.96	0.13	-0.05	-0.84	0.36	-0.26
Silt	**0.96**	-0.13	0.05	**0.84**	-0.36	0.26
Clay	0.24	0.41	0.43	0.40	-0.15	0.07
CaCO3	-0.84	0.39	0.02	-0.85	0.31	-0.33

Parameters	Arasalar Component			Thirumalairajanar Component		
	Factor 1	Factor 2	Factor 3	Factor 1	Factor 2	Factor 3
OM	**0.88**	-0.39	-0.11	**0.99**	0.03	0.00
Fe	**0.94**	-0.15	-0.19	**0.91**	-0.02	-0.25
Mn	**0.94**	-0.24	-0.12	**0.94**	0.06	-0.05
Cr	**0.85**	0.20	-0.06	**0.86**	0.13	-0.07
Cu	**0.94**	0.13	-0.08	**0.91**	0.28	-0.07
Pb	0.31	-0.08	**0.76**	**0.85**	0.13	-0.30
Zn	-0.03	-0.16	**0.88**	**0.86**	0.21	-0.13
Co	0.10	**0.79**	-0.33	0.26	-0.20	**0.84**
Ni	0.07	0.66	-0.43	**0.89**	0.26	-0.17
Living	0.13	0.54	0.61	0.004	**0.94**	0.07
Dead	0.42	**0.81**	0.05	0.05	**0.93**	0.32
Total pop.	0.39	**0.86**	025	0.04	**0.94**	0.26
Total individuals	0.08	0.61	-0.12	-0.68	0.09	0.10

CONCLUSION

In the study area various sediments characteristics affect the distribution and diversity of Foraminifera. Fine sand, clay, silt, organic matter and calcium carbonate of the sediments are the important determining factors of abundance and species richness of foraminifera. Several authors reported that sediments rich in fine sand and clay have higher density and diversity of foraminifera than the coarse grain sands. In the present study, very high abundance and diversity of foraminifera found at eastern part of the Arasalar estuary and eastern part of the Thirumalairajanar estuary are correlated to the presence of high amount of fine sand and clay in the sediment composition. Organic matter plays an important role in controlling the distribution of foraminifera. Calcium carbonate also influences the distribution of foraminifera. In the present study also, a direct relationship was observed between the calcium content and abundance and diversity of foraminifera. The species belonging to sub-orders Miliolina and Rotaliina were abundant on the east coast of these two estuaries, where calcium content was more. It is evident for the fine sand, clay, organic matter and calcium carbonate play an important role in controlling the diversity and distribution of foraminifera in each estuaries.

The geochemistry of water samples and sediment samples collected from the Arasalar and Thirumalairajanar estuaries in south east coast of India. Various analyses like Enrichment analysis, Contamination factor and various statistical analyses had been done. The results for enrichment factor shows indications of pollution. In view of the rapid urbanization and industrialization effluents in to these study regions the elevated concentrations are not surprising. The results also indicate that this region is more contaminated with Co and Pb. Based on the comparative levels of metals in the studied area to other areas, the levels of metals content are lower than in other areas. The extractable metals indicate that the sediments of these study area were relatively unpolluted. But compared to the Arasalar estuary, Thirumalairajanar estuary is slightly polluted with reference to some locations. The spatial distributions of trace metals are also controlled by other factors like geochemical, precipitation and flocculation of particulate substances in these regions.

Reference

ALVE, E. (1995) Benthic foraminiferal responses to estuarine pollution: a review. J. Foram. Res., v. 25, pp.190-203.

ACEVEDO-FIGUEROA, D., JIMENEZ, B. D. and RODRGUEZ-SIERRA, C. J. (2006) Trace metals in sediments of two estuarine lagoons from Puerto Rico. Environ. Pollution, v. 141, pp.336-342

BASAHAM, A.S. and EL-SAYED, M.A. (1998) Distribution and phase association of some major and trace elements in the Arabian Gulf sediments, Estuar. Coastal Shelf Sci., v. 46, pp.185-194.

COQUERY M. and WELBOURN P. (1995) The relationship between metal concentration and organic matter in sediments and metal concentration in the aquatic macrophyte Eriocaulonseptangulare, Water Res., v. 29, 2094–2102.

DEMASION, G.J. and MOORE, G.T. (1980) Anoxic environments and oil source bed genesis. Org. Geochem., v. 2, 9-31.

GHOSH, A. (2012) Estuarine Foraminifera from the Gulf of Cambay. J. Geo. Soc. Ind., v. 80, pp.65-74.

GHOSH, A., SAHA, S., SARASWAT, P. K., BANERJEE, S. and BURLE, S. (2009) Intertidal Foraminifera in the Macrotidal Estuaries of the Gulf of Cambay: implications for interpreting sea level change in palaeo-estuaries. Marine Petroleum Geol., v. 26, pp.1592-1599.

HAKANSON, L. (1980) An ecological risk index for aquatic pollution control. A sedimentological approach. Water Res., v.14, pp.975-1001.

JAYARAJU, N. and REDDY, K.R. (1995) Foraminiferal ecosystem in relation to coastal and estuarine Sediments of Kovalam-Tuticorin, South India. J. Geo. Soc. Ind., v. 46, pp.565 573.

KUMAR, A.V., PATIL, R.S., NAMBI, K.S.V. (2001) Source apportionment of suspended particulate matter at two traffic junctions in Mumbai, India. Atmospheric Environment, v. 35, pp.4245-4251.

LANGSTON, W.J. (1982) The distribution of mercury in British estuarine sediments and its availability to deposit-feeding bivalves. J. Mar. Bio. Assoc. U. K, v.62, pp.667-684.

LIBES, S. M. (1992) An introduction to biochemistry. John Wiley and Sons, Inc., New York, 73p.

LOEBLICH, A. R. and TAPPAN, H. (1987). Foram. genera and their classification. Von Nostrand Reinhold, New York, pp. 970.

LOSKA, K.,WIECHULA, D. and KORUS, I. (2004) Metal contamination of farming soils affected by industry. Environment International, v.30, 159-165.

MULLER, W.A. and LEE, J.J. (1969) Apparent indispensability of bacteria in foraminiferan nutrition. Jour. Protozool., v. 16, pp.471-478.

MURRAY, J. (2006) Ecology and application of benthic foraminifera. *Cambridge University Press,* 426p.

MUTHURAJ, S. and JAYAPRAKASH, M. (2007) Distribution and enrichment of trace metals in marine sediments of Bay of Bengal, off Ennore, south-east coast of India. Environ. Geology, v.56, pp.207-217.

NASR, S.M., OKBAH, M.A. and KASEM, S.M. (2006) Environmental assessment of heavy metal pollution in bottom sediment of Aden port, Yemen. Inter. J. Ocean Oceanography, v.1, pp.99-109.

NIGAM, R., PANCHANG, R. and BANERJEE, P. (2005). Foraminifera in surface sediments of Mandovi River estuary: indicators for mining pollution and high sea stand in Goa, India, J. Coastal Res., v. 21.pp 1-24.

PARPOKARI, A.L. (1979) Distribution of organic carbon in sediments of the northwestern continental shelf of India. Indian J. Mar. Sci., v. 8, pp.127-129.

PEKEY, H., KARAKAŞ, D., AYBERK, S., TOLUN, L. and BAKOĞLU, M. (2004) Ecological risk assessment using trace elements from surface sediments of İzmit Bay (Northeastern Marmara Sea) Turkey. Marine Pollution Bull., v.48, pp.946–953.

RAO and BALASUBRAMANIAN (1996) studied the distribution of foraminifera in the Cochin estuary. J. Mar. Bio. Assoc. India, v.38, pp.50-57.

RIDGWAY, J. and SHIMMIELD, G. (2002) Estuaries as repositories of historical contamination and their impact on shelf seas. Estuarine, Coastal and Shelf Science, v.55, pp.903-928.

RUBIO, B., NOMBELA, M.A. and VILAS, F. (2000) Geochemistry of major trace elements in sediments of the Ria de vigo (NW Spain) an assessment of metal pollution, Marine Pollution Bull., v. 40, pp.968-980.

SALOMONS, W. and FORSTNER, U. (1984) Metals in Hydrocycle. Springer-Verlag, New York, pp.63-98.

SARASWAT R., KURTARBAR, S. R., MAZUMDER, A. and NIGAM, R. (2004) Foraminifers as indicators of marine pollution: a culture experiment with *Rosalina leei*. Marine Pollution Bull. v.48, pp.91-96.

SAVVIDES, C., PAPADOPOULOS, A., HARALAMBOUS, K.J. and LOIZIDOU, M. (1995) Sea sediments contaminated with heavy metals, metal speciation and removal. Water Science Tech., v.32, pp.65-73.

STUMM, W. and MORGAN, J.J. (1996) Aquatic chemistry: chemical equilibria and rates in natural waters, 3rd edn., Wiley, New York.

SZEFER, P., GLASBY, G.P., PEMPKOWIAK, J. and KALISZAN, R. (1995) Extraction studies of heavy-metal pollutants in surface sediments from southern Baltic Sea off Poland. Chem. Geol., v.120, pp.111-126.

TAYLOR, S.R. and Mc-LENNAN, S.M. (1995) The geochemical evolution of the continental crust. Geophy. Review, v.33, pp.264-271.

TURNER, A. (2000) Trace metal concentration in sediments from U.K. Estuaries: an empirical evaluation of the role of hydrous iron and manganese oxides, Estuar. Coast. Shelf. Sci., v.50, 355-371.

WALTON, W.R. (1952). Techniques for recognition of living foraminifera. Contri. Cushman Found. Foram. Res., v.3, pp.56-60

WEDEPOHL, K.H. (1995) The composition of the continental crust. Geochem. Cosmochem. Acta. v.5, pp.1217-1232.

ZWOLSMAN, J., VAN ECK, G. and BURGER, G. (1996) Spatial and temporal distribution of trace metals in sediments from the Scheldt estuary, south-west Netherlands, Estuarine Coastal Shelf Sci.,v.43, pp.55-79.

Micropaleontology and its Applications
Pradeep K. Kathal, Rajiv Nigam & Abu Talib (Editors)
Scientific Publishers (India), 2018, 265-284 pp.

Foraminifera as Bio-Indicators of Pollution: A Review of Research over the Last Decade

Thejasino Suokhrie,* R. Saraswat and R. Nigam

Micropaleontology Laboratory, CSIR-National Institute of Oceanography, Goa - 403004, India
*Email: theja.geo12@gmail.com

Abstract

This paper provides a synthesis of research carried out in the last decade (2005 till present) on application of foraminifera in pollution monitoring. Over the years, various characteristics of foraminifera have been applied for ecotoxicology and pollution monitoring. Rapid changes in abundance are often used as indicators of stressed environment. Presence of stress tolerant taxa with less tolerant or sensitive taxa has been reported by many workers from the most polluted sites. *Ammonia tepida* has been invariably reported from a majority of the polluted sites, suggesting it to be a potential bio-indicator species. A few workers, however, suggest that the preference of *A. tepida* for fine grained organic carbon rich sediments may be the reason for its dominance in polluted regions. Tests with morphological deformities like twinned or reduced chambers, curled or uncoiled shell, surface protrusions, irregular suture lines, and even dissolution are common in foraminifera living in polluted regions. A large increase in laboratory culture studies to understand the response of foraminifera to specific pollutants has been noted in the last decade. Laboratory culture studies have led to better discrimination between natural and anthropogenic changes in the environment.

Keywords: Foraminifera, bio-indicators, marine pollution.

INTRODUCTION

The final destination for a majority of the pollutants is the nearby ocean and thus the pollutants adversely affect marine ecosystem. Cousteau (1971) rightly states that "there is only one pollution because every single thing, every single chemical whether in the air or on land ends up in the ocean". United Nations Convention on the Law of the Sea [1982, Article 1(4)] defines pollution as "The introduction by man, directly or indirectly, of substances or energy into the marine environment, including estuaries, which results or is likely to result in such deleterious effects as harm to living resources and marine life, hazards to human health, hindrance to marine activities, including fishing and other legitimate uses of the sea, impairment of quality for use of sea water and reduction of amenities". Polluting effluents reach the marine ecosystem either from domestic or industrial sources. The threat of pollutants in marine life comes in various forms such as highly persistent pollutants like polycyclic aromatic hydrocarbons (PAHs), polychlorinated biphenyls (PCB's), pesticides (dichloro diphenyl trichloroethane), toxic metals (Cd, Hg, Ag, Co, Cr, Ni, Pb, Zn, Cu), and others. The coastal areas of the world are a major recipient of these pollutants, hampering the commercial fishery industry. The pollutants ultimately find their way across the food chain reaching even up to higher trophic levels (Binelli and Provini, 2003, Sarkar et al., 2006). As microorganisms form the lowest level in the food chain, the initial effects are first reflected in them. Among the vast variety of microorganisms, calcareous test bearing group called foraminifera have shown potential for effective pollution biomonitoring, apart from many other applications. These unicellular microorganisms are effective environmental indicators as they respond quickly to the slightest of environmental change. Besides, foraminifera are abundant and preserve the changes in their tests, making it possible to study even the past environmental status.

Most of the studies addressing the pollution status by utilising foraminifera do so by evaluating the abundance of tolerant taxa and a decrease in sensitive taxa. Presence or absence of key species has also been statistically correlated with the contaminant in the environment. Earlier, Nigam et al. (2006) reviewed the studies on the application of foraminifera in ecotoxicology, carried out from 1979 onwards till 2004 and they opined that, "more and more results be published in widely circulated journals dealing with environmental/pollution aspects". They further added that the applications of foraminifera in pollution monitoring would be more effective if foraminiferal culture studies are supplemented with advance crystallographic and molecular studies. Following a similar pattern, this review is an update on research from 2005 till date. We have summarized the pollutants into those that source from sewage sludge disposal and those from industrial and commercial sources including anthropogenic pollutants. Application of foraminifera for temporal bio-monitoring as well as culture based experiments to understand the effect of pollutants on foraminifera has also been reviewed.

EFFECT OF SEWAGE OUTFALL

The quick and perceptible response is the key for effective application of organisms as pollution bio-indicators. The response of sewage sludge disposal on benthic foraminifera was compared with that on macrofauna by Mojtahid et al. (2008) who noted that benthic foraminifera were relatively more affected than macrofauna at the disposal site. The differential response was attributed to low pH that acts as additional stressor for foraminifera, thus suggesting the effectiveness of benthic foraminifera as pollution indicator. The effective application of benthic foraminifera in pollution monitoring also depends on understanding the specific response of benthic foraminifera to a particular pollutant. Teodoro et al. (2010) combined geochemical and foraminiferal data to classify sewage submarine outfalls with a different mix of pollutants. *Ammonia tepida* and *Galvinopsis praegeri* were the most abundant species at all affected sites. The sites polluted with sewage rich in toxic metals had low foraminiferal abundance, high percentages of *A. tepida*, low percentage of epiphytic species and more deformed fauna. The sites more polluted with organic pollutants were characterized by framboids of pyrite and pyritised fauna (Caruso et al., 2011). The health of Rodrigo de Freitas Lagoon, in Brazil was evaluated by Gutterres et al. (2011) by using foraminiferal abundance. At the highest impacted area, *A. tepida – Elphidium excavatum* assemblage exhibited low diversity with weak, dwarf and corroded tests. Agglutinated species like *Ammobaculites dilatatus* and *Ammotium salsum* were abundant at less impacted area. Koukousioura et al. (2011) also identified three assemblages reflecting various environmental conditions. *Amphistegina lobifera* assemblage suggested good water quality and oligotrophic environmental conditions. Increase in miliolid and rotalid foraminifera suggested abundant food, whereas low abundance of *A. lobifera* and dominance of stress tolerant taxa *Ammonia* spp. and *Elphidium* spp., reflected the influence of sewage pollution. Eichler et al. (2012) noted higher foraminiferal diversity at the mouth of the estuarine system, which was under the influence of the domestic sewage discharged from a submarine outfall.

EFFECT OF INDUSTRIAL AND ANTHROPOGENIC POLLUTANTS

The range of industrial and anthropogenic pollutants include mining effluents, heavy metal contamination from industries, oil refineries, organic pollutants particularly from agriculture, thermal pollution and oil spills. The adverse influence of mining activity was evident from a drastic decrease in foraminiferal abundance from 1994 to 2001, as a result of continued increase in suspended load due to mining activities in the catchment of Mandovi estuary in Goa, India (Nigam et al., 2005). Interestingly, in the same year, Panchang et al (2005) exhibited an improvement in the environment of another adjacent estuary, namely Zuari estuary, based on an increase in foraminiferal abundance and attributed it to decreased suspended load indicating less mining activity. In general, an increase in pollutants was marked by a decrease in species diversity with increased abundance of stress tolerant species and high percentage of abnormalities (Ferraro et al., 2006; Bergin et al., 2006; Romano et al., 2008; Valenti et al., 2008; Bergamin et al., 2009; Cherchi et al., 2009; Frontalini et al., 2009; Teodoro et al., 2010; Jayaraju, 2011; Cherchi et al., 2012; Brunner et al., 2013; Capotondi et al., 2015; Raja et al., 2016) (Table 1). These studies also report the presence of stress tolerant species like *Ammonia tepida*, *Haynesina germanica*, *Elphidium* spp. (Bergin et al., 2006;

Frontalini and Coccioni, 2008; Valenti et al., 2008; Coccioni et al., 2009; Caruso et al., 2011; Elshanawany, 2011). The presence of heavy metals Fe and Cu caused disappearance of fauna (Romano et al., 2008). Foraminifera at the most polluted oil disposal sites were less dense with reduced standing stock and shortened depth of habitation (DOH). The adverse response of foraminifera at these sites was evident from the absence of living specimens, increased abundance of tolerant species like *Ammonia tepida, Chilostomella oolina,* bolivinids and buliminids or opportunistic taxa and disappearance of more sensitive taxa. Additionally, aberrant specimens were also observed (Mojtahid et al., 2006; Jorissen et al., 2009; Brunner et al., 2013; Capotondi et al., 2015; Schintu et al., 2016). Agriculture and aquaculture drainage with pesticides gave rise to twinned and reduced chambers in the foraminifera. *H. germanica* was a dominant high tolerant species in such areas (Jayaraju and Reddy, 2008, Gabriella et al., 2009). The effects of thermal pollution on foraminifera from power plant and heat plumes have also been studied (Nina et al., 2011; Titelboim et al., 2016). The adverse effect of oil pollution on foraminifera was evident from increase in mortality and an accelerated reproduction rate (Ernst et al., 2006). The crude oil spill in "Herbei Spirit" led to breakages of the chambers in *Ammonia beccarii* and *Elphidium subincertum* species caused by decalcification because of low pH (6.98 on average) (Gyu et al., 2014). Another oil spill study by Li et al. (2015) suggested less diverse but high abundance of total fauna, living fauna and abnormal specimens. *Bucella frigida, Elphidium macellum, Cribrononion asiaticum* and *Verneuilinulla advena* were pollution resistant taxa while *Ammoglobigerina globigeriniformis, Cribononion incertum* and *Quinquiloculina ungeriana* were reported as opportunistic taxa.

Table 1. Studies (since 2005) mentioning application of Foraminifera for pollution monitoring.

Author	Pollutants (source/type)	Remarks
Nigam et al., 2005	Mining effluents	The drastic decline in foraminiferal abundance was attributed to increasing suspended load due to mining activities.
Panchang et al., 2005	Mining effluents	The increase in foraminiferal number, total species number, a decline in total suspended matter and reduced mining activity indicated an improvement in the health of the Zuari estuary.
Bergin et al., 2006	Heavy metal pollution	Benthic foraminiferal abundance decreases at the most contaminated area. *Ammonia tepida* was found to be a tolerant species. Test deformities were also observed.
Ferraro et al., 2006	Heavy metal contamination	A decrease in foraminiferal richness and diversity with an increase in Pb, Zn, Ni and Hg in three docks of a harbour in Naples.
Mojtahid et al., 2006	Oil disposal site	Within 70 m of the discharge points, foraminifera were less dense. Between 70m to 250m density was high with species like *Bulimina aculeata, Bulimina marginata, Textularia sagittula, Trifarina bradyi* and *Bolivina* spp. A pollution Index was developed which correlated benthic foraminifera to the distance of the disposal site.

Author	Pollutants (source/type)	Remarks
Burone et al., 2007	Urban sewage outfalls	*Ammonia tepida* is more stress tolerant than *Buliminella elegantissima.*
Carnahan et al., 2008	Heavy metals (Cu, Zn, Cr, Hg, Pb, and Ni)	*Ammonia* and *Cribroelphidium* were found to be stress-tolerant genera.
Frontalini and Coccioni, 2008	Heavy metal	Deformities with variety of malformations observed throughout metal polluted area. The Foraminiferal Abnormality Index (FAI) of 4.7 with 2 on an average reflected heavy metal concentration. *Ammonia parkinsoniana* preferred clean and unpolluted environment and thus deeply affected by heavy metal pollution even in low concentrations. More tolerant species were *Ammonia tepida, Aubignyna perlucida, Eggerella scabra*, and *Nonionella turgida.*
Jayaraju and Reddy, 2008	Untreated industrial waste with heavy metals, agriculture and aquaculture drainage water with pesticides	Faunal deformities like corrosion, cavity development, broken peripheries and reduction in overall growth associated with heavy metal pollution. Agricultural wastes had lesser effect as evident from twinned and reduced chambers.
Mojtahid et al., 2008	Organic matter pollution from sewage sludge disposal	Abundance and biodiversity of fauna was minimal at the site of severe pollution where only tolerant species were found.
Romano et al., 2008	Industrial pollution, Polycyclic Aromatic Hydrocarbons (PAH)	Strong Pb and Zn pollution decreased the faunal abundance. Fe and associated Cu led to disappearance of fauna. The number of deformed tests of *Miliolinella subrotunda* statistically related to PAH's, Mn, Zn.
Valenti et al., 2008	Heavy metal concentration	*Adelosina* spp. and *Quinqueloculina* spp.were sensitive to overall metal pollution while *Elphidium* spp. shows an "opportunistic" behaviour, increasing its abundance in the presence of higher levels of total normalised contamination.
Bergamin et al., 2009	Cu, Hg, Pb, Zn, and PAHs probably due to the presence of sunken vessels	High concentration of PCBs (up to 144 ng g^{-1} d.w.) resulted in decreased foraminiferal diversity and density, increased abundance of "pollution -tolerant species" and abnormal specimens.
Carboni et al., 2009	Urbanisation induced pollution	Variation in foraminiferal assemblages from more stressed to less stressed area.
Carnahan et al., 2009	Urbanisation	Increased abundance of stress-tolerant taxa, especially *Ammonia* spp.

Author	Pollutants (source/type)	Remarks
Cherchi et al., 2009	Industrial pollution	Increasing pollution caused low species diversity, low population density and more frequent abnormal specimens.
Coccioni et al., 2009	Heavy trace element concentration (Cd, Cu, Ni, Mn, Pb, Zn and Hg) due to industrial pollution	The heaviest populated industrial zones are marked by greater proportion of abnormal specimens. *Ammonia tepida*, *Hyanesina germanica* are dominant in stressed environments and the other species include *Cribroelphidium oceanensis. Aubignyna perlucida, Ammonia parkinsoniana* and *Bolivina striatula.*
Debenay and Fernandez, 2009	Heavy metal rich terrigenous particles from an open cast mining	Increase in species richness, diversity, density, and decreasing percentages of *Ammonia tepida* with decreasing impact of freshwater and pollution.
Frontalini et al., 2009	Industrial wastes mostly trace element (Cr, Cu, Hg, Ni, Pb and Zn) and agricultural and domestic effluent	The innermost part of the lagoon having the highest load of trace element reveals lower diversity and higher percentages of abnormalities when compared to the outer parts of the lagoon. Nano particles of Fe and S, Ba and S, and La, Ce, Nd and S in abnormal specimens of *Ammonia tepida*, is also observed which leads to interference of the crystalline arrangement of carbonate tests.
Gabriella et al., 2009	Human impact especially agriculture	Dominance of high tolerant species *Hyanesina germanica* was attributed to pollutant inputs in the inner part. The absence of the inner and intermediate lake assemblages in the silty sediments of Lungo Lake suggested strong environmental disturbance related to high human impact.
Hyams-Kaphzan et al., 2009	Sewage sludge outfall	Foraminifera are less abundant and diverse with *Ammonia tepida* being dominant.
Jorissen et al., 2009	Pollutants from oil exploration activities	Increase in tolerant or opportunistic taxa (*Chilostomella oolina*, bolivinid and buliminid) and disappearance of more sensitive taxa (*Uvigerina peregrina, Cancris. auriculus* and *Cribrostoimoides subglobosus*). Large size taxa showed more sensitivity than small sized taxa.
Martinez-Colon et al., 2009	Heavy metals Potentially Toxic Elements (PTE)	Reviewed research on PTEs in coastal waters and opined on the requirement of genomic studies on foraminiferal bio-indicators.
Rumolo et al., 2009	Cd, Co, Cu, Li, Ni, Pb, V and Zn from harbour	*Ammonia tepida* abundance was 2 to 100 times higher than that in unpolluted Mediterranean seawater.
Denoyelle et al., 2010	Pollutants from offshore oil drilling sites	Most polluted sites characterised by poor fauna and strong dominance of tolerant taxa. The foraminiferal index (FIEI) was more discriminative than the macrofaunal index (ITI).
Martins et al., 2010	Heavy metal contaminants (Al, Cd, Co, Cu, Fe, Pb, Mn Zn)	Low diversity with an increase in test abnormalities.

Author	Pollutants (source/type)	Remarks
Souder et al., 2010	Environmental stress	Seven different types of morphological abnormalities and five different surface texture anomalies were observed.
Teodoro et al., 2010	Domestic sewage outfall	Low species diversity and richness as well as dominance of detritivore feeder species.
Caruso et al., 2011	Sewage outfalls and toxic metals (Cu, Cr, Hg, Pb, Zn)	Pollutants modify the biogeochemical parameters of bottom water in coastal marine areas, which can have a strong impact on benthic foraminiferal assemblages, distribution and test deformations.
Elshanawany, 2011	Heavy metals from industries and domestic and agricultural effluents	Pollution opportunists like *A. tepida*, *Quinqueloculina lata* and *Porosononion* spp. were dominant. Pollution sensitive taxa include *Quinqueloculina vulgaris, Elphidium spp., Asterigerinata mamilla, Rosalina macropora, Ammonia beccarii, Triloculina trigonula, Peneroplis pertusus,* and *Quinqueloculina* spp.
Frontalini and Coccioni, 2011	Review of pollution studies from Italian research	Possibility of qualitative examination on "the degree of sensitivity of selected taxa to different sources of pollution (organic vs. chemical pollution). Confirm that benthic foraminifera are sensitive bio-indicators and successful candidates for inclusion in an integrated pollution-monitoring program.
Gutterres et al., 2011	Heavy metals (Pb and Zn) and untreated sewage (Cu), Polycyclic Aromatic Hydrocarbons (PAH)	The environmental stress is severe in the inner part of the lagoon. Low abundance and sterility of five samples in this area was attributed to high levels of heavy metals and organic matter.
Jayaraju, 2011	Iron ore tailing pollution	*Ammonia, Elphidium* and *Quinqueloculina* were dominant.
Koukousioura et al., 2011	Sewage outfall	Stress-tolerant taxa *Ammonia* spp. and *Elphidium* spp., observed in the vicinity of the primary sewage outfall. *Amphistegina* could reflect nitrification-induced decline in the water quality of coastal environments.
Nina et al., 2011	Thermal pollution	The total foraminiferal abundance and species richness and diversity were low at the thermally polluted site. *Textularia agglutinans* has very low resilience to elevated SST and was completely absent from the most thermally polluted stations. The miliolids especially *Lachlanella sp.*, was tolerant to increased SST and survived at the most thermally polluted stations.
Celia Magno et al., 2012	Industrial pollution, oil refineries and petrochemicals	Observed that the high tolerance of species (*Ammonia tepida*) to pollution could be due to their preference for fine sediments which are the main receptor of contaminants.
Cherchi et al., 2012	Heavy metal	High levels of heavy metals in the shell associated with a decrease in foraminiferal richness, population density and biodiversity with the presence of morphologically abnormal specimens.

Author	Pollutants (source/type)	Remarks
Choi, 2012	Organic pollutants and heavy metal enrichment (Cu, Zn, and Pb)	Low abundance of benthic foraminifera and dominance of *Cribroelphidium excavatum* and *Trochammina hadai* appears to indicate high levels of heavy metals and organic matter.
Donnici et al., 2012	Heavy metal concentration (Cu, Pb, Zn, Cr, Ag and Cd, Fe, Mn, Ni) due to industrial sectors, river discharge, urban settlements	*Ammonia beccarii* positively correlated with Fe, Mn and Ni. Presence of *Buliminella elegantissima* indicated anthropogenic pollution including Cd and Ag. Lack of foraminiferal colonisation and presence of sterile sediments indicated unliveable condition.
Eichler et al., 2012	Domestic sewage	The benthic fauna observed to be degraded due to sand accumulation.
Foster et al., 2012	Heavy metals (Cu and Cd) from industries and agricultural sources	*Quinqueloculina bicostata* was positively related to the toxicity of the sediment.
Brunner et al., 2013	Oil pollution from well blowout based on Total Polycyclic Aromatic Hydrocarbons (TPAH)	Highest oil concentration reduced standing stock and shortening the depth of habitation (DOH). At less heavily oiled site, foraminifera responded with a shallower DOH, but with a boom in standing stock. Deformed, dead foraminifera occurred in all heavily oiled cores.
Dijkstra et al., 2013	Pollution from offshore petroleum and oil exploration	Sessile epifaunal species, *Lobatula lobatula, Trifarina angulosa ,Cassidulina laevigata, Melonis barleeanus, Pullenia bulloides* and *Nonionella auricular* were identified as good bio-indicators.
Martins et al., 2013	Heavy metals (Zn, Pb, Cu, Cr, As, Co, Hg)	*Bolivina ordinaria, Bolivina pseudoplicata* and *Bulimina elongata/gibba*are tolerant species and more tolerant species include *Hyanesina germanica, Ammonia tepida* and *Quinqueloculina seminulum.* Foraminifera were more tolerant to Zn in any phase than Cu.
Melis and Covelli, 2013	Agricultural activities, clam farming with high Hg concentration	*Ammonia tepida* is a stress tolerant species. High percentage of abnormalities was found in *Elphidium gunteri* and *Hyanesina germanica* at highest eutrophication. Abnormalities decreased at highest Hg concentrated site as Hg is in sulphide form.
Popadić et al., 2013	Industrial pollutants	Stress-tolerant species (*Ammonia* and *Haynesina* group) prevailed in fine-grained sediments with increased concentration of heavy metals. *E. exigua* and *B. marginata* are confirmed to be tolerant to heavy metal concentration.
Baz, 2014	Heavy metal pollution	X-ray microanalysis reveals that deformed specimens contain higher levels of heavy metals (Mn, Cu, and Hg) than non-deformed ones.
Gandhi and Nathan, 2014	Mn, Cr, Cu, Ni, Pb and Zn concentration	Morphological abnormalities in *Spiroloculina, Quinqueloculina, Elphidium.*

Author	Pollutants (source/type)	Remarks
Gyu et al., 2014	Oil spill	Breakages of the chambers in the *Ammonia beccarii* and *Elphidium subincertum* species.
Li and Li, 2014	Trace metal (Co, Cr, Cu, Ni, Pb and Zn, Ba, Sr, V and Zr)	*Ammonia tepida, Haplophragmoides canariensis* and *Elphidium nakanokawaense* are tolerant to metal enrichment.
Capotondi et al., 2015	Pollutants from harbour and oil refinery	Higher abundance of opportunistic, low-oxygen tolerant taxa along the coastline was observed. Near the oil refinery, living specimens were absent and the diversity was low associated with the dominance of opportunistic species. Aberrant species were also found. *Ammonia parkinsoniana* and *Quinqueloculina* seem to be the most sensitive taxa.
Debenay et al., 2015	Untreated effluents (normal-saline effluents, nutrient input and microbial stimulation) from semi-intensive shrimp farms	Pollution indicator *Ammonia tepida* found near the outflow area.
Lee and Nikraz, 2015	Industrial effluents, domestic and agricultural waste, land reclamation	High *Ammonia – Elphidium* Index (AEI) with low FORAM Index at more anthropogenic activity while high foraminiferal diversity with low abundances and high FORAM Index in less anthropologically impacted region.
Martin and Nesbitt, 2015	Toxic sediments	Low species diversity and strongly dominated by species tolerant of various contaminants and dysoxia.
Li et al., 2015	Trace metal pollution	*Ammonia tepida* is the most abundant species in the polluted area. High diversity corresponds to heavy pollution.
Youssef, 2015	Heavy metal pollution (Fe, Mn, Zn, Cu, Pb, Ni, Cr, and Cd)	Some foraminiferal tests display abnormalities in coiling, general shape of chambers and apertures. *Sorites marginalis* and *Peneroplis planatus* showed significantly high concentrations of Fe, Mn, Pb and Cu.
Laroche et al., 2016	Pollution from offshore oil drilling site	Eight foraminiferal molecular operational taxonomic units were identified that have potential to be used as bio-indicator taxa.
Raja et al., 2016	Heavy metal concentration (Co, Cr, Cu, Fe,Mn, Ni, and Pb)	Low diversity of fauna and predominance of opportunistic species. Increase in abnormal species.
Sabbagh, 2016	Domestic and industrial effluents	Pollution-sensitive smaller species (*Ammonia parkinsoniana, Discorbinella bertheloti* and *Pararotalia nipponica)* represented low $CaCO_3$, low heavy metal concentration and organic matter. *Amphistegina lessonii, Peneroplis pertusus, Neorotalia calcar,* and *Sorites orbiculus,* as well as smaller species *Elphidium crispum, Glabratella patelliformis,* and *Discorbis turbo* represented more protected site. Rich organic matter and heavy metals, particularly Cu and Zn resulted in deformities.

Author	Pollutants (source/type)	Remarks
Schintu et al., 2016	Heavy metal from oil refinery	Living assemblages were characterized by low diversity in samples collected close to the port areas. *Ammonia tepida* and bolivinids are stress tolerant.
Titelboim et al., 2016	Thermal pollution	The most thermally tolerant species was *Pararotalia calcariformata.*
Yahiya and Miranda, 2016	Industrial pollution from factory and fish landing centre	Pollution tolerant species like *Navicula*, and *Discorbis* sp. were found at the industrially polluted site.

ABNORMAL TESTS AND DEFORMITIES AS POLLUTION INDICATORS

Apart from using assemblages and presence or absence of species, morphological abnormalities were also considered as pollution bio-indicators, as the percentage of deformed or abnormal tests increases noticeably at the polluted site. Several workers reported deformities in fauna along with malformations or abnormal specimens from polluted sites (Bergin et al., 2006; Jayaraju and Reddy, 2008; Frontalini and Coccioni, 2008; Bergamin et al., 2009; Martins et al., 2010; Cherchi et al., 2012). At the heaviest populated zones, fauna showed greater proportion of morphologically abnormal specimens (Cherchi et al., 2012). Jayaraju and Reddy (2008), Melis and Covelli (2013) found that morphological deformities in foraminifera depended on the nature of the pollutant. Jayaraju and Reddy (2008), also noted that heavy metal pollution caused corrosion, cavity development with broken peripheries while pollutants from agricultural wastes caused twinned or reduced chamber in fauna. Further, heavy metals like Fe, Mn, Zn, Cu in foraminiferal tests displayed abnormalities in coiling, shape of chambers and aperture (Youssef, 2015). Increased instances of curled, asymmetrical, and uncoiled shells, irregular suture lines, surface protrusions, and breakage/repair along with surface pits, dissolution features, microborings, microbial biofilm were also observed (Souder et al., 2010). High polychlorinated biphenyls (PCB) and Polycyclic Aromatic Hydrocarbons (PAH) in the sediment caused deformities in *Haynesia orbicular* and *Miliolinella subrotunda* (Scott, 2005; Romano et al., 2008). Frontalini and Coccioni (2008) reported that the Foraminiferal Abnormality Index (FAI) in low heavy metal concentration had a value of 4.7. Further studies observed that in the most polluted areas, foraminiferal tests were weak, dwarf, corroded (Gutterres et al., 2011). The abnormalities in tests were attributed to the disruption of crystal lattice as evident from higher levels of Mn, Cu and Hg in deformed specimens as compared to the non-deformed ones (Baz, 2014).

APPLICATIONS OF TEMPORAL DISTRIBUTION OF FORAMINIFERA IN POLLUTION MONITORING

Foraminifera have an added advantage over other pollution monitoring techniques because of its preservation potential thereby making it possible to study past environmental variations, even in the absence of baseline data. Some of the major findings based on application of temporal variations in foraminifera to reconstruct pollution history are discussed here (Table 2). The deformities in *Haynesina orbicular* occurred simultaneously with high polychlorinated biphenyls (PCB's) in the sediments. Tsujimoto et al. (2006a; b; 2008) analyzed a sediment core deposited during the past 200 years and identified *Trochammina hadai-Eggerella advena* assemblage which proliferated during the periods of intense nutrient discharge leading to eutrophication. A subsequent restriction on the discharge led to a decrease in this assemblage. The study of a 0.50 m core in the Suances estuary covering a period of 130 years suggested a "change in foraminiferal assemblage and defaunation during the period of more intense pollution" (Irabien et al., 2008). Similarly, rapid change in foraminiferal assemblage in the inner and intermediate areas of the inshore fringing reefs as compared to the outer area located away from the land run-off was observed by Uthicke et al. (2012) in eleven sediment cores. Dijkstra et al. (2015) carried out baseline survey in SW Barents Sea where petroleum industry related activities have been projected to increase in coming decades. It was found that though the metal concentrations correspond to predicted no effect levels, a strong correlation between metal concentration and sediment properties, and the strong influence of water masses on sediment distribution, suggests that change in oceanographic conditions might influence the pollutant deposition in the region in future. Li et al. (2015) studied samples from an oil spill site in Bohai Sea and observed that larger foraminifera were more sensitive to oil, and the small foraminifera showed high mortality and reproduction. The Foraminiferal Index of Environmental Impact (FIEI) increased from low to high oil-polluted station and from deep layer to surface sediment reflecting the impact of oil pollution.

CULTURE STUDIES TO UNDERSTAND THE EFFECT OF SPECIFIC POLLUTANTS ON FORAMINIFERA

The adverse effects of different pollutants on foraminifera are well documented. The increasing emphasis now is to document specific response to particular pollutants. Laboratory culture studies can help to understand the effect of specific pollutants on benthic foraminifera. Sewage derived particulate organic matter did not affect the diet of *A. becarii* and *H. germanica* as observed in a microcosm experiment (Topping et al., 2006). The response of foraminifera to oil spills is seen in the form of statistically significant deformities in the foraminiferal tests. Deformed specimens of *Miliammina fusca* at heavily oil polluted sites were reported by Sabean et al. (2009). The adverse effect of oil pollution was tested under controlled conditions in the laboratory (Ernst et al., 2006). The living benthic foraminifera were subjected to saturated seawater mix (SSW) and a water accommodated fraction of the oil (WAF). During the first two weeks of the experiment, there was general decline in density with exception in some of the mesocosms where density increased. It was hence concluded that "there is a dual response to oil-induced pollution: foraminiferal faunas may respond by a strongly increased mortality, and/or by an accelerated reproduction of some of the taxa". De Nooijer et al. (2007) subjected benthic foraminifera to different concentrations of copper and observed that the Cu partitioning was constant over a wide range of Cu/Ca ratios for both

Ammonia tepida and *Heterostegina depressa*. A gradual introduction of mercury in the ambient environment of *Rosalina leei* caused decreased growth rate, irregular reproduction, deformed chambers and abnormalities (Nigam et al., 2009). A similar response was also observed in *Ammonia parkinsoniana* that responded to high levels of mercury by proliferation of lysosomes and even leading to cell death at the highest concentrations (Frontalini et al., 2016). Increased instances of abnormalities in benthic foraminifera subjected to heavy metal pollution were further validated in laboratory culture when cadmium caused a change in the coiling direction in *Pararotalia nipponica* apart from the presence of other abnormalities (Linshy et al., 2013). Increase in the concentration of heavy metals like Ni, Cu and Mn led to more incorporation of these metals into the tests of *Ammonia tepida* (Munsel et al., 2010).

Table 2. Applications of temporal distribution of foraminifera in pollution monitoring.

Author	**Pollutants (source/type)**	**Remarks**
Scott, 2005	Industrial and Organic Carbon from domestic sources	Deformities in *Haynesina orbicular*. Presence of more organic carbon tolerant taxa with increase in harbour activities.
Di Leonardo et al., 2007	Polycyclic aromatic hydrocarbon (PAH) and Hg concentration	Reduced foraminiferal abundance and increased abnormalities.
Irabien et al., 2008	Al, Zn, Pb, Cd, Cu, Ni, Cr and As	Foraminiferal assemblages reached an afaunal stage in recent decade.
Tsujimoto et al., 2006a, 2006b, 2008	Eutrophication sourced from industrial development	*Trochammina hadai-Eggerella advena* assemblage reflects eutrophication.
Debenay and Fernandez, 2009	High easily oxidized material (EOM) from aquaculture	Exceptional abundances of abnormal tests indicate high accumulation of EOM.
Uthicke et al., 2012	Agricultural runoff from settlement areas	Change in foraminiferal assemblage due to enhanced agricultural run-off.
Dijkstra et al., 2015	Metal concentration from petroleum industry	Variation in metal concentration and foraminifera was assessed at a site where petroleum industry activity was projected to increase.
Li et al., 2015	Oil spill (PAH, oils, sulphides)	Abnormal specimens were positively correlated to oils or PAH's.

Table 3. Culture studies.

Author	**Pollutants (source/type)**	**Remarks**
Topping et al., 2006	Sewage derived Particulate Organic Matter (POM)	POM did not have a direct effect on the food source and diet of foraminifera.
Ernst et al., 2006	Oil spills	In oil induced pollution, foraminifera respond by increasing mortality and/or by an accelerated reproduction of some of the taxa.

Author	Pollutants (source/type)	Remarks
Cadre, 2006	Copper	Copper concentration higher than 200μgl^{-1} was lethal to pollutant tolerant *Ammonia* sp. Delay in reproduction and deformities in new chamber formation was observed. Also, the number of juveniles reduced.
Nigam et al., 2009	Mercury	Presence of mercury in the ambient environment leads to decreased growth rate, deformed chambers, irregular rate of reproduction and abnormalities in *Rosalina leei.*
Sabean et al., 2009	Oil spill	Dramatic increase in deformities in *Milliamina fusca.*
Munsel et al., 2010	Ni, Cu, Mn	The concentration of these elements in the carbonate shell of *Ammonia tepida* increased with its concentration in the culture solutions. Decreased incorporation of Ni and Cu was attributed to chemical competition effects. Mn concentration was found to increase 20-fold due to antagonistic effects with Cu
Prazeres et al., 2011	Zinc	Exposure of *Amphistegina lessonii* to high Zn concentrations led to mortality, visual alterations (white spots and dark-brown areas in the test), oxidative stress biomarkers and metallothionein-like proteins. No significant mortality was observed as an activation of some components of the antioxidant system counteracted the oxidative stress induced by Zn exposure.
Prazeres et al., 2012	Cd, Cu, Zn	Ca^{+2} - ATPase activity in *Amphistegina lessonii* found to be a good biomarker of metal exposure.
Denoyelle et al., 2012	Cd, Fuel oil no.2, Non Aqueous Base Mud (NABM), Water base mud (WBM)	Foraminifera were sensitive to all four tested contaminants after 30 days incubation. NABW was found to be more toxic than WBM
Hess et al., 2013	Water based drill cutting	Reduction in abundance and species richness. Affected reproduction in surviving populations of *B. marginata* and *Nonionellina labradorica* and microhabitat of *N. labradorica. Stainforthia fusiformis* species is the most tolerant to the water-based drill cuttings.
Linshy et al., 2013	Cadmium	Increased Cd in *Pararotalia nipponica* affected normal growth, caused morphological abnormalities and a peculiar change in coiling direction.
Frontalini et al., 2015	Lead	Numerical increase of lipid, proliferation of residual bodies, a thickening of the organic lining, mitochondrial degeneration, autophagosome proliferation and development of inorganic aggregates in *Ammonia parkinsoniana.*
Frontalini et al., 2016	Mercury	Increased number of lipid droplets, promoted the proliferation of lysosomes and cell death at the highest levels in *Ammonia parkinsoniana.*

CONCLUSION

We have synthesized the research carried out from 2005 till present on the application of foraminifera in pollution monitoring. Widespread application of foraminifera as a potential bio-indicator has been observed, wherein observations were made from varied pollution sites like industrial effluents, human settlements, oil refineries, offshore petroleum drilling sites, and others. A majority of the field studies reported presence or absence of characteristic benthic foraminifera and increased instances of abnormalities. The field based observations were also substantiated by laboratory culture studies. Advancements in culture studies are noted as molecular techniques have been used to understand the mechanisms through which foraminifers respond to pollution. We conclude that benthic foraminifera are a reliable bio-indicator for pollution monitoring. However, much work is still to be done to identify characteristic response of benthic foraminifera to specific pollutants.

Acknowledgements: Authors are grateful to the Director, Council of Scientific and Industrial Research-National Institute of Oceanography, Goa, India for permission and support and Department of Science and Technology (Project DST/CCP/NCC&CV/130/2017(G) under National "Network Programme on Climate Change and Coastal Vulnerability"). TS acknowledges University Grants Commission, New Delhi, for financial support in the form of Senior Research Fellowship.

REFERENCES

BAZ, S.M.EL. (2014) Benthic foraminifera as bioindicators of heavy metal pollution in Lake Burullus, Egypt. Arab. J. Geosci., v.8, pp.5491-5509 doi:10.1007/s12517-014-1611-8

BERGAMIN, L., ROMANO, E., FINOIA, M.G., VENTI, F., BIANCHI, J., COLASANTI, A. and AUSILI, A. (2009) Benthic foraminifera from the coastal zone of Baia (Naples, Italy): Assemblage distribution and modification as tools for environmental characterisation. Mar. Pollut. Bull., v.59, pp.234–244. doi:10.1016/j.marpolbul.2009.09.015

BERGIN, F., KUCUKSEZGIN, F., ULUTURHAN, E. and BARUT, I.F. (2006) The response of benthic foraminifera and ostracoda to heavy metal pollution in Gulf of Izmir (Eastern Aegean Sea). Estuar. Coast. Shelf Sci., v.66, pp.368–386. doi:10.1016/j.ecss.2005.09.013

BINELLI, A. and PROVINI, A. (2003) POPs in edible clams from different Italian and European markets and possible human health risk. Mar. Pollut. Bull., v.46, pp.879–886. doi:10.1016/S0025-326X(03) 00043-2

BRUNNER, C.A., YEAGER, K.M., HATCH, R., SIMPSON, S., KEIM, J., BRIGGS, K.B. and LOUCHOUARN, P. (2013) Effects of Oil from the 2010 Macondo Well Blowout on Marsh Foraminifera of Mississippi and Louisiana, USA. Environ. Sci. Technol., dx.doi.org/10.1021/es401943y

BURONE, L., PIRES-VANIN, P., SOUSA, S., MAHIQUES, M. and BRAGA, E. (2007) Benthic foraminiferal variability on a monthly scale in a subtropical bay moderately affected by urban sewage. Sci. Mar., v.71, pp.775–792. doi:10.3989/scimar.2007.71n4775

CADRE, VALERIE LE (2006) Morphological and cytological responses of Ammonia (foraminifera) to copper contamination: Implication for the use of foraminifera as bioindicators of pollution. Environ. Pollut., v.143, pp.304–317. doi:10.1016/j.envpol.2005.11.033

CAPOTONDI, L., BERGAMI, C., ORSINI, G. and RAVAIOLI, M. (2015) Benthic foraminifera for environmental monitoring: a case study in the central Adriatic continental shelf. Environ. Sci. Pollut. Res., v.22, pp.6034–6049. doi:10.1007/s11356-014-3778-7

CARBONI, M.G., SUCCI, M.C., BERGAMIN, L., BELLA, L. DI, FREZZA, V. and LANDINI, B. (2009) Benthic foraminifera from two coastal lakes of southern Latium (Italy). Preliminary evaluation of environmental quality. Mar. Pollut. Bull., v.59, pp.68–280. doi:10.1016/j.marpolbul.2009.08.010

CARNAHAN, E.A., HOARE, A.A.M., HALLOCK, P., LIDZ, B.H. and REICH, C.D. (2008) Distribution of heavy metals and foraminiferal assemblages in sediments of Biscayne Bay, Florida, USA. J. Coast. Res., v.24, pp.159–169. doi:10.2112/06-0666.1

CARNAHAN, E.A., HOARE, A.M., HALLOCK, P., LIDZ, B.H. and REICH, C.D. (2009) Foraminiferal assemblages in Biscayne Bay, Florida, USA: Responses to urban and agricultural influence in a subtropical estuary. Mar. Pollut. Bull., v.59, pp.221–233. doi:10.1016/j.marpolbul.2009.08.008

CARUSO, A., COSENTINO, C., TRANCHINA, L., BRAI, M., CARUSO, A., COSENTINO, C., TRANCHINA, L. and BRAI, M. (2011) Response of benthic foraminifera to heavy metal contamination in marine sediments (Sicilian coasts, Mediterranean Sea). Chem. Ecol., v.27, pp.9-30. doi:10.1080/02757540.2010.529076

CELIA MAGNO, M., BERGAMIN, L., FINOIA, M.G., PIERFRANCESCHI, G., VENTI, F. and ROMANO, E. (2012) Correlation between textural characteristics of marine sediments and benthic foraminifera in highly anthropogenically-altered coastal areas. Mar. Geol., v.315-318, pp.143–161. doi:10.1016/j.margeo.2012.04.002

CHERCHI, A., BUOSI, C., ZUDDAS, P., GIUDICI, G. DE and PARIS, C. (2012) Bioerosion by microbial euendoliths in benthic foraminifera from heavy metal-polluted coastal environments of Portovesme (south-western Sardinia, Italy). Biogeosciences., v.9, pp.4607–4620. doi:10.5194/bg-9-4607-2012

CHERCHI, A., PELO, S. DA, IBBA, A., MANA, D., BUOSI, C. and FLORIS, N. (2009) Benthic foraminifera response and geochemical characterization of the coastal environment surrounding the polluted industrial area of Portovesme (South-Western Sardinia, Italy). Mar. Pollut. Bull. J., v.59, pp.281–296. doi:10.1016/j.marpolbul.2009.09.016

CHOI. (2012) High benthic foraminiferal diversity in polluted busan north port (Korea). J. Foram. Res., v.42, pp.327–339.

COCCIONI, R., FRONTALINI, F., MARSILI, A. and MANA, D. (2009) Benthic foraminifera and trace element distribution: A case-study from the heavily polluted lagoon of Venice (Italy). Mar. Pollut. Bull., v.59, pp.257–267. doi:10.1016/j.marpolbul.2009.08.009

COUSTEAU, J. (1971) Our oceans are dying. New York Times, 14 November

DEBENAY, J.P., DELLA PATRONA, L., HERBLAND, A. and GOGUENHEIM, H. (2009) The impact of easily oxidized material (EOM) on the meiobenthos: Foraminifera abnormalities in shrimp ponds of New Caledonia; implications for environment and paleoenvironment survey. Mar. Pollut. Bull., v.59, pp.323–335. doi:10.1016/j.marpolbul.2009.08.012

DEBENAY, J.P. and FERNANDEZ, J.M. (2009) Benthic foraminifera records of complex anthropogenic environmental changes combined with geochemical data in a tropical bay of New Caledonia (SW Pacific). Mar. Pollut. Bull., v.59, pp.311–322. doi:10.1016/j.marpolbul.2009.09.014

DEBENAY, J.P., MARCHAND, C., MOLNAR, N., ASCHENBROICH, A. and MEZIANE, T. (2015) Foraminiferal assemblages as bioindicators to assess potential pollution in mangroves used as a natural biofilter for shrimp farm effluents (New Caledonia). Mar. Pollut. Bull., v.93, pp.103–120. doi:10.1016/j.marpolbul.2015.02.009

DE NOOIJER, L. J., REICHART, G. J., DUEÑAS-BOHÓRQUEZ, A., WOLTHERS, M., ERNST, S. R., MASON, P. R. D., and VAN DER ZWAAN, G. J. (2007) Copper incorporation in foraminiferal calcite: results from culturing experiments. Biogeosciences Discussions., v.4, pp.961-991.

DENOYELLE, M., GESLIN, E., JORISSEN, F.J. and CAZES, L. (2012) Innovative use of foraminifera in ecotoxicology: A marine chronic bioassay for testing potential toxicity of drilling muds. Ecol. Indic., v.12, pp.17–25. doi:10.1016/j.ecolind.2011.05.011

DENOYELLE, M., JORISSEN, F.J., MARTIN, D., GALGANI, F. and MINÉ, J. (2010) Comparison of benthic foraminifera and macrofaunal indicators of the impact of oil-based drill mud disposal. Mar. Pollut. Bull., v.60, pp.2007–2021. doi:10.1016/j.marpolbul.2010.07.024

DI LEONARDO, R., BELLANCA, A., CAPOTONDI, L., CUNDY, A. and NERI, R. (2007) Possible impacts of Hg and PAH contamination on benthic foraminiferal assemblages: An example from the Sicilian coast, central Mediterranean. Sci. Total Environ., v.388, pp.168–183. doi:10.1016/j.scitotenv.2007.08.009

DIJKSTRA, N., JUNTTILA, J., CARROLL, J., HUSUM, K., HALD, M., ELVEBAKK, G. and GODTLIEBSEN, F. (2013) Baseline benthic foraminiferal assemblages and habitat conditions in a sub-Arctic region of increasing petroleum development. Mar. Environ. Res., v.92, pp.178–196. doi:10.1016/j.marenvres.2013.09.014

DIJKSTRA, N., JUNTTILA, J., HUSUM, K., CARROLL, J. and HALD, M. (2015) Natural variability of benthic foraminiferal assemblages and metal concentrations during the last 150 years in the Ingøydjupet trough, SW Barents Sea. Mar. Micropaleontol., v.121, pp.16–31. doi:10.1016/j.marmicro. 2015.09.005

DONNICI, S., SERANDREI-BARBERO, R., BONARDI, M. and SPERLE, M. (2012) Benthic foraminifera as proxies of pollution: The case of Guanabara Bay (Brazil). Mar. Pollut. Bull., v.64, pp.2015–2028.

EICHLER, P.P.B., EICHLER, B.B., SEN, B. and RÖSCH, A. (2012) Foraminifera as indicators of marine pollutant contamination on the inner continental shelf of southern Brazil. Mar. Pollut. Bull., v.64, pp.22–30. doi:10.1016/j.marpolbul.2011.10.032

ELSHANAWANY, R. (2011) Anthropogenic impact on benthic foraminifera, Abu-qir bay,. J. Foram. Res., v.41, pp.326–348.

ERNST, S.R., MORVAN, J., GESLIN, E., LE BIHAN, A., JORISSEN, F.J. (2006) Benthic foraminiferal response to experimentally induced Erika oil pollution. Mar. Micropaleontol., v.61, pp.76–93. doi:10.1016/j.marmicro.2006.05.005

FERRARO, L., SPROVIERI, M., ALBERICO, I., LIRER, F., PREVEDELLO, L. and MARSELLA, E. (2006) Benthic foraminifera and heavy metals distribution: A case study from the Naples Harbour (Tyrrhenian Sea, Southern Italy). Environ. Pollut., v.142, pp.274–287. doi:10.1016/j.envpol.2005.10.026

FOSTER, W.J., ARMYNOT DU CHÂTELET, E. and ROGERSON, M. (2012) Testing benthic foraminiferal distributions as a contemporary quantitative approach to biomonitoring estuarine heavy metal pollution. Mar. Pollut. Bull., v.64, pp.1039–1048. doi:10.1016/j.marpolbul.2012.01.021

FRONTALINI, F., BUOSI, C., DA PELO, S., COCCIONI, R., CHERCHI, A. and BUCCI, C. (2009) Benthic foraminifera as bio-indicators of trace element pollution in the heavily contaminated Santa Gilla lagoon (Cagliari, Italy). Mar. Pollut. Bull., v.58, pp.858–877. doi:10.1016/j.marpolbul.2009.01.015

FRONTALINI, F. and COCCIONI, R. (2008) Benthic foraminifera for heavy metal pollution monitoring: A case study from the central Adriatic Sea coast of Italy. Estuar. Coast. Shelf Sci., v.76, pp.404–417. doi:10.1016/j.ecss.2007.07.024

FRONTALINI, F. and COCCIONI, R. (2011) Benthic foraminifera as bioindicators of pollution: A review of Italian research over the last three decades. Rev. Micropaleont. v.54, pp.115-127.

FRONTALINI, F., CURZI, D., CESARINI, E., CANONICO, B., GIORDANO, F.M., DE MATTEIS, R., BERNHARD, J.M., PIERETTI, N., GU, B., ESKELSEN, J.R., JUBB, A.M., ZHAO, L., PIERCE, E.M., GOBBI, P., PAPA, S. and COCCIONI, R. (2016) Mercury-pollution induction of intracellular lipid accumulation and lysosomal compartment amplification in the benthic foraminifer ammonia parkinsoniana. PLoS One., v.11, pp.1–14. doi:10.1371/journal.pone.0162401

FRONTALINI, F., CURZI, D., GIORDANO, F.M., BERNHARD, J.M., FALCIERI, E. and COCCIONI, R. (2015) Effects of lead pollution on *Ammonia parkinsoniana* (foraminifera): Ultrastructural and micro-analytical approaches. Eur. J. Histochem., v.59, pp.1–11. doi:10.4081/ejh.2015.2460

GABRIELLA, M., CRISTINA, M., BERGAMIN, L., DI, L., FREZZA, V. and LANDINI, B. (2009) Benthic foraminifera from two coastal lakes of southern Latium (Italy). Preliminary evaluation of environmental quality. Mar. Pollut. Bull., v.59, pp.268–280. doi:10.1016/j.marpolbul. 2009.08.010

GANDHI, M.S. and NATHAN, A.S. (2014) Benthic foraminifera and geochemical studies with influence on pollution studies along the coast of Cuddalore, Tamil Nadu-ITS, India. Arab. J. Geosci., v.7, pp.917–925. doi:10.1007/s12517-012-0775-3

GUTTERRES, C., SILVA, D., ANTONIO, J., NETO, B., OLINDO, R. and JR, G. (2011) Benthic foraminifera distribution in a tourist lagoon in Rio de Janeiro, Brazil: A response to anthropogenic impacts. Mar. Pollut. Bull., v.62, pp.2055–2074. doi:10.1016/j.marpolbul.2011.07.023

GYU, Y., KIM, S., UN, D., SICK, J., JUN, H., WOO, M., HAK, B., HYUN, M. and HO, Y. (2014) Decalcification of benthic foraminifera due to '"Hebei Spirit"' oil spill, Korea. Mar. Pollut. Bull., v.87, pp.276–285.

HESS, S., ALVE, E., CECILIE, H. and NORLING, K. (2013) Benthic foraminiferal responses to water-based drill cuttings and natural sediment burial: Results from a mesocosm experiment. Mar. Micropaleontol., v.101, pp.1–9.

HYAMS-KAPHZAN, O., ALMOGI-LABIN, A., BENJAMINI, C. and HERUT, B. (2009) Natural oligotrophy vs. pollution-induced eutrophy on the SE Mediterranean shallow shelf (Israel): Environmental parameters and benthic foraminifera. Mar. Pollut. Bull., v.58, pp.1888–1902. doi:10.1016/ j.marpolbul.2009.07.010

IRABIEN, M.J., CEARRETA, A., LEORRI, E., GÓMEZ, J. and VIGURI, J. (2008) A 130 year record of pollution in the Suances estuary (southern Bay of Biscay): Implications for environmental management. Mar. Pollut. Bull., v.56, pp.1719–1727. doi:10.1016/j.marpolbul.2008.07.006

JAYARAJU, N. (2011) Impact of Iron Ore Tailing on Foraminifera of the Uppateru River Estuary, East Coast of India. J. Environ. Prot., (Irvine,. Calif). v.02, pp.213–220. doi:10.4236/jep.2011.23025

JAYARAJU, N. and REDDY, B.C.S.R. (2008) The response of benthic foraminifera to various pollution sources: A study from Nellore Coast, East Coast of India. Environ. Monit. Assess., v.142, pp.319–323. doi:10.1007/s10661-007-9931-8

JORISSEN, F.J., BICCHI, E., DUCHEMIN, G., DURRIEU, J. and GALGANI, F. (2009) Deep-Sea Research II Impact of oil-based drill mud disposal on benthic foraminiferal assemblages on the continental margin off Angola. Deep. Res. Part II Top. Stud. Oceanogr., v.56, pp.2270–2291. doi:10.1016/j.dsr2.2009.04.009

KOUKOUSIOURA, O., DIMIZA, M.D., TRIANTAPHYLLOU, M. V. and HALLOCK, P. (2011) Living benthic foraminifera as an environmental proxy in coastal ecosystems: A case study from the Aegean Sea (Greece, NE Mediterranean). J. Mar. Syst., v.88, pp.489–501. doi:10.1016/j.jmarsys. 2011.06.004

LAROCHE, O., WOOD, S., TREMBLAY, L., ELLIS, J., LEJZEROWICZ, F., PAWLOWSKI, J., LEAR, G., ATLAS, J. and POCHON, X. (2016) First evaluation of foraminiferal metabarcoding for monitoring benthic impacts at an offshore oil drilling site. Mar. Environ. Res., v.120, pp.225–235. doi:http://dx.doi.org/10.1016/j.marenvres.2016.08.009

LEE, A. and NIKRAZ, H. (2015) BOD: COD Ratio as an Indicator for River Pollution. Int. Proc. Chem. Biol. Environ. Eng., v.51, pp.139–142. doi:10.7763/IPCBEE

LI, T., LI, X., SUN, G., ZHOU, Y., CHEN, F., ZHONG, H., YANG, C., LUO, W. and YAO, Y. (2015) Assessment of the impacts of trace metals on benthic foraminifera in surface sediments from the northwestern Taiwan Strait. Mar. Pollut. Bull., v.98, pp.78–94. doi:10.1016/j.marpolbul.2015.07.010

LI, T., XIANG, R. and LI, T. (2014) Influence of trace metals in recent benthic foraminifera distribution in the Pearl River Estuary. Mar. Micropaleontol., v.108, pp.13–27.

LI, Y., GANG, T., BI, H., LIN, W., PENG, W., YE, J. and CHUN, C. (2015) Responses of benthic foraminifera to the 2011 oil spill in the Bohai Sea. Mar. Pollut. Bull., v.96, pp.245–260.

LINSHY, V.N., SARASWAT, R., KURTARKAR, S.R. and NIGAM, R. (2013) Experiment to decipher the effect of heavy metal cadmium on coastal benthic foraminifer *Pararotalia nipponica* (ASANO). J. Palaeontol. Soc. India., v.58, pp.205–211.

MARTIN, R.A. and NESBITT, E.A. (2015) Foraminiferal evidence of sediment toxicity in anthropogenically influenced embayments of Puget Sound, Washington, U.S.A. Mar. Micropaleontol., v.121, pp.97–106.

MARTINEZ-COLON, M., HALLOCK, P. and GREEN-RUIZ, C. (2009) Strategies for Using Shallow-Water Benthic Foraminifers As Bioindicators of Potentially Toxic Elements: a Review. J. Foram. Res., v.39, pp.278–299. doi:10.2113/gsjfr.39.4.278

MARTINS, V., FERREIRA, E., SEQUEIRA, C., ROCHA, F. and DUARTE, A.C. (2010) Estuarine, Coastal and Shelf Science Evaluation of the ecological effects of heavy metals on the assemblages of benthic foraminifera of the canals of Aveiro (Portugal). Estuar. Coast. Shelf Sci., v.87, pp.293–304. doi:10.1016/j.ecss.2010.01.011

MARTINS, V.A., FRONTALINI, F., TRAMONTE, K.M., FIGUEIRA, R.C.L., MIRANDA, P., SEQUEIRA, C., FERNÁNDEZ-FERNÁNDEZ, S., DIAS, J.A., YAMASHITA, C., RENÓ, R., LAUT, L.L.M., SILVA, F.S., ANTONIETA, M., RODRIGUES, C., BERNARDES, C., NAGAI, R., SOUSA, S.H.M., MAHIQUES, M., RUBIO, B., BERNABEU, A., REY, D. and ROCHA, F. (2013) Assessment of the health quality of Ria de Aveiro (Portugal): Heavy metals and benthic foraminifera. Mar. Pollut. Bull., v.70, pp.18–33.

MELIS, R. and COVELLI, S. (2013) Distribution and morphological abnormalities of recent foraminifera in the Marano and Grado Lagoon (North Adriatic Sea, Italy). Mediterr. Mar. Sci., v.14, pp.432–450. doi:10.12681/mms.351

MOJTAHID, M., JORISSEN, F., DURRIEU, J., GALGANI, F., HOWA, H., REDOIS, F. and CAMPS, R. (2006) Benthic foraminifera as bio-indicators of drill cutting disposal in tropical east Atlantic outer shelf environments. Mar. Micropaleontol., v.61, pp.58–75. doi:10.1016/j.marmicro.2006.05.004

MOJTAHID, M., JORISSEN, F. and PEARSON, T.H. (2008) Comparison of benthic foraminiferal and macrofaunal responses to organic pollution in the Firth of Clyde (Scotland). Mar. Pollut. Bull., v.56, pp.42–76. doi:10.1016/j.marpolbul.2007.08.018

MUNSEL, D., KRAMAR, U., DISSARD, D., NEHRKE, G., BERNER, Z., BIJMA, J., REICHART, G.J. and NEUMANN, T. (2010) Heavy metal incorporation in foraminiferal calcite: Results from multi-element enrichment culture experiments with *Ammonia tepida*. Biogeosciences., v.7, pp.2339–2350. doi:10.5194/bg-7-2339-2010

NIGAM, R., LINSHY, V.N., KURTARKAR, S.R. and SARASWAT, R. (2009) Effects of sudden stress due to heavy metal mercury on benthic foraminifer *Rosalina leei*: Laboratory culture experiment. Mar. Pollut. Bull., v.59, pp.362–368. doi:10.1016/j.marpolbul.2009.08.014

NIGAM, R., PANCHANG, R. and BANERJEE, P. (2005) Foraminifera in Surface Sediments of Mandovi River Estuary: Indicators for Mining Pollution and High Sea Stand in Goa, India. J. Coast. Res., v.21, pp.853–859. doi:10.2112/03-0061.1

NIGAM, R., SARASWAT, R. and PANCHANG, R. (2006) Application of foraminifers in ecotoxicology: Retrospect, perspect and prospect. Environ. Int., v.32, pp.273–283. doi:10.1016/j.envint.2005.08.024

NINA, R., ALMOGI-LABIN, A., ABRAMOVICH, S. and HERUT, B. (2011) The effect of thermal pollution on benthic foraminiferal assemblages in the Mediterranean shoreface adjacent to Hadera power plant (Israel). Mar. Pollut. Bull., v.62, pp.1002–1012. doi:10.1016/j.marpolbul.2011.02.036

POPADIĆ, A., VIDOVIĆ, J., ĆOSOVIĆ, V., MEDAKOVIĆ, D., DOLENEC, M. and FELJA, I. (2013) Impact evaluation of the industrial activities in the Bay of Bakar (Adriatic Sea, Croatia): Recent benthic foraminifera & heavy metals. Mar. Pollut. Bull., v.76, pp.333–348. doi:10.1016/j.marpolbul.2013.09.039

PRAZERES, M. DE F., MARTINS, S.E. and BIANCHINI, A. (2011) Biomarkers response to zinc exposure in the symbiont-bearing foraminifer *Amphistegina lessonii* (Amphisteginidae, Foraminifera). J. Exp. Mar. Bio. Ecol., v.407, pp.116–121. doi:10.1016/j.jembe.2011.07.015

PRAZERES, M.D.F., MARTINS, S.E., BIANCHINI, A. (2012) Impact of metal exposure in the symbiont-bearing foraminifer *Amphistegina lessonii*. Proc. 12th Int. Coral Reef Symp., 13–16.

PANCHANG, R., NIGAM, R., BAIG, N. and NAIK, G.N. (2005) A foraminifera testimony for the reduced adverse effects of mining in Zuari estuary, Goa. Int. J. Env. Study, v.62, pp.579-591.

RAJA, B.C.S., JAYARAJU, N., SREENIVASULU, G., SURESH, U. and REDDY, A.N. (2016) Heavy metal pollution monitoring with foraminifera in the estuaries of Nellore coast, East coast of India. Mar. Pollut. Bull., v.113, pp.542–551.

ROMANO, E., BERGAMIN, L., GRAZIA, M., GABRIELLA, M., AUSILI, A. and GABELLINI, M. (2008) Industrial pollution at Bagnoli (Naples, Italy): Benthic foraminifera as a tool in integrated programs of environmental characterisation. Mar. Pollut. Bull., v.56, pp.439–457. doi:10.1016/j.marpolbul.2007.11.003

RUMOLO, P., MANTA, D.S., SPROVIERI, M., COCCIONI, R., FERRARO, L. and MARSELLA, E. (2009) Heavy metals in benthic foraminifera from the highly polluted sediments of the Naples harbour (Southern Tyrrhenian Sea, Italy). Sci. Total Environ., v.407, pp.5795–5802. doi:10.1016/j.scitotenv.2009.06.050

SABBAGH, E. (2016) Benthic foraminiferal proxies for pollution monitoring in almukalla coastal area, hadramout governate, Republic of Yemen. J. Foram. Res., v.46, pp.369–392.

SABEAN, J.A.R., SCOTT, D.B., LEE, K., VENOSA, A.D. (2009) Monitoring oil spill bioremediation using marsh foraminifera as indicators.Mar.Pollut.Bull.,v.59,pp.352–361. doi:10.1016/j.marpolbul.2009.08. 013

SARKAR, A., RAY, D., SHRIVASTAVA, A.N. and SARKER, S. (2006) Molecular Biomarkers: Their significance and application in marine pollution monitoring. Ecotoxicology, v.15, pp.333–340. doi:10.1007/s10646-006-0069-1

SCHINTU, M., MARRUCCI, A., MARRAS, B., GALGANI, F., BUOSI, C., IBBA, A. and CHERCHI, A. (2016) Heavy metal accumulation in surface sediments at the port of Cagliari (Sardinia, western Mediterranean): Environmental assessment using sequential extractions and benthic foraminifera. Mar. Pollut. Bull., v.111, pp.45–56. doi:10.1016/j.marpolbul. 2016.07.029

SCOTT, D.B. (2005) Pollution Monitoring in Two North American Estuaries: Historical Reconstructions Using Benthic Foraminifera. J. Foram. Res., v.35, pp.65–82. doi:10.2113/35.1.65

SOUDER, H.C., MCCLOSKEY, B., HALLOCK, P. and BYRNE, R. (2010) Shell anomalies observed in a population of Archaias angulatus (Foraminifera) from the Florida Keys (USA) sampled in 1982-83 and 2006-07. Mar. Micropaleontol., v.77, pp.71–81. doi:10.1016/j.marmicro.2010.07.005

TEODORO, A.C., DULEBA, W., GUBITOSO, S., PRADA, S.M., LAMPARELLI, C.C. and BEVILACQUA, J.E. (2010) Analysis of foraminifera assemblages and sediment geochemical properties to characterise the environment near Araçá & Saco da Capela domestic sewage submarine outfalls of São Sebastião Channel, São Paulo State, Brazil. Mar. Pollut. Bull., v.60, pp.536–553. doi:10.1016/j.marpolbul.2009.11.011

TITELBOIM, D., ALMOGI-LABIN, A., HERUT, B., KUCERA, M., SCHMIDT, C., HYAMS-KAPHZAN, O., OVADIA, O., and ABRAMOVICH, S. (2016) Selective responses of benthic foraminifera to thermal pollution. Mar. Pollut. Bull., v.105, pp.324-336.

TOPPING, J.N., MURRAY, J.W. and POND, D.W. (2006) Sewage effects on the food sources and diet of benthic foraminifera living in oxic sediment: A microcosm experiment. J. Exp. Biol., v.329, pp.239–250. doi:10.1016/j.jembe.2005.09.013

TSUJIMOTO, A., NOMURA, R., YASUHARA, M., YAMAZAKI, H. and YOSHIKAWA, S. (2006a) Impact of eutrophication on shallow marine benthic foraminifers over the last 150 years in Osaka Bay, Japan. Mar. Micropaleontol., v.60, pp.258–268. doi:10.1016/j.marmicro.2006.06.001

TSUJIMOTO, A., NOMURA, R., YASUHARA, M. and YOSHIKAWA, S. (2006b) Benthic foraminiferal assemblages in Osaka Bay, southwestern Japan: faunal changes over the last 50 years. Paleontol. Res., v.10, pp.141–161. doi:10.2517/prpsj.10.141

TSUJIMOTO, A., YASUHARA, M., NOMURA, R., YAMAZAKI, H., SAMPEI, Y., HIROSE, K. and YOSHIKAWA, S. (2008) Development of modern benthic ecosystems in eutrophic coastal oceans: The foraminiferal record over the last 200 years, Osaka Bay, Japan. Mar. Micropaleontol., v.69, pp.225–239. doi:10.1016/j.marmicro.2008.08.001

UNITED NATIONS CONVENTION ON THE LAW OF THE SEA, Adopted Montego Bay, Jamaica, 10 December 1982.

UTHICKE, S., PATEL, F. and DITCHBURN, R. (2012) Elevated land runoff after European settlement perturbs persistent foraminiferal assemblages on the Great Barrier Reef. Ecology, v.93, pp.111–121. doi:10.1890/11-0665.1

VALENTI, D., TRANCHINA, L., BRAI, M., CARUSO, A., COSENTINO, C. and SPAGNOLO, B. (2008) Environmental metal pollution considered as noise: Effects on the spatial distribution of benthic foraminifera in two coastal marine areas of Sicily (Southern Italy). Ecol. Modell., v.3, pp.449–462. doi:10.1016/ j.ecolmodel.2008.01.023

YAHIYA, M. and MIRANDA, M.T.P. (2016) Planktons as Indicators of Heavy Metal Pollution along South West Coast of India. Int. J. Sci. Res., v.5, pp.977–982.

YOUSSEF, M. (2015) Heavy metals contamination and distribution of benthic foraminifera from the Red Sea coastal area. Oceanologia., v.57, pp.236–250

Micropaleontology and its Applications
Pradeep K. Kathal, Rajiv Nigam & Abu Talib (Editors)
Scientific Publishers (India), 2018, 285-300 pp.

Coralline Algae from the Kopili Formation (Late Eocene), Jaintia Hills, Meghalaya, India

S. Kishore[1], A. K. Jauhri[2], A. P. Singh[3], R. P. Tiwari[3#*] and S. K. Singh[4]

[1]Department of Botany, University of Lucknow, Lucknow, India.
[2]Department of Geology, University of Lucknow, Lucknow, India.
[3]Department of Geology, Mizoram University, Aizawl, India
[#]Presently Vice Chancellor, Dr. Harisingh Gour Vishwavidyalaya, Sagar, M. P., India
[4]Department of Botany, Charak Institute of Education, Lucknow, India.
***Email**: rptm@rediffmail.com

Abstract

The late Eocene Kopili Formation succession exposed between Borghat and Lama in the Jaintia Hills has been studied for the taxonomic composition of coralline algae. Though not as diversified as in the Lakadong and the Prang formations of the present area, it is characterized by six species belonging to six genera of coralline algae. These taxa are first time recorded from the Kopili Formation. These include: *Hydrolithon lemonei, Harveylithon samöense, Phymatolithon calcareum, Mesophyllum alternans, Sporolithon lugeoni* and Geniculates *sensu lato.* The studied coralline taxa and associated larger benthic foraminifera suggest that the depositional environment of the lower carbonate part of the Kopili Formation was characterized by of warm, shallow-water conditions in a bathymetric zone of 20-30m under moderate-low hydrodynamic energy.

Keywords: Coralline algae, Foraminifera, late Eocene, Kopili Formation, Meghalaya.

MEMORIAL

Professor Anil Kumar Jauhri, a retired Professor in the Centre of Advanced Study in Geology, University of Lucknow, Lucknow and the working Chief Editor, Journal of the Paleontological Society of India, Lucknow did not live to see the publication of this paper. We remember him as a dedicated teacher and a scientist, who had a distinguished career in the field of Micropaleontology and dedicated himself to repay an obligation to his profession by enriching it with new knowledge. He encouraged us enough to complete the draft of this paper, a lasting tribute to his devotion to Micropaleontology.

RPT

INTRODUCTION

As limited exposures of the fossiliferous bands of the Kopili Formation are available in the Shillong Plateau, very few micropalaeontological studies have been made from this lithounit, e.g. Garo Hills (Samanta, 1964, 1965, 1968, 1969), Khasi Hills (Nagappa, 1959) and Jaintia Hills (Matsumaru and Sarma, 2010). In particular, the observations on the coralline algae have scarcely been made. Sarma et al. (2014) were first to present their findings on four taxa of this group from the Lumshnong area of the Jaintia Hills. Their work records presence of coralline algae in the Kopili Formation and indicates scope of work on this group of microfossils in the Kopili deposits. Though the coralline algal diversity is poor in comparison to the Lakadong and Prang formations (Misra et al., 2002; Sarma and Ghosh, 2006; Kishore et al., 2007, 2009), its presence in this formation is to be mentioned with sufficient details.

The present study reports the occurrence of coralline algae associated with larger benthic foraminifera from a section of the Kopili Formation exposed between the villages Borghat and Lama of the Jaintia Hills, Meghalaya (Fig. 1). The purpose of the present paper is to provide some taxonomic details of the recorded forms and discuss their palaeoenvironmental implications during deposition of the succession.

GEOLOGICAL SETTING AND STRATIGRAPHY

The study area is a part of the Shillong Plateau, which comprises the Assam shelf, Mikir Hills and the Upper Assam valley. The Shillong Plateau appears to be an uplifted part of the basement bounded by nearly east-west aligned faults towards northern and southern boundaries of the plateau (Murty, 1983). The plateau witnessed the marine transgression during the Late Cretaceous when platform conditions were established to result in the deposition of the basal sandstone. The deposition continued to occur during the Palaeogene (Palaeocene-Eocene) in shallow marine to lagoonal environment. The sedimentation was controlled by the tectonics of convergence of the Indian and Burmese (Myanmar) continents. During the phases of the least supply of terrigenous clastic material in the early Palaeogene, the carbonates were extensively deposited in this region.

Medlicott (1971), Evans (1932), Ghosh (1940), Wilson and Metre (1953), Nagappa (1959), Biswas (1962), Mathur and Evans (1964), Murthy et al. (1976), Dutta and Jain (1980), Dasgupta (1977), Ghose (1976), Samanta and Raychaudhary (1983), Jauhri and Agarwal (2001), and others contributed to the general stratigraphy of the area (Table 1).

Palaeogene carbonate sequence exposed in the Jaintia Hills is grouped as the Sylhet Limestone Group (=Sylhet Limestone Stage of Wilson and Metre, 1953; Nagappa, 1959; Dasgupta, 1977), which includes three mappable units (the Lakadong Formation, the Umlatdoh Formation and the Prang Formation), each with a carbonate member with distinctive larger foraminiferal assemblages. These units include two sandstone interbeds, one between the Lakadong Limestone and the Umlatdoh Limestone and the other between the Umlatdoh Limestone and Prang Formation (Table 1). The sequence ranges in age from the late Palaeocene to the middle Eocene and overlies the upper Palaeocene Therria Formation characterized by hard, compact sandstone. It is overlain by the Kopili Formation mainly characterized by thinly bedded shale-sandstone alternations.

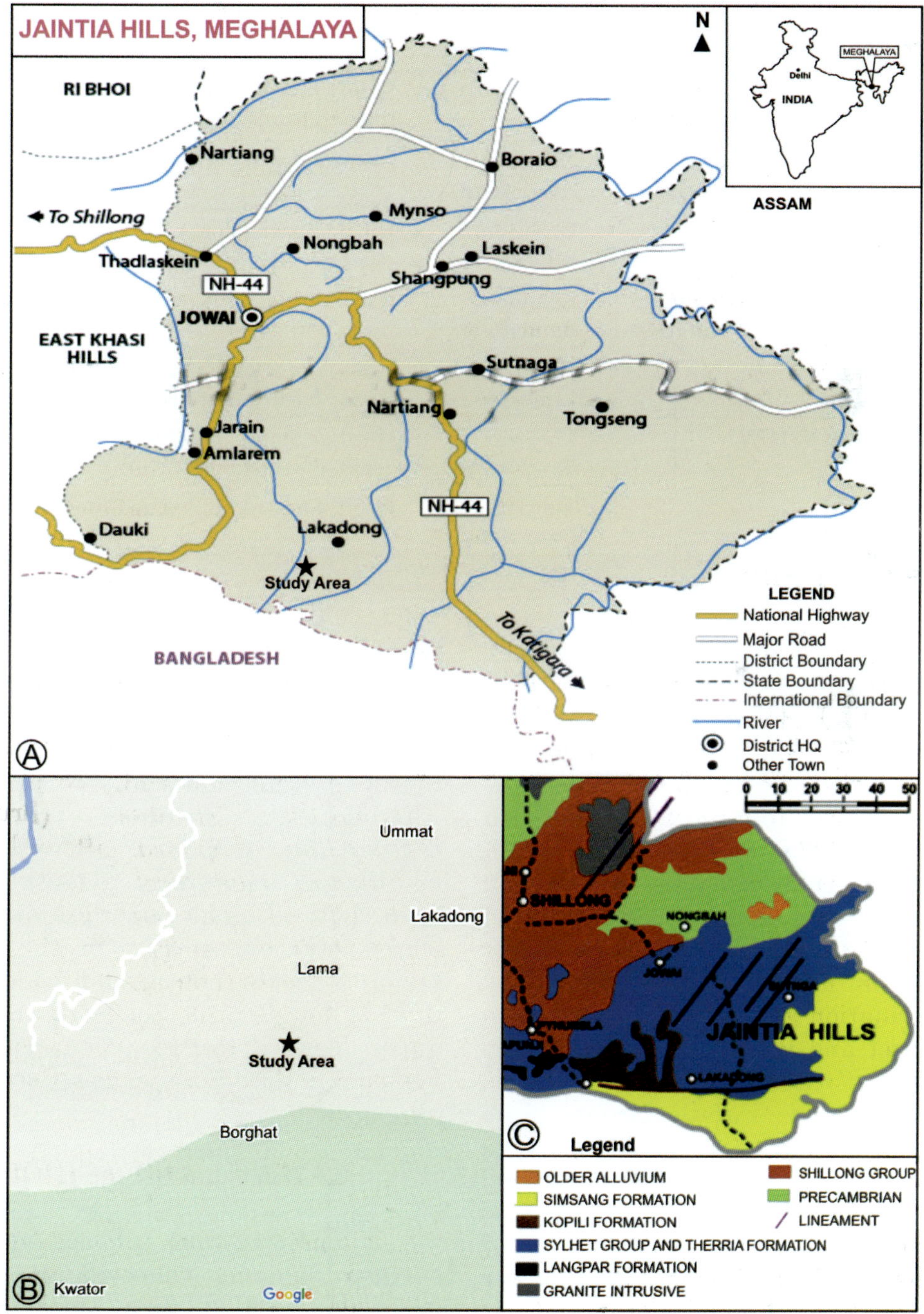

Fig.1. Location (A and B) and geological map (C) of the study area in the Jaintia Hills, Meghalaya. Inset shows position of Meghalaya in India. Star (*) indicates location of the sampling locality where the Kopili Formation is exposed (based on maps prepared by Department of Mining and Geology, Govt. of Meghalaya).

Table 1. Lithostratigraphic succession in the study area

KOPILI FORMATION

SYLHET LIMESTONE GROUP	Prang Formation		Richly fossiliferous, light to dark grey limestone with large nummulitids (50 m)
	Umlatdoh Formation	Nurpuh Sandstone	Arkosic, ferruginous sandstone (20 m)
		Umlatdoh Limestone	Hard, massive foraminiferal limestone (40 m)
	Lakadong Formation	Lakadong Sandstone	Soft, friable, light coloured sandstone with coaly horizons (12 m)
		Lakadong Limestone	Hard, compact, fossiliferous limestone rich in larger foraminifera (100 m)
THERRIA FORMATION		Therria Sandstone	Hard, compact, burrowed, coarse to medium-grained sandstone (30 m)
		Therria Limestone	Hard, compact, unbedded limestone (100 m)

LANGPAR FORMATION

The Kopili Formation was named by Evans (1932) after the Kopili River that runs through Meghalaya and Assam. This formation includes beds forming the uppermost horizons of the late Palaeocene-Eocene sequence. It is best developed in Khorungma (25°28': 92°05') in the North Cachar Hills. This formation is also well developed in the southeastern part of Shillong Plateau, *i.e.* Jowai-Badarpur Road section in the Jaintia Hills. The Kopili Formation gradually thins out from east to west and is conformable with the underlying Sylhet Limestone. It is overlain by the Barail Group. The lithology of this formation mainly consists of the alternations of fine-grained sandstone and grey and carbonaceous shale with subordinate limestone or calcareous bands and marls. The lower portion of the Kopili Formation is dominated by limestone and marl. Based on its fossil content, *e.g.* foraminifera characterized by species of *Pellatispira*, *Nummulites*, *Discocyclina* and planktic foraminifera (Pl. 2, figs. 1-9), it has been considered equivalent of Priabonian, *i.e.* Zone P15-P16 (Nagappa, 1959; Mehrotra and Banerji, 1973; Samanta, 1968, 1969). In the present material, the presence of *Nummulites striatus* (Bruguiere), *Discocyclina dispansa* (Sowerby) and *Pellatispira madaraszi* (Hantken) have been noticed. The stratigraphic distribution of these species in the Tethyan region suggests their age equivalence with the Shallow Benthic Zones 19 and 20, i.e. upper Zone P15-P16, corresponding to Priabonian (late Eocene) age (Serra-Kiel et al., 1998).

MATERIAL AND METHODS

The present work is based on nineteen outcrop samples collected at different intervals from a 32m thick sequence exposed in a road section beween Borghat and Lama villages of the Jaintia Hills. The samples were processed for the study of coralline algae and foraminifera in random thin section as they could not be disaggregated to isolate the specimens. Seventy five thin sections were prepared and effort was made to obtain the desired orientation by sectioning samples along and across the bedding planes. These microfossils were studied under light microscope.

RESULTS

Six coralline algal species have been determined: one sporolithacean algae, two hapalidiacean and three corallinacean algae. These have been assigned to six genera. Taxonomic observations were made in light of taxonomy developed through studies by Woelkerling (1988), Rasser and Piller (1999), Harvey et al. (2003), Bassi et al. (2007), Iryu et al. (2009) and Aguirre et al. (2012). In the present study, the taxonomic criteria suggested by Le Gall and Saunders (2007), Lee Gall et al. (2010) and Rösler et al. (2016) have been followed.

SYSTEMATIC DESCRIPTION

Division RHODOPHYTA Wittstein, 1901

Class FLORIDEOPHYCEAE Cronquist, 1960

Subclass CORALLINOPHYCIDAE Le Gall and Saunders, 2007

Order SPOROLITHALES Le Gall et al., 2010

Family SPOROLITHACEAE Verheij, 1993

Genus SPOROLITHON Heydrich, 1897

Sporolithon lugeoni (Pfender) Moussavian and Kuss 1990 (Pl. 1, fig. 1)

Archaeolithothamnium lugeoni Pfender, 1926, p. 324, pls. 8,9. - Johnson, 1966, p.263, pl.2. - Moussavian and Kuss, 1990, p. 934-936.

Description: Growth form encrusting to warty. Core filaments not preserved. Cell shape rectangular, without growth rhythms and trichocytes, cell layer irregular, cell fusions present. Cell length 10-18 μm, cell diameter 10-12 μm. Epithallial cells not preserved.

Sporangial compartments arranged in sori. Individual sporangial compartments with rounded to elliptical or ovoid corners in longitudinal section; sporangia 60-70 μm in height and 30-40 μm in width. Sori usually arise from a layer of elongated cells. 1-2 filaments (paraphyses) are interspersed between the sporangial compartments. (*Sample No.*: KOP-9; *Slide No.*: KF-9)

Remarks: Following Woelkerling (1988), Moussavian and Kuss (1990) who established the priority of *Sporolithon* Heydrich, 1897 over *Archaeolithothamium* Rothpletz, 1891, the present specimen is assigned to *Sporolithon*. Recently, Lee Gall *et al.* (2010) separated Sporolithaceae family from the order Corallinales and placed it under a separate order Sporolithales on the basis of multiple phlylogenetic analysis. The present species resembles *Sporolithon lugeoni* mainly in shape, size and their arrangement in sporangial compartments from the basal layer. Johnson (1966) reported *Sporolithon* (*Archaeolithothamnium*) *lugeoni* from the Palaeocene of Borneo. Misra et al. (2011) also reported it from the Prang Formation (middle-late Eocene), Meghalaya, India.

Order CORALLINALES Silva and Johansen, 1986

Family HAPALIDIACEAE Gray, 1864

Subfamily MELOBESIOIDEAE Bizzozero, 1885

Genus *MESOPHYLLUM* Lemoine, 1928

Mesophyllum alternans (Foslie, 1907) Cabioch and Mendoza, 1998
(Pl. 1, figs. 2-4)

Lithothamnion philippi f. *alternans* Foslie, 1907, p. 17. - Cabioch & Mendoza, p. 209, figs. 1-20. - Hrabovský et al., 2015, p. 9, fig. 9.

Description: Growth form encrusting to warty. Thallus organization monomerous. Core filaments non-coaxial to coaxial (indicated by an arrow: Pl. 1, fig. 3), core portion usually 130-150 μm thick, cell fusions present. Cells 16-22 μm in length and 10-12 μm in width. Peripheral filaments well developed. Peripheral cells rectangular in shape, 10-12 μm in length and 8-10 μm in width. Cell fusions present. Tetra/biosporangial conceptacles multiporate with sterile filaments. Conceptacles 170-180 μm in height and 410-650 μm in width (*Sample No.*: KOP-7, 9, 10; *Slide No.*: KF-7; KF-9; KF-10).

Remarks: The coaxial to non-coaxial core filaments, multiporate conceptacles and cell fusions helped in identification of the present specimens as *Mesophyllum*. These specimens are comparable with *Mesophyllum alternans* mainly in shape, size and arrangement of sterile filaments in the conceptacles. Hrabovský et al. (2015) reported this species from the early Langhian (early middle Miocene), Czech Republic.

Genus PHYMATOLITHON Foslie, 1898

Phymatolithon calcareum (Pallas) Adey and McKibbin, 1970
(Pl. 1, fig. 5)

Phymatolithon calcareum (Pallas) Adey and McKibbin, 1970, p. 100, fig. 1. - Rebelo *et al.*, 2014, p. 376, figs. 4 A-D.

Description: Growth form encrusting with a thallus thickness of usually 600-700 μm. Thallus organization monomerous. Core portion non-coaxial, 140-160 μm thick, cell fusions present. Cells 21-25 μm in length and 10-12 μm in width. The peripheral region of encrusting portion restricted to the dorsal part of the thallus. Cells of the peripheral region 18-20 μm in length and 8-12 μm in width. Cell fusions present. Epithallial cells rounded, present on the top of the conceptacle. Tetra/ bisporangial conceptacles multiporate, 150-160 μm in height and 450-490 μm in width (Sample No.: KOP-9; Slide No.: KF-9).

Remarks: The non-coaxial core filament, multiporate conceptacles, cell fusions and shape of the epithallial cells indicate that the present form is similar to *Phymatolithon calcareum*. Rebelo *et al.* (2014) reported this species from the late Miocene of Santa Maria Island (Azores, NE Atlantic) and Braga and Aguirre (2001) recorded it from the Pliocene of Spain. Kishore et al. (2015) reported it from the Pleistocene of Neil Island of the south Andaman, India.

Family CORALLINACEAE Lamouroux, 1812

Subfamily HYDROLITHOIDEAE Kato and Baba in Kato et al., 2011

Genus *HYDROLITHON* (Foslie) Foslie, 1909

Hydrolithon lemoinei (Miranda, 1935) Aguirre et al., 2011
(Pl. 1, figs. 6-8)

Melobesia lemoinei Miranda, 1935, p. 284, fig. 3A, B, pl. 38, fig. 1. - Aguirre et al., 2011, p. 282, fig. 6A-C. - Hrabovský et al., 2015, p. 15, fig. 16.

Description: Growth form encrusting. Specimens encrust on substrate or biogenic remains, and frequently on other coralline algae. Thallus organization dimerous, with dorsiventral organization. Primigenous filament contains only single-layered palisade tissue (indicated by an arrow: Pl. 1, fig. 6). Cell length 25-30 μm, cell diameter 15-20 μm. Postigenous filaments considerably variable in thickness, they may be single layered or multilayered. Cells of the postigenous filaments also vary in their shape and size. Cell fusions are present. Cells are 50-55μm in length and 20-25 μm in width. Epithallial cells not observed. Tetra/ bisporangial uniporate conceptacles, usually completely raised above the thallus surface. Conceptacle height 450-500 μm, diameter 650-680 μm. Cell filaments around conceptacles not preserved (Sample No.: KOP-2, 7,9,10; Slide No.: KF-9; KF-10).

Remarks: Dimeours thallus organization is a characteristic feature of the *Hydrolithon lemoinei*. Besides this, cell fusions, shape and size of postigenous and primigenous filaments and uniporate nature of conceptacles also suggest its closeness with this taxon. Though it is one of the most dominant species in the Kopili Formation, sometimes its morphology indicates resemblance with *Lithoporella melobesioides* when it has preserved only the primigenous filaments. Aguirre et al. (2011) have reported this species from the Oligocene of Spain. Hrabovský et al. (2015) have reported this species from the early Miocene of Czech Republic.

Subfamily METAGONIOLITHOIDEAE (Johansen) Rösler et al., 2016

Genus HARVEYLITHON Rösler et al., 2016

Harveylithon samöense (Foslie) Rösler et al., 2016
(Pl. 1, fig. 9)

Lithophyllum samöense Foslie, 1906, p. 20. - Keats and Chamberlain, 1994, p. 15. - Rösler et al., 2016, p. 425.

Description: Growth form encrusting to warty, with a thallus thickness of encrusting portion usually 130-150 μm. Thallus organisation monomerous. Core filaments non-coaxial, core portion usually 50-60 μm thick. Cell fusions present. Cell length 10-12 μm, cell diameter 8-10 μm. The cells of peripheral region restricted to the dorsal part of the thallus. Cell length 6-8 μm, cell diameter 6-8 μm. The epithallial cells not preserved. Cells of adjacent filaments joined by cell fusions; secondary pit- connections not observed. Trichocytes occurring singly on the thallus surface (indicated by arrows: Pl. 1, fig. 9), commonly becoming buried within the thallus. Tetra/bisporangial conceptacles uniporate (unipore indicated by an arrow: Pl. 1, fig. 9). Conceptacle height 160-170 μm, diameter 300-320 μm (Sample No.: KOP-10; Slide No.: KF-10).

Remarks: The thallus morphology, presence of cell fusions, trichocytes and uniporate conceptacles indicate affinity of the present specimen with *Harveyliton samöense*. Thallus organization and the shape and size of the cells also suggest closeness with this taxon. Harvey *et al.* (2006) reported it from the south-eastern Australia. Malakar et al. (2016) reported this species from the shallow reef of the South Andaman, India. Recently, Rösler et al. (2016) transferred this taxon (*Hydrolithon samöense*) to *Harveylithon*, a new genus created by them in the Subfamily Metagoniolithoideae.

Family CORALLINACEAE Lamouroux, 1812

Subfamily CORALLINOIDEAE (Areschoug) Foslie, 1908

Geniculates *sensu lato* Bassi, 1998 (Pl. 1, fig. 10)

Geniculates *sensu lato* Bassi, 1998, p.21 pl. 10, fig. 8.

Description: Plant geniculate, fragments of thalli about 160-170 μm in length and 60-65 μm in diameter. Genicula not preserved. Core cells of intergenicula 24-26 μm in length and 5-8 μm in diameter. Conceptacles not preserved (Sample No.: KOP-2; Slide No.: KF-2).

Remarks: The present form is comparable with Geniculates *sensu lato* Bassi in its morphological characters. Bassi (1998) reported it from the late Eocene of Northern Italy. Misra et al. (2011) also reported this form from the Prang Formation (middle-late Eocene), Jaintia Hills, Meghalaya, India.

DISCUSSION AND CONCLUSIONS

The Kopili Formation succession exposed between Borghat and Lama in the Jaintia Hills is characterized by occurrence of some coralline algae, which is not as diversified as in the Lakadong and Prang formations of this area. The latter are exceptionally rich in benthic foraminifera and coralline algae. In the present paper, six species belonging to six genera of coralline algae are recorded from this formation. All these taxa are first time reported from the study area. Out of these, the occurrence of two species (*Hydrolithon lemonei* and *Mesophyllum alternans*) is reported for the first time from India, two species (*Harveylithon samöense* and *Phymatolithon calcareum*) are known from the South Andaman, and two species (*Sporolithon lugeoni* and Geniculates *sensu lato*) are known to occur in the Prang Formation in the Jaintia Hills.

The examination of thin sections of the samples reveals presence mainly of larger benthic foraminifera and coralline algae. Among the former, the species of *Nummulites, Discocyclina* and *Pellatispira* are commonly observed, e.g. *Discocyclina dispansa* (Sowerby), *Discocyclina* sp., *Nummulites striatus* (Bruguière) and *Pellatispira madaraszi* (Hantken). Coralline algae include species of *Sporolithon, Hydrolithon, Harveylithon, Mesophyllum, Phymatolithon* and one indeterminate fragment of geniculate coralline algae.

The coralline algal forms start appearing as fragments in the sample nos. KOP-1 and 2 from the lower part of the succession but disappear in the overlying sample nos. KOP-3, 4 and 5 (Fig. 2). They reappear as fragments in the sample nos. KOP-6 and 7 of the lower part of the succession. In contrast, the lower part is characterized by rich LBF assemblage consisting of nummulitids and discocylininids in abundance. However, coralline algae are noticed in large number in the middle part (sample nos. KOP-9 and 10) of the succession. Thereafter, they decline and finally disappear in the upper part of the succession where rich LBF assemblage replaces them (Fig.2).

Their poor, fragmented development or their absence in some intervals of the lower part of the succession seems to be related with high-energy environment which does not allow their colonization. In this environment, coralline algae are hard competitors and attempt to settle either on foraminiferal shells or on other skeletal pieces as encrusting forms. They gradually enlarge the area of colonization and

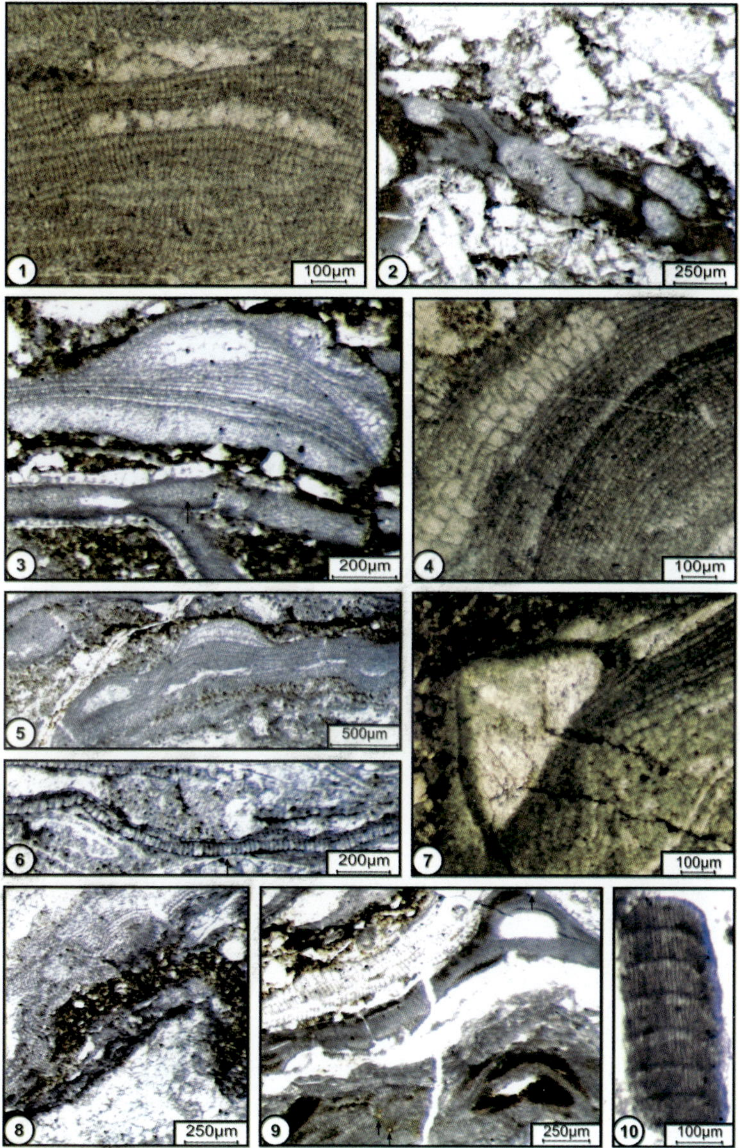

Plate1.1. *Sporolithon lugeoni*: showing peripheral filaments and sporangial compartments. **2-4.** *Mesophyllum alternans*. **2.** Showing peripheral filaments with many multiporate conceptacles. **3.** Showing non-coaxial to coaxial core filaments (indicated by an arrow) and multiporate conceptale. **4.** Enlarged view of peripheral filaments and multiporate conceptacles. **5.** *Phymatolithon calcareum*: showing non-coaxial core filaments, peripheral filaments, a row of rounded epithallial cells above the conceptacle and multiporate conceptale. **6-8.** *Hydrolithon lemonei*. **6.** Showing a single layer of primgenous (indicated by an arrow) and postigenous filaments. **7.** Enlarged view of postigenous filaments and uniporate conceptacle. **8.** Showing a multilayered postigenous filaments with variable cell morphology. **9.** *Hydrolithon samöense*: showing non-coaxial core filaments, peripheral filaments, trichocytes (indicated by arrows) and uniporate conceptale (unipore indicated by an arrow). **10.** Geniculates *sensu lato* Bassi, 1998: Enlarged view of the intergenicula of geniculate coralline algae.

Plate2.1. *Pellatispira* sp., axial section encrusted by coralline algae (KOP-9). **2**. *Discocyclina* sp., surrounded by coralline algal filaments (KOP-9). **3**. Coralline algae encrusting on *Discocyclina* sp. (KOP-10). **4**. Planktonic form, possibly *Turborotalia* sp. (KOP-9). **5**. *Pellatispira* sp., equatorial section (KOP-7). **6**. *Discocyclina* sp. (upper left), *Nummulites* sp. (centre) and coralline algae (lower left) (KOP-10). **7**. Discocyclinids, Ornamented rotaliids (referable to *Pellatispira* sp. - centre left and lower (KOP-9). **8**. Coralline algae (upper) encrusting on *Discocyclina* sp. (lower) (KOP-10). **9**. Rotaliids, discocyclinids, and coralline algal fragments (KOP-9).

register increase in their diversity in response to improved (moderate to low-energy) environment. The coralline algae of the lower part include mainly the fragments of *Hydrolithon lemonei*, fragments of melobesioids and geniculate coralline algae. The middle part of the succession is marked by the well-developed coralline algae characterized by *Hydrolithon, Harveyliton* and *Mesophyllum* in abundance and poor development of *Sporolithon* in some intervals. In the upper part, discocyclinids appear in abundance and are accompanied by subordinate nummulitids and few rotaliids. It seems that much deeper ramp environment may have prevented the occurrence of corallines.

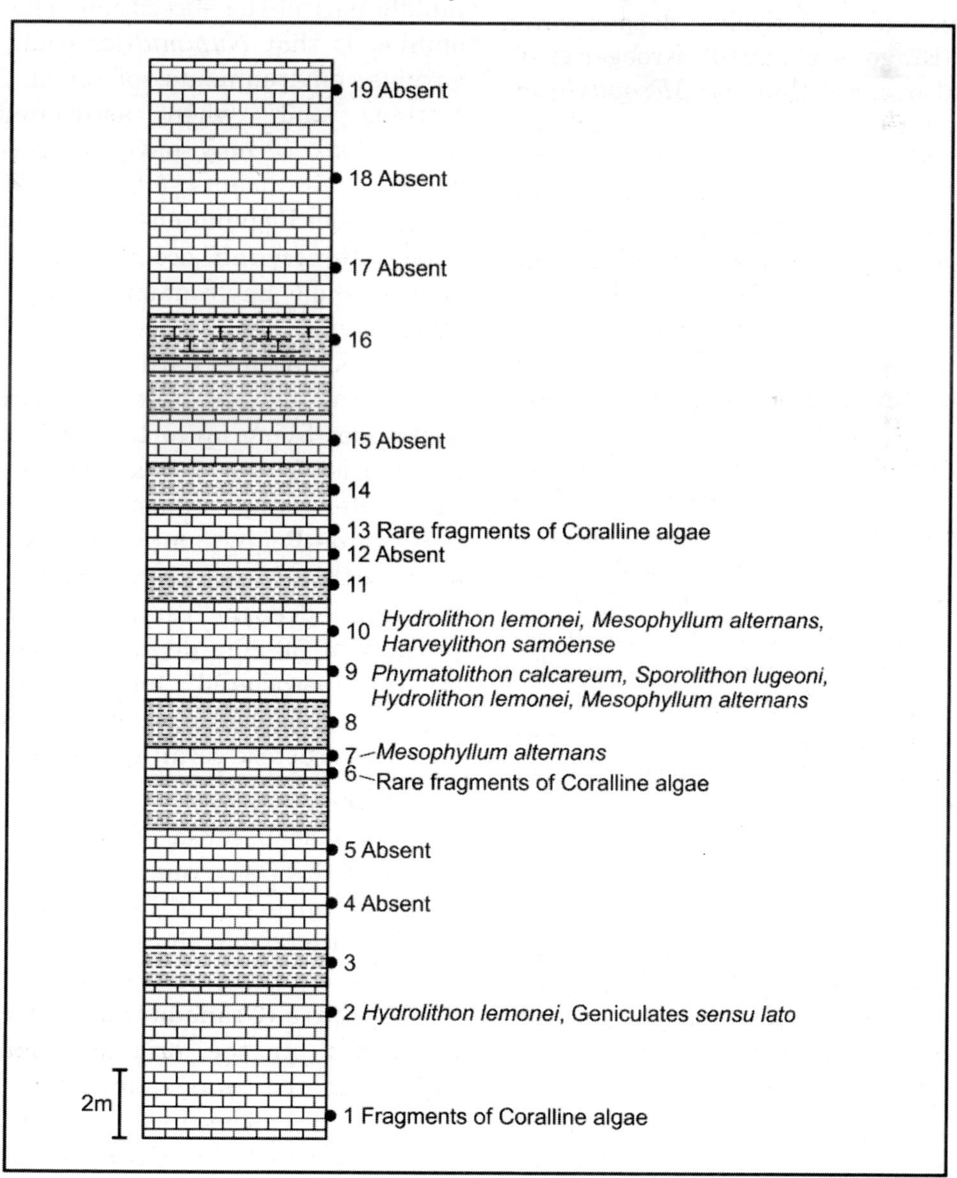

Fig.2. Distribution of coralline algae in the lithocolumn.

The presence of *Hydrolithon, Harveyliton* and *Mesophyllum* indicates tropical climate (Braga et al., 2010). While *Hydrolithon*, classified under Hydrolithoideae related to Mastophoroideae in habit, and *Harveylithon*, assigned to Metagoniolithoideae, characterize the shallow-water carbonate environment of the photic zone, the dominance of *Mesophyllum* is suggestive of relatively deeper-water settings (Braga et al., 2010). Kroeger et al. (2006) also note that is *Mesophyllum*-dominated association occurs in the lower part of the photic zone (30-40m) in warm-water environments. However, *Hydrolithon lemoinei* is the most abundant species in the present material. Next come in rank *Mesophyllum*, which is followed by *Harveylithon* and *Phymatolithon*, the poorly represented forms in the present material. The latter is adapted to deeper water environment like that of *Mesophyllum*. This order of abundance suggests that the depositional environment was situated near the shallower limit of lower photic zone, possibly in a depth zone between 30m-40m, where the moderate-low hydrodynamic conditions may have allowed corallines to develop encrusting to warty growth forms. Most of the algal crusts in the present material are relatively thinner indicating their colonization in the lower photic zone (Rasser and Piller, 1997).

The presence of a geniculate coralline form as fragments, *i.e.* Geniculates *sensu lato* (a shallow-water form) in the present material is not suitable for palaeoenvironmental interpretation. It seems that the presence of corallines of both shallow and deep photic zones suggests a tropical/ subtropical environment with bathymetric range from 30 to 40m in moderate-low hydrodynamic conditions suitable for the growth for encrusting to warty growth-forms.

Foraminiferal associations studied in these samples comprise assemblages dominated by *Nummulites*, *Discocylina*, *rotaliids* and *Pellatispira*. These forms prominently occur in association with coralline algae in the samples from the middle part of the succession. The general opinion is that *Nummulites* could live in various environments of open, marine settings and, when associated with *Discocylina*, it may have occupied deep-water environments (40-80 m) of shelf (Geel, 2000; Beavington-Penny and Racey, 2004). *Discocylina* occupies a relatively deeper shelf environment, extending to 100m water depth (Geel, 2000). *Pellatispira* which prefers hard bottoms, occupies outer to deeper shelf environment, sometimes living up to 130m (Geel, 2000). Summarising the faunal association on carbonate ramps during the Eocene, Beavington-Penny and Racey (2004) suggest that *Nummulites* and *Discocylina* live on deeper ramp environments, possibly at 50-80m depth. Ecological data of the associated foraminiferal taxa support palaeobathymetric observations based on crustose coralline algae.

Acknowledgements: The authors express a deep sense of gratitude to the Heads of the Departments of Geology and Botany, University of Lucknow, Lucknow for encouragement, library and laboratory facilities. Authors are grateful to Prof. Julio Aguirre (Granada, Spain) for helpful comments on the original manuscript. This research was supported by the Council of Scientific and Industrial Research, New Delhi (Grant No. 24 (0279)/ 05/EMR-II).

REFERENCES

ADEY, W.H. and MCKIBBIN, D.L. (1970) Studies on the maerl species *Phymatolithon calcareum* (Pallas) nov. comb. and *Lithothamnium corallioides* Crouan in the Ria de Vigo. Botanica Marina, v.(13), 100-106.

AGUIRRE, J., BRAGA, J.C. and BASSI, D. (2011) Taxonomic assessment of coralline algal species (Rhodophyta; Corallinales and Sporolithales) described by Pfender, Lemoine and Miranda from northern Spain type localities. Ann. Naturhist. Mus. Wien, Serie A, v.(113), 267-289.

AGUIRRE, J., BRAGA, J.C., DE REVIERS, B. and WOELKERLING, W.J. (2012) Reassessment of Lemoine's newly discovered types of fossil corallines (Corallinales, Rhodophyta) preserved at the Museum national d'histoire naturelle, Paris. Cryptogamie, Algologie, v.(33/3), 289-326.

BASSI, D (1998) Coralline red algae (Corallinales, Rhodophyta) from the upper Eocene Calcare di Nago (Lake Garda, Northern Italy). Annali dell' Universita di Ferrara, v.(7), 1-50.

BASSI, D., HOTTINGER, L. and NEBELSICK, J.H. (2007) Larger foraminifera from the Upper Oligocene of the Venetian area, North-East Itlay. Palaeontology, v.(50/4), 845-868.

BEAVINGTON-PENNY, S.J. and RACEY, A. (2004) Ecology of extant nummulitids and other larger benthic foraminifera: applications in palaeoenvironmental analysis. Earth-Science Reviews, v.(67), 219-265.

BISWAS, B. (1962) Stratigraphy of the Mahadeo, Langpar, Cherra and Tura formations, Assam, India. Bull. Geol. Min. Met. Soc. India, v.(25), 1-48.

BRAGGA, J.C. and AGUIRRE, J.M. (2001) Coralline algal assemblage in upper Neogene reef and temperate carbonates in southern Spain. Palaeogeo., Palaeoclimat., Palaeoeco., v.(175), 27-41.

BRAGA, J.C., BASSI, D. and PILLER, W.E. (2010) Palaeoenvironmental signi□cance of Oligocene-Miocene coralline algae. *In*: Carbonate systems during the Oligocene-Miocene climatic transition (Mutti, M., Piller, W.E. and Betzler, C., Eds.), Special Publication Number 42(IAS), Wiley-Blackwell, p. 165-182.

CABIOCH, J. and MENDOZA, M.L. (1998) *Mesophyllum alternans* (Foslie) comb. nov. (Corallinales, Rhodophyta), a Mediterraneo-Atlantic species, and new considerations on the *Lithothamnion philippii* Foslie complex. Phycologia, v.(37/3), 208-221.

DASGUPTA, A.B. (1977) Geology of the Assam-Arakan region. Quart. Jour. Min. Metal. Soc. India, v.(49), 1-54.

DUTTA, S.K. and JAIN, K.P. (1980) Geology and Palynology of the area around Lumshnong, Jaintia Hills. Meghalaya, India. Biol.Mem., v.(5/1), 56-81.

EVANS, P. (1932) Explanatory notes to accompany a table showing the Tertiary succession in Assam. Trans. Min. Metal. Soc. India, v.(27), 155-260.

FOSLIE, M. (1906) Algologiske notiser II. Det Kongelige Norske Videnskabers Selskabs Skrifter, v.(2), 1-28.

FOSLIE, M. (1907) Algologiske notiser IV. Det Kongelige Norske Videnskabers Selskabs Skrifter, v.(6), 1-30.

GEEL, T. (2000) Recognition of stratigraphic sequences in carbonate platform and slope deposits: empirical models based on microfacies analysis of Palaeogene deposits in Southeastern Spain. Palaeogeo., Palaeoclimat., Palaeoeco., v.(155), 211-238.

GHOSH, A.M.N. (1940) The stratigraphical position of the Cherra Sandstone, Assam. Rec. Geol. Surv. India, v.(75), 17-18.

GHOSE, B.K. (1976) Palaeogene reef-complex of Meghalaya Plateau and its oil potentialities. Sci. Cult., v.(42), 248-253.

HARVEY, A.S., BROADWATER, S.T., WOELKERLING, W.J. and MITROVSKI, P.J. (2003) *Choreonema* (Corallinales, Rhodophytra): 18s rDNA and resurrection of the Hapalidiaceae for the subfamilies Choreonematoideae, Austrolithoideae and Melobesioideae. Jour. Phycol., v.(39), 988-998.

IRYU, Y., BASSI, D. and WOELKERLING, W.J. (2009) Re-assessment of the type collections of fourteen Corallinalean species (Corallinales, Rhodophyta) described by W. Ishijima (1942-1960). Palaeontology, v.(52/2), 401-427.

JAUHRI, A.K. and AGARWAL, K.K. (2001) Early Palaeogene in the South Shillong Plateau, NE India: local biostratigraphic signals of global tectonic and oceanic changes. Palaeogeo., Palaeoclimat., Palaeoeco., v.(168), 187-203.

JOHNSON, J.H. (1966) Tertiary Red algae from Borneo. Bull. Brit. Mus. (Nat. Hist.), Geology, v.(11), 255-280.

HRABOVSKÝ, J., BASSO, D. and DOLÁKOVÁ, N. (2015) Diagnostic characters in fossil coralline algae (Corallinophycidae: Rhodophyta) from the Miocene of southern Moravia (Carpathian Foredeep, Czech Republic). Jour. Syst. Palaeont., http://dx.doi.org/10.1080/ 14772019.2015.1071501.

KATO, A., BABA, M. and SUDA, S. 2011. Revision of the Mastophoroideae (Corallinales, Rhodophyta) and polyphyly in nongeniculate species widely distributed on Pacifc coral reefs. J. Phycol., v.(47), 662-672.

KEATS, D.W. and CHAMBERLAIN, Y.M. (1994). Three species of *Hydrolithon* (Rhodophyta, Corallinaceae): *Hydrolithon onkodes* (Heydrich) Penrose and Woelkerling, *Hydrolithon superficiale*sp. nov. and *H. samöense* (Foslie) comb. nov. from South Africa. S. African J. Bot., v.(60), 8-21.

KISHORE, S., JAUHRI, A.K., SINGH, S.K., MALAKAR, B. and MISRA, P.K. (2015). Coralline algae from Neill West Coast Formation (Pleistocene), Neil Island, South Andaman, India. J. Pal. Soc. India, v.(60/2), 57-69.

KISHORE, S., SINGH, A.P., JAUHRI, A.K., MISRA, P.K., SINGH, S.K. and LYNGDOH, B.C. (2007) Coralline algae from the Prang Formation (Eocene) of the South Khasi Hills, Meghalaya, India. Rev. de Paléobiol., Genève, v.(26/2), 615-623.

KISHORE, S., SINGH, A.P., SINGH, S.K., TIWARI, R.P., JAUHRI, A.K. and MISRA, P.K. (2009) Geniculate Coralline algae from the Prang Formation (Eocene) of the Jaintia Hills, Meghalaya, India. J. Pal. Soc. India, v.(54/2), 189-195.

KROEGER, K.F., REUTER, M. and BRACHERT, T.C. (2006) Palaeoenvironmental reconstruction based on non-geniculate coralline red algal assemblages in Miocene limestone of Central Crete. Facies, v.52(3), 381-409.

LE GALL, L., PAYRI, C. E., BITTNER, L. and SAUNDERS, G.W. (2010) Multigene phylogenetic analyses support recognition of the Sporolithales ord. nov. Mol. Phylogen. Evol., v.(54), 302-305.

MALAKAR, B., SINGH, D., KISHORE, S., VENU, S. and SINGH, S.K. (2016) *Hydrolithon* (Foslie) Foslie (Hydrolithoideae, Corallinaceae) from shallow reef of South Andaman, India. Inter. Jour. Appl. Biol. Pharma. Tech., USA, v.(7/3), 121-129.

MATHUR, L.P. and EVANS, P. (1964). Oil in India. *In*: Proc. 18th Int. Geol. Congr., New Delhi. p.1-85.

MATSUMARU, K. and SARMA, A. (2010) Larger foraminiferal biostratigraphy of the lower Tertiary of Jaintia Hills, Meghalaya, NE India. Micropaleont., v.(50/6), 539-565.

MEDLICOTT, H.B. (1871) Geological sketch of the Shillong Plateau in north-eastern Bengal. Mem. Geol. Surv. India, v.(7), 151-207.

MEHROTRA, K.K. and BANERJI, R.K. (1973) Middle-Upper Eocene Biostratigraphy of Khasi and Jaintia Hills based on planktonic and larger foraminifera. J. Pal. Soc. India, v.(18), 22-26.

MIRANDA, F. (1935) Algas coralináceas fósiles del Terciario de San Vicente de la Barquera (Santander). Boletín de la Soc. Esp. de Hist. Nat., v.(35), 279-287.

MISRA, P.K., JAUHRI, A.K., KISHORE, S. and SINGH, S. K. (2002) Calcareous algae from the Lakadong Formation of the South Shillong Plateau, NE India. Rev. Paléobiol., Genève, v.(21/2), 717-734.

MISRA, P.K., JAUHRI, A.K., TIWARI, R.P., KISHORE, S., SINGH, A.P. and SINGH, S. K. (2011) Coralline algae from the Prang Formation (middle-late Eocene) of the Lumshnong area, Jaintia Hills, Meghalaya, India. J. Geol. Soc. India, v.(78/10), 355-364.

MOUSSAVIAN, E. and KUSS, J. (1990) Typification and status of *Lithothamnium aschersonii* Schwager, 1883 (Corallinaceae, Rhodophyta) from Palaeocene limestone of Egypt. A contribution to the priority of the genera *Archaeolithothamnium* Rothpletz and *Sporolithon* Heydrich. Berliner geowiss. Abh., v.(120/2), 929-942.

MURTY, K.N. (1983) Geology and hydrocarbon prospects of Assam shelf: recent advances and present status. *In*: L.L. BHANDARI, et al. (Eds.), Petroliferous Basin of India. Petrol. Asia Jour., v.1, 1-14.

MURTHY, M.V.N., CHAKARABORTY, C. and TALUKDAR, S.C. (1976) Stratigraphic revision of the Cretaceous-Tertiary sediments on the Shillong Plateau. Rec. Geol. Surv. India, v.(107), 80-90.

NAGAPPA, Y. (1959) Foraminiferal biostratigraphy of Cretaceous-Eocene succession in the India-Pakistan-Burma region. Micropaleontology, v.(5/2), 141-181.

PFENDER, J. (1926) Les Mélobésiées dans les calcaires crétacés de la Basse-Provence. Mém. Soc. Géol. France, n. sér., v.(3/6), 32.

RASSER, M. and PILLER,W.E. (1997) Depth distribution of calcareous associations in the Northern Red Sea (Safaga, Egypt) and their geological implications. Proc. 8th Int. Coral. Reef. Symp. 1., p. 743-748.

RASSER, M. and PILLER,W.E. (1999) Application of neontological taxonomic concepts to the Late Eocene coralline algae (Rhodophyta) of the Austrian Molasse Zone. Jour. Micropalaeont., v.(18), 67- 80.

REBELO, A.C., RASSER, M.W., RIOSMENA-RODÍGUEZ, R., NETO, A.I. and ÁVILA, S.P. (2014) Rhodolith forming coralline algae in the Upper Miocene of Santa Maria Island (Azores, NE Atlantic): a critical evaluation. Phytotaxa, v.(190/1), 370-382.

Rösler, A., Perfectti, F., Peña, V. and Braga, J.C. (2016) Phylogenetic relationships of Corallinaceae (Corallinales, Rhodophyta): Taxonomic implications for reef-building Corallines. J. Phycol. v.(52), 412-431.

ROTHPLETZ, A. (1891) Fossile Kalkalgen aus den Familien der Codiaceen und der Corallineen. Zeitsch. Deutsch. Geol. Gesell., v.(43), 295-322.

SAMANTA, B.K. (1964) The occurrence of Indo-Pacific *Discocyclina* in eastern India. Micropaleontology, v.(10), 339-353.

SAMANTA, B.K. (1965) *Discocyclina* from the Upper Eocene of Assam, India. Micropal., v.(11), 415-430.

SAMANTA, B.K. (1968) *Nummulites* (Foraminifera) from the Upper Eocene Kopili Formation of Assam, India. Palaeontology, v.(5), 669–682.

SAMANTA, B.K. (1969) Eocene planktonic foraminifera from the Garo Hills, Assam, India. Micropalaeontology, v.(15/3), 325-350.

SAMANTA, B.K. and RAYCHOUDHURY, A.K. (1983) A revised lithostratigraphic classification of the Cretaceous-Lower Tertiary shelf sediments of the Eastern Khasi and Jaintia Hills, Meghalaya. Quar. J. Min. Metal. Soc. India, v. (55/3), 101-129.

SARMA, A. and GHOSH, A.K. (2006) A new record of calcareous algae from Shella Formation (Jaintia Group) of South Jaintia Hills, Meghalaya, India. Curr. Sci., v.(90/9), 1276-1281.

SARMA, A., GHOSH, A.K. and SARKAR, S. (2014) First record of Coralline Red Algae from the Kopili Formation (late Eocene) of Meghalaya, N-E India. Natl. Acad. Sci. Lett., v.(37/6), 503–507.

SERRA-KIEL, J., HOTTINGER, L., CAUS, E., DROBNE, K., FERRANDEZ, C., JAUHRI,

A. K., LESS, G., PAVLOVEC, R., PIGNATTI, J., SAMSO, J. M., SCHAUB, H., SIREL, E., STROUGO, A., TAMBAREAU, Y., TOSQUELLA, J. AND ZAKREVSKAYA, E. (1998) Larger foraminiferal biostratigraphy of the Tethyan Paleocene and Eocene. Bull. de la Soc. Geol. de France, v.(169/2), 281-299.

WILSON, C.E. and METRE, W.B. (1953) Assam and Arakan. *In*: V.C. Illing, (Ed.), World's Oil Fields. The Eastern Hemisphere. The Science of Petroleum. London, Oxford University Press, 6, pp.119-123.

WOELKERLING, W.J. (1988) The Coralline Red algae: An analysis of the genera and subfamilies of nongeniculate Corallinaceae. Brit. Mus. (Nat. Hist.), Oxford University Press, London and Oxford, 268p.

Micropaleontology and its Applications
Pradeep K. Kathal, Rajiv Nigam & Abu Talib (Editors)
Scientific Publishers (India), 2018, 301-309 pp.

The Significance of Spinosity in Foraminifera – A New Perspective

N. Rajeshwara Rao[1*] and S. M. Hussain[2]

[1]Department of Applied Geology, University of Madras, Guindy, Chennai - 600 025, India
[2]Department of Geology, University of Madras, Guindy, Chennai - 600 025, India
***Email:** raonandamuri@gmail.com

Abstract

Spinose ornamentation has been recorded in several benthic genera, and spines have always been the prime possession of all that remain of the extant planktic species, except, those of the genus *Globorotalia*. In the planktic counterparts, the significance of spinosity has been rather well established, but little is known about benthic taxa that grow spines; little else is known about whether there are any ecological drivers that influence or even promote the growth of spines. It is to be noted, however, that this study does not concern deep-water foraminiferal genera (*Bulimina*, *Uvigerina*, *Trifarina*, *Mimosina* etc), but refers to the inner continental shelf, wherein a particular species, *Ammonia dentata* (=??? *Asterorotalia dentata*) (Parker and Jones, 1865) exhibits a morphological signature that could actually provide a clue to paleo-offshore depth. It has been observed across several transects in the south-western Bay of Bengal from off Chennai in the north to off Karaikal in the south, that spines in *A. dentata* reflect fairly reliable ecological conditions in the inner continental shelf. While the short, blunt spines in this species characterize near-shore, relatively shallower waters influenced by the effects of tides, waves and currents, with more coarser substrates, morphological variants with long, slender spines typify comparatively deeper waters that are more serene, with substrates that are much finer, with relatively higher mud (silt + clay) content.

Key words:. Foraminifera, spinosity, environmental significance.

INTRODUCTION

The presence of spines has been observed and recorded for the past several decades in both benthic and planktic foraminifera. With regard to the latter, however, there are only ~40 living taxa in all the oceans and seas put together. They may be less in number, but they have been, and are, the most sensitive group of microscopic protists that have either evolved or become extinct in the geologic record. Among the extant planktic species, majority of them are spinose, except for a couple of commonly and abundantly occurring taxa - *Globorotalia menardii* and *G. tumida.* On the other hand, there are innumerable benthic counterparts that grow spines, which constitute an integral part of the adult ornamentation, but have never been within any generic or even suborder-level constraint which, probably, needs to be addressed in the classification of foraminifera.

Hilbrecht and Thierstein (1996) observed interesting benthic behavioural activities in planktic foraminifera in laboratory culture experiments, such as re-orientation, crawling and burrowing. *Globigerinella siphonifera* created well-organized burrows and excavated sediment on a circular pile. They found that *Globorotalia menardii* and *G. siphonifera* have specific crawling and burrowing orientations, with crawling speeds of up to 1 mm.min^{-1} and burrowing speeds of 0.5–2.0 mm.hr^{-1}. Four other species – *Globigerinoides ruber, G. sacculifer, Hastigerina pelagica* and *Orbulina universa* – did not exhibit any benthic activity and died on the sediment substrates. They concluded that the ability to live in a benthic life mode may be a strategy in species adapted to the deep chlorophyll maximum, or neritic environments, where they exploit floating snow aggregates (marine snow) as a pseudo-benthic habitat and as a source of food.

In cultures without sediment substrates, the spinose species – *G. siphonifera, G. sacculifer, H. pelagica* and *O. universa* (terminal stage) – recovered their spines and floated, while the non-spinose *G. menardii* settled on the glass bottom and kept crawling on its umbilical side for several days. Such culture experiments, without sediment substrates, allowed reproduction of well-known behaviours in planktic foraminifera (Hemleben et al., 1985). In unkeeled, spinose planktic species, the globular chamber walls serve as mechanical support for the radial spines (Hottinger, 2006). The latter, in their turn, support the clothes-lines produced by the pseudopods to expose the symbionts during daytime to light and free gas exchange in the open water (Hemleben et al., 1989). Gary and Healy-Williams (1986) applied an automated imaging system to examine the morphology of three species of *Bolivina* from the north-western Gulf of Mexico. They opined that overall test shape appeared to be the most important discriminator of depth, but changes in the periphery (i.e. spinosity) of the foraminifera were also noted in one species. Their studies emphasized the fact that quantification of such changes offers the possibility of precise and rapid depth determinations with an accuracy and resolution not possible with many other techniques.

According to Furbish and Arnold (1997), planktic foraminifera adjust their buoyancies to compensate for gravitational settling by manufacturing, in part, low-density lipids or gases. They also opined that the process of growing acicular spines, as in *Orbulina universa* and *Globigerinoides sacculifer*, results in two contrasting consequences: both increasing and decreasing the settling velocity, due to increased weight and fluid drag, respectively. Erez (2003) opined that in spinose planktic foraminifera, the shell also serves as basis for the spines, which allow for the construction of a wide pseudopodial "spider web" that help to catch their prey (small copepods). This is important particularly in turbulent shallow water where they are found. The spines may also assist in dispersing the symbiotic algae in such a way as to provide CO_2 and to reduce oxygen toxicity by diffusion. A partitioning of finite spine mass into many moderate to short spines is less effective in producing drag than one involving fewer long spines.

Long spines are, however, more susceptible to mechanical breakage due to the torque that viscous forces apply to them. However, most of the spines in planktic foraminifera are discarded or broken during separation of the tests from the sediment samples, or even while mounting them for SEM studies (Huber, 2006). Planktic foraminifera can also be adapted to different ocean layers of particular temperatures and densities. They have fewer buoyancy problems in cooler denser water, so the species living their have a lower test-porosity than forms in warmer or shallower waters. These warm forms improve buoyancy with test-porosity and prominent spines (Becker, 2006). In cultured specimens of *Hastigerina pelagica*, masses of gametes were observed to flow along its spines in cytoplasmic strands in the early to middle stages of gradual release (Ketten and Edmond, 1979). Among the smaller algae (zooxanthellae) associated with planktic foraminifera, *Globigerinoides ruber*, *Gs. sacculifer* and *Orbulina universa*, Bé et al. (1977) observed a circadian periodicity of ingress into the test in the evening and egress to the rhizopodial network and the distal parts of the spines during the day (Köhler-Rink and Kühl, 2005). According to Hohenegger (1994), *Calcarina gaudichaudii* is able to withstand higher hydrodynamic energy due to its robust spines as compared to *Neorotalia calcar*. According to observations made by Hull et al. (2011) in Monterey Bay, California, there were numerous copepods in various stages of ingestion on the spines of *Hastigerinella digitata* and a conspicuous lack of detritus attached to the spine network, providing support for the hypothesis that this species is an obligate carnivore. Spero (1988) collected pre-sphere, trochospiral tests of *Orbulina universa* from the surface waters of the Pacific Ocean off Santa Catalina island, California, and documented that approximately 24 hours before the release of gametes, the spines were resorbed and a veneer of calcite was deposited on the surface of the test. Growth of thin, elongate spines provides one way for larger-sized planktic foraminifera to increase fluid drag and reduce their settling speed (Huber, 2006).

Van Hengstum (2008) carried out paleoenvironmental analysis using thecamoebians and foraminifera in Mexican anchialine caves and observed that all the seven samples constituting an assemblage from a mean depth of 4.6 m contained more specimens of the thecamoebian strain *Centropyxis aculeata* "aculeata" (with spines) than the strain without spines, *C. aculeata* "discoides" (without spines). Earlier, Reinhardt et al. (1998) divided *C. aculeata* and *C. constricta* into two ecophenotypes or 'strains" based on the presence or absence of spines, which Van Hengstum (2008) endorsed by concluding that shells with the presence or absence of spines are, indeed, different ecophenotypes of the same species. We have evidence of this relationship from elsewhere in the globe even in non-tropical settings: from the salt marshes of British Columbia (Riveiros et al., 2007) and the U.K. (Gehrels et al., 2001). These results have indicated that in environments with marginal-salinity, the expression of spines on the thecamoebian test is related to salinity. However, since spines are also expressed on thecamoebians in lacustrine environments, other ecological or environmental factors such as pH, percent moisture etc) may control the expression of spines on the phenotype (Van Hengstum, 2008).

We presume that there are only few authentic records about the functions of spines in either benthic or planktic foraminifera, especially from the Bay of Bengal or even the Arabian Sea. Perhaps the earliest observation about the presence of variable spinosity in a benthic foraminiferal species, *Ammonia dentata* (Parker and Jones), was by Ragothaman (1974), wherein he noted that specimens of this species were morphologically of two types: one with short, stout, blunt spines; and the other, with long, slender spines, an important observation that went unnoticed for at least 25 years, until a similar trend in the species was observed by Rajeshwara Rao (1998), in his studies on foraminiferal taxa off Karikkattukuppam, near Chennai. Subsequent studies from various offshore regions, all south of Chennai, confirm the occurrence of two morphological types of *A. dentata*. Ghose (1966), however, had earlier made several morphometric measurements on *Asterorotalia trispinosa* (Thalmann), and opined that the spines in this species probably aided its presence in suspension just above the sediment-water interface.

MATERIALS AND METHODS

This study encompasses recorded data that has been compiled from studies on benthic foraminifera from off Karikkattukupam (near Chennai), Pondicherry, Cuddalore, Nagapattinam and Karaikal. All the samples examined are surface samples collected from various transects off the coastal towns mentioned above. It is to be noted that the gradient of the continental shelf decreases from Chennai through Karaikal, which is the southernmost onshore port in the study area. This obviously implies that depths to the water column decrease, in general, with latitudes due south, due to progradation of the Cauvery Delta (in recent geological history) and the break-up of the continents (in the geological past). The Cauvery Delta is the southern-most of the four major east coast deltas. With its apex at about 30 km inland west of Tanjavur (Tanjore), the river flows eastward through its five distributaries. Studies of Meijrink (1971) and Sambasiva Rao (1982) revealed a number of abandoned distributary channels and relict beach ridges. Both these studies suggested that the earliest shoreline representing the Holocene maximum transgression was in the south-eastern side of the delta and that there was a continuous counter-clockwise shift in the delta lobes towards north-east. The Cauvery Delta at that point of time could be surmized as a cuspate delta with its mouth somewhere in the present offshore region (Nageswara Rao et al., 2005).

All the surface sediment samples were collected with a Peterson grab sampler, and a part of each sediment sample was preserved in a buffered solution of 10% formaldehyde. A pinch of $CaCO_3$ was added to sustain the pH of the preservative. This procedure was undertaken to differentiate the "living" from the dead foraminiferal populations; Rose Bengal staining technique was used for this purpose. It has been well established that the rose Bengal technique, albeit overestimating "living" populations when compared to the "Sudan Black" technique, is still the widely used technique (Murray, 1991). Moreover, staining in tests of agglutinated species is easily recognized if rose Bengal is used (Bernhard, 1988).

The bottom sediment samples were collected off Karikkattukuppam (Rajeshwara Rao, 1998); off Pondicherry and Cuddalore (Ramesh, 2005); off Nagapattinam and Karaikal, (Stephen, 2005). Table 1 presents the data pertaining to the depth of the water column.

RESULTS AND DISCUSSION

Eggins et al. (2004) observed that spines in the planktic foraminiferal species, *Orbulina universa* d'Orbigny, 1839, dampen buoyancy-driven vertical migration, are used to catch prey, and provide a framework on which to host algal symbionts (Hemleben, 1989). According to Erez (2003), in spinose planktic foraminifera, the shell also serves as basis for the spines, which allow for the construction of a wide pseudopodial "spider web" that help to catch their prey (small copepods). This is important particularly in turbulent shallow water where they are found. The spines may also assist in dispersing the symbiotic algae in such a way as to provide CO_2 and to reduce oxygen toxicity by diffusion.

Table 1. Variations (in %) in *Ammonia dentata* with water depth off different regions in the Bay of Bengal.

Water depth (m)	Locations	*A. dentata* (SBS) in %	*A. dentata* (LSS) in %
7.0	Off Karikattukuppam	99.01	0.99
10.0	Off Karikattukuppam	97.11	2.89
14.0	Off Karikattukuppam	97.26	2.74
18.0	Off Karikattukuppam	97.15	2.85
21.0	Off Karikattukuppam	98.01	1.99
24.0	Off Karikattukuppam	96.17	3.83
28.0	Off Karikattukuppam	90.17	9.83
34.0	Off Karikattukuppam	76.29	23.71
38.0	Off Karikattukuppam	43.19	56.81
41.0	Off Karikattukuppam	29.16	70.84
45.0	Off Karikattukuppam	23.23	76.77
48.0	Off Karikattukuppam	10.11	89.89
51.0	Off Karikattukuppam	9.72	90.28
53.0	Off Karikattukuppam	7.67	92.33
55.0	Off Karikattukuppam	5.18	94.82

Water depth (m)	Locations	*A. dentata* (SBS) in %	*A. dentata* (LSS) in %
6.1	Off Pondicherry	98.08	1.92
8.7	Off Pondicherry	98.03	1.97
10.5	Off Pondicherry	97.26	2.74
11.7	Off Pondicherry	97.01	2.99
12.4	Off Pondicherry	97.14	2.86
16.8	Off Pondicherry	95.13	4.87
6.0	Off Cuddalore	98.98	1.02
11.5	Off Cuddalore	97.77	2.23
15.5	Off Cuddalore	95.19	4.81
24.7	Off Cuddalore	90.18	9.82
30.3	Off Cuddalore	47.26	52.74
34.9	Off Cuddalore	39.15	60.85
33.6	Off Cuddalore	39.01	60.99
20.3	Off Nagapattinam	94.19	5.81
22.5	Off Nagapattinam	93.22	6.78
25.6	Off Nagapattinam	93.26	6.74
27.3	Off Nagapattinam	90.19	9.81
29.1	Off Nagapattinam	90.02	9.98
28.5	Off Nagapattinam	90.33	9.67
29.3	Off Karaikal	22.77	77.23
25.8	Off Karaikal	90.11	9.89
17.4	Off Karaikal	94.43	5.57
12.9	Off Karaikal	97.14	2.86
7.8	Off Karaikal	99.04	0.96
6.1	Off Karaikal	98.17	1.83
7.2	Off Karaikal	97.79	2.21

Growth of thin, elongate spines provides one way for larger sized planktic foraminifera to increase fluid drag and reduce their settling speed. As Huber (2006) remarked, "While the option to grow elongate spines did not exist for Cretaceous planktic foraminifera, since spinose shells did not evolve until the early Danian, we have to assume that Cretaceous planktic foraminifera may have had the ability to manufacture lipids and/or gases, and thus could have offset some of the negative buoyancy effects of calcite addition to the shell".

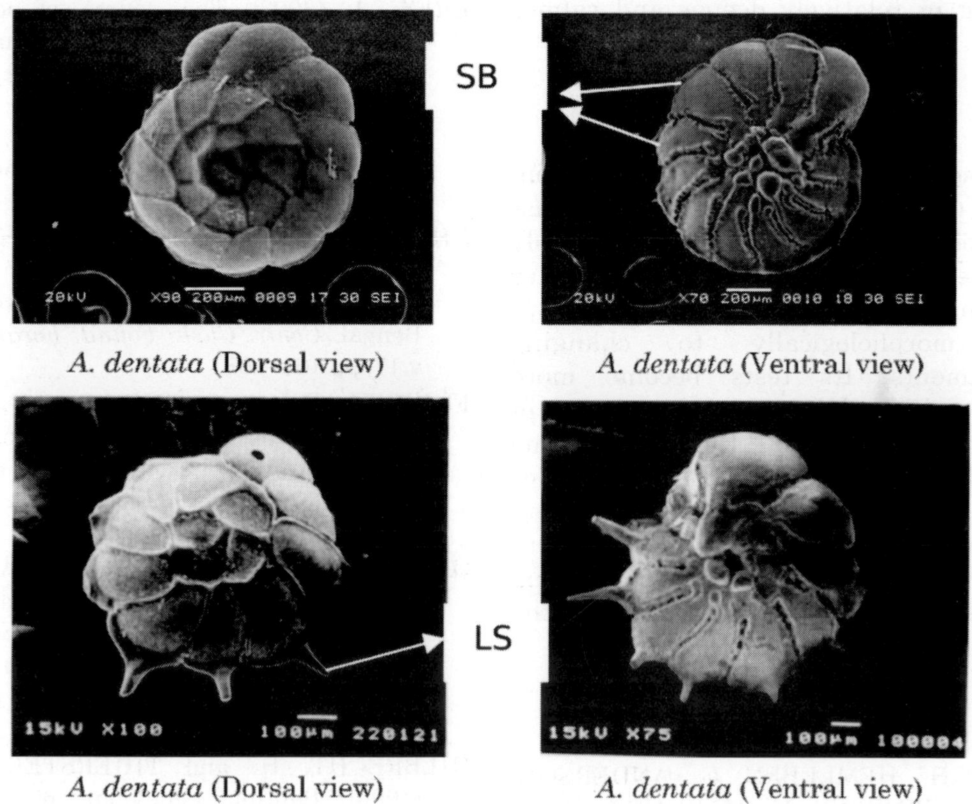

A. dentata (Dorsal view) *A. dentata* (Ventral view)

A. dentata (Dorsal view) *A. dentata* (Ventral view)

Fig.1. *Ammonia dentata* (a) with SB, dorsal view (b) ventral view (c) with LS (d) ventral view.

The present study has revealed an interesting fact. Spines do offer the opportunity of buoyancy or other functions, but in a species such as *Ammonia dentata* (Parker and Jones, 1865), they seem to modify themselves differently to shallow-water, relatively more turbulent conditions (short, blunt spines of more robust tests) as opposed to comparatively calmer conditions at greater water depths, where the short, blunt spines seem to evolve into long, slender spines, perhaps abetting better buoyancy. This might also be a pointer to variable feeding habits of the same species with morphological variations in the form of the nature of its spines.

Table 1 presents data assimilated from various sites off the south-east coast of India with regard to the correlation of the nature of spinosity in *Ammonia dentata* in relation to depth. It is well established that the sediment grain size generally decreases with increase in water depth offshore, so it is presumed that the ability of this species to produce long, slender spines in deeper habitats is related to its feeding strategy (suspension feeder?). Culture experiments should throw more light on this aspect in the future. Fig. 1 shows the difference in morphology between specimens possessing short, blunt spines in shallower waters and those having long, slender spines dwelling in relatively deeper and calmer waters.

CONCLUSION

Based on the data generated from several offshore regions off the south-east coast of India, it can be reasonably concluded that *Ammonia dentata* Parker and Jones, 1865 has the ability to adapt itself morphologically to changing environments. Its tests become more robust with short, blunt spines in shallower waters, while the tests become less robust while growing long, slender spines in comparatively deeper waters (~>30 m water depth). This would be very useful in inferring paleodepths utilizing this species and its spinosity provided the spines are well preserved.

REFERENCES

BÉ, A.W.H., HEMLEBEN, C., ANDERSON, O.R., SPINDLER, M., HACUNDA, J. and TUNTIVATE-CHOY, S. (1977) Laboratory and field observations of living planktic foraminifera. *Micropal.,* v.23(2), pp.155-179.

BECKER, S. (2006) Modern foraminifera: Biological and ecological basics. www.geo.tufreiberg.de/oberseminar/os05_06/sindy_becker.pdf

BERNHARD, J.M. (1988) Postmortem vital staining in benthic foraminifera: duration and importance in population and distribution studies. *J. Foram. Res.*, v.18, pp. 143–146.

EGGINS, S.M., SADEKOV, A. and De DECKKER, P. (2004) Modulation and daily banding of Mg/Ca in *Orbulina universa* tests by symbiont photosynthesis and respiration: a complication forseawater thermometry? *Earth Planetary Sci. Letters*, v. 225, pp. 411– 419.

EREZ, J. (2003) The source of ions for biomineralization in foraminifera and their implications for paleoceanographicproxies, in biomineralization. *In*: DOVE, P., DE YOREO, J. J. and WEINER, S. (Eds.), *Mineral. Soc. America*, Washington, D.C., pp.115–149.

GHOSE, B.K. (1966) *Asterorotalia trispinosa* (Thalmann). A spinose rotaliid foraminifera from Digha Beach, southern Bengal. *Contr. Cush. Found. Foram. Res.,* v.17, pp.104–108.

FURBISH, D.J. and ARNOLD, A.J. (1997) Hydrodynamic strategies in the morphological evolution of spinose planktic foraminifera. *Bull. Geol. Soc. America*, v.109(8), pp.1055–1072.

GEHRELS, W.R, ROE, H.M. and CHARMAN, D.J. (2001) Foraminifera, testate amoebae and diatoms as sea level indicators in UK salt marshes: a quantitative multiproxy approach. *J. Quarter. Sci.,* v.16(3), pp.201–220.

HILBRECHT, H. and THIERSTEIN, H.R. (1996) Benthic behaviour of planktic foraminifera. *Geology*, v.24, pp.200–202.

HEMLEBEN, C. (1989) Modern Planktic Foraminifera, Springer Verlag, New York: 363 p.

HEMLEBEN, C., SPINDLER, M., BREITINGER, I. and DEUSER, W. G. (1985) Field and laboratory studies on the ontogeny and ecology of some globorotaliid species from the Sargasso Sea off Bermuda. *J. Foram. Res.,* v.15, pp.1505–1512.

HEMLEBEN, C., SPINDLER, M. and ANDERSON, O. R. (1989) *Modern Planktic Foraminifera*. Springer, New York.

HOTTINGER, L. (2006) The “face” of benthic foraminifera. Bollettino della Società Paleontologica Italiana, v.45(1), 2006, pp.75–89.

HUBER, B.T. (2006) Convergent evolutionary patterns of shell ornamentation in Cretaceous planktic foraminifera: Was there method to the madness? *Anuario do Instituto de Geociencias*, v. 29(1), pp.68–69.

HULL, P.M., OSBORN, K.J., NORRIS, R.D. and ROBISON, B.H. (2011) Seasonality and depth distribution of a mesopelagic foraminifer, *Hastigerinella digitata*, in Monterey Bay, California. *Limno. Oceanogr.*, v.56(2), pp.562–576.

KÖHLER-RINK, S. and KÜHL, M. (2005) The chemical environment of the symbiotic planktic foraminifer *Orbulina universa. Marine Bio. Res.*, v.1, pp.68–78.

MEIJRINK, A.M.J. (1971) Reconnaissance of Quaternary geology of Cauvery Delta. *J. Geol. Soc. India*, v.2, pp.113–124.

MURRAY, J.W. (1991) *Paleoecology of Benthic Foraminifera.* Longman Group, London, 397 p.

NAGESWARA RAO, K., TAKAYASU, K., SADAKATA, N. and HEMA MALINI, B. (2005) Studies on the Holocene evolution of the east coast deltas of India: present status and future prospects. In: Chen, Z., Saito, Y. and Goodbred S. L. (Eds.), *Mega Deltas of Asia – Geol. Evolution Human Impact*, China Ocean Press, pp. 190–199.

RAGOTHAMAN, V. (1974) *The study of foraminifera from off Porto Novo, Tamil Nadu State.* Ph.D. thesis, University of Madras, Chennai, 246 p.

RAJESHWARA RAO, N. (1998) Recent foraminifera from inner shelf sediments of the Bay of Bengal, off Karikkattukuppam, near Chennai, India. Ph.D. thesis, University of Madras, Chennai, 283 p.

RAMESH, R.K. (2005) Recent benthic foraminifera from the inner shelf sediments off Pondicherry and Cuddalore, south-east coast of India. Ph.D. thesis, University of Madras, Chennai, 224 p.

SAMBASIVA RAO (1982) Morphology and evolution of modern Cauvery Delta, Tamilnadu, India. *Transactions of the Indian Institute of Geography*, v.4(1), pp.67–78.

SPERO, H. J. (1998) Ultrastructural examination of chamber morphogenesis and biomineralization in the planktic foraminifer *Orbulina universa. Marine Bio.*, v.99(1), pp.9–20.

STEPHEN, V. P. (2005) Recent foraminifera, geochemistry and sedimentology of the inner shelf sediments off Nagapattinam, Tamilnadu, south-east coast of India, Ph.D. thesis, University of Madras, Chennai, 242 p.

VAN HENGSTUM, P.J. (2008) Paleoenvironmental analysis using thecamoebians and foraminifera in Mexican anchialine caves: a focus on Aktun Ha (Carwash) Mexico. M.Sc. thesis, McMaster University, Ontario, Canada, 112 p.

Micropaleontology and its Applications
Pradeep K. Kathal, Rajiv Nigam & Abu Talib (Editors)
Scientific Publishers (India), 2018, 311-330 pp.

Experimental Studies on Benthic Foraminifera: an Update on the Changing Trends and Challenges over the Past Decade

Sujata R. Kurtarkar*[1], Linshy V. N.[2], Amrata Kaithwar[1], Rajeev Saraswat[1], and Rajiv Nigam[1]

[1]Micropalaeontology Lab, National Institute of Oceanography, Dona Paula, Goa - 403004, India
[2]408, Technocrats Appt., Plot No.25, Sec.56, Gurgaon - 122011, India
[*]**Email**: sujatak2@gmail.com

Abstract

The increasing application of foraminiferal proxies in various fields of paleontological research led to an urge to explore the fundamental biological processes of foraminifera. As a result, geologists and paleontologists started venturing in to the less travelled roads of living foraminiferal studies. There are several inherent properties that render foraminifera as a group more difficult to handle experimentally than numerous other popular laboratory protozoa. But it didn't stop the researchers from attempting all possible techniques to reveal whatsoever possible on the biological processes of foraminifera and to align this knowledge to refine many field based results. The outcome was an enormously abundant set of information and knowledge from different corners of the world in the form of books, manuscripts, reports and reviews from time to time. We had reviewed the experimental studies carried out on benthic foraminifera, with the aim to refine applications of benthic foraminifera, in 2007. This review is a continuation of the same covering the last decade from 2006 to 2016 with compilations ranging from studies, which refined the understanding of factors that bring out changes in benthic foraminiferal abundance, diversity, morphology, isotopic and chemical composition of the test. We report that increasing efforts are now made to understand the effect of ambient parameters on trace element partitioning in benthic foraminifera. Additionally, the effect of ocean acidification on benthic foraminifera has also been studied in detail, under controlled laboratory conditions. The species specific response in a majority of the cases necessitates continuous efforts to comprehensively understand the factors affecting benthic foraminifera.

Keywords: Culture, benthic foraminifera, element, isotope, acidification, pollution, abundance, morphology.

INTRODUCTION

Among protozoan, foraminifera are widely used as a proxy for paleoceanography/ paleoclimate studies. The sensitivity of foraminifera to the environment and capability to incorporate the ambient signatures in the test makes it a very efficient tool to study the past climate. A majority of the foraminiferal proxies were developed based on samples collected from field. Off late, researchers are venturing into the laboratory experiments on foraminifera to validate field based findings. Foraminifera being exclusively marine organism, it is often difficult to recreate the field conditions in laboratory. Even though, only basic laboratory equipment is required to establish and maintain a thriving culture of the benthic foraminifera from a single asexually reproducing specimen (Seears and Wade, 2013), advance techniques are required for complicated experiments. In spite of the limitations, many researchers nationally as well as internationally have ventured in the less travelled road of foraminiferal culture studies. The laboratory culture studies help in understating the response of foraminiferal morphology, diversity, abundance to one or a combination of parameters.

Linshy et al. (2007) compiled the laboratory studies carried out on benthic foraminifera from the year 1935 to 2007. Based on the thorough review, it was suggested that the proper understanding of parameters affecting the morphology as well as the exact mechanism that causes deformities in benthic foraminifera, is still lacking. They further observed that the factors affecting the isotopic composition of benthic foraminifera were not explored fully as well as the calibration equations to reconstruct quantitative paleoclimatic information from benthic foraminifera, have to be developed. A comprehensive review on the morphology based biodiversity and trophic diversity of deep sea benthic foraminifera was provided by Gooday et al. (2008). Later, Martinez-Colon et al. (2009), du Châtelet and Debenay (2010) and Frontalini and Coccioni (2011) compiled the studies covering the application of benthic foraminifera as pollution bio-indicators. Subsequently, Kontakiotis et al. (2016) compiled the studies to assess the reliability of foraminiferal Mg/Ca thermometry by comparing field-samples and culture experiments. This review is a continuation of the laboratory culture studies on foraminifera, from 2007 to present. With the advancement of technology, laboratory culture studies have increased profusely. The International (Table 1) and National studies (Table 2) are discussed separately to compare the gaps effectively.

Table 1. Foraminiferal culture studies on benthic foraminifera (International Scenario).

Sr. No.	Authors and Year of Publication	Study Details
1	Kohoa et al., 2008	Reported the response of benthic foraminifera to phytodetritus, either directly or indirectly due to enhanced bacterial activity; the response was measured as an increase in total standing stalk of foraminifera.
2.	Mc Corkle et al., 2008	Difference in the offset of $\delta^{13}C$ and $\delta^{18}O$ was noticed between the cultured specimens and the water in which the specimens were growing.
3.	Rollion-Bard et al., 2008	Studied influence of seawater carbonate chemistry and temperature on the intra-shell oxygen isotope ratio in *Amphistegina.*
4.	Dissard et al., 2009	Effect of the fluorescence indicator calcein on Mg and Sr incorporation into foraminiferal calcite.
5.	Kuroyanagi et al., 2009	Growth rate, shell weight and the number of chambers, generally decreased with decerasing pH, after 10 weeks of culture in asexually produced individually of *Marginopora kudakajimensis.*
6.	Nomaki et al., 2009	In situ feeding experiment using ^{13}C-labelled unicellular algae, showed microbial degradation of ^{13}C labeled algal material and the production of bacterial biomass within 2 days. The biomass produced was gradually turned over by respiration or predation within 6 days.
7.	de Nooijer et al., 2009a	Physiological controls on seawater uptake and calcification in the benthic foraminifer *Ammonia tepida.*
8.	de Nooijer et al., 2009b	Foraminifera promote calcification by elevating their intracellular pH.
9.	Pucci et al., 2009	Experimental results show that all dominant foraminiferal taxa from the sixteen short sediment cores from a 35 m deep site in the Adriatic Sea survive strong hypoxic conditions.
10.	Sabean et al., 2009	Monitoring of oil spill using marsh foraminifera as indicators.
11.	Alve, 2010	Studied the response of benthic foraminiferal taxa to food for two years under laboratory conditions.
12.	Allison et al., 2010	Evaluated the effect of pH, $\Delta[CO_3^{2-}]$ on Mg/Ca in *Elphidium williamsoni.*
13.	Alve and Goldstein, 2010	Propagules of certain species are sufficiently resilient to survive transport, remain dormant for two years, and then start growing and reproducing once conditions are favourable.
14.	Baras et al., 2010	Studied the calibration of $\delta^{18}O$ of cultured benthic foraminiferal calcite as a function of temperature.
15.	Dissard et al., 2010a	Studied the impact of salinity on Mg/Ca and Sr/Ca ratio in *Ammonia tepida.*

Sr. No.	Authors and Year of Publication	Study Details
16.	Dissard et al., 2010b	Impact of seawater pCO_2 on calcification and Mg/Ca and Sr/Ca ratio in *Ammonia tepida*.
17.	Flipsson et al., 2010	Calibrated paleotemperature proxies from $\delta^{18}O$ and Mg/Ca.
18.	Munsel et al., 2010	Studied the effect of heavy metal Ni, Cu and Mn in *Ammonia tepida*.
19.	Raitzsch et al., 2010	Investigated the effect of calcite saturation state on Mg and Sr incorporation into benthic foraminifera *Heterostegina depressa* (high Mg calcite, symbiont-bearing) and *Ammonia tepida* (low Mg calcite, symbiont-barren).
20.	Allison et al., 2011	Effect of seawater pH and calcification rate on test Mg/Ca and Sr/Ca in *Elphidium williamsoni*.
21.	Dueñas-Bohórquez et al., 2011	Studied the impact of carbonate ion concentration on Mg and Sr incorporation in *Ammonia tepida* and *Heterostegina depressa*.
22.	Enge et al., 2011	Studied the response of individual benthic foraminifera to a simulated phytodetritus pulse.
23.	Goldstein and Alve, 2011	Studied viability of foraminiferal communities from coastal propagule banks.
24.	Geslin et al., 2011	Observed the oxygen respiration rates of benthic foraminifera.
25.	Haynert et al., 2011	Effect of high pCO_2 concentration on benthic foraminifera *Ammonia aomoriensis*.
26.	Hikami et al., 2011	Studied the response of two reef foraminifera harboring different algal symbionts to ocean acidification.
27.	Mojtahid et al., 2011	Studied grazing of intertidal benthic foraminifera on bacteria by using pulse-chase radio-tracing.
28.	Prazeres et al., 2011	Studied the effect of zinc (Zn) on symbiont-bearing foraminifer *Amphistegina lessonii*.
29.	Toyofuku et al., 2011	Evaluated the effect of different salinity and temperature on Mg/Ca and $\delta^{18}O$.
30.	Denoyelle et al., 2012	Evaluated the application of benthic foraminifera as potential bio-indicator.
31.	Frontalini and Coccioni, 2012	Studied the response of benthic foraminiferal assemblages to Copper.
32.	Glas et al., 2012	pH within extracellular microenvironment decreases during calcification in symbiont-free benthic foraminifera *Ammonia sp.*
33.	Prazeres et al., 2012	Observed the impact of metal exposure in symbiont-bearing foraminifer *Amphistegina lessonii*.
34.	Uthicke and Fabricus, 2012	Studied productivity, respiration, and abundance of symbiont bearing foraminifer *Marginopora vertebralis* near natural CO_2 seeps, as well as under artificially enhanced pCO_2.

Sr. No.	Authors and Year of Publication	Study Details
35.	Vogel and Uthicke, 2012	Investigated the impact of ocean acidification on calcification rates of large benthic foraminifera (*Amphistegina radiata* and *Heterostegina depressa* and dinoflagellate-bearing *Marginopora vertebralis*).
36.	Wit et al., 2012	Assessed the effect of vital processes on Mg/Ca in benthic foraminifer *Bulimina marginata*.
37.	Hess et al., 2013	Effect of burial by water-based drill cuttings and natural test sediment were investigated in a mesocosm experiment.
38.	Ishida, et al., 2013	Observed the effects of CO_2 on benthic biota.
39.	Keul, et al., 2013	Confirms the strong potential of *Ammonia* sp. foraminiferal size-normalized weights (SNW) as a [CO_3^{2-}] proxy.
40.	Keul, et al., 2013a	Confirms the strong potential of foraminiferal U/Ca as a [CO_3^{2-}] proxy.
41.	Langlet et al., 2013	Observed the effect of experimentally induced anoxia in insitu experiment on benthic foraminifera.
42.	McIntyre-Wressnig et al., 2013	Effect of ocean acidification on the growth and survival of symbiont-bearing benthic foraminifer *Amphistegina gibbosa*.
43.	Seears and Wade, 2013	A note on the laboratory culture of benthic foraminifer *Cornuloculina balkwilli*.
44.	Enge et al., 2014	Role of benthic foraminifera in uptake of phytodetritus in the core regions of the Indian margin OMZ.
45.	Geslin et al.,2014	The survival rate, growth and reproduction capacities of *Ammonia tepida*, *Melonis barleeanus* and *Bulimina marginata* were studied under strongly oxygen depleted conditions.
46.	Haynert et al., 2014	Culture experiment was conducted on benthic foraminifera to observe its response to ocean acidification in its natural sediment environment.
47.	de Nooijer et al., 2014	Studied Mg/Ca and Sr/Ca variability in clones of benthic foraminifera *Ammonia tepida*.
48.	Nardelli et al., 2014	Reported that benthic foraminifera not only survive, but also calcify under anoxic conditions.
49.	Diz et al., 2015	Described an experimental set up to carry out ocean acidification experiments with non-symbiotic benthic foraminifera (*Miliolinella* spp.).
50.	Frontalini et al., 2015	The response of *Ammonia parkinsoniana* to different concentrations of Pb was evaluated at the cytological level.
51.	Prazeres et al., 2015	Assessed the biochemical and morphological impact of different pH on *Amphistegina lessonii* and *Marginopora vertebralis*.
52.	Vigier et al., 2015	Suggested lithium isotope of foraminifera as a novel proxy for ocean dissolved inorganic carbon (DIC).

Sr. No.	Authors and Year of Publication	Study Details
53.	Wollenburg et al., 2015	Evidence for methane seepage-related change in $\delta^{13}C$ of *Cibicides wuellerstorfi*.
54.	Brouwer, et al., 2016	Reported diet shift and population dynamics of estuarine foraminifera under experimentally induced hypoxia.
55.	Cesbron et al., 2016	Reported the uptake of oxygen in intertidal sediments which is almost five times more than the maximum recorded by other authors.
56.	Enge et al., 2016	Foraminifera dominate the short-term processing of phytodetritus in the OMZ core but are less important in the lower OMZ boundary region of the Indian margin.
57.	Frontalini et al., 2016	Examined the response of *Ammonia parkinsoniana*, to different concentration of Hg.
58.	Yu et al., 2016	Observed the potential application of *Ammonia aomoriensis* mean proloculus size as a salinity index.
59.	van Dijk et al., 2017b	Impact of pH and [CO_3^{2-}] on Zn incorporation.
60.	van Dijk et al., 2017a	Observed the element incorporation in hyaline and porcelaneous foraminifera as a function of pCO_2.

Table 2. Foraminiferal culture studies on benthic foraminifera (National Scenario)

Sr. No.	Authors and Year of Publication	Study Details
1	Nigam et al., 2008	Response of *Rosalina leei* to different salinity and temperature.
2	Nigam et al., 2009	Effect of sudden stress due to heavy metal mercury on benthic foraminifer *Rosalina leei*.
3	Saraswat et al., 2011	Reported that the optimum temperature for growth and reproduction of *Rosalina globularis* are different.
4	Kurtarkar et al., 2011	*Rosalina leei* can recover from adverse effect of short-term salinity changes.
5	Kurtarkar et al., 2013	Effect of temperature and salinity on stable isotopic composition of shallow water benthic foraminifera *Rosalina* sp. and *Pararotalia nipponica*.
6	Linshy et al., 2013	Observed the effect of heavy metal cadmium on benthic foraminifera *Pararotalia nipponica*.
7	Linshy et al., 2014	Specimens of *Ammonia tepida* and *Bolivina variabilis* were subjected to the marked green algae *Dunaliella tertiolecta* for 2, 7, 21 and 42 days and were analyzed for carbon signals.
8	Saraswat et al., 2015	Reported that salinity induced pH changes affect calcification as well as reproduction in benthic foraminifera *Rosalina globularis*.
9	Kurtarkar et al., 2015	Assessed the effect of calcein incorporation on physiological processes of benthic foraminifera.

EFFECT OF PHYTODETRITUS PULSE ON BENTHIC FORAMINIFERA

Food is one of the essential factors for growth and other activities of all living organisms. Foraminifera mostly feed upon organic matter, including small organisms, mainly diatoms, bacteria, coccolithophores, dinoflagellates, and parts of other plants and animals. It is believed that the type and amount of phytodetritus can alter foraminiferal population. The effect of *Thalassiosira pseudonana* (species of marine centric diatom) on benthic foraminifera was studied by Koho et al. (2008) and it was found that, several foraminiferal taxa, namely *Melonis barleeanum, Bigenerina cylindrica, Chilostomella oolina,* responded to the diatom bloom and its abundance increased. The vertical distribution of these infaunal species was, however not significantly influenced by the additional food. The abundance of only two infaunal species, viz. *Pullenia sp.* and *Trochammina sp.*, increased in surface sediments. Foraminifera have diverse food preferences. Based on long-term laboratory monitoring of benthic foraminifera under different phytodetritus regimes, benthic foraminifera can be grouped into three categories. The Category 1 included taxa dependent on seasonal supply of fresh phytodetritus to maintain their population. As compared to the fresh phytodetritus dependent taxa, Category 2 included taxa that do not require regular supply of fresh organic material and instead utilize organic material and degradation products. Surprisingly, Category 3 included taxa that were able to maintain their population for two years without any supply of fresh organic material (Alve, 2010). Benthic foraminifera immensely control the fate of various lipid biomarkers derived from primary producers and subsequent carbon cycling at the deep-sea floor (Nomaki et al., 2009). The grazing of intertidal calcareous benthic foraminifera (*Haynesina germanica* and *Ammonia beccarii,* and the single-chambered agglutinated species *Psammophaga sp.*,) on bacteria (*Halomonas sp.*), was observed by ^{3}H and ^{14}C labelling technique (Mojtahid et al., 2011). Both *Ammonia tepida* and *Bolivina variabilis*, showed fast and strong ingestion of the added food after 2 days, followed by a decreasing isotopic signal in the biomass with time (Linshy et al., 2014). The work suggested radioisotope labeling technique as a promising tool to study foraminifera prey interactions and to understand the trophic pathways in foraminifera and the fate of ingested carbon under laboratory condition.

CALCIFICATION AND REPRODUCTION

A majority of benthic foraminifera are calcareous in nature. The secretion of calcareous test under a diverse set of conditions is a complicated phenomenon. The ocean acidification is expected to adversely affect calcification. Therefore, several attempts have been made to understand the calcification process. The relation between seawater up take and calcification in juvenile specimens of *Ammonia tepida* was observed under controlled laboratory condition and it was found that endocytosis of seawater is a part of the calcification process (de Nooijer et al., 2009a). Benthic foraminiferal abundance depends on reproduction efficiency and frequency under different ambient conditions. Benthic foraminiferal propagules (numerous small juveniles) can survive transport, remain "dormant" for two years, and then start growing and reproducing once conditions are favourable (Alve and Goldstein, 2010). Additionally, the propagule banks provide a tool for documenting changes in coastal assemblages that potentially result from warming or cooling climates (Goldstein and Alve, 2011).

EFFECT OF OXYGEN ON BENTHIC FORAMINIFERA

Oxygen is another important parameter that defines the microhabitat of benthic foraminifera. Depending upon the climate change and eutrophication, the concentration of oxygen may vary. The response of benthic foraminifers to different oxygen concentration varies from species to species. *Nouria polymorphinoides* and *Nonionella turgida* show a clear tendency to move to the oxygen deficient sediment surface in the nitrogen-bubbled cores, whereas, *Bulimina* spp. and *Eggerella scabra* did not show such a migrational response (Pucci et al., 2009). The differential response was attributed to the nutrition concentration at the sediment water interface. Observation of oxygen respiration rates and fluxes for seventeen species of benthic foraminifera by using microelectrodes suggest that foraminifera contribute more to anaerobic organic matter mineralization than to aerobic organic matter mineralization (Geslin et al., 2011). The relative foraminiferal contribution to aerobic and anaerobic carbon remineralization, however varies from location to location. Aerobic carbon remineralization attain 7% of diffusional oxygen uptake in intertidal sediments, and is almost five times more than the maximum reported by Geslin et al. (2011) (Cesbron et al., 2016). Due to the differential response to oxygen availability, both the calcareous as well as agglutinated benthic foraminifera can survive up to 10 months of anoxia with co-occurring hydrogen sulfides (Langlet et al., 2013). Foraminiferal standing stocks however, decreases with sampling time in an irregular manner. *Ammonia tepida* and *Melonis barleeanus* showed a very high survival and growth rate under strongly oxygen-depleted conditions alternating with short periods of anoxia in all experimental conditions (Geslin et al., 2014). The survival rate of *Bulimina marginata* was comparatively much lower. In another study, however, both *Ammonia tepida* and *Bulimina marginata* as well as *Cassidulina laevigata* not only survived but were also able to calcify at different depths, under anoxic conditions with or without nitrates (Nardelli et al., 2014). The benthic foraminiferal population density was more at sediment surface and decreased gradually with depth, thus confirming field results (Enge et al., 2011). Due to the capability to thrive under oxygen depleted condition, foraminifera may play an important role in short-term carbon cycling in the oxygen minima zone (OMZ) core region on the Indian margin as observed from the uptake of large amounts of organic matter by a few species (Enge et al., 2014). It should however be noted that even though the foraminifera dominate the short-term processing of phytodetritus in the core region of the oxygen minimum zone but are less important in the lower oxygen minimum zone boundary region of the Indian margin as biological interactions and species distribution of foraminifera change with depth and oxygen levels (Enge et al., 2016). The differential effect of diet and induced winter hypoxia on the population structure of benthic foraminifera was further confirmed and it was found that benthic foraminifera generally responded inversely to progressive stages of ecosystem recovery (Brouwer et al., 2016).

EFFECT OF OCEAN ACIDIFICATION ON BENTHIC FORAMINIFERA

The ambient pH is very critical for all calcifying organisms, including foraminifera. The open ocean pH varies from 7.9 to 8.3. As the concentration of CO_2 is increasing in the atmosphere, in near future, it is expected that the pH of the ocean, which is said to be one of the biggest sinks for this atmospheric CO_2, will decrease. A decrease in pH will adversely affect calcifying organisms. The effect of ocean acidification may vary from organism to organism. Realizing the need of the hour, many researchers are trying to understand the effect of decrease in pH, due to increase in CO_2 in atmosphere, on calcifying organisms. Proper experimental design is required for carrying out ocean acidification experiments, especially the contamination free media, as benthic foraminifera may not survive due to ciliate infestation (Diz et al., 2015). The laboratory culture studies suggest that, ongoing ocean acidification can result in decrease of calcite production by foraminifera (de Nooijer et al., 2009b). The results are further supported by another study wherein it was observed that ≥929 µatm CO_2, caused reduced growth and shell dissolution in *Ammonia aomoriensis* (Haynert et al., 2011). In an interesting study, it was suggested that benthic foraminifera can't survive and thus likely to be extinct in the near future under pCO_2 exceeding 700 µatm, (Uthicke and Fabricus, 2012). The findings were later contested by several studies. No lethal effect on the fitness and survival of *Amphistegina gibbosa* was observed under elevated pCO_2 or the associated decrease in pH as well as the calcite saturation state of the seawater (McIntyre-Wressnig et al., 2013). The growth and mortality of living *Ammonia aomoriensis* were not affected by elevated pCO_2 levels, but ~30% of empty tests of *A. aomoriensis* were subjected to dissolution at high pCO_2 of 3247 µatm and Ω_{calc} undersaturation. The tests of subdominant species *Elphidium incertum*, however remained intact (Haynert et al., 2014). No significant effect of elevated pCO_2 on benthic foraminifera *Amphistegina radiata* and *Heterostegina depressa* net production and dark respiration was observed after long and short term exposures (Vogel and Uthicke, 2012). The shell weight of *Marginopora kudakajimensis* was most closely dependent upon pH, suggesting that fossil shell weight can be more useful for reconstruction of paleo-pH changes (Kuroyanagi et al., 2009). Yet in another study, Dissard et al. (2010b), reported that *Ammonia tepida* shell weight decreases with decreasing $[CO_3^{2-}]$, and also with increasing temperature. Increase in size normalized weight and growth rate of *Ammonia* sp. was found under high $[CO_3^{2-}]$, which confirmed that size normalized weight can be used as a proxy for $[CO_3^{2-}]$ (Keul et al., 2013). The growth in *Amphisorus* and *Calcarina* sp. showed opposite response to the carbonate species (carbonate ion and CO_2) which may be because of the different symbiotic algae (Hikami et al., 2011). The main factor influencing the response to changes in pH levels in *Amphistegina lessonii* and *Marginopora vertebralis* is the difference in biochemical pathways of calcification (Prazeres et al., 2015). These species are also capable to regulate biochemical functions to cope with short-term increase in acidity. Growing emphasis is on conducting community level experiments to understand differential interaction between groups. The elevated CO_2 concentration may result in increased nanobenthos density (Ishida et al., 2013). The calcification during chamber formation of *Ammonia sp.* strongly influences the extracellular pH of the microenvironment (Glas et al., 2012). The laboratory culture studies suggest a very large pH tolerance range of benthic foraminifera and a clear species specific response to ocean acidification.

FACTORS AFFECTING STABLE ISOTOPIC COMPOSITION OF BENTHIC FORAMINIFERA

The stable isotopic ratio (Oxygen and Carbon) of calcareous foraminifera are extensively used to reconstruct past hydrographic changes. Therefore, continuous efforts have been made to understand the effect of various ambient parameters on stable isotopic fractionation in foraminifera. Laboratory culture studies showed that the carbonate ion composition of seawater, growth rate, developmental stage, light levels affect the stable oxygen and carbon isotopic composition of benthic foraminifera. The $\delta^{18}O$ in *Amphistegina lobifera* decerases with increasing pH and a change of 0.2‰ was observed with every °C change in temperature, in line with the response of planktic foraminifera (Rollion-Bard et al., 2008). A similar (−0.22‰°C^{-1}) difference in $\delta^{18}O$ of *Bulimina marginata* cultured over a wide range of temperature (4 to 19°C) was also observed confirming that the temperature dependency of foraminiferal $\delta^{18}O$ was parallel to the thermodynamically determined changes observed in inorganically precipitated calcite and thus further validating the use of $\delta^{18}O$ in paleoceanographic studies (Barras et al., 2010). A comparison of $\delta^{18}O$ and $\delta^{13}C$ of cultured benthic foraminifera with water chemistry and field specimens suggests differences due to species-dependent vital effects, ontogenetic variations in shell chemistry; and the aqueous carbonate chemistry ([CO_3^{2}] or pH) of the experimental system (McCorkle et al., 2008). In another study, it was confirmed that the combination of $\delta^{18}O$ and Mg/Ca of *Ammonia beccarii* can be applied to reconstruct $\delta^{18}O_{seawater}$ as well as the salinity (Toyofuku et al., 2011). The effect of ontogenetic variations on $\delta^{18}O$ was also observed in *Bulimina aculeata* and *Bulimina marginata*, thus suggesting a consistency with patterns observed in earlier culture studies (Filipsson et al., 2010). As compared to oxygen, limited attempts have been made to understand the factors affecting carbon isotopic fractionation in benthic foraminifera. The persistent methane emanation also impacts the carbon isotopic composition of deep-sea benthic foraminifera (Wollenburg et al., 2015). In addition to O and C, a few other stable isotopes have also been measured to develop potential new proxies. Recently, it was reported that Li and its isotopes may serve as a new complementary proxy for the past ocean carbonate chemistry (Vigier et al., 2015).

FACTORS AFFECTING ELEMENTAL COMPOSITION OF BENTHIC FORAMINIFERA

Elemental composition of foraminiferal test is also used for paleoceanographic and paleoclimatic interpretations. Trace elements get incorporated into the test from the seawater during the test secretion. Laboratory culture studies have helped evaluate the factors affecting elemental compostion of several benthic foraminiferal species. A majority of culture studies have been conducted on benthic formainifera to understand the effect of ambient conditions on Mg/Ca and Sr/Ca (Dissard et al., 2009; 2010a; 2010b; Duenas-Bohorquez et al., 2011; de Nooijer et al., 2014). A precise evaluation of the effect of ambient conditions on trace element ratio requires to delineate the earlier formed calcite from that secreted under controlled laboratory conditions. A fluorescent dye calcein was suggested as potential technique to track newly formed calcite. The effect of flourescent dye calcein on benthic foraminifera *Ammonia tepida* was investigated and it was found that the presence of calcein does not impact Mg and Sr incorporation into biologically and inorganically precipitated calcium carbonate (Dissard et al., 2009). The impact of salinity on Mg/Ca and Sr/Ca ratio was evaluated from the change in partition coefficients (D_{Me}=(Me/Ca calcite)/(Me/Ca seawater); where (Me: i.e. Sr and Mg)). D_{Mg} per unit salinity is 2.8% at 10°C and 3.3% at 15°C, for the salinity interval 20-40 psu. It implies that a salinity increase of 2 psu results in enhanced Mg incorporation equivalent to 1°C increase in temperature. As compared to D_{Mg}, D_{Sr} increase per unit salinity was 0.8% at 10°C and 1.3% at 15°C, for the salinity interval 20–40 psu (Dissard et al., 2010a). It should however be noted that the partition coefficient of trace elements is species specific. Allison et al., 2010 also suggested that a correlation between ambient water Δ [CO_3^{2-}] and test Mg/Ca is not ubiquitous in all calcareous foraminiferal species. *Heterostegina depressa* and *Ammonia tepida* species responded differently to the seawater chemistry (Raitzsch et al., 2010). They observed that, the partition coefficient of Mg (D_{Mg}) in *A. tepida* significantly decreases with increasing [Ca^{2+}] and, therefore calcite saturation state (Ω), while D_{Sr} in the same benthic species does not vary. On the other hand, D_{Mg} of *H. depressa* shows only a minor decrease with increasing [Ca^{2+}] and thus (Ω), while D_{Sr} increases considerably with [Ca^{2+}] and also (Ω). Interestingly, changes in [CO_3^{2-}] or dissolved inorganic carbon, does not significantly affect Mg distribution coefficient. Contrary to that, Sr incorporation increases with increasing [CO_3^{2-}] (Dissard et al., 2010b). The effect of seawater carbonate ion concentration ([CO_3^{2-}]), and thereby calcite saturation state (Ω), on Mg and Sr incorporation into calcite of *Ammonia tepida* and *Heterostegina depressa* has also been observed. It was found that the ions involved in biomineralization (i.e. Ca^{2+} and dissolved inorganic carbon [DIC]) are processed by separate cellular transport mechanisms. Further based on similar response of Mg and Sr incorporation, it was proposed that differences in the Ca^{2+} transport mechanism only affect divalent cation partitioning (Dueñas-Bohórquez et al., 2011). *Ammonia tepida* was cultured under constant conditions while determining its growth rates, size-normalized weights and single-chamber Mg/Ca and Sr/Ca (de Nooijer et al., 2014). No detectable ontogenetic control on the incorporation of these elements in the species was observed. Difference between the groups was similar to the intraclone group variability in elemental composition, suggesting that any genetic differences between the studied clone groups do not affect trace element partitioning. As benthic foraminiferal species abundance varies regionally, continuous efforts are made to explore trace element partitioning coefficients in multiple species. *Bulimina marginata* was found to have a low sensitivity for trace element incorporation, and thus is not ideal for reconstructing paleotemperature due to associated large uncertainties and large sample size needed for an accurate temperature reconstruction (Wit et al., 2012). No significant correlation was observed between Mg/Ca and Sr/Ca, either within single tests or between tests cultured at the same pH, thus suggesting additional control on incorporation of these elements in foraminiferal tests (Allison et al., 2011). Laboratory culturing of benthic foraminifer *Ammonia* sp. at a range of carbonate chemistry manipulation treatments suggests a strong potential of foraminiferal U/Ca as a [CO_3^{2-}] proxy (Keul et al., 2013a). Recently, it was suggested that foraminiferal Zn/Ca might be useful when combined with other [CO_3^{2}] proxies to reconstruct past nutrient profiles (van Dijk et al., 2017a). Additionally, combining foraminiferal Zn/Ca with other independent carbonate system proxy (for pH), enables reconstruction of the complete carbon system and thereby, past atmospheric CO_2. In another experiment, van Dijk et al. (2017b) observed a higher incorporation of Zn and Ba when pCO_2 increases from 350 to 1200 ppm. They also observed a contrasting trend in element incorporation in both hyaline and porcelaneous foraminifers whereas, a similar trend was observed in element incorporation as a function of seawater carbonate chemistry. Similar studies with metals like Ni, Cu and Mn were also carried out on benthic foraminifera *Ammonia tepida* and partition coefficients for Cu and Ni (0.14±0.02 and 1.0±0.5, respectively) were calculated. These coefficients can thus be used in environmental and palaeo studies to more accurately determine the Cu/Ca and Ni/Ca ratios of sea water (Munsel et al., 2010). The culture studies are thus exploring novel elemental proxies for better paleoclimatic reconstruction.

EFFECTS OF SPECIFIC POLLUTANTS ON FORAMINIFERA

Anthropogenic activities severely affect the marginal marine regions. Being abundant, benthic foraminifera have been often used to understand the effect of pollutants on marine ecosystem. Laboratory culture studies help in understanding specific effect of particular pollutant on benthic foraminifera. Sabean et al. (2009) observed the effect of residual oil on benthic foraminifera *Miliammina fusca* and reported that the percentage of deformed tests in foraminifera can be an excellent indicator of oil pollution. Lead contamination can cause numerical increase in lipid droplets, proliferation of residual bodies, thickening of the organic lining, mitochondrial degeneration, autophagosome proliferation and the development of inorganic aggregates in *Ammonia parkinsoniana* (Frontalini et al., 2015). The presence of Hg at the basal part of pores, in the organic linings of the foramen/septa, and as cytoplasmic accumulations in the foraminiferal test of *Ammonia parkinsoniana,* was observed in response to different concentrations of Hg (Frontalini et al., 2016). Prazeres et al. (2011) observed the effect of zinc on benthic foraminifera *Amphistegina lessonii* and suggested that an activation of some components of the antioxidant system occurred in this species to counteract the oxidative stress induced by Zn exposure, and therefore, a possible complete loss of the symbiont was prevented. In another study, the Ca^{2+}-ATPase activity was evaluated in adult holobionts of *Amphistegina lessonii* and the results indicate that holobiont Ca^{2+}-

Biometry and dissolution features of the benthic foraminifer *Ammonia aomoriensis* at high pCO_2. Mar. Ecol. Prog. Ser., v. 432, pp. 53–67, doi:10.3354/meps09138.

HAYNERT, K., SCHÖNFELD, J., SCHIEBEL, R., WILSON, B. and THOMSEN, J. (2014) Response of benthic foraminifera to ocean acidification in their natural sediment environment: a long-term culturing experiment. Biogeosciences, v.11, pp.1581– 1597.

HESS, S., ALVE, E., TRANNUM, H.C. and NORLING, K. (2013) Benthic foraminiferal responses to water-based drill cuttings and natural sediment burial: Results from a mesocosm experiment. Mar. Micropaleontol., v.101, pp.1–9. doi:10.1016/ j.marmicro

HIKAMI, M., USHIE, H., IRIE, T., FUJITA, K., KUROYANAGI, A., SAKAI, K., NOJIRI, Y., SUZUKI, A. and KAWAHATA, H. (2011) Contrasting calcification responses to ocean acidification between two reef foraminifers harboring different algal symbionts. Geophys. Res. Lett., v.38, pp.1–5. doi:10.1029/ 2011GL048501

ISHIDA, H., GOLMEN, L.G., WEST, J., KRÜGER, M., COOMBS, P., BERGE, J.A., FUKUHARA, T., MAGI, M. and KITA, J. (2013) Effects of CO_2 on benthic biota: An in situ benthic chamber experiment in Storfjorden (Norway). Mar. Pollut. Bull., v.73, pp.443–451. doi:10.1016/j.marpolbul

KEUL, N., LANGER, G., DE NOOIJER, L.J. and BIJMA, J. (2013) Effect of ocean acidification on the benthic foraminifera *Ammonia* sp. is caused by a decrease in carbonate ion concentration. Biogeosciences, v.10, pp.6185–6198. doi:10.5194/ bg-10-6185-2013

KEUL, N., LANGER, G., DE NOOIJER, L.J., NEHRKE, G., REICHART, G.J. and BIJMA, J. (2013a) Incorporation of uranium in benthic foraminiferal calcite reflects seawater carbonate ion concentration. Geochemistry, Geophys. Geosystems, v.14, pp.102–111. doi:10.1029/2012GC004330

KOHO, K.A., LANGEZAAL, A.M., VAN LITH, Y.A., DUIJNSTEE, I.A.P. and VAN DER ZWAAN, G.J., (2008) The influence of a simulated diatom bloom on deep-sea benthic foraminifera and the activity of bacteria: A mesocosm study. Deep Res. Part I Oceanogr. Res. Pap., v.55, pp.696–719. doi:10.1016/j.dsr

KONTAKIOTIS, G., MORTYN, G.P., ANTONARAKOU, A. and DRINIA, H. (2016) Assessing the reliability of foraminiferal Mg/Ca thermometry by comparing fieldsamples and culture experiments: A review. Geol. Quaterly, v.60, pp.547-560. doi: 10.7306/gq.1272.

KUROYANAGI, A., KAWAHATA, H., SUZUKI, A., FUJITA, K. and IRIE, T. (2009) Impacts of ocean acidification on large benthic foraminifers: Results from laboratory experiments. Mar. Micropaleontol., v.73, pp.190–195. doi:10.1016/ j.marmicro

KURTARKAR, S.R., NIGAM, R., SARASWAT, R. and LINSHY, V.N. (2011) Regeneration and abnormality in benthic foraminifer *Rosalina leei*: Implications in reconstructing past salinity changes. Riv. Italiana di Paleontolog Stratigraf, v.117, pp.189–196.

KURTARKAR, S.R., NIGAM, R., LINSY, V.N. and SARASWATI, R. (2013) Effect of temperature and salinity of stable isotopic composition of benthic traminiferae tests: Culture experiments on shallow water species. Spc. vol.V. Geol. Soc. Int., pp.245-252

KURTARKAR, S.R., SARASWAT, R., NIGAM, R., BANERJEE, B., MALLICK, R., NAIK, D.K. and SINGH, D.P. (2015) Assessing the effect of calcein incorporation on physiological processes of benthic foraminifera. Mar. Micropaleontol., v.114, pp.36–45. doi:10.1016/j.marmicro.2014.10.001

LANGLET, D., GESLIN, E., BAAL, C., METZGER, E., LEJZEROWICZ, F.,

RIEDEL, B., ZUSCHIN, M., PAWLOWSKI, J., STACHOWITSCH, M. and JORISSEN, F.J. (2013) Foraminiferal survival after long-term in situ experimentally induced anoxia. Biogeosciences, v.10, pp.7463–7480. doi:10.5194/bg-10-7463-2013

LINSHY, V.N., NIGAM, R. and HEINZ, P. (2014) Response of shallow water benthic foraminifera to a 13C-labelled food pulse in the laboratory. *In*: Experimental Approaches in Foraminifera: Collection, Maintenance and Experiments, edited by: Kitazato, H. and Bernhard, J. M., Springer, Berlin, pp. 115-131.

LINSHY, V.N., RANA, S.S., KURTARKAR, S., SARASWAT, R. and NIGAM, R. (2007) Appraisal of laboratory culture experiments on benthic foraminifera to assess/develop paleoceanographic proxies, Indian J. Mar. Sci., v.36, pp.301–321.

LINSHY, V.N., SARASWAT, R., KURTARKAR, S.R. and NIGAM, R. (2013) Experiment to decipher the effect of heavy metal cadmium on coastal benthic foraminifer *Pararotalia nipponica* (ASANO). J. Palaeontol. Soc. India, v.58, pp.205–211.

MARTINEZ-COLON, M., HALLOCK, P. and GREEN-RUIZ, C. (2009) Strategies for using shallow water benthic foraminifers as bioindicators of potentially toxic elements: a review. J. foraminifer. res., v. 39, pp. 278–299. doi:10.2113/gsjfr.39.4.278

MCCORKLE, D.C., BERNHARD, J.M., HINTZ, C.J., BLANKS, J.K., CHANDLER, G.T. and SHAW, T.J. (2008) The carbon and oxygen stable isotopic composition of cultured benthic foraminifera. Geol. Soc. London, Spec. Publ., v.303, pp.135–154. doi:10.1144/sp303.10

MCINTYRE-WRESSNIG, A., BERNHARD, J.M., MCCORKLE, D.C. and HALLOCK, P. (2013) Non-lethal effects of ocean acidification on the symbiont-bearing benthic foraminifer *Amphistegina gibbosa*. Mar. Ecol. Prog. Ser., v.472, pp.45–60. doi:10.3354/meps09918

MOJTAHID, M., ZUBKOV, M. V., HARTMANN, M. and GOODAY, A.J. (2011) Grazing of intertidal benthic foraminifera on bacteria: Assessment using pulse-chase radiotracing. J. Exp. Mar. Bio. Ecol., v.399, pp.25–34. doi:10.1016/j.jembe

MUNSEL, D., KRAMAR, U., DISSARD, D., NEHRKE, G., BERNER, Z., BIJMA, J., REICHART, G.J. and NEUMANN, T. (2010) Heavy metal incorporation in foraminiferal calcite: Results from multi-element enrichment culture experiments with *Ammonia tepida*. Biogeosciences, v.7, pp.2339–2350. doi:10.5194/bg-7-2339-2010

NARDELLI, M.P., BARRAS, C., METZGER, E., MOURET, A., FILIPSSON, H.L., JORISSEN, F. and GESLIN, E. (2014) Experimental evidence for foraminiferal calcification under anoxia. Biogeosciences, v.11, pp.4029–4038. doi:10.5194/bg-11-4029-2014

NIGAM, R., KURTARKAR, S.R., SARASWAT, R., LINSHY, V.N. and RANA, S.S. (2008) Response of benthic foraminifera *Rosalina leei* to different temperature and salinity, under laboratory culture experiment v.88, pp.699–704. doi:10.1017/S0025315408001197

NIGAM, R., LINSHY, V.N., KURTARKAR, S.R. and SARASWAT, R. (2009) Effects of sudden stress due to heavy metal mercury on benthic foraminifer *Rosalina leei*: Laboratory culture experiment. Mar. Pollut. Bull., v.59, pp.362–368. doi:10.1016/j.marpolbul.

NOMAKI, H., OHKOUCHI, N., HEINZ, P., SUGA, H., CHIKARAISHI, Y., OGAWA, N.O., MATSUMOTO, K. and KITAZATO, H. (2009) Degradation of algal lipids by deep-sea benthic foraminifera: An in situ tracer experiment. Deep Res. Part I Oceanogr. Res. Pap., v.56, pp.1488–1503. doi:10.1016/j.dsr.

PRAZERES, M. DE F., MARTINS, S.E. and BIANCHINI, A. (2011) Biomarkers response to zinc exposure in the symbiont-bearing foraminifer *Amphistegina lessonii*

(Amphisteginidae, Foraminifera). J. Exp. Mar. Bio. Ecol., v.407, pp.116–121. doi:10.1016/j.jembe.

PRAZERES, M.D.F., MARTINS, S.E. and BIANCHINI, A. (2012). Impact of metal exposure in the symbiont-bearing foraminifer *Amphistegina lessonii*. Proc. 12th Int. Coral Reef Symp., pp. 13–16.

PRAZERES, M., PRAZERES, M., UTHICKE, S. and PANDOLFI, J.M. (2015) Ocean acidification induces biochemical and morphological changes in the calcification process of large benthic foraminifera. Proc. R. Soc. B, 282: 20142782. doi:10.1098/rspb.2014.2782

PUCCI, F., GESLIN, E., BARRAS, C., MORIGI, C., SABBATINI, A., NEGRI, A. and JORISSEN, F.J. (2009) Survival of benthic foraminifera under hypoxic conditions: Results of an experimental study using the cell tracker green method. Mar. Pollut. Bull., v.59, pp.336–351. doi:10.1016/j.marpolbul.2009.08.015

RAITZSCH, M., DUEÑAS-BOHÓRQUEZ, A., REICHART, G.-J., DE NOOIJER, L.J. and BICKERT, T. (2010) Incorporation of Mg and Sr in calcite of cultured benthic foraminifera: Impact of calcium concentration and associated calcite saturation state. Biogeosciences, v.7, pp. 869–881. doi:10.5194/bg-7-869-2010

ROLLION-BARD, C., EREZ, J. and ZILBERMAN, T. (2008) Intra-shell oxygen isotope ratios in the benthic foraminifera genus *Amphistegina* and the influence of seawater carbonate chemistry and temperature on this ratio. Geochim. Cosmochim. Acta, v.72, pp.6006–6014. doi:10.1016/j.gca.2008.09.013

UTHICKE, S. and FABRICUS, K. E. (2012) Productivity gains do not compensate for reduced calcification under near future ocean acidification in the photosynthetic benthic foraminifer species *Marginopora vertebralis*. Global Change Biology, v.18, pp.2781–2791, doi: 10.1111/j.1365-2486

SABEAN, J.A.R., SCOTT, D.B., LEE, K. and VENOSA, A.D. (2009) Monitoring oil spill bioremediation using marsh foraminifera as indicators. Mar. Pollut. Bull., v.59, pp.352–361. doi:10.1016/j.marpolbul.2009.08.013

SARASWAT, R., KOUTHANKER, M., KURTARKAR, S.R., NIGAM, R., NAQVI, S.W.A. and LINSHY, V.N. (2015) Effect of salinity induced pH/alkalinity changes on benthic foraminifera: A laboratory culture experiment, Estuarine. Coastal Shelf Sci., v.153, pp.96–107.

SARASWAT, R., NIGAM, R. and PACHKHANDE, S. (2011) Difference in optimum temperature for growth and reproduction in benthic foraminifer *Rosalina globularis*: Implications for paleoclimatic studies. J. Exp. Mar. Bio. Ecol., v. 405, pp. 105–110. doi:10.1016/j.jembe.2011.05.026

SEEARS, H. A, WADE, C.M. (2013) A note on the laboratory culture of the benthic foraminifer *Cornuloculina balkwilli* (MacFadyen). BMC Res. Notes, v.6, 369. doi:10.1186/1756-0500-6-369

TOYOFUKU, T., SUZUKI, M., SUGA, H., SAKAI, S., SUZUKI, A., ISHIKAWA, T., DE NOOIJER, L.J., SCHIEBEL, R., KAWAHATA, H. and KITAZATO, H. (2011) Mg/Ca and $\delta^{18}O$ in the brackish shallow-water benthic foraminifer *Ammonia "beccarii."* Mar. Micropaleontol., v.78, pp.113–120. doi:10.1016/j.marmicro

VAN DIJK, I., DE NOOIJER, L.J. and REICHART, G.-J. (2017a) Trends in element incorporation in hyaline and porcelaneous foraminifera as a function of pCO_2. Biogeosciences, v.14, pp. 497–510. doi:10.5194/bg-14-497-2017

VAN DIJK, I., DE NOOIJER, L.J., WOLTHERS, M. and REICHART, G.J. (2017b) Impacts of pH and $[CO_3^{2-}]$ on the incorporation of Zn in foraminiferal calcite. Geochim. Cosmochim. Acta, v.197, pp.263–277. doi:10.1016/j.gca

VIGIER, N., ROLLION-BARD, C., LEVENSON, Y. and EREZ, J. (2015) Lithium isotopes in foraminifera shells as a novel proxy for the ocean dissolved inorganic carbon (DIC). Comptes Rendus Geosci., v.347, pp.43–51. doi:10.1016/j.crte

VOGEL, N. and UTHICKE, S. (2012) Calcification and photobiology in symbiont-bearing benthic foraminifera and responses to a high CO_2 environment. J. Exp. Mar. Bio. Ecol., v.424, pp.15–24. doi:10.1016/j.jembe

WIT, J.C., DE NOOIJER, L.J., BARRAS, C., JORISSEN, F.J. and REICHART, G.J. (2012) A reappraisal of the vital effect in cultured benthic foraminifer *Bulimina marginata* on Mg/Ca values: Assessing temperature uncertainty relationships. Biogeosciences, v. 9, pp. 3693–3704. doi:10.5194/bg-9-3693-2012

WOLLENBURG, J.E., RAITZSCH, M. and TIEDEMANN, R. (2015) Novel high-pressure culture experiments on deep-sea benthic foraminifera - Evidence for methane seepage-related $\delta^{13}C$ of *Cibicides wuellerstorfi*. Mar. Micropaleontol., v.117, pp.47–64. doi:10.1016/j.marmicro

YU, ZHOUFEI, LEI, YANLI, LI and TIEGANG, (2016) Mean proloculus size as a salinity index in benthic foraminifera *Ammonia aomoriensis*: based on culture and seasonal studies. J. Palaeontol. Soc. India. v. 61(2), pp. 215-223.

Micropaleontology and its Applications
Pradeep K. Kathal, Rajiv Nigam & Abu Talib (Editors)
Scientific Publishers (India), 2018, 331-342 pp.

Potential Applicability of Benthic Foraminiferal Microhabitats in Deciphering Ocean- Bottom Oxygenation History of the Eastern Arabian Sea

S. R. Bharti[1,2], K. Verma[1] and A. D. Singh[1]*

[1] Center of Advanced Study in Geology, Banaras Hindu University, Varanasi - 221005, India.
[2]Paleontology Division I CHQ, Geological Survey of India, 15 A & B Kyd Street, Kolkata – 700016, India
***Email:** arundeosingh@yahoo.com

Abstract

Benthic foraminifera show species-specific microhabitat preferences depending on their adaptation capabilities to the changing ocean bottom environments governed by the nutrient (organic carbon) and dissolved oxygen conditions. Microhabitats of benthic foraminifera are classified according to depth at which they live in sediments. We investigated benthic foraminiferal assemblages of a core collected from the upper continental slope off Goa at 1230 m water depth in the eastern Arabian Sea and determined succession of predominant foraminiferal habitat types on millennial time scale in the last 70 kyr BP. We suggest that the percentage distribution of two major microhabitat categories (epifauna and infauna) have potential for tracing changes in bottom water oxygenation. High abundance of epifaunal group in the core appears to be a measure for the oxic and oligotrophic environment and the abundance maxima of infaunal group are indicative of oxygen-poor and eutrophic environment. This study reveals that the bottom oxygen condition in the eastern Arabian Sea has varied in the past on millennial scale.

Keywords: Benthic foraminifera, microhabitat, paleoceanography, Arabian Sea.

INTRODUCTION

The Arabian Sea is characterised by a persistent Oxygen Minimum Zone (OMZ) basin wide at water depths between 150 and 1250 m (Wyrtki, 1973; Schulte, et al., 1999). The depletion of oxygen in the water column results from the combined effects of high primary productivity induced large flux of organic matter (Wyrtki, 1973; Sarma, 2002), moderate consumption rates and poor ventilation (Olson et al., 1993). Benthic foraminifera are found to be reliable indicators of these oceanographic changes as they respond rapidly to dissolved-oxygen concentration and organic carbon flux (Quinterno and Gardner, 1987; Sen Gupta and Machhain-Castillo, 1993; Loubere et al., 1993; Kahio, 1994; Bernhard et al., 1997; Jannink et al., 1998).Variations in composition of benthic foraminiferal assemblages in response to the changes in paleoceanographic conditions are observed in terms of species diversity, abundances of environment sensitive species and group and microhabitat categories etc. (Hermelin and Shimmield, 1990, 1995; Schmiedl and Leuschner, 2005; Nigam et al., 2007; Gupta et al., 2008; Singh et al., 2015).

Previous distribution studies of benthic foraminifera have demonstrated various morphotypes based on the shape of the test, the nature of test coiling and the presence or absence of the surface pores; and they are related to the nature of the substrate, bottom oxygen environment and the depth within sediment these taxa live (Corliss, 1985; Corliss et al., 1886; Gooday, 1986; Corliss and Chen, 1988; Bernhard, 1989; Corliss and Fois, 1990; Jorissen et al., 1995). Microhabitats of benthic foraminifera are classified according to depth at which they live in sediments. The main microhabitat categories of benthic foraminifera are epifaunal, shallow infaunal and deep infaunal. Epifaunal taxa live on top of the sediment, shallow infauna in the top 2 cm and deep infauna below 2 cm within sediment (Murray, 1991; Barmawidjaya et al., 1992; Buzas et al., 1993). Studies on modern benthic foraminifera indicate that among the various environmental factors, trophic and oxygen conditions are the main factors controlling the foraminiferal microhabitats (Gooday, 1986; Mackensen and Douglas, 1989; Corliss and Chen, 1990; Corliss and Emerson, 1990; Barmawidjaya et al., 1992; Jorissen et al., 1992; Rathburn and Corliss, 1994; Nigam et al., 2007). The respective importance of each of these two factors in controlling the microhabitats of foraminifera, however, is difficult because these environmental parameters are generally closely coupled, especially within the OMZ. In recent years, attempts have been made to study past changes of the OMZ intensity, deep water circulation and bottom water oxygenation in the Arabian Sea using benthic foraminiferal assemblage records from the marginal sediments (eg. Hermelin and Shimmield, 1995; von Rad et al., 1999; den Dulk et al., 1998, 2000; Schmiedl and Leuschner, 2005). There have been a few studies on fossil benthic foraminiferal assemblages from the Indian margin sediment cores in order to reconstruct late Quaternary history of the OMZ intensity variation in the eastern Arabian Sea (eg. Singh et al., 2014; Verma et al., *in press*). Most of the previous studies used diversity, abundance and morphotype variations of benthic foraminifera in paleoceanographic reconstructions. In this study, we investigated benthic foraminiferal assemblages for specific morphologic characteristics associated with two broad microhabitat categories (epifauna vs infauna), in a sediment core collected off Goa, to record millennial scales changes in sea-bottom oxygen condition in the eastern Arabian Sea.

MATERIAL AND METHODS

A 9.65 m long piston core (MD 76-131) was recovered from the upper continental slope off Goa (at about 1230 m depth; Lat.-15°31.8'N; Long.-72°34.1' E) [Fig. 1]. The core location lies at the base of the present day oxygen minimum zone (150-1250 m water depth; Wyrtki, 1973), thus ideal to record temporal variation in the OMZ intensity and sea-bottom environmental condition. Foraminiferal assemblages were analysed at close sampling intervals (2 - 4 cm). For separation of foraminiferal tests, sediment samples were processed following the conventional micropaleontological techniques. Samples were washed through wet sieving over a 63 µm screen. Dry residue larger than 63 µm was sieved again over a 125 µm screen. Benthic foraminiferal census counts were made on the coarse fraction (>125 µm). For identification and taxonomic classification, we followed mainly the criteria of Loeblich and Tappan (1988, 1992) and Sen Gupta (2002). We studied morphologic characteristics of benthic foraminifera tests following concepts of Phleger (1951), Bernhard (1986), Corliss and Chen (1988), Corliss and Fois (1990) and Živkovic and Babić (2003). The microhabitat categories (epifauna and infauna) were determined based on the modern distribution data and test morphology of benthic foraminifera (references as in Table 1). Based on the census data, the relative abundance (%) of each species and microhabitat groups were estimated. We also compared our faunal record with the proxy record of the OMZ variation (aragonite %; Naidu et al., 2014). A chronological framework of the core MD76-131 developed by Singh et al (2011) was adopted. The core represents a record of 5 to 70 kyr BP.

RESULTS

The benthic foraminiferal assemblages in the examined core are composed of buliminids, cassidulinids, uvigerinids, cibicidids, miliolids and fursenkoinids and other important taxa such as *Sphaeroidina bulloides*, *Osangularia bengalensis*, *Hoeglundina elegans*, *Bolivina robusta* and *Hyalinea balthica* (in decreasing order of relative abundance). Earlier studies have suggested a good correlation between microhabitat preference and morphologic characteristic of an individual taxon (eg. Corliss and Fois, 1990). The benthic foraminifera with different habitats show distinct morphological features (eg. Corliss, 1985; Corliss and Emerson, 1990). The main microhabitat categories are epifaunal, shallow infaunal and deep infaunal. Epifaunal taxa live on top of the sediment, shallow infauna in the top 2 cm and deep infauna below 2 cm within sediment (Murray, 1991; Barmwidjaja et al., 1992; Buzas et al., 1993). Because, it is difficult to draw a line between epifaunal and shallow infaunal microhabitats, we described these two categories together as epifaunal microhabitat and deep infaunal as infaunal microhabitat (Jorissen et al., 1995). The constituent species of these two benthic foraminiferal microhabitat groups are listed in Table 1. Epifaunal species generally have rounded trochospiral, plano-convex, milioline or biconvex, trochospiral tests. Whereas, infaunal taxa are characterised by planispiral, flattened, tapered and cylindrical, spherical or ooid tests (Corliss and Chen, 1988). The abundance pattern of epifaunal species

Table 1. Microhabitat preferences for important benthic foraminifera encountered in this study.

Epifauna		Infauna	
Karrerialla subrotundata	14[b]	*Chillostomella oolina*	1,4,6,8,11,14
Miliolids	1,2,3,6,11	*Chillostomella ovoidea*	1,2,3,10
Ceratobulimina pacifica	2,3,10	*Bulimina aculeata*	(1 SI), 13
Hoeglundina elegans	1,4,6,11,14	*Bulimina marginata*	1, 2,3,10
Epistominella exigua	2,8,11,12,13	*Bulimina striata*	1, 2,3,10
Sphaeroidina bulloides	14	*Bulimina costata*	1, 2,3,10
Cibicides refulgens	1,2,3,10	*Globobulimina pacifica*	1, 2,3,10
Cibicides wuellerstorfi	1,2,3,8, 10, 15[b]	*G. pupoides*	1, 2,3,10
Cibicides marialana gigas	1,2,3,10	*Uvigerina proboscidea*	4,10
Cibicides mediocris	1,2,3,10	*Uvigerina peregrina*	1,2,3,4,6,8,11,14
Cibicides robertsonianus	15[b]	*Uvigerina interrupta-costata*	1,2,3,10
Ehrenbergina pacifica	1,2,3,10	*Fursenkoina bradyi*	1,2,3,10
Oridorsalis umbonatus	2,4,8, (14, E/SI)	*Fursenkoina bramletti*	1,2,3,10
Gyroidinoides orbicularis	2,7,11	*Fursenkoina cornuta*	1,2,3,10
Gavelinopsis lobatulus	(4,E/SI),11,(14[b] SI)	*Bolivina robusta*	1,2,3,10,15
Osangularia bengalensis	1,2,3,10	*Cassidulina carinata*	3[b],10
Hyalinea balthica	1,2,3,10	*Cassidulina crassa*	1,2,3,10
		Cassidulina subglobosa	1,2,3,10 1,2,3,10
		Pullenia bulloides	1,2,3,10
		Pullenia quadriloba	1,2,3,10
		Pullenia quinqueloba	13
		Rotaliatinopsis semiinvoluta	1,2,3,10
		Melonis barleeanum	

[a] The categories were based on the following references: 1. Corliss (1985); 2. Corliss and Chen (1988); 3. Corliss and Fois (1990); 4. Rathburn and Corliss (1984); 5. Barmawidjaja et al. (1992); 6. Buzas et al. (1993); 7. Linke and Lutze (1993); 8. Gooday (1986); 9. Kitazato (1994); 10. Rathburn et al. (1996); 11. Guichard (1997); 12. de Stigter (1996); 13. Jannink et al. (1998); 14. Jorissen et al. (1992, 1995); 15. Kaiho (1994).

[b] Indicates references classifying comparable species, or species differently named. In some cases the references are between brackets, and added abbreviations reflecting the categories the taxa were assigned to by these authors: E, Epifauna; SI, Shallow Infauna; I, Infauna.

reflects bottom water condition of sediment/water interface, whereas, abundances of infaunal taxa are indicative of the environmental characteristics within sediment at deeper depth. Epifaunal taxa are known to prefer high-oxygen bottom water condition (Kaiho, 1994). Infaunal taxa are considered to prefer relatively lower-oxygen habitat because of a decrease in dissolved-oxygen content downward in sediment (Fenchel, 1969; Corliss and Chen, 1988 and Kaiho, 1994). The quantitative benthic foraminiferal data enabled us to determine succession of predominant benthic foraminiferal habitat types prevailed at the examined core-site, in response to changes in bottom environmental conditions during the last 70 kyr BP. Fig. 2 shows the relative abundances of epifaunal and infaunal components in the calcareous benthic foraminiferal community. The microfaunal record shows major variation in abundances of microhabitat categories down-core on millennial scale. Predominance of epifaunal group in benthic population is recorded during 10-12.5 kyr BP, 15-17.5 kyr BP and 23-24 kyr BP (Fig. 2). The abundance of epifaunal taxa remained relatively high between 25 and 37 kyr BP with small scale fluctuation. The epifaunal benthic foraminifera show rapid and significant increase in its abundance during 38 - 39 kyr BP (Fig. 2). The period between 40 and 45 kyr BP is characterised by a major decline in abundance of epifaunal group in benthic assemblage, preceding a brief period (46-47 kyr BP) of its predominance. There has also been a considerable increase in abundance of epifaunal taxa between 58 and 60 kyr BP. The faunal record reveals that the pattern of temporal variation in abundance of infaunal group is opposite to the variation pattern of epifaunal group with peak abundances of former corresponding to the abundance minima of later (Fig. 2).

DISCUSSION

The ocean bottom oxygen condition and trophic level are the main environmental factors controlling the foraminiferal microhabitats (Gooday, 1986; Corliss and Chen, 1988). The dissolved oxygen condition in the eastern Arabian Sea waters at intermediate depths is mainly controlled by the strength of the oxygen minimum zone (OMZ), which has been attributed to the monsoonal wind induced primary productivity, organic carbon flux and deep water ventilation by externally sourced waters (Altabet et al., 1995; Boyle et al., 1995; Schulte et al., 1999 and Reichart et al., 2002). As, the examined core site is located present day at the base of the OMZ, it is likely that the changes in sea-bottom environment to be related with the fluctuations in the OMZ intensity. Earlier paleoceanographic studies reveal that the OMZ intensity in the eastern Arabian Sea has varied significantly in the past 70 kyr BP (Singh et al., 2006; Naidu et al., 2014). The temporal variations in relative abundance of epifaunal and infaunal microfaunal groups in the examined core exhibit changes in bottom oxygen condition coupled with the trophic level (organic carbon flux). The peak abundances of epifaunal component indicate high levels of bottom water oxygenation. Furthermore, the predominance of epifaunal taxa in our record may also be related to the low surface productivity condition (low organic carbon flux), as both the oxygen level and export flux of organic carbon to sea-floor is closely coupled. However, den Dulk et al (2000) opined this group is more sensitive to oxygen variations than the export productivity in the northern Arabian Sea. The infaunal taxa are known to flourish in oxygen-poor environment with high organic carbon content, where the trophic level is high (Gooday, 1986). Therefore, it can be suggested that abundance maxima of epifaunal group occurring during 10-12.5 kyr BP, 15-17.5 kyr BP, 23-24 kyr BP, 28-30 kyr BP, 38-39 kyr BP, 46-47 kyr BP and 58-60 kyr BP are linked to the weakening of the OMZ, which would have led to oxygenated environment at the sea-bottom. Interestingly, these periods of epifaunal maxima correspond to the cold North Atlantic events [Younger Dryas (YD) and Heinrich (H1 to H6)]. Previous studies have suggested a weak OMZ in the eastern Arabian during the cold YD and Heinrich events, attributed to the weak monsoonal wind, low primary productivity related organic carbon flux and enhanced incursion of oxygenated southern ocean sourced glacial mode and intermediate waters (Singh et al., 2006, Singh, 2007 and Naidu et al., 2014). When, we compared our faunal record with the record of aragonite content (a proxy for the OMZ intensity variation, Singh, 2007), we found a close correspondence between epifaunal abundance maxima and aragonite maxima during H1, H2 and H3 events. This further supports that the OMZ was significantly weak during these cold periods. It is intriguing to note that aragonite record does not show significant increase during the periods corresponding to the YD, H4 and H5 events. This inconsistency is probably related to low production of aragonitic pteropods during these specific time intervals. Nevertheless, $\delta^{15}N$ isotope record from the same core (Ivanochko, 2004), which reflects

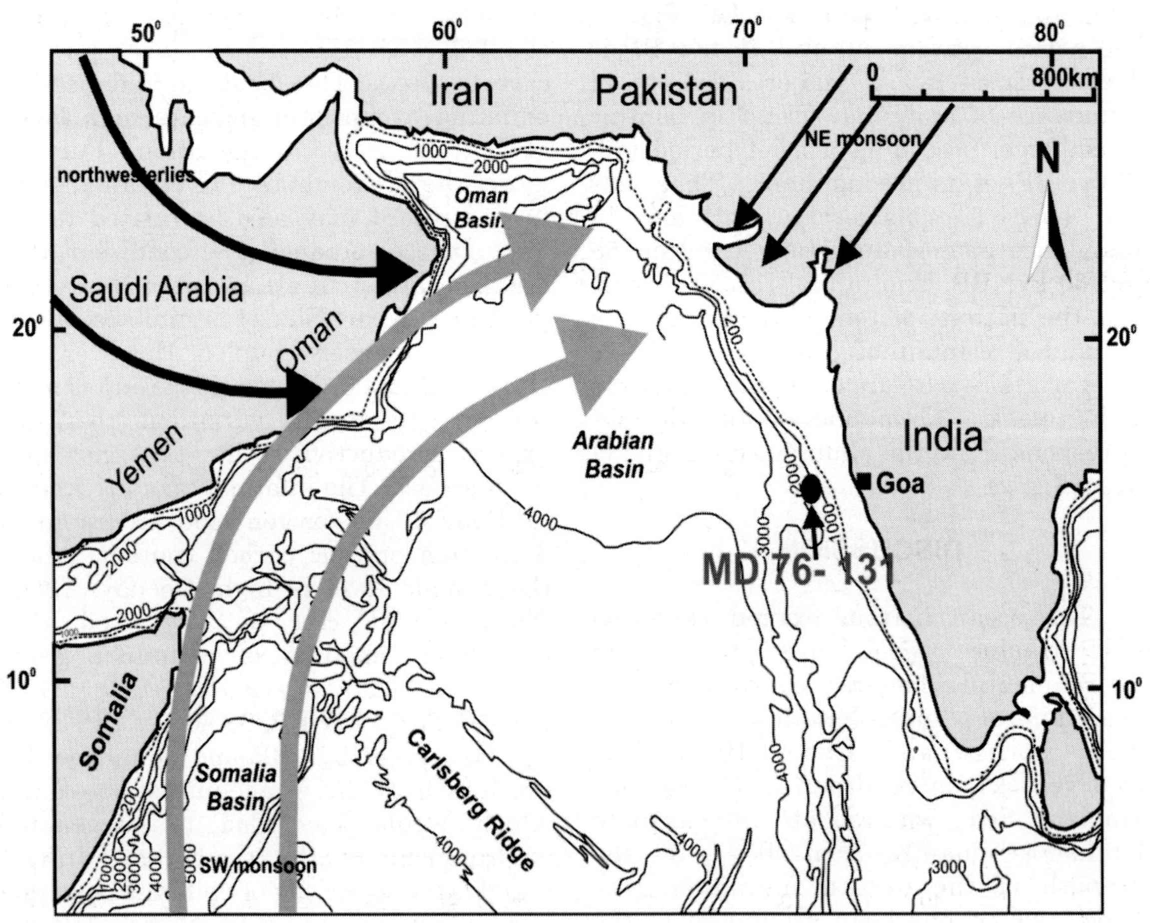

Fig.1. Map showing location of Core MD 76-131 and schematic presentation of the seasonal monsoon winds (modified after Prell and Van Campo, 1986; Sirocko *et al.*, 1993; Clemens *et al.*, 1996; Ivanova, 1999).

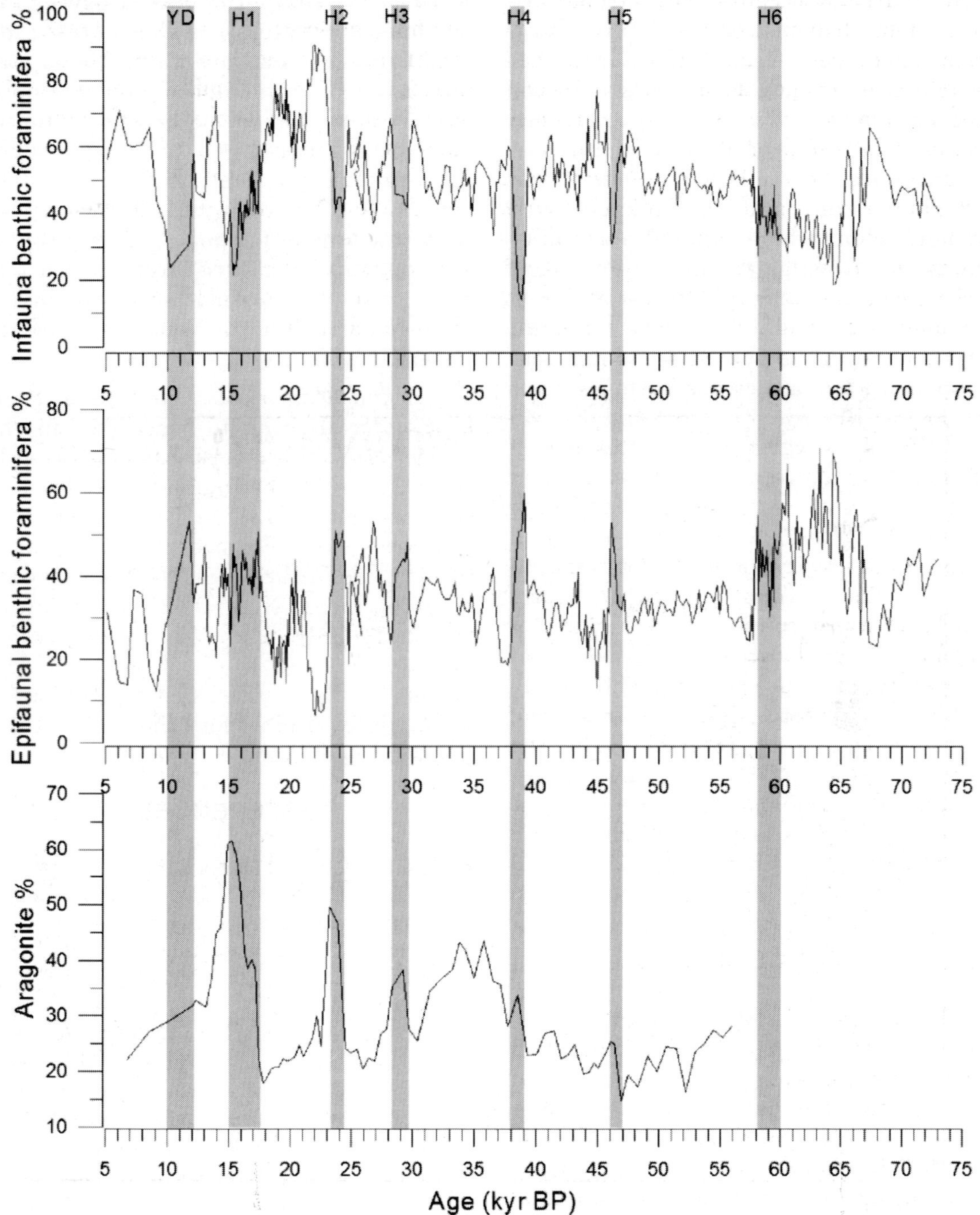

Fig.2. Temporal variations in relative abundances of epifaunal and infaunal benthic foraminifera (>125 μm) and aragonite content (Naidu et al., 2014). Shaded intervals showing YD (Younger Dryas), H1-H6 (Heinrich events) are the cold North Atlantic events.

denitrification associated with strength of the OMZ, demonstrates oxygenated environment and weak OMZ during YD and Heinrich events. Past changes in the oxygenation history of the northern Arabian Sea in response to the changes in monsoon climate induced productivity on millennial scale and its linkage with northern Hemisphere climate instability has been investigated in recent years (Reichart et al., 1998; von Rad et al., 1999 and den Dulk et al., 2000). The temporal variation pattern of abundance of infaunal group shows major increase during 5-9 kyr BP, 14 kyr BP, 18-23 kyr BP and 44 - 45 kyr BP. We suggest these intervals of infaunal abundance maxima to be related to the oxygen-poor environment and high trophic condition attributed to high surface productivity induced export flux to the sea-floor coupled with poorly ventilated water column. Several lines of evidences suggest increased monsoon wind induced primary productivity and poorly ventilated water column resulting intensified OMZ at the core-site during the early Holocene and the periods corresponding to the Bølling/Allerød warm event (~14 kyr BP) and last glacial maximum (~18-23 kyr BP) [Ivanochko, 2004 and Naidu et al., 2014].

CONCLUSION

The benthic foraminiferal microhabitat patterns (epifaunal-infaunal) in the examined core reveal that dissolved oxygen level of underlying waters in the eastern Arabian Sea has considerably varied in the last 70 kyr BP. The periods representing predominance of epifaunal group are characterised by oxic and oligotrophic bottom environment associated with weaker monsoon circulation related surface productivity and enhanced incursion of southern sourced oxygenated intermediate waters. On the contrary, the abundance maxima of infaunal group are indicative of oxygen-poor and eutrophic environment linked with the increased monsoon wind induced primary productivity and poorly ventilated intermediate waters. Our microfaunal record suggests that the major change in sea-bottom environment of eastern Arabian Sea took place during the periods corresponding to the cold north Atlantic Younger Dryas and Heinrich events.

Acknowledgements: We thank P. K. Kathal for inviting to contribute in this special issue. This work was supported by the grants from the Indian Space Research Organisation (ISRO-GBP) and the Department of Science and Technology (Project no. SR/S4/ES-30/2002), Government of India, New Delhi. First author thanks to CSIR fellowship which enabled him to carry out this work in BHU. KV thanks to the University Grants Commission, New Delhi for the BSR UGC-RFSMS Fellowship.

REFERENCES

ALTABET, M.A., FRANCOIS, R., MURRAY, W.D. and PRELL, W.L. (1995) Climate-related variations in denitrification in the Arabian Sea from sediment $^{15}N/^{14}N$ ratios. Nature, v.373, pp.506-509. doi:10.1038/373506a0

BARMAWIDJAJA, D.M., JORISSEN, F.J., PUSKARIC, S. and Van Der ZWAAN, G.J., (1992) Microhabitat selection by benthic foraminifera in the northern Adriatic Sea. J. Foram. Res., v.22(4), pp.297-317.

BERNHARD, J.M. (1989) The distribution of benthic foraminifera with respect to oxygen concentration and organic carbon levels in shallow-water Antarctic sediments. Limnology and Oceanography, v.34, pp.1131-1141.

BERNHARD, J.M., SEN GUPTA, B.K. and BORNE, P.F. (1997) Benthic foraminifera proxy to estimate dysoxic bottom-water oxygen concentrations: Santa Barbara Basin, U.S. Pacific continental margin. J. Foram. Res., v.27, pp.301-310.

BOYLE, E.A., LABEYRIE, L. and DUPLESSY, J., (1995) Calcitic foraminiferal data confirmed by cadmium in aragonitic Hoeglundina: application to the last glacial maximum in the northern Indian Ocean. Palaeoceanography, v.10, pp.881-900.

BUZAS, M.A. CULVER, S.J. and JORISSEN, F.J. (1993) A statistical evaluation of the microhabitats of living (stained) infaunal benthic foraminifera. Mar. Micropaleontol. v.20, pp.311-320.

CLEMENS, S., MURRAY, D. and PRELL, W. (1996) Nonstationary phase of the Plio-Pleistocene Asian monsoon. Science, v.274, pp.943-948.

CORLISS, B.H. (1985) Microhabitats of benthic foraminifera within deep-sea sediments. Nature, v.314, pp. 435-438.

CORLISS, B.H. and CHEN, C. (1988) Morphotype patterns of Norwegian Sea deep-sea benthic foraminifera and ecological implications. Geology, v.16, pp.716-719.

CORLISS, B.H. and EMERSON, S. (1990) Distribution of Rose Bengal stained deep-sea benthic foraminifera from the Nova Scotian continental margin and Gulf of Maine. Deep-Sea Res. v.37, pp.381-400.

CORLISS, B.H. and FOIS, E. (1990) Morphotype analysis of deep sea benthic foraminifera from the northwestern Gulf of Mexico. Palaios, v.6, pp.589-605.

CORLISS, B.H., MARTINSON, D.G. and KEFFER, T. (1986) Late Quaternary deep ocean circulation. Geol. Soc. America, Bull., v. 97, pp. 1106-1121.

den DULK, M., REICHART, G.J., van HEYST, S., ZACHARIASSE, W.J. and van der Zwaan, G.J. (2000) Benthic foraminifera as proxies of organic matter flux and bottom water oxygenation? A case history from the northern Arabian Sea. Palaeogeogr. Palaeoclimatol. Palaeoecol. v.16, pp.1 337-359.

den DULK, M., REICHART, G.J., MEMON, G.M., ROELOFS, E.M.B., ZACHARIASSE, W.J. and van Der ZWAAN, G.J. (1998) Benthic foraminiferal response to variations in surface water productivity and oxygenation in the northern Arabian Sea. Mar. Micropaleontol. v.35, pp.43-66.

de STIGTER, H.C. (1996) Recent and fossil foraminifera in the Adriatic Sea: distribution patterns in relation to organic carbon flux and oxygen concentration at the seabed. Geologica Ultraiectina., v.144, pp.254.

FENCHEL, T.M. (1969) The ecology of marine microbenthos, IV Structure and function of the benthic ecosystem. Its chemical and physical factors and microfauna communities with special reference to the ciliated Protozoa. Ophelia., v.6, pp.1-82.

GOODAY, A.J., (1986) Meiofaunal foraminiferans from the bathyal Porcupine Seablight (northeast Atlantic): size structure, standing stock, taxonomic composition, species diversity and vertical distribution in the sediment. Deep-Sea Res., v.33, pp.1345-1373.

GUICHARD, S., (1997) Thesis Evolution de l'environment dans la zone de l'upwelling maturitanien pendant les derniers 236,000 ans, le temoignage de l'ecosysteme benthique. Thesis. l'Universite Bordeaux I, N° d'ordre: 1799.

GUPTA, A.K., DAS, M., CLEMENS, S.C. and MUKHERJEE, B. (2008) Benthic foraminiferal faunal and isotopic changes as recorded in Holocene sediments of the northwest Indian Ocean. Paleoceanography, v.23 (2), doi:10.1029/2007PA 001546

HERMELIN, J.O.R. and SHIMMIELD, G.B. (1990) The importance of the Oxygen Minimum Zone and sediment geochemistry in the distribution of Recent benthic foraminifera in the northwest Indian Ocean. Mar. Geol., v.91, pp.1-29.

HERMELIN, J.O.R. and SHIMMIELD, G.B. (1995) Impact of productivity events on benthic foraminiferal fauna in the Arabian Sea over the last 150,000 years. Paleoceanography, v.10, pp.85-116.

IVANOCHKO, T.S. (2004) Sub – orbital scale variation in the intensity of the Arabian Sea monsoon, Unpublished Ph. D. Thesis, University of Edinburgh, Scotland, pp.230.

IVANOVA, E.M. (1999) Late quaternary Monsoon history and paleoproductivity of the western Arabian Sea, Unpublished Ph.D. Thesis, Free University, The Netherlands, pp.159.

JANNINK, N.T., ZACHARIASSE, W.J. and van der Zwaan, G.J. (1998) Living (Rose Bengal stained) benthic foraminifera from the Pakistan continental margin (northern Arabian Sea). Deep-Sea Res., v.I (45), pp.1483-1513.

JORISSEN, F.J., BARMAWIDJAJA, D.M., PUSKARIC, S. and van der ZWAAN, G.J. (1992) Vertical distribution of benthic foraminifera in the northern Adriatic Sea: The relation with the organic flux. Mar. Micropaleont., v.19, pp.131-146.

JORISSEN, F.J., de STIGTER, H.C. and WIDMARK, J.G.V. (1995) A conceptual model explaining benthic foraminiferal microhabitats. Mar. Micropaleontol., v.26, pp.3-15.

KAIHO, J. (1994) Benthic foraminifera dissolved-oxygen index and dissolved-oxygen levels in the modern ocean. Geology, v.22, pp.719-722.

KITAZATO, H. (1994) Foraminiferal microhabitats in four marine environments around Japan. Mar. Micropaleont., v.24, pp.29-41.

LINKE, P. and LUTZE G.F. (1993) Microhabitat preferences of benthic foraminifera-a static concept or a dynamic adaptation to optimize food acquisition? Mar. Micropaleontol., v.20, pp.215-234.

LOEBLICH, A.R. JR. and TAPPAN, H. (1988) Foraminiferal Genera and their Classification. Van Nostrand Rienhold Company, New York, v.2V, pp.1182.

LOEBLICH, A.R. JR. and TAPPAN, H. (1992) Present status of foraminiferal Classification. *In*: Studies in Benthic Foraminifera (eds) Takayanagi Y and Saito T Tokai University Press, Tokyo, pp.93-102.

LOUBERE, P. (1993) GARY A. and LAGOE, M., Benthic foraminiferal microhabitats and the generation of a fossil assemblage: theory and preliminary data. Mar. Micropaleont., v.20, pp.165-181.

MACKENSEN, A. and DOUGLAS, R.G. (1989) Down-core distribution of live and dead deep-water benthic foraminifera in box cores from the Weddell Sea and the California continental borderland. Deep-Sea Res., v.36, pp.879-900.

MURRAY, J.W. (1991) Ecology and Paleoecology of Benthic Foraminifera. Longman, Harlow, pp.397.

NAIDU, P.D., Singh, A.D., Ganeshram, R.S. and Bharti, S.K. (2014) Abrupt climate-induced changes in carbonate burial in the Arabian Sea: Causes and consequences. Geochem. Geophys. Geosyst., v.15(1), pp. 398-1406. doi:org/10.1002/2013GC005065

NIGAM, R., MAZUMDER, A., HENRIQUES, P.J. and SARASWAT, R. (2007) Benthic foraminifera as proxy for oxygen-depleted conditions off the central west coast of India. J. Geol. Soc. India, v.70, pp.1047-1054.

OLSON, D.B., HITCHCOCK, G.L., FINE, R.A. and WARREN, B.A. (1993) Maintenance of the low-oxygen layer in the central Arabian Sea. Deep-Sea Res., Part II 40, pp.673–685.

PHLEGER, F.B. (1951) Ecology of foraminifera, northwest Gulf of Mexico, Part 1: Foraminifera distribution. Geol. Soc. Amer. Mem., v.46, pp.1-88.

PRELL, W.L. and van CAMPO, E. (1986) Coherent response of the Arabian Sea upwelling and pollen transport to late Quaternary monsoonal winds. Nature, v.323, pp 526-528.

QUINTERNO, P.J. and GARDNER, J.V. (1987) Benthic foraminifers on the continental shelf and upper slope, Russian River area, northern California. J. Foram. Res., v. 17, pp. 132-152.

RATHBURN, A.E. and CORLISS, B.H. (1994) The ecology of living (stained) deep-sea benthic foraminifera from the Sulu Sea. Paleoceanography, v.9, pp.87-150.

RATHBURN, A.E., CORLISS, B.H., TAPPA, K.D. and LOHMANN, K.C. (1996) Comparisons of the ecology and stable isotopic compositions of living (Stained) benthic foraminifera from the Sulu Sea and South China Sea. Deep-Sea Res. I, v.43, pp.1617-1646.

REICHART, G.L., LOURENS, L.J. and ZACHARIASSE, W.J. (1998) Temporal variability in the northern Arabian Sea oxygen minimum zone (OMZ) during the last 225,000 years. Paleoceanography, v.13, pp. 07-621.

REICHART, G.J., SCHENAU, S.J., de LANGE, G.J. and ZACHARIASSE, W.J. (2002) Synchroneity of oxygen minimum zone intensity on the Oman and Pakistan Margins at sub Milankovitch time scales. Mar. Geol., v. 185, pp. 403-415.

SARMA, V.V.S.S. (2002) An evaluation of physical and biogeochemical processes regulating perennial suboxic conditions in the water column of the Arabian Sea. Global Biogeochem. Cycles, v.16, pp. 9.1-29.11.

SCHMIEDL, G. and LEUSCHNER, D.C. (2005) Oxygenation changes in the deep western Arabian Sea during the last 190,000 years: Productivity versus deep water circulation. Paleocenography 20 PA2008, doi: 10.1029/2004PA001044.

SCHULTE, S., ROSTEK, F., BARD, E., RULLKÖTTER, J. and MARCHAL, O. (1999) Variations of oxygen-minimum and primary productivity recorded in sediments of the Arabian Sea. Earth Planet. Sci. Lett., v.173, pp.205-221.

SEN GUPTA, B.K. (2002) Systematics of modern foraminifera. In: Modern Foraminifera (ed.) Sen Gupta BK, Kluwer Academic Publishers Dordrech, pp.7-36.

SEN GUPTA, B.K. and Machain-Castillo, M.L. (1993) Benthic foraminifera in oxygen - poor habitats. Mar. Micropaleontol., v.20, pp.183-201.

SINGH, A.D. (2007) Episodic preservation of pteropods in the eastern Arabian Sea: Monsoonal change, oxygen minimum zone intensity and aragonite compensation depth. Ind. J. Mar. Sci., v. 36, pp. 378-383.

SINGH, A.D., JUNG, S.J.A., DARLING, K., GANESHRAM, R., IVANOCHKO, T.S. and KROON, D. (2011) Productivity collapses in the Arabian Sea during glacial cold phases. Paleoceanography, v.26, PA3210, doi: 10.1029/2009PA001923.

SINGH, A.D., KROON, D. and GANESHRAM, R.S. (2006) Millennial scale variations in productivity and OMZ intensity in the eastern Arabian Sea. J. Geol. Soc. India, v.68, pp.369-377.

SINGH, A.D., RAI, A.K., VERMA, K., DAS, S. and BHARTI, S.K. (2015) Benthic foraminiferal diversity response to the climate induced changes in the eastern Arabian Sea oxygen minimum zone during the last 30 ka BP. Quat. Int., v.374, pp.118-125, doi.org/10.1016/j.quaint.2014. 11.052.

SIROCKO, F., SARNTHEIN, M., ERLENKEUSER, H., LANGE, H., ARNOLD, M. and DUPLESSY, J.C. (1993) Century scale events in monsoon climate over the past 24,000 years. Nature, v.364, pp.322-324.

VERMA, K., BHARTI, S.K. and SINGH, A.D. (2017) Late glacial – Holocene record of benthic foraminiferal morphotypes from the eastern Arabian Sea OMZ: paleoenvironmental implications. J. Earth. Syst. Sci., (*in press*)

von RAD, U., SCHULZ, H., RIECH, V., den DULK, M., BERNER, U. and SIROCKO, F. (1999) Multiple monsoon - controlled breakdown of oxygen - minimum conditions during the past 30,000 years documented in laminated sediments of Pakistan. Palaeogeogr. Palaeoclimatol. Palaeoecol., v.152, pp.129-161.

WYRTKI, K. (1973) Physical oceanography of the Indian Ocean. In: The biology of the Indian Ocean (eds.) Zeitschel B and Gerlach SA Springer Verlag, Berlin, pp.18–36.

ŽIVKOVIĆ, S. and BABIĆ, L. (2003) Palaeoceanographic implications of smaller benthic and planktonic foraminifera from the Eocene Pazin Basin (Coastal Dinarides, Croatia). Facies, v.49, pp.49-60.